CONVERSION FACTORS

Physical Quantity	Symbol	Conversion Factor
Area	A	$1\ ft^2 = 0.0929\ m^2$ $1\ in.^2 = 6.452 \times 10^{-4}\ m^2$
Density	ρ	$1\ lb_m/ft^3 = 16.018\ kg/m^3$ $1\ slug/ft^3 = 515.379\ kg/m^3$
Energy	Q or W	$1\ Btu = 1055.1\ J$ $1\ cal = 4.186\ J$ $1\ (ft)(lb_f) = 1.3558\ J$ $1\ (hp)(hr) = 2.685 \times 10^6\ J$
Force	F	$1\ lb_f = 4.448\ N$
Heat flow rate	q	$1\ Btu/hr = 0.2931\ W$ $1\ Btu/sec = 1055.1\ W$
Heat flux	q''	$1\ Btu/(hr)(ft^2) = 3.1525\ W/m^2$
Heat generation per unit volume	q_G'''	$1\ Btu/(hr)(ft^3) = 10.343\ W/m^3$
Heat transfer coefficient	h_c	$1\ Btu/(hr)(ft^2)(°F) = 5.678\ W/m^2 \cdot K$
Length	L	$1\ ft = 0.3048\ m$ $1\ in. = 2.54\ cm = 0.0254\ m$ $1\ mile = 1.6093\ km = 1609.3\ m$
Mass	m	$1\ lb_m = 0.4536\ kg$ $1\ slug = 14.594\ kg$
Mass flow rate	$\dot{m}$	$1\ lb_m/hr = 0.000126\ kg/s$ $1\ lb_m/sec = 0.4536\ kg/s$
Power	$\dot{W}$	$1\ hp = 745.7\ W$ $1\ (ft)(lb_f)/sec = 1.3558\ W$ $1\ Btu/sec = 1055.1\ W$ $1\ Btu/hr = 0.293\ W$

CIRCLE AREA : $\pi R^2 = \dfrac{\pi d^2}{4}$

  CIRCUMFERENCE : $2\pi R = \pi D$

CYLINDER SURFACE AREA : $2\pi R h + ENDS$

  CYLINDER VOLUME : $\pi R^2 H$

SPHERE SURFACE AREA : $4\pi R^2$

SPHERE VOLUME : $4\pi R^3/3$

CONVERSION FACTORS (continued)

PHYSICAL QUANTITY	SYMBOL	CONVERSION FACTOR
Pressure	P	1 $lb_f/in.^2 = 6894.8$ N/m^2 1 $lb_f/ft^2 = 47.88$ N/m^2 1 atm $= 101{,}325$ N/m^2
Specific heat capacity	c	1 Btu/(lb$_m$)(°F) $= 4{,}187$ J/kg·K
Specific energy	Q/m	1 Btu/lb$_m = 2326.1$ J/kg
Temperature	T	$T(°R) = (9/5)T(K)$ $T(°F) = [T(°C)](9/5) + 32$ $T(°F) = [T(K) - 273.15](9/5) + 32$
Thermal conductivity	k	1 Btu/(hr)(ft)(°F) $= 1.731$ W/m·K
Thermal diffusivity	α	1 ft^2/sec $= 0.0929$ m^2/s 1 ft^2/hr $= 2.581 \times 10^{-5}$ m^2/s
Thermal resistance	R_t	1 (hr)(°F)/Btu $= 1.8958$ K/W
Velocity	V	1 ft/sec $= 0.3048$ m/s 1 mph $= 0.44703$ m/s
Viscosity, dynamic	μ	1 lb$_m$/(ft)(sec) $= 1.488$ N·s/m^2 1 centipoise $= 0.00100$ N·s/m^2
Viscosity, kinematic	ν	1 ft^2/sec $= 0.0929$ m^2/s 1 ft^2/hr $= 2.581 \times 10^{-5}$ m^2/s
Volume	V	1 ft$^3 = 0.02832$ m^3 1 in.$^3 = 1.6387 \times 10^{-5}$ m^3 1 gal (U.S. liq.) $= 0.003785$ m^3

BASIC HEAT TRANSFER

Basic Heat Transfer

Frank Kreith

Solar Energy Research Institute
and
University of Colorado

William Z. Black

Georgia Institute of Technology

1817

HARPER & ROW, PUBLISHERS, New York
Cambridge, Hagerstown, Philadelphia, San Francisco,
London, Mexico City, São Paulo, Sydney

Sponsoring Editor: Charlie Dresser
Project Editor: Penelope Schmukler
Designer: T. R. Funderburk
Production Manager: Marion A. Palen
Compositor: Science Typographers
Printer and Binder: Murray Printing
Art Studio: J & R Technical Services

BASIC HEAT TRANSFER

Library of Congress Cataloging in Publication Data

Kreith, Frank.
  Basic heat transfer.
  Includes index.
  1. Heat--Transmission.   I.   Black, William Z.,
1940-   joint author.   II.   Title.
QC320.K69   536'.2   79-9190
ISBN 0-700-22518-8

CONTENTS

Preface   xi
Nomenclature   xiii

1 PRINCIPLES OF HEAT TRANSFER   1

1-1   Introduction   1
1-2   Conduction Heat Transfer   2
      Plane wall   3
      Plane walls in series   5
      Electric analog for conduction   7
      Contact resistance   10
      Thermal conductivity   11
1-3   Convection Heat Transfer   14
1-4   Radiation Heat Transfer   19
1-5   Combined Heat-Transfer Mechanisms   21
1-6   Dimensions and Units   25
1-7   Dimensional Analysis   26
      Primary dimensions and dimensional formulas   27
      Buckingham π theorem   27
      Determination of dimensionless groups   29
References   32
Problems   34

Sections marked with an asterisk (*) may be omitted without loss of continuity.

v

2 STEADY STATE CONDUCTION 42

2-1   Introduction 42
2-2   Conduction Equation 43
        Rectangular coordinates 44
        Dimensionless form 46
        Cylindrical coordinates 49
        Spherical coordinates 50
2-3   Steady, One-dimensional Conduction Without Generation 51
        Rectangular coordinates 52
        Cylindrical coordinates 54
        Spherical coordinates 58
        Overall heat-transfer coefficient 59
        Critical insulation thickness for a cylinder 62
*2-4 Effect of Variable Thermal Conductivity 64
2-5   Steady, One-dimensional Conduction with Generation 68
        Rectangular coordinates 68
        Cylindrical coordinates 70
2-6   Heat Transfer from Fins 74
        Fin efficiency 79
2-7   Steady, Two-, and Three-dimensional Conduction 84
        Analytical methods 85
        Graphical methods 87
        Analog methods 93
        Numerical methods 94
        Relaxation techniques 100
        Matrix techniques 106
        Iteration techniques 112
References 118
Problems 119

3 TRANSIENT CONDUCTION 136

3-1   Introduction 136
3-2   Transient Conduction with Negligible Internal Resistance 137
3-3   Transient Conduction in a Semi-infinite Solid 142
3-4   Chart Solutions to Transient Conduction Problems 147
        One-dimensional solutions 147
        Two- and three-dimensional solutions 160
3-5   Numerical Solutions to Transient Conduction Problems 165
        Explicit method 165
        Graphical interpretation of the numerical method 174
        Implicit method 176
References 189
Problems 191

4 ANALYSIS OF CONVECTION HEAT TRANSFER   *197*

4-1   Introduction   *197*
4-2   The Conservation of Mass, Momentum, and Energy Equations for Laminar Flow over a Flat Plate   *202*
4-3   The Integral Momentum and Energy Equations for a Laminar Boundary Layer   *206*
4-4   Evaluation of Heat-Transfer and Friction Coefficients in Laminar Flow   *208*
4-5   Analogy Between Heat and Momentum Transfer in Turbulent Flow over a Flat Surface   *212*
4-6   Reynolds Analogy for Trubulent Flow over a Flat Plate   *218*
4-7   Laminar Forced Convection in a Tube   *221*
4-8   Reynolds Analogy for Turbulent Flow in a Tube   *226*
References   *229*
Problems   *230*

5 ENGINEERING RELATIONS FOR CONVECTION HEAT TRANSFER   *237*

5-1   Introduction   *237*
5-2   Dimensionless Parameters for Correlating Convection Data   *237*
      Correlation of data   *239*
5-3   Convection Heat Transfer in Flow Through Tubes and Ducts   *240*
      Turbulent flow in tubes and ducts   *241*
      Laminar flow in tubes and ducts   *243*
      Liquid metals   *246*
      Forced convection in transition flow   *248*
5-4   Forced-Convection Heat Transfer in External Flow   *248*
      Flat plates   *248*
      Single cylinders and spheres   *249*
      Tube banks   *253*
5-5   Natural Convection   *255*
      Vertical planes and cylinders   *259*
      Horizontal cylinders, spheres, and blocks   *259*
      Free convection in enclosed spaces   *261*
*5-6   Combined Free and Forced Convection   *262*
      Combined free and forced convection in horizontal circular tubes   *263*
      Effects of superimposed free convection in vertical circular tubes   *264*
*5-7   Heat Transfer in High-Speed Flow   *265*
References   *269*
Problems   *272*

6 RADIATION 282

6-1 Introduction 282
6-2 Physics of Radiation 284
Concept of a blackbody 284
Planck's law 284
Wien's displacement law 285
Stefan-Boltzmann law 287
Radiation functions 287
6-3 Radiation Properties 291
Total radiation properties 292
Kirchhoff's law 293
Monochromatic radiation properties 295
Concept of a gray body 299
Directional radiation properties 301
6-4 Radiation Shape Factor 307
Shape-factor algebra 311
Crossed-string method 312
6-5 Radiative Exchange Between Black Surfaces 315
Refractory surfaces 320
6-6 Radiative Exchange Between Gray Surfaces 325
Two gray surfaces forming an enclosure 327
Three gray surfaces forming an enclosure 328
6-7 Matrix Methods 333
Surfaces with known temperatures 334
Surfaces with known net heat flux 337
*6-8 Radiation Through Absorbing, Transmitting Media 346
*6-9 Radiative Properties of Gases 350
*6-10 Solar Radiation 358
Tilted surfaces 368
Solar radiation on earth 372
References 378
Problems 380

7 HEAT EXCHANGERS 395

7-1 Introduction 395
7-2 Basic Types of Heat Exchangers 396
7-3 The Overall Heat-Transfer Coefficient 401
Fouling factors 403
7-4 The Log-Mean Temperature Difference 405
7-5 Heat-Exchanger Effectiveness 413
*7-6 Solar Energy Collectors 422
Energy balance for a flat-plate collector 423

        Collector-heat-loss conductance   *424*
        Collector efficiency factor   *432*
        Collector-heat-removal factor   *433*
*7-7   Heat Transfer and Flow in Packed Beds   *436*
*7-8   Heat Pipes   *438*
        Sonic limitation   *442*
        Entrainment limitation   *444*
        Wicking limitation   *444*
        Boiling limitations   *447*
References   *448*
Problems   *450*

8 CONDENSATION, BOILING, AND MASS TRANSFER   *454*

8-1   Introduction   *454*
8-2   Condensation Heat Transfer   *454*
        Filmwise condensation   *455*
        Effect of turbulence in the film   *460*
        Effect of high vapor velocity   *462*
        Condensation of superheated vapor   *463*
        Dropwise condensation   *463*
        Mixtures of vapors and noncondensable gases   *464*
8-3   Boiling Heat Transfer   *464*
        Correlation of boiling-heat-transfer data   *467*
        Pool boiling   *468*
        Nucleate boiling with forced convection   *471*
        Maximum heat flux in nucleate boiling   *475*
        Boiling and vaporization in forced convection   *475*
        Transition boiling and film boiling   *479*
*8-4   Mass Transfer   *480*
        Mass transfer by diffusion   *481*
        Mass transfer by convection   *486*
References   *491*
Problems   *494*

APPENDIXES

Appendix A
THE INTERNATIONAL SYSTEM OF UNITS   *497*

Appendix B
VECTOR OPERATIONS   *502*

Appendix C
HYPERBOLIC FUNCTIONS *504*

Appendix D
GAUSS ERROR FUNCTION *506*

Appendix E
THERMODYNAMIC PROPERTIES OF SOLIDS *508*

Appendix F
THERMODYNAMIC PROPERTIES OF LIQUIDS *514*

Appendix G
THERMODYNAMIC PROPERTIES OF GASES *520*

Appendix H
THERMODYNAMIC PROPERTIES OF LIQUID METALS *528*

Appendix I
NORMAL EMISSIVITIES *531*

Appendix J
MASS DIFFUSIVITIES *533*

Appendix K
TEMPERATURE CONVERSIONS *535*

Appendix L
DIMENSIONLESS GROUPS *536*

Appendix M
SUBPROGRAM FOR MATRIX INVERSION *538*

Appendix N
HEAT-TRANSFER PROGRAMS *540*

Answers to Odd-Numbered Problems *542*
Index *549*

PREFACE

This book presents an introductory treatment of engineering heat transfer. It is a text designed for a one-semester or a one-quarter course at the junior or senior level for engineering students in any of the classic disciplines. A background in physics, thermodynamics, and ordinary differential equations is assumed; some familiarity with fluid mechanics will be useful, but is not essential for understanding the material in the book.

The presentation follows the classical lines established in the original text of the senior author, but emphasizes applications to engineering problems and use of the computer. Throughout the book, emphasis has been placed on a physical understanding of the processes by which heat is transferred and on how to make appropriate assumptions and simplification in real situations to obtain an engineering answer. The entire book is presented in the SI system of units, which will be universally adopted all over the world in the near future.

Conduction heat transfer is treated from both the analytical and numerical viewpoints, but emphasis is placed on computer solutions. In the treatment of convection, solutions for laminar and turbulent flow are first presented from an analytical viewpoint, and then empirical equations for geometries significant in engineering practice are presented. The number of correlation equations have been reduced to a minimum to simplify the task of the reader in obtaining a numerical answer. Particular emphasis has been placed on defining the appropriate reference temperature to be used in empirical equations for each important geometric configuration. Radiation heat transfer is presented from the point of view of the network method, which is also used to integrate all the heat-transfer processes into a unified treatment.

In the chapter on heat exchangers, the log-mean temperature difference and the effectiveness approaches are presented, and the advantages and disadvantages of each method are pointed out. In addition, the heat-exchanger chapter presents separate sections dealing with modern heat-exchange devices, such as heat pipes, packed beds, and solar collectors.

The problems at the end of each chapter are graduated by increasing complexity. The first few problems are rather routine in nature, whereas other problems require the students to make engineering assumptions and take imaginative approaches to obtain a solution.

The appendixes contain a survey of thermal properties useful in solving the problems. The property tables are, however, merely intended to supplement the discussion in the book and to provide a handy source of data for solving the problems. Whenever answers to problems are provided, they have been obtained with the physical properties listed in the appendixes. For more complete information on physical properties, the reader will, however, have to consult an appropriate reference book.

We have tried to avoid overspecialization in the presentation of the subject matter, and to present material common to many facets of engineering. We make no claim to originality but hope that the method of presentation will make it easy for students to learn and for teachers to teach.

Frank Kreith

William Black

NOMENCLATURE

a	Acceleration	F	Force
a	Velocity of sound	$F_{1 \to 2}$	Shape factor for radiation for area 1 to area 2
A	Area (cross-section)		
A_s	Surface area	g	Acceleration of gravity
c	Specific heat capacity, general	g_c	Inertia proportionality factor
c	Velocity of light	G	Mass flux density ($G = \rho V$)
c	Mole concentration	G	Irradiation (Incident radiant flux)
c_p	Specific heat capacity, constant pressure	h	Planck constant
c_v	Specific heat capacity, constant volume	h	Surface heat transfer coefficient
C	Heat capacity, general	i	Specific enthalpy
C_f	Friction coefficient	I	Radiant intensity
C	Molar concentration	I	Electric current
D_h	Hydraulic diameter	j	Heat transfer factor
D	Diameter	J	Mechanical equivalent of heat
D	Mass diffusion coefficient, mass diffusivity	J	Radiosity
		k	Boltzmann's constant
E	Energy	k	Thermal conductivity
E	Emissive power of a radiating body or gas	l	Length
		l	Prandtl mixing length
E	Electric potential	L	Length
f	Friction factor	m	Mass

xiii

$\dot{m}$	Mass flow rate	y	Coordinate dimension, usually perpendicular to surface
$\mathfrak{M}$	Molecular weight, molar mass		
M	Mach number		
N	Number of molecules	z	Coordinate dimension
N	Mass transfer rate	α	Absorptivity, absorptance
p	Perimeter	α	Thermal diffusivity
p	Pressure	β	Volumetric thermal expansion coefficient
q	Heat transfer rate		
q''	Heat flux	γ	Ratio of specific heats
q'''	Rate of heat generation per unit volume	δ	Film thickness
		δ	Velocity boundary layer thickness
Q	Heat		
$\dot{Q}$	Volumetric flow rate $(\dot{Q} = VA)$	δ_t	Thermal boundary layer thickness
r	Radius	Δ	Change, finite difference
r	Recovery factor	ϵ	Emissivity, emittance
R	Fouling factor	ϵ	Turbulent diffusivity
R	Gas constant	ϵ_H	Eddy diffusivity of heat
R_u	Universal gas constant	ϵ_M	Eddy viscosity
R	Thermal resistance	η	Dynamic (absolute) viscosity
S	Conduction shape factor	η	Efficiency, effectiveness
t	Time	θ	Normal angle
t	Thickness	θ	Temperature difference
T	Temperature	λ	Wavelength
u	Local velocity parallel to surface	μ	Dynamic (absolute) viscosity
		ν	Kinematic viscosity
U	Internal energy	ν	Frequency
U	Overall heat transfer coefficient	ξ	Dimensionless coordinate
		π	Dimensionless group
v	Local velocity perpendicular to surface	ρ	Density
		ρ	Reflectivity, reflectance
v	Specific volume	σ	Stefan-Boltzmann constant
V	Velocity	σ	Surface tension
V	Volume	τ	Dimensionless time
x	Coordinate dimension, usually parallel to surface	τ	Transmissivity, transmittance
		τ	Shear stress
x	Mole fraction	ϕ	Angle
X	Mole fraction	ω	Solid angle
		ω	Mass fraction

BASIC HEAT TRANSFER

Chapter 1

PRINCIPLES OF HEAT TRANSFER

1-1 INTRODUCTION

Whenever a temperature difference exists in the universe, energy will be transferred from the region of higher temperature to the region of lower temperature. According to thermodynamic concepts, the energy that is transferred as a result of a temperature difference is called *heat*. Although the laws of thermodynamics deal with energy transfer, they can only treat systems that are in equilibrium. They can, therefore, be used to predict the amount of energy required to change a system from one equilibrium state to another, but they cannot predict how fast these changes will occur. The science of heat transfer supplements the first and second laws of classical thermodynamics by providing methods of analyses that can be used to predict rates of energy transfer.

To illustrate the different types of information that can be obtained from a thermodynamic and a heat-transfer analysis, consider the heating of a steel rod placed into hot water. Thermodynamic laws can be used to predict the final temperatures after the two systems have reached equilibrium, and the quantity of energy transferred between the initial and final equilibrium states, but they cannot tell us what the rate of heat transfer and the temperature of the rod will be after a given time or how long it will take to obtain a desired temperature. A heat-transfer analysis, on the other hand, can predict the rate of heat transfer from the water to the steel rod, and from this information we can then calculate the temperature of the rod as well as the temperature of the water as a function of time.

For a complete heat-transfer analysis it is necessary to deal with three different mechanisms: conduction, convection, and radiation. The design and analysis of heat-exchange and energy-conversion systems requires familiarity with each of these mechanisms of heat transfer, as well as their interactions. In this chapter we will consider the basic principles of heat transfer and some simple applications. In subsequent chapters each heat-transfer mode will be treated in detail.

1-2 CONDUCTION HEAT TRANSFER

Conduction is the only heat-transfer mode in opaque solid media. When a temperature gradient exists in such a body, heat will be transferred from the higher- to the lower-temperature region. The rate at which heat is transferred by conduction, q_k, is proportional to the temperature gradient, dT/dx, times the area through which heat is transferred, A [Fig. 1-1(a)], or

$$q_k \propto A \frac{dT}{dx} \tag{1-1}$$

where

$$T = \text{temperature}$$
$$x = \text{direction of heat flow}$$

The actual rate of heat flow depends on the thermal conductivity, k, a physical property of the medium. The rate equation can therefore be quantitatively expressed as

$$q_k = -kA \frac{dT}{dx} \tag{1-2}$$

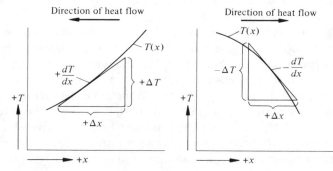

Figure 1-1 (a) Sketch illustrating sign convention for conduction heat flow.

The minus sign is a consequence of the second law of thermodynamics, which requires that heat *must* flow in the direction of lower temperature. The gradient, as shown in Fig. 1-1(b), will be negative if the temperature decreases with increasing values of x. If we designate that heat transferred in the positive direction is to be a positive quantity, the negative sign must be inserted in the right-hand side of Eq. 1-2.

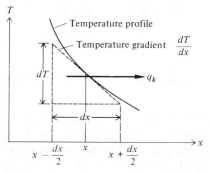

Figure 1-1 (b)   Direction of heat conduction.

Equation 1-2 is *Fourier's law of heat conduction** and serves to define the thermal conductivity k. If the area is in square meters, the temperature in degrees Kelvin, x in meters, and the rate of heat transfer in watts, k has the units of watts per meter per degree Kelvin ($W/m \cdot K$).

Plane Wall

A simple illustration of Fourier's law is the case of heat transfer through a plane wall shown in Fig. 1-2. When both surfaces of the wall are at uniform, but different, temperatures heat will flow only in one direction, perpendicular to the wall surfaces. If the thermal conductivity is uniform, integration of Eq. 1-2 gives

$$q_k = -\frac{kA}{L}(T_2 - T_1) = \frac{kA}{L}(T_1 - T_2) \tag{1-3}$$

where

$$L = \text{thickness of wall}$$
$$T_1 = \text{temperature at left surface } (x = 0)$$
$$T_2 = \text{temperature at right surface } (x = L)$$

*Fourier was a French mathematician (1768–1830) who made important contributions to the analytic treatment of conduction (see Ref. 1).

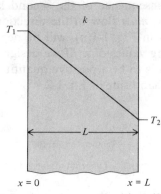

$$x = 0 \qquad\qquad x = L$$

Figure 1-2 Conduction through a plane wall with uniform thermal conductivity.

Example 1-1. A glass window in a storefront has an area of 12 m² and a thickness of 1 cm. The thermal conductivity of the glass is 0.8 W/m·K. On a cold day the outside surface temperature of the glass is 272 K (−1°C)* and the inside surface is at 276 K (3°C). Determine (a) the heat-transfer rate through the glass, and (b) the temperature at a plane midway between the inside and outside glass surfaces.

Solution

a. The heat-transfer rate through the glass is

$$q_k = \frac{kA(T_1 - T_2)}{L} = \frac{0.8 \times 12 \times 4}{0.01} = 3840 \text{ W}$$

b. The temperature of the midplane (T at $L/2$) is 274 K, which is the mean of the two surface temperatures because the temperature profile is linear in the glass.

For many materials the thermal conductivity is not uniform but varies with temperature. In many cases it is possible to approximate this variation over certain ranges of temperature as a linear function of temperature:

$$k(T) = k_0(1 + \beta T) \tag{1-4}$$

where

$k_0 =$ value of conductivity at reference temperature

$\beta =$ empirically determined constant

---

*Temperatures substituted into the Fourier law may be measured in °C or K, because even though the magnitude of the two temperature scales differ by 273, the *size* of the temperature difference is the same. That is, a temperature gradient of 1° C/m is equal to a temperature gradient of 1 K/m.

As shown in more detail in Chapter 2, in such cases integration of Eq. 1-2 gives

$$q_k = \frac{k_0 A}{L} \left[(T_1 - T_2) + \frac{\beta}{2} (T_1^2 - T_2^2) \right] \tag{1-5}$$

or

$$q_k = \frac{k_m A}{L} (T_1 - T_2) \tag{1-6}$$

where k_m is the value of k at the average temperature $(T_1 + T_2)/2$.

Plane Walls in Series

If heat is conducted through several walls in good thermal contact, as for example in the multilayer construction used in most houses, the analysis is only slightly more difficult. In the steady state the rate of heat flow through all the sections must be the same. However, as shown in Fig. 1-3 for a three-layer system, the gradients in the layers are different. The heat-transfer rate can be written for each section and set equal to one another, or

$$q_k = \left(\frac{kA}{L} \right)_A (T_1 - T_2) = \left(\frac{kA}{L} \right)_B (T_2 - T_3) = \left(\frac{kA}{L} \right)_C (T_3 - T_4) \tag{1-7}$$

Eliminating the intermediate temperatures T_2 and T_3 in Eq. 1-7, the rate of

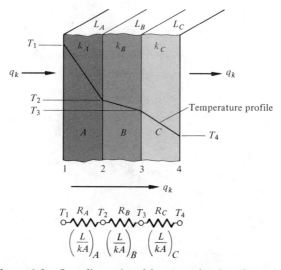

Figure 1-3 One-dimensional heat conduction through a composite wall and electric analog.

heat flow can be written in the form

$$q_k = \frac{T_1 - T_4}{(L/kA)_A + (L/kA)_B + (L/kA)_C} \tag{1-8}$$

For a multilayered slab of N layers in perfect thermal contact, the rate of heat flow is

$$q_k = \frac{T_i - T_{i+1}}{(L/kA)_i} = \frac{T_1 - T_{N+1}}{\displaystyle\sum_{i=1}^{i=N} (L/kA)_i} \tag{1-9}$$

where T_1 is the surface temperature of layer 1 and T_{N+1} is the surface temperature of layer N.

Example 1-2. A furnace wall (see Fig. 1-4) consists of a 1.2-cm-thick stainless steel inner layer covered by a 5-cm-thick outer layer of asbestos board insulation. The temperature of the inside surface of the stainless steel is 800 K and the outside surface of the asbestos is 350 K. Determine the heat-transfer rate through the furnace wall per unit area and the temperature of the interface between the stainless steel and the asbestos. The thermal conductivities for the steel and the asbestos are, respectively:

$$k_1 = 19 \text{ W/m·K}$$
$$k_2 = 0.7 \text{ W/m·K}$$

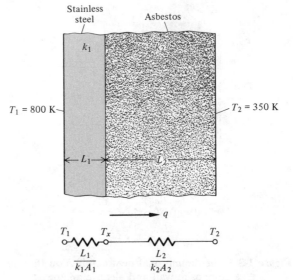

Figure 1-4 Furnace wall for example 1-2.

Solution: The heat-transfer rate is

$$q_k = \frac{T_1 - T_2}{L_1/k_1 A + L_2/k_2 A}$$

The heat-transfer rate per unit area is

$$\frac{q_k}{A} = \frac{T_1 - T_2}{L_1/k_1 + L_2/k_2} = \frac{800 - 350}{0.012/19 + 0.05/0.7} = 6245 \text{ W/m}^2$$

The interface temperature, T_x, is determined from the equation

$$\frac{q_k}{A} = \frac{T_1 - T_x}{L_1/k_1}$$

Solving for T_x gives

$$T_x = T_1 - \frac{q_k}{A}\left(\frac{L_1}{k_1}\right) = 800 - 6245\left(\frac{0.012}{19}\right) = 796 \text{ K}$$

The temperature drop across the stainless steel is therefore only about 4 K; the temperature drop across the asbestos is 446 K.

Electric Analog for Conduction

This is a convenient starting point to introduce a different viewpoint for the analysis of heat transfer which can be applied to more complex problems and will be followed up in later chapters. The new approach makes use of concepts developed in electric-circuit theory and is often called the *analogy between the flow of heat and electricity*. If the heat-transfer rate is considered to be analogous to the flow of electricity, the combination L/kA is viewed as a resistance, and the temperature difference as analogous to a potential difference, Eq. 1-2 can be written in a form similar to *Ohm's law* in electric-circuit theory:

$$q_k = \frac{\Delta T}{R_k} \tag{1-10}$$

where

$$\Delta T = T_1 - T_2, \text{ a thermal potential}$$

$$R_k = \frac{L}{kA}, \text{ a thermal resistance}$$

The reciprocal of the thermal resistance is referred to as the *thermal conductance*, and k/L, the thermal conductance per unit area, is called the *unit thermal conductance* for conduction heat flow. Similarly, Eq. 1-8 can be extended to heat flow through three sections in series, as shown in Fig. 1-4, in the form

$$q_k = \frac{\Delta T}{R_A + R_B + R_C} \quad \text{SERIES} \tag{1-11}$$

where

$$\Delta T = T_1 - T_4$$

$$R_A = \left(\frac{L}{kA}\right)_A$$

$$R_B = \left(\frac{L}{kA}\right)_B$$

$$R_C = \left(\frac{L}{kA}\right)_C$$

The electric-analog approach can also be used to solve more complex problems. For example, in many situations conduction occurs in materials placed in parallel. Figure 1-5 shows a slab consisting of two materials of area A_1 and A_2 in parallel; the corresponding thermal circuit is shown to the right of the physical system. To solve this problem note that for a given temperature difference across the slab each layer of the composite can be analyzed separately, provided the conditions necessary for one dimensional conduction through each of the two sections are met. If the temperature difference between the adjacent materials is small, the heat flow parallel to the layers will dominate over any heat flow normal to the layers and the problem may be treated as one dimensional without any serious loss of accuracy.

Since heat can flow through the two materials along separate paths, the total rate of heat flow is the sum of the two:

$$q_k = q_1 + q_2$$

$$= \frac{T_1 - T_2}{(L/kA)_1} + \frac{T_1 - T_2}{(L/kA)_2} = \left(\frac{1}{R_1} + \frac{1}{R_2}\right)(T_1 - T_2) \qquad (1\text{-}12)$$

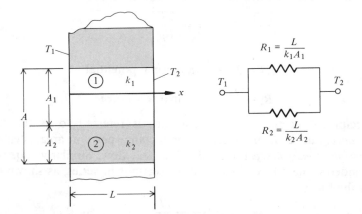

Figure 1-5 Heat conduction through a wall with two sections in parallel.

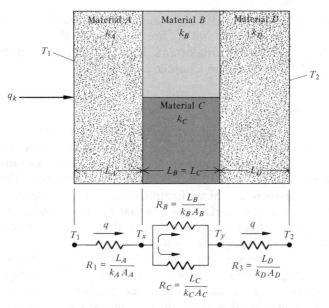

Figure 1-6 Series/parallel thermal circuit, rectangular coordinates.

Observe that the total heat-transfer area is the sum of the two individual areas and that the reciprocal of the total resistance equals the sum of the reciprocals of all the individual resistances. As shown in Fig. 1-5, the thermal circuit for this problem is a parallel arrangement of two resistances, R_1 and R_2.

A more complex application of the thermal network approach is illustrated in Fig. 1-6, where heat is transferred through a composite structure involving thermal resistance in series and in parallel. For this system the resistance of the middle layer, R_2 in Fig. 1-6, becomes

$$R_2 = \frac{R_B R_C}{R_B + R_C} \tag{1-13}$$

and the rate of heat flow is

$$q_k = \frac{\Delta T_{\text{overall}}}{\sum\limits_{n=1}^{n=N} R_n} = \text{WATTS} \tag{1-14}$$

where

$$N = \text{number of layers in series}$$
$$R_n = \text{thermal resistance of } n\text{th layer}$$
$$\Delta T_{\text{overall}} = \text{temperature difference across two outer surfaces}$$

The preceding circuit analysis assumes that the heat flow is one-dimensional. If the resistances R_B and R_C are significantly different, two-dimensional effects can become important. Such two-dimensional conduction problems will be discussed in Chapter 2.

Contact Resistance

When different conducting surfaces are placed in contact as shown in Fig. 1-4, a thermal resistance is often present at the interface of the solids. The interface resistance, frequently called the *contact resistance*, is developed when two materials will not fit tightly together and a thin layer of fluid is trapped between them. Examination of an enlarged view of the contact between the two surfaces shows that the solids touch only at peaks in the surface, and that the valleys in the mating surfaces are occupied by a fluid, possibly air, a liquid or a vacuum.

The interface resistance is primarily a function of surface roughness, the pressure holding the two surfaces in contact, the interface fluid, and the interface temperature. At the interface, the mechanism of heat transfer is complex. Conduction takes place through the contact points of the solid, while heat is transferred by convection and radiation across the trapped interface fluid.

If the heat flux through two solid surfaces in contact is q/A and the temperature difference across the fluid gap separating the two solids is ΔT_i, the interface resistance R_i is defined by

$$R_i = \frac{\Delta T_i}{q/A} \tag{1-15}$$

When two surfaces are said to be in *perfect thermal contact*, the interface resistance approaches zero and there is no temperature difference across the interface. For *imperfect thermal contact*, a temperature difference occurs at the interface.

Most of the problems at the end of the chapter do not consider interface resistance, even though it exists whenever solids are mechanically joined. Regardless of this fact, we should always be aware of the existence of the interface resistance and the resulting temperature difference across the interface. Particularly with rough surfaces and low bonding pressures, the temperature drop across the interface can be significant and should not be ignored.

The subject area of interface resistance is a complex one, and no single theory or set of empirical data accurately describes the interface resistance for surfaces of engineering importance. References 2 and 3 should be consulted for a more detailed discussion of this subject.

Example 1-3. The wall of a house consists of a layer of common brick ($L_1 = 0.100$ m, $k = 0.70$ W/m·K) and a layer of gypsum plaster ($L_2 = 0.038$ m, $k = 0.48$ W/m·K). Compare the rate of heat transfer through this wall with another which has between the brick and gypsum an interface resistance of 0.1 K/W.

Solution: The rate of heat transfer through the idealized wall per square meter area and per degree kelvin temperature difference is

$$\frac{q_k}{A(T_i - T_0)} = \frac{1}{L_1/k_1 + L_2/k_2} = \frac{1}{0.100/0.70 + 0.038/0.48} = 4.50 \text{ W/m}^2\cdot\text{K}$$

The interface will add a third resistance in series, and the rate of heat transfer will be reduced to

$$\frac{q_k}{A(T_i - T_0)} = \frac{1}{R_1 + R_2 + R_{inter}} = \frac{1}{0.222 + 0.1} = 3.11 \text{ W/m}^2\cdot\text{K}$$

Thermal Conductivity

The *thermal conductivity* is a material property defined by Eq. 1-2. Except for gases at low temperatures it is not possible to predict this property analytically. Available information about the thermal conductivity of materials is therefore largely based on experimental measurements. In general, the thermal conductivity of a material varies with temperature, but in many practical situations a constant value based on the average temperature of the system will give satisfactory results. Table 1-1 lists

Table 1-1 Thermal Conductivities of Some Metals, Nonmetallic Solids, Liquids, and Gases

MATERIAL	THERMAL CONDUCTIVITY AT 300 K (W/m·K)
Copper	386.
Aluminum	204.
Carbon steel	54.
Glass	0.75
Plastics	0.2–0.3
Water	0.6
Ethylene glycol	0.25
Engine oil	0.15
Freon (liquid)	0.07
Hydrogen	0.18
Air	0.026

typical values of the thermal conductivities for some metals, nonmetallic solids, liquids, and gases to illustrate the order of magnitude to be expected in practice. Additional information is presented in Appendixes E through H.

The mechanism of thermal conduction in gases can be explained qualitatively by the kinetic theory. All molecules in a gas are in random motion and exchange energy and momentum when they collide with one another. However, since higher temperatures are associated with molecules possessing more kinetic energy, when a molecule from a high-temperature region moves into a region of lower temperature, it transports kinetic energy on a molecular scale to the lower-temperature region. Upon impact with a molecule of lower kinetic energy, an energy transfer occurs which is seen as a transfer of heat from a macroscopic viewpoint. The physical mechanics of conduction in liquids is qualitatively similar, but since

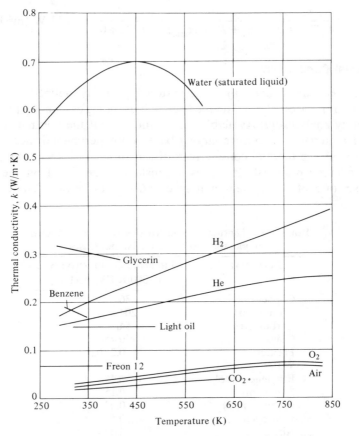

Figure 1-7 Variation of thermal conductivity with temperature for various gases and liquids.

molecules in liquids are more closely spaced and their force fields play a significant role in the energy transfer during collision, the picture is even more complex than in gases.

Figure 1-7 shows how the thermal conductivity of some gases varies with temperature. The thermal conductivity of gases is almost independent of pressure, except near the critical point. According to a simplified analysis based on a kinetic exchange model, the thermal conductivity of gases will increase as the square root of the absolute temperature.

Figure 1-7 also shows the thermal conductivity of some liquids as a function of temperature. It can be seen that except for water, the thermal conductivity of liquids decreases with increasing temperatures, but the

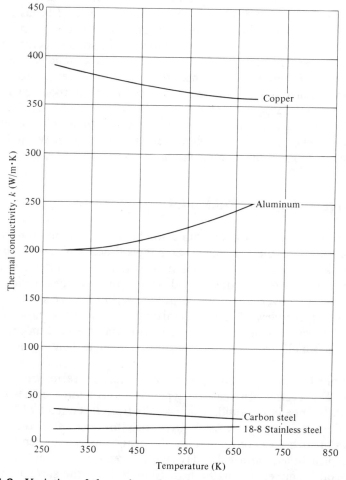

Figure 1-8 Variation of thermal conductivity with temperature for some metals.

change is so small that in most practical situations the thermal conductivity may be assumed constant at some average temperature; there is no appreciable dependence on pressure in liquids.

Figure 1-8 shows the thermal conductivities of some metals and non-metallic solids. In solids, thermal energy is transported by free electrons and by vibrations in the lattice structure. In general the movement of free electrons is the more important mode, and since in good electrical conductors a large number of free electrons move within the lattice structure, good electrical conductors are also good heat conductors (e.g., copper, silver, and aluminum). On the other hand, good electrical insulations are also good thermal insulators (e.g., glass and plastics). The best types of thermal insulators, however, rely for their insulating effectiveness on trapping a gas within a porous structure. In those materials the transfer of heat may occur by several modes: conduction through a fibrous or porous solid structure; conduction and/or convection through air trapped in the void spaces; and radiation between portions of the solid structure, which is especially important at high temperatures or in evacuated enclosures. Special types of superinsulation materials have been developed for cryogenic applications at very low temperatures, down to about 25 K. These kinds of superinsulators consist of several layers of highly reflective materials, separated by evacuated spaces to minimize conduction and convection, and can achieve effective conductivities as low as 0.02 W/m·K. More complete information on superinsulation is given in References 4 and 5.

1-3 CONVECTION HEAT TRANSFER

When a fluid comes in contact with a solid surface at a different temperature, the resulting thermal-energy-exchange process is called *convection heat transfer*. This process is a common experience, but a detailed description of the mechanism is complicated. In this introductory section we will not attempt to cover analytical procedures, but rather concentrate on presenting an overview of the mechanism and present the basic equations that can be used to calculate the rate of convection heat transfer in those subsystems which are important parts of complete heating and cooling systems.

There are two kinds of convection processes: natural or *free convection* and *forced convection*. In the first type the motive force comes from the density difference in the fluid, which results from its contact with a surface at a different temperature and gives rise to buoyant forces. Typical examples of such free convection are the heat transfer between the wall or the roof of a house on a calm day, the convection in a tank in which a heating coil is immersed, or the heat transfer from the surface of a solar collector when there is no wind blowing.

Forced convection occurs when an outside motive force moves a fluid past a surface at a higher or lower temperature than the fluid. Since the fluid velocity in forced convection is larger than in free convection, more heat can be transferred at a given temperature difference. The price to be paid for this increase in the rate of heat transfer is the work required to move the fluid past the surface. But regardless of whether the convection is free or forced, the rate of heat transfer, q_c, can be written in the form of *Newton's law of cooling*:

$$q_c = \bar{h}_c A (T_s - T_{f, \infty}) \qquad (1\text{-}16)$$

where

$\bar{h}_c$ = unit thermal convective conductance, or average convection heat-transfer coefficient, at liquid-to-solid interface, $W/m^{2} \cdot K$

A = surface area in contact with fluid, m^2

T_s = surface temperature, K

$T_{f, \infty}$ = temperature of undisturbed fluid far away from heat-transfer surface, K

Equation 1-16 serves only as a definition of $\bar{h}_c$. The numerical value of $\bar{h}_c$ must be determined analytically or experimentally. The SI units for $\bar{h}_c$ are watts per square meter per degree Kelvin and Table 1-2 lists some approximate values of convection-heat-transfer coefficients, including boiling and condensation, usually considered to be a part of the area of convection.

**Table 1-2 Approximate Values
of Convection-Heat-Transfer Coefficients**

CONVECTION MODE AND FLUID	$\bar{h}_c$ (W/m²·K)
Free convection, air	5–25
Free convection, water	20–100
Forced convection, air	10–200
Forced convection, water	50–10,000
Boiling water	3,000–100,000
Condensing water vapor	5,000–100,000

Example 1-4. Water at 300 K flows over one side of a plate of 1×2 m in area, maintained at 400 K. If the convection-heat-transfer coefficient is 200 $W/m^{2} \cdot K$, calculate the rate of heat transfer by convection from the plate to the water:

Solution: Using Eq. 1-16, the rate of heat transfer is

$$q_c = \bar{h}_c A (T_s - T_{f, \infty}) = 200 \times 2 \times (400 - 300) = 40,000 \text{ W}$$

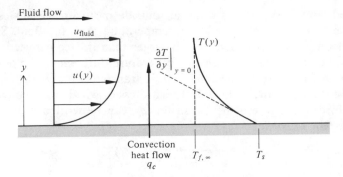

Figure 1-9 Velocity and temperature distributions
for forced convection over a heated plate.

Methods for calculating a heat-transfer coefficient are taken up in
Chapters 4 and 5. Here we shall merely examine the process and qualitatively relate the convection of heat to the flow of the fluid. Figure 1-9
shows a heated flat plate cooled by a stream of air flowing over it. Also
shown are the velocity and the temperature distributions. The first point to
note is that the velocity, $u(y)$, decreases in the direction toward the surface
as a result of viscous forces. Since the velocity of the fluid layer adjacent to
the wall is zero, the heat transfer per unit area* between the surface and
this fluid layer must be by conduction alone:

$$\frac{q_c}{A} = q_c'' = -k_f \frac{\partial T}{\partial y}\bigg|_{y\to 0} = \bar{h}_c(T_s - T_{f,\infty}) \qquad (1\text{-}17)$$

Although this viewpoint suggests that the process can be viewed as
conduction, the temperature gradient at the surface, $(\partial T/\partial y)|_{y=0}$, is de-
termined by the rate at which the fluid farther from the wall can transport
the energy into the mainstream. Thus the temperature gradient at the wall
depends on the flow field, with higher velocities able to produce larger
temperature gradients and higher rates of heat transfer. At the same time,
however, the thermal conductivity of the fluid plays a role. For example,
the value of k_f for the water is an order of magnitude larger than that of
air; thus, as shown in Table 1-2, the convection-heat-transfer coefficient
for water is larger than the coefficient for air.

The situation is quite similar in free convection, as shown in Fig. 1-10.
The principal difference is that in forced convection the velocity approaches the free-steam value imposed by an external force, whereas in
free convection the velocity at first increases with increasing distance from

---

*In this text a prime superscript indicates a quantity per unit length, a double prime is that
quantity per unit area, and a triple prime signifies the quantity per unit volume.

PARALLEL CIRCUIT

$$R = \frac{R_1 R_2}{R_1 + R_2}$$

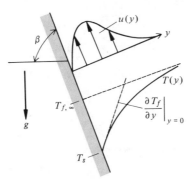

Figure 1-10 Velocity and temperature distributions for free convection over a heated plate inclined on angle β from the horizontal.

the plate because the action of viscosity diminishes rather rapidly while the density difference decreases more slowly. Eventually, however, the buoyant force also decreases as the fluid density approaches the value of the surrounding fluid; this will cause the velocity to reach a maximum and approach zero far away from the heated surface. The temperature fields in free and forced convection have similar shapes, and in both cases the heat-transfer mechanism at the fluid/solid interface is conduction.

The preceding discussion indicates that the convection-heat-transfer coefficient will depend on the density, viscosity, and velocity of the fluid as well as on its thermal properties (thermal conductivity and specific heat). Whereas in forced convection the velocity is usually imposed on the system by a pump or a fan and can be directly specified, in free convection the velocity will depend on the temperature difference between the surface and the fluid, the coefficient of thermal expansion of the fluid (it determines the density change per unit temperature difference) and the force field, which in systems located on earth is simply the gravitational force.

Also, convection heat transfer can be treated within the framework of a thermal resistance network. From Eq. 1-16 the thermal resistance in convection heat transfer is given by

$$R_{CONDUCTION} = \frac{L}{KA} \qquad R_c = \frac{1}{\bar{h}_c A} \qquad (1\text{-}18)$$

and this resistance at a surface to fluid interface can easily be incorporated into a network. For example, the heat transfer from the interior of a room at T_i through a wall to atmosphere outside at T_o is shown in Fig. 1-11. The rate of heat transfer is given by

$$q = \frac{T_i - T_o}{\displaystyle\sum_{i=1}^{i=3} R_i} = \frac{T_i - T_o}{R_1 + R_2 + R_3} = \frac{\Delta T_{overall}}{R_{TOTAL}} \qquad (1\text{-}19)$$

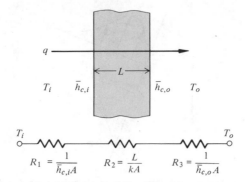

Figure 1-11 Thermal network for heat transfer through a plane wall with convection over both surfaces.

where

$$R_1 = \frac{1}{\bar{h}_{c,i}A}$$

$$R_2 = \frac{L}{kA}$$

$$R_3 = \frac{1}{\bar{h}_{c,o}A}$$

Example 1-5. A 0.1 m-thick brick wall ($k = 0.7$ W/m·K) is exposed to a cold wind at 270 K through a convection-heat-transfer coefficient of 40 W/m²·K. On the other side is calm air at 330 K, with a free convection-heat-transfer coefficient of 10 W/m²·K. Calculate the rate of heat transfer per unit area (i.e., the heat flux).

Solution: From Eq. 1-19 the three resistances are

$$R_1 = \frac{1}{\bar{h}_{c,o}A} = \frac{1}{40 \times 1} = 0.025 \text{ K/W}$$

$$R_2 = \frac{L}{kA} = \frac{0.1}{0.7 \times 1} = 0.143 \text{ K/W}$$

$$R_3 = \frac{1}{\bar{h}_{c,i}A} = \frac{1}{10 \times 1} = 0.10 \text{ K/W}$$

Thus the heat transfer rate per unit area is

$$\frac{q}{A} = \frac{330 - 270}{\left(0.025 + 0.143 + 0.10\right)_l} = 224 \text{ W/m}^2$$

$$\sigma = 5.67 \times 10^{-8} \ W/m^2 \cdot K^4$$

1-4 RADIATION HEAT TRANSFER

Whereas conduction and convection heat transfer can only take place through a material medium, radiation can transport heat even through a perfect vacuum. In the *radiative mode of heat transfer* the energy is transported in the form of electromagnetic waves which travel at the speed of light. There are many different electromagnetic radiation phenomena (e.g. x rays), but here we will consider only thermal radiation that transports energy as heat.

The quantity of energy leaving a surface as radiant heat depends upon the absolute temperature and the nature of the surface. A perfect radiator, or *blackbody*,* emits radiant energy from its surface at a rate q_r given by

BLACK → ~~BERK~~
$$q_r = \sigma A T^4 \tag{1-20}$$

The rate of heat flow by radiation, q_r, will be in watts if the surface area A is in square meters (m^2), the surface temperature T in K, and the dimensional constant σ, called the *Stefan-Boltzmann constant*, is taken at its SI value of $5.67 \times 10^{-8} \ W/m^2 \cdot K^4$.

An inspection of Eq. 1-20 shows that any black surface radiates heat at a rate proportional to the fourth power of the absolute temperature. While the rate of emission is independent of the conditions of the surroundings, the evaluation of a net transfer of radiant heat requires a difference in the surface temperature of two or more bodies between which the exchange is taking place. If a blackbody radiates to an enclosure that completely surrounds it, whose surface is also black (i.e., absorbs all the radiant energy incident upon it), the net rate of radiant heat transfer is given by

BLACK → BLACK
$$q_r = \sigma A_1 (T_1^4 - T_2^4) \tag{1-21}$$

where T_1 is the temperature of the blackbody in degrees Kelvin (K) and T_2 is the surface temperature of the enclosure in degrees Kelvin (K).

Real bodies do not meet the specifications of an ideal radiator but emit radiation at a lower rate than do blackbodies. If they emit, at a temperature equal to that of a blackbody, a constant fraction of blackbody emission at each wavelength, they are called *gray bodies*. A gray body emits radiation at the rate $\epsilon \sigma A T^4$. The net rate of heat transfer from a gray body at a temperature T_1 to a black surrounding body at T_2 is

BLACK → GRAY $$q_r = \sigma A_1 \epsilon_1 (T_1^4 - T_2^4) \tag{1-22}$$

GRAY → BLACK

*A detailed discussion of the meanings of these terms is presented in Chapter 6.

$$\epsilon = \frac{Emmission \ \ GRAY \ @ \ T_o}{Emmission \ \ BLACK \ @ \ T_o}$$

where ϵ_1 is the emittance of the gray surface and is equal to the ratio of emission from the gray surface to the emission from a perfect radiator at the same temperature.

Example 1-6. Calculate the rate of heat loss into space by radiation from the upper surface of a horizontal square flat plate, 2×2 m in area, at a temperature of 500 K, with an emittance of 0.6.

Solution: From Eq. 1-20 the rate at which radiation is emitted from a blackbody at 500 K is

$$q_r = 5.67 \times 10^{-8} \times 4 \times (500)^4$$
$$= 14,180 \text{ W}$$

However, since the emittance is 0.6 for the surface, the actual heat loss will be $0.6 \times 14,180 = 8508$ W.

If neither of two bodies is a perfect radiator and if the two bodies possess a given geometrical relationship to each other, the net heat transfer by radiation between them is given by

$$q_r = \sigma A_1 \mathscr{F}_{1-2}(T_1^4 - T_2^4) \tag{1-23}$$

where $\mathscr{F}_{1-2}$ is a modulus which modifies the equation for perfect radiators to account for the emittances and relative geometries of the actual bodies.

In many engineering problems, radiation is combined with other modes of heat transfer. The solution of such problems can often be simplified by using a thermal resistance, R_r, for radiation. The definition of R_r is similar to that of the thermal resistance for convection and conduction. If the heat transfer by radiation is written

$$q_r = \frac{T_1 - T_2'}{R_r} \tag{1-24}$$

the resistance, by comparison with Eq. 1-23, is given by

$$R_r = \frac{T_1 - T_2'}{\sigma A_1 \mathscr{F}_{1-2}(T_1^4 - T_2^4)} \tag{1-25}$$

Also a unit thermal conductance can be defined for radiation $\bar{h}_r$ by

$$\bar{h}_r = \frac{1}{R_r A_1} = \frac{\sigma \mathscr{F}_{1-2}(T_1^4 - T_2^4)}{T_1 - T_2'} \tag{1-26}$$

where T_2' is any convenient reference temperature, whose choice is often dictated by the convection equation.

Example 1-7. Calculate the radiation unit thermal conductance for a small spherical thermocouple junction located in a large black pipe carrying air. The pipe temperature is 300 K, the thermocouple temperature is 500 K, and the emittance of the thermocouple surface is 0.3.

Solution: From Eqs. 1-22 and 1-26, assuming that the reference temperature is the pipe temperature T_2, we get

$\epsilon \Rightarrow \mathfrak{F}_{1-2}$

$$\bar{h}_r = \frac{\sigma \epsilon_1}{T_1 - T_2}(T_1^4 - T_2^4)$$

$$\boxed{\bar{h}_r = \sigma \epsilon_1 (T_1^2 + T_2^2)(T_1 + T_2)}$$ $*$ (1-26)

$$= 5.67 \times 10^{-8} \times 0.3(500^2 + 300^2)(800)$$

$$= 4.63 \text{ W/m}^2 \cdot \text{K}$$

1-5 COMBINED HEAT-TRANSFER MECHANISMS

In practice, heat is usually transferred in steps through a number of different series-connected sections and heat transfer frequently occurs by two mechanisms in parallel. The transfer of heat from the products of combustion in the chamber of a rocket motor through a thin wall to a coolant flowing in an annulus over the outside of the wall will illustrate such a case (Fig. 1-12).

Products of combustion contain gases, such as CO, CO_2, and H_2O, which emit and absorb radiation. In the first section of this system, heat is therefore transferred from the hot gas to the inner surface of the wall of the rocket motor by the mechanisms of convection and radiation acting in parallel. The total rate of heat flow q to the surface of the wall some

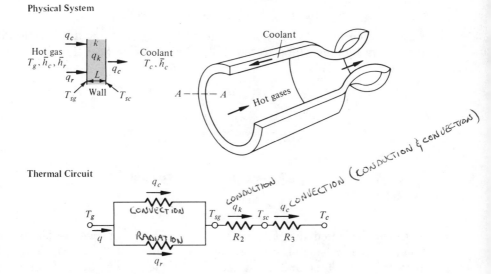

Figure 1-12 Heat transfer in a rocket motor.

distance from the nozzle is

HOT GAS -> WALL SURFACE

$$q = q_c + q_r$$

$$= \bar{h}_c A(T_g - T_{sg}) + \bar{h}_r A(T_g - T_{sg})$$

or

$$q = (\bar{h}_c A + \bar{h}_r A)(T_g - T_{sg})$$

$$= \frac{T_g - T_{sg}}{R_1} \tag{1-27}$$

where

T_g = temperature of hot gas

T_{sg} = temperature at inner surface of wall

$R_1 = \dfrac{1}{(\bar{h}_r + \bar{h}_c)A}$, combined thermal resistance of first section

In the steady state, heat is conducted through the shell, the second section of the system, at the same rate as to the surface and

$$q = q_k = \frac{kA}{L}(T_{sg} - T_{sc}) \qquad \text{O.S. WALL SURFACE} \rightarrow$$

$$\text{INSIDE WALL SURFACE}$$

$$= \frac{T_{sg} - T_{sc}}{R_2} \tag{1-28}$$

where

T_{sc} = surface temperature at wall on coolant side

R_2 = thermal resistance of second section

After passing through the wall, the heat flows through the third section of the system by convection to the coolant. The rate of heat flow in the last step is

WALL SURFACE -> COOLANT

$$q = q_c = \bar{h}_c A(T_{sc} - T_c)$$

$$= \frac{T_{sc} - T_c}{R_3} \tag{1-29}$$

where

T_c = temperature of coolant

$R_3 = $ thermal resistance in third section of system

It should be noted that the symbol $\bar{h}_c$ stands for convection-unit-surface conductance in general, but the numerical values of the convection coefficients in the first and third sections of the system depend on many factors, and will in general be different. Also, the areas of the three heat-flow sections are not equal. But since the wall is very thin, the change in the heat-flow area is so small that it can be neglected in this system.

In practice, often only the temperatures of the hot gas and the coolant are known. If intermediate temperatures are eliminated by algebraic addition of Eqs. 1-27, 1-28, and 1-29, the rate of heat flow is

$$q = \frac{T_g - T_c}{R_1 + R_2 + R_3} = \frac{\Delta T_{\text{total}}}{R_1 + R_2 + R_3} \qquad (1\text{-}30)$$

where the thermal resistances of the three series-connected sections or heat-flow steps in the system are defined in Eqs. 1-27, 1-28, and 1-29.

In Eq. 1-30 the rate of heat flow is expressed only in terms of an overall temperature potential and the heat-transfer characteristics of individual sections in the heat-flow path. From these relations it is possible to evaluate quantitatively the importance of each individual thermal resistance in the path. An inspection of the order of magnitudes of the individual terms in the denominator often indicates the means of simplifying a problem. When one or the other term dominates quantitatively, it is sometimes permissible to neglect the rest. As we gain facility in the techniques of determining individual thermal resistances and conductances, there will be numerous occasions where such approximations will be illustrated. There are, however, certain types of problems, notably in the design of heat exchangers, where it is convenient to simplify the writing of Eq. 1-30 by combining the individual resistances or conductances of the thermal system into one quantity, called the overall unit conductance, the overall transmittance, or the overall coefficient of heat transfer, U. The use of an overall coefficient is a convenience in notation, and it is important not to lose sight of the significance of the individual factors that determine the numerical value of U.

Writing Eq. 1-30 in terms of an overall coefficient gives

$$q = UA\Delta T_{\text{total}} \qquad (1\text{-}31)$$

where

$$UA = \frac{1}{R_1 + R_2 + R_3} = \frac{1}{R_{\text{total}}} \qquad (1\text{-}32)$$

The overall coefficient U may be based on any chosen area. To avoid misunderstandings, the area basis of an overall coefficient should therefore always be stated. Additional information about the overall heat transfer coefficient U will be presented in later chapters.

The overall heat-transfer coefficient will be found useful primarily in problems involving thermal systems consisting of several series-connected sections. The analysis of heat flow at boundaries of complicated geometry and in unsteady-state conduction problems can be simplified by using a combined unit-thermal-surface conductance $\bar{h}$. The combined unit-thermal-surface conductance, or *unit-surface conductance* for short, combines the effects of heat flow by convection and radiation between a surface and a fluid and is defined by

$$\bar{h} = \bar{h}_c + \bar{h}_r \tag{1-33}$$

The unit-surface conductance specifies the average total rate of heat flow per unit area between a surface and a fluid per unit temperature difference. Its units are $W/m^2 \cdot K$.

Example 1-8. A 0.5-m-diameter pipe ($\epsilon = 0.9$) carrying steam has a surface temperature of 500 K. The pipe is located in a room at 300 K and the convection-heat-transfer coefficient between the pipe surface and the air in the room is 20 $W/m^2 \cdot K$. Calculate the combined unit surface conductance and the rate of heat loss per meter of pipe length.

Solution: This problem may be idealized as a small object (the pipe) inside a large black enclosure (the room). Thus the radiation-heat-transfer coefficient is $h_r = 5.67 \times 10^{-8} (.9)(500^2 + 300^2)(500 + 300) = 13.88$

$$h_r = \sigma\epsilon(T_1^2 + T_2^2)(T_1 + T_2) = 13.9 \ W/m^2 \cdot K$$

The combined unit surface conductance is

$$h = h_c + h_r = 20 + 13.9 = 33.9 \ W/m^2 \cdot K$$

and the rate of heat loss per meter is

$$q = \pi D L h (T_{pipe} - T_{air}) = \pi \times 0.5 \times 1 \times 33.9 \times 200 = 10{,}650 \ W$$

Example 1-9. In the design of a heat exchanger for aircraft application (Fig. 1-13), the maximum wall temperature is not to exceed 800 K. For the conditions tabulated below, determine the maximum permissible unit thermal resistance per square-meter area of the metal wall between hot gas on the one side and cold gas on the other.

Hot-gas temperature $= T_g = 1300$ K

Unit-surface conductance on hot side, $\bar{h}_1 = 200 \ W/m^2 \cdot K$

Unit-surface conductance on cold side, $\bar{h}_c = 400 \ W/m^2 \cdot K$

Coolant temperature $= T_c = 300$ K

Solution: In the steady state we can write

(q/A) from gas to hot side of wall

$\qquad = (q/A)$ from hot side of wall through wall to cold gas

Physical System

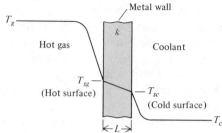

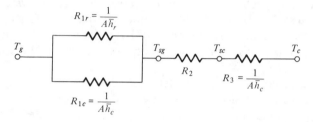

Detailed Thermal Circuit

Simplified Circuit

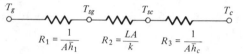

Figure 1-13 Physical system and thermal circuit for example 1-9.

or

$$\frac{q}{A} = \frac{T_g - T_{sg}}{R_1} = \frac{T_g - T_c}{R_1 + R_2 + R_3} = \frac{1300 - 800}{1/200} = \frac{1300 - 300}{(1/200) + R_2 + (1/400)}$$

where T_{sg} is the hot surface temperature. Substituting numerical values for the unit thermal resistances and temperatures yields

$$\frac{1300 - 800}{0.005} = \frac{1300 - 300}{R_2 + 0.0075}$$

Solving for R_2 gives

$$R_2 = 0.0025 \ \text{m}^2 \cdot \text{K/W}$$

A unit thermal resistance larger than 0.0025 $\text{m}^2 \cdot \text{K/W}$ would raise the inner wall temperature above 800 K.

1-6 DIMENSIONS AND UNITS

A *dimension* is a name given to any measurable quantity. For example, the space occupied by an object is qualified by the dimension called the volume. The distance between two points is qualified by the dimension called the length. Common dimensions used in a heat-transfer course are length, time, mass, force, heat, and temperature.

Before numerical calculations can be made, each dimension must be quantified with a defined, reproducible *unit*. Units are the arbitrary names that specify the magnitude of each dimension. For example, the meter is a unit for the dimension of length. Other units of length have also been used to quantify the dimension of length. Some of these are foot, yard, mile, millimeter, centimeter, and kilometer.

Several different unit systems are presently in use throughout the world. In industry and research and development, the SI system (Système International d'Unités) is fast becoming the most widely used system of units. The SI system has been adopted by the International Organization for Standardization and is recommended by a large number of national standard organizations. For these reasons we will use SI units throughout this book.

The units used in the SI system are described in Appendix A. A complete list of conversion factors between the SI system and the engineering system of units that is frequently used in the United States today appears in Table A-6. For convenience, a condensation of this table appears on the inside cover of the text.

The units assigned to the SI system and other commonly used systems are summarized in Table 1-3.

Table 1-3   Base Units and Derived Units for Several Systems

DIMENSION	SYSTEM			
	SI	MKS	CGS	ENGINEERING
Length	m	m	cm	ft
Time	s	s	s	sec
Force	N	N	dyne	lb_f
Mass	kg	kg	g	lb_m
Temperature	K	°C	°C	°F
Heat	J	kcal	cal	Btu

1-7 DIMENSIONAL ANALYSIS

The inconvenience of changing from one set of units to another can often be avoided by using dimensionless parameters whose values are the same in any set of units. The process of determining appropriate dimensionless numbers is called *dimensional analysis*. This method not only combines several variables into dimensionless groups which are independent of the system of units, but it facilitates interpretation of experimental data.

The most serious limitation of dimensional analysis is that it gives no information about the nature of a phenomenon. In fact, to apply dimensional analysis it is necessary to know beforehand what variables influence the phenomenon, and the success or failure of the method depends on the proper selection of these variables. It is therefore important to have at least a preliminary theory or a thorough physical understanding of a phenomenon before a dimensional analysis can be performed. However, once the pertinent variables are known, dimensional analysis can be applied to most problems by a routine procedure which is outlined below.*

Primary Dimensions and Dimensional Formulas

The first step is to select a system of primary dimensions. The choice of the primary dimensions is arbitrary, but the dimensional formulas of all pertinent variables must be expressible in terms of them. In the SI system the primary dimensions of length L, time θ, temperature T, and mass M are used.

The dimensional formula of a physical quantity follows from definitions or physical laws. For instance, the dimensional formula for the length of a bar is $[L]$ by definition.† The average velocity of a fluid particle is equal to a distance divided by the time interval taken to traverse it. The dimensional formula of velocity is therefore $[L/\theta]$, or $[L\theta^{-1}]$ (i.e., a distance or length divided by a time). The units of velocity could be expressed in meters per second, feet per second, or miles per hour, since they all are a length divided by a time. The dimensional formulas and the symbols of physical quantities occurring frequently in heat-transfer problems are given in Table 1-4

Buckingham π Theorem

To determine the number of independent dimensionless groups required to obtain a relation describing a physical phenomenon, the Buckingham π (pi) theorem may be used.‡ According to this rule, the required number of independent dimensionless groups that can be formed by combining the physical variables pertinent to a problem is equal to the total number of these physical quantities n (e.g., density, viscosity, heat-transfer coefficient)

---

*The algebraic theory of dimensional analysis will not be developed here. For a rigorous and comprehensive treatment of the mathematical background, Chapters 3 and 4 of Reference 7 are recommended.

†Square brackets [ ] denote that the quantity has the dimensional formula stated within the brackets.

‡A more rigorous rule, proposed by van Driest (Ref. 6), shows that the π theorem holds as long as the set of simultaneous equations formed by equating the exponents of each primary dimension to zero is linearly independent. If one equation in the set is a linear combination of one or more of the other equations (i.e., if the equations are linearly dependent), the number of dimensionless groups is equal to the total number of variables n minus the number of independent equations.

Table 1-4 Some Physical Quantities with Associated Symbols and Dimensions

QUANTITY	SYMBOL	DIMENSIONS IN $ML\theta T$ SYSTEM
Length	L, x	L
Time	t	θ
Mass	M	M
Force	F	ML/θ^2
Temperature	T	T
Heat	Q	ML^2/θ^2
Velocity	V	L/θ
Acceleration	a, g	L/θ^2
Work	W	ML^2/θ^2
Pressure	p	$M/\theta^2 L$
Density	ρ	M/L^3
Internal energy	u	L^2/θ^2
Enthalpy	h	L^2/θ^2
Specific heat	c	$L^2/\theta^2 T$
Absolute viscosity	μ	$M/L\theta$
Kinematic viscosity	$\nu = \mu/\rho$	L^2/θ
Thermal conductivity	k	$ML/\theta^3 T$
Thermal diffusivity	α	L^2/θ
Thermal resistance	R	$T\theta^3/ML^2$
Coefficient of expansion	β	$1/T$
Surface tension	σ	M/θ^2
Shear per unit area	τ	$M/L\theta^2$
Unit-surface conductance	$\bar{h}_c$	$M/\theta^3 T$
Mass flow rate	$\dot{m}$	M/θ

minus the number of primary dimensions m required to express the dimensional formulas of the n physical quantities. If we call these groups $\pi_1, \pi_2, \ldots$, the equation expressing the relationship among the variables has a solution of the form

$$F(\pi_1, \pi_2, \pi_3, \ldots) = 0 \tag{1-34}$$

In a problem involving five physical quantities and three primary dimensions, $n - m$ is equal to 2 and the solution either has the form

$$F(\pi_1, \pi_2) = 0 \tag{1-35}$$

or the form

$$\pi_1 = f(\pi_2) \tag{1-36}$$

Experimental data for such a case can be presented conveniently by plotting π_1 against π_2. The resulting empirical curve reveals the functional relationship between π_1 and π_2 which cannot be deduced from dimensional analysis.

For a phenomenon that can be described in terms of three dimensionless groups (i.e., if $n - m = 3$), Eq. 1-34 has the form

$$F(\pi_1, \pi_2, \pi_3) = 0 \tag{1-37}$$

but can also be written as

$$\pi_1 = f(\pi_2, \pi_3) \tag{1-38}$$

For such a case, experimental data can be correlated by plotting π_1 against π_2 for various values of π_3. Sometimes it is possible to combine two of the π's in some manner and to plot this parameter against the remaining π on a single curve, as shown in Chapter 5.

Determination of Dimensionless Groups

A simple method for determining dimensionless groups will now be illustrated by applying it to a conduction-heat-transfer problem and to a problem in fluid flow.

Example 1-10. Determine dimensionless parameters to relate the maximum temperature, T_m, in a slab of thickness L, with thermal conductivity k if heat is generated uniformly at the rate q_G''' per unit volume and one surface is maintained at temperature T_1 while the other surface is insulated.

Solution: We begin by writing π as a product of the variables, each raised to an unknown power

$$\pi = T_m{}^a T_1{}^b k^c L^d q_G'''^e$$

and then substitute the dimensional formulas from Table 1-4:

$$\pi = [T]^a [T]^b \left[\frac{ML}{\theta^3 T} \right]^c [L]^d \left[\frac{M}{\theta^3 L} \right]^e$$

For π to be dimensionless, the exponents of each primary dimension must separately add up to zero. Equating the sum of the exponents of each primary dimension to zero, we obtain the set of equations

$$a + b - c = 0 \qquad \text{for } T$$

$$c + e = 0 \qquad \text{for } M$$

$$-3c - 3e = 0 \qquad \text{for } \theta$$

$$c + d - e = 0 \qquad \text{for } L$$

Evidently, any set of values of a, b, c, d, and e that simultaneously satisfies this set of equations will make π dimensionless. There are five unknowns but only four equations. We should notice that the balances for M and θ produce the same equation, so there are only three independent equations rather than four. We can therefore choose values for two of the exponents in each of the dimensionless groups. The only restriction on the choice of the exponents is that each of the selected exponents be independent of the others. An exponent is independent if the determinant formed with the coefficients of the remaining terms does not vanish (i.e., is not equal to zero).

Since T_m is the variable we eventually want to evaluate, set its exponent, a, equal to 1. Since we want q_G''' to be the independent variable, we do not want to combine it with T_m and therefore set its exponent $e = 0$. This gives the equations

$$1 + b - c = 0$$
$$c + 0 = 0$$
$$c + d - 0 = 0$$

Solving the equations simultaneously we get $b = -1$, $c = 0$, $d = 0$, and the first dimensionless group is

$$\pi_1 = \frac{T_m}{T_1}$$

the ratio of the maximum temperature to the surface temperature.

For π_2 we let a be zero, so T_m will not appear again, and $e = 1$, to have the independent variable appear to the first power in π_2. Simultaneous solution of the equations with these choices yields $c = -1$, $b = -1$, $d = 2$; thus

$$\pi_2 = \frac{L^2 q_G'''}{k T_1}$$

This problem can be expressed in terms of Eq. 1-36 as

$$\frac{T_m}{T_1} = f\left(\frac{L^2 q_G'''}{k T_1} \right)$$

Dimensionless analysis cannot reveal the nature of the functional relation between π_1 and π_2, but in Chapter 2 it will be shown that

$$\frac{T_m}{T_1} = 1 + \frac{L^2 q_G'''}{2 k T_1}$$

Example 1-11. Find dimensionless parameters to correlate data for the pressure drop Δp in a pipe of diameter D and length L when a fluid of viscosity μ and density ρ is flowing through the pipe at an average velocity V.

Solution: There are six variables in this problem, but only three dimensions leading to three independent equations. Thus, three variables must be selected in each evaluation of a dimensionless π parameter. The variables of the problem, their dimensions, and their exponents are tabulated below.

VARIABLE	DIMENSION	EXPONENT
L	$[L]$	a
Δp	$[M/L\theta^2]$	b
ρ	$[M/L^3]$	c
D	$[L]$	d
μ	$[M/L\theta]$	e
V	$[L/\theta]$	f

Next we write

$$\pi = [L]^a \left[\frac{M}{L\theta^2}\right]^b \left[\frac{M}{L^3}\right]^c [L]^d \left[\frac{M}{L\theta}\right]^e \left[\frac{L}{\theta}\right]^f$$

For the first dimensionless group we let $a=1$, $b=0$, and $c=0$ and obtain the following equations:

$$L: \quad 1+d-e+f=0$$
$$M: \quad e=0$$
$$\theta: \quad -e-f=0$$

Thus $d=-1$ and

$$\pi_1 = \frac{L}{D}$$

For the second dimensionless group we let $a=0$, $b=1$, and $d=0$. This yields, upon simultaneous solution of the pertinent equations,

$$\pi_2 = \frac{\Delta p}{\rho V^2}$$

Similarly, if we let $e=1$, $b=0$, and $a=0$, the third dimensionless group is

$$\pi_3 = \frac{\mu}{\rho VD}$$

To correlate experimental data one could plot π_2 versus π_3 for various values of π_1. However, in engineering practice it has been found convenient to use the quotient of π_2 and π_1, called the *friction factor*, f. Moreover, it can be shown analytically that

$$f = \pi_2/\pi_1 = \frac{\Delta p D}{\rho V^2 L} = \phi(\pi_3)$$

Figure 1-14 shows experimental results, plotted as π_2/π_1 versus the reciprocal of π_3. As discussed in Chapter 4, if $1/\pi_3$ is less than about 2300, the flow is laminar and

$$f = \pi_2/\pi_1 = 64\pi_3$$

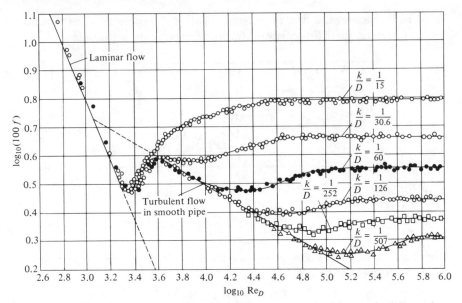

Figure 1-14 Relation between friction factor, f, and reynolds number, Re_D, for smooth and artificially roughened pipes [$\log_{10}(\pi_1\pi_2)$ versus $\log_{10}(1/\pi_3)$]. From J. Nikuradse, "Gesetzmassigkeiten der turbulenten Stromung in glatten Rohren," *VDI Forschungsheft*, vol. 356, 1932; "Stromungsgesetze in rauhen Rohren," *VDI Forschungsheft*, vol. 361, 1933.

But when $1/\pi_3 > 2300$, the functional relationship between f and π_3 changes because the flow undergoes a change from "laminar" to "turbulent" flow. Dimensional analysis cannot, of course, predict this physical phenomenon. Moreover, in turbulent flow not only π_3, but also the surface roughness, affects the value of f. This is clearly illustrated in Fig. 1-14, where experimental results for π_2/π_1 versus π_3^{-1} are plotted in the laminar- and turbulent-flow regimes. In turbulent flow the data show a pronounced dependence on the ratio of the average height of surface roughness elements, k, to the pipe diameter, D. For smooth pipes, however, the empirical relation

$$\pi_2/\pi_1 = f = 0.046\pi_3^{0.2} = 0.046/\text{Re}_D^{0.2}$$

correlates the experimental data in the turbulent-flow regime over a wide range of π_3 (Ref. 8).

REFERENCES

1. J. B. Fourier, *Théorie analytique de la chaleur*, Paris, 1822; A. Freeman, trans., Dover Publications, Inc., New York, 1955.
2. W. M. Rohesenow and J. P. Hartnett, eds. *Handbook of Heat Transfer*, Sec. 3 (by P. J. Schneider), McGraw-Hill Book Co., New York, N.Y., 1973.

3. T. N. Veizirogen, "Correlation of Thermal Contact Conductance Experimental Results," Prog. Astron. Aero., *20*, Academic Press, Inc., New York, 1967.
4. R. W. Vance and W. M. Duke, eds., *Applied Cryogenic Engineering*, John Wiley & Sons, New York, 1962.
5. R. Barron, *Cryogenic Systems*, McGraw-Hill Book Co., New York, 1967.
6. E. R. Van Driest, "On Dimensional Analysis and the Presentation of Data in Fluid Flow Problems," *J. Appl. Mech.*, vol. 13, 1940.
7. H. L. Langhaar, *Dimensional Analysis and Theory of Models*, John Wiley & Sons, Inc., New York, 1951.
8. J. Nikuradse, "Gesetzmässigkeiten der turbulenten Strömung in glatten Rohren," *VDI Forschungsheft*, vol 356, 1932; "Strömungsgesetze in rauhen Rohren," *VDI Forschungsheft*, vol. 361, 1933.

PROBLEMS

The problems in this chapter are organized in the manner shown in the table.

PROBLEM NUMBERS	SECTIONS	SUBJECT
1-1 to 1-18	1-2	Conduction heat transfer
1-19 to 1-25	1-3	Convection heat transfer
1-31 to 1-34	1-4	Radiation heat transfer
1-35 to 1-37	1-5	Combined modes
1-38 to 1-41	1-7	Dimensional analysis

1-1 Determine the heat-transfer rate per unit surface area through a brick wall ($k = 0.3$ W/m·K) when one surface of the brick is at 25°C and the other surface is at -10°C. The thickness of the wall is 10 cm.

1-2 A furnace wall is constructed of silica brick, $k = 1.1$ W/m·K, and the inside surface of the brick is 450°C. The thickness of the wall is 30 cm and the exterior surface of the brick is at a temperature of 55°C. Determine the heat-transfer rate through the brick per unit area of the wall.

1-3 The heat-transfer rate through a plane wall is 1000 W/m². One surface of the wall is maintained at 100°C. The thermal conductivity of the wall is 28 W/m·K and it is 25 cm thick. Determine the temperature of the second surface of the wall.

1-4 The temperature distribution in a plane wall with a thermal conductivity of 2.0 W/m·K is

$$T(x) = 100 + 150x$$

where T is in °C and x is in meters measured from one surface of the wall. Determine the heat flux conducted through the wall. Is the heat flow in the positive or negative x direction?

1-5 The heat-conduction rate through a 1-cm-thick piece of Plexiglas ($k = 0.195$ W/m·K) is 300 W. The area of the Plexiglas is 2 m². One surface of the Plexiglas is maintained at a temperature of 30°C. Determine the temperature of the other surface and the temperature of the midplane of the Plexiglas.

1-6 The thermal resistance of a wall of a residential dwelling is 9.0 K/W. Determine the heat-transfer rate through 30 m² of wall surface when the temperature difference across the wall is 30°C.

1-7 The R factor frequently used to describe the resistance to heat flow of insulations is defined as

$$R_{\text{factor}} = \frac{L}{k}$$

where L is the thickness of the insulation. Calculate the R factor for 10-cm thicknesses of the following materials: Fiberglas, plaster, plywood, and common brick. Use the thermal conductivity values given in Table E-3.

1-8 Considering heat-transfer principles, is stainless steel a good material to use in making cooking utensils? What factors in addition to heat transfer must be considered when designing equipment, such as cookware, that is purchased by the general public.

1-9 Several rods 1 cm in diameter and 10 cm long are insulated around their periphery. One end of the rods is at 100°C while the other is at 0°C, so the heat flows axially by conduction. Determine the heat transfer rates through rods made of (a) copper, (b) aluminum, (c) stainless steel, (d) asbestos, (e) cardboard, and (f) fiberglass. Use the thermal conductivity values given in Appendix E.

1-10 A thin, flat heater with a temperature of 200°C and area of 0.2 m² is to be sandwiched between two layers of insulation with $k = 0.35$ W/m·K. The energy is dissipated by the heater at a rate of 1000 W. Calculate the thickness of insulation required to ensure that its exterior surface is at a temperature less than 50°C.

1-11 The thermal conductivity of a material varies with temperature according to the relationship

$$k = 2.2 + (4 \times 10^{-4})T$$

where k is measured in W/m·K and T is in K. Determine the heat-transfer rate when two planes separated by 40 cm of this material are maintained at temperatures of 100°C and 200°C. The cross-sectional area of the material is 1.8 m².

1-12 Integrate Eq. 1-2 for the case of a plane surface with thermal conductivity that varies linearly with temperature and verify Eq. 1-6.

1-13 Determine the heat flux through a plane wall that has a thermal conductivity that varies quadratically as

$$k = k_0(1 + BT + CT^2)$$

Express your answer in terms of k_0; B; C; the temperatures of both surfaces of the wall, T_1 and T_2; and the width of the wall, L. Calculate the heat flux through the wall when

$$T_1 = 200°C \qquad T_2 = 500°C$$
$$L = 15 \text{ cm} \qquad k_0 = 15 \text{ W/m·K}$$
$$B = 10^{-4}K^{-1} \qquad C = 10^{-8}K^{-2}$$

1-14 A metallic refrigerator wall is to be covered with a rigid foam insulation that has a thermal conductivity of 0.03 W/m·K. The interior of the refrigerator is to be maintained at $-20°C$. The cooling capacity of the refrigerator is 2 kW and the surface area of the refrigerator wall is 100 m². Determine the minimum insulation thickness needed so that condensation will not occur on the exterior surface of the insulation, assuming that the dew-point temperature of the air outside the refrigerator is 15°C.

1-15 Assume one-dimensional heat transfer through the composite wall shown in the figure:

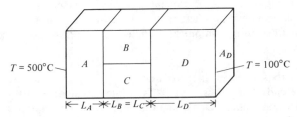

(a) Draw the thermal circuit for the wall, labeling all resistances and known potentials with appropriate symbols.
(b) Determine the heat-transfer rate through the wall.
(c) Determine the temperature of the left-hand face of material D.

$$k_A = 75 \text{ W/m·K} \qquad L_A = 20 \text{ cm}$$
$$k_B = 60 \text{ W/m·K} \qquad L_B = L_C = 25 \text{ cm}$$
$$k_C = 58 \text{ W/m·K} \qquad L_D = 40 \text{ cm}$$
$$k_D = 20 \text{ W/m·K} \qquad A_A = A_D = 2m^2$$
$$A_B = A_C$$

1-16 At steady state the temperature profile in a laminated system is as shown. Which material has the higher thermal conductivity? Show sufficient work and reasoning to justify your answer.

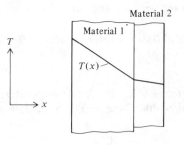

1-17 Consider the composite wall shown in the figure. The left surface of the wall is submerged in water that has an ambient temperature of 70°C and the convective-heat-transfer coefficient on that surface is 60 W/m²·K. Determine the value for k_x.

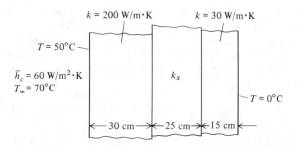

1-18 At steady state the temperature profile in a three-material composite body is shown in the figure. Determine the correct conclusion:

(a) $k_A > k_B > k_C$.
(b) $k_A > k_C > k_B$.
(c) $k_B > k_A > k_C$.
(d) $k_C > k_A > k_B$.
(e) $k_C > k_B > k_A$.

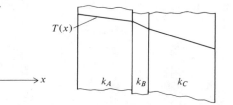

1-19 On a calm day the convective-heat-transfer coefficient for the roof of a building is 6 W/m²·K. Determine the convective-heat-transfer rate from the roof of the building if the exterior surface of the roof is at 15°C and the ambient air is at −5°C. The surface area of the roof is 400 m². Calculate the heat-transfer rate if $\bar{h}_c$ increases to 85 W/m²·K due to a breeze.

1-20 Determine the convective-heat-transfer rate from a 1-cm-diameter ball bearing that has a temperature of 200°C submerged in oil that has an ambient temperature of 100°C. Assume that $h = 1000$ W/m²·K.

1-21 Calculate the convective-heat-transfer rate from a plate at 200°C to air with an ambient temperature of 30°C. The convective-heat-transfer coefficient between the plate and air is 30 W/m²·K and the surface area of the plate is 10 m².

1-22 A 1000-W electric heater with a surface area of 0.1 m² is exposed to a 20°C fluid. Calculate the surface temperature of the heater for the following situations:

(a) The heater is surrounded by air with $\bar{h}_c = 30$ W/m²·K.
(b) The heater is surrounded by still water with $\bar{h}_c = 500$ W/m²·K.
(c) The heater is surrounded by agitated water with $\bar{h}_c = 5000$ W/m²·K.

1-23 A 100-W electric heater is surrounded by 20°C air with $\bar{h}_c = 50$ W/m²·K. What is the minimum surface area of the heater such that the surface temperature of the heater will not exceed 60°C?

1-24 Air at 20°C is blown over the top surface of a 10-cm-thick pure iron horizontal plate. The convective-heat-transfer coefficient is 20 W/m²·K and the top surface of the plate gains 350 W/m² by radiation from the surroundings. The lower surface of the plate transfers 200 W/m² to the surroundings. Draw the thermal circuit for this situation and calculate the steady-state temperatures of both surfaces of the iron plate.

1-25 One surface of a flat wall is exposed to a fluid with an ambient temperature of 20°C. The surface of the wall adjacent to the fluid is covered with a 4-cm-thick insulation that has a thermal conductivity of 0.5 W/m·K. The temperature of the plane separating the wall and insulation is maintained at 500°C. Compute the value for the convective-heat-transfer coefficient which must be maintained on the outer surface of the insulation so that the temperature of the outer surface of the insulation does not exceed 50°C. Calculate the heat-transfer rate through the insulation per m² of surface area.

1-26 A composite wall shown in the figure is comprised of two different materials. One surface of the wall is kept at 20°C while the other surface is exposed to air with an ambient temperature of 150°C:

$q = 738.9 \text{ W/m}^2$

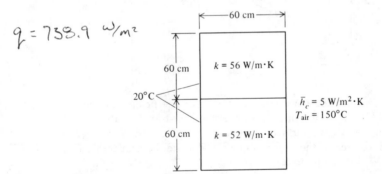

(a) Draw the thermal circuit for this problem.
(b) Calculate all thermal resistances.
(c) Determine the heat-transfer rate through the wall per unit depth of wall.
(d) Calculate the surface temperature of the wall that is exposed to the air.

1-27 Two rough metal surfaces are pressed together. One surface has a thickness of 15 cm and a thermal conductivity of 45 W/m·K while the other surface has a thickness of 25 cm and $k = 70$ W/m·K. The interface resistance per unit area between the two surfaces is 10^{-2} K·m²/W. The overall temperature difference across the exterior surfaces of the two surfaces is 400°C:
(a) Draw the thermal circuit for this system.

(b) Determine the heat-transfer rate per unit surface area.

(c) Calculate the temperature difference across the interface resistance.

(d) Determine the heat-transfer rate per unit area assuming no interface resistance.

1-28 A small transistor is rated at 250 mW. It is to be cooled by attaching an array of aluminum fins that have a total surface area of 10 cm². The ambient air temperature surrounding the fins is 25°C and the convective-heat-transfer coefficient on the surface of the fins is 12 W/m²·K. Because the fins are made from aluminum, they have a low thermal resistance and they may be assumed to be isothermal over their entire surface. Determine the steady-state operating temperature of the transistor.

1-29 In Problem 1-28 there is an interface resistance between the body of the transistor and the array of fins. The interface resistance is estimated to be 60 K/W. Draw a thermal circuit for this problem. Identify all resistances. Estimate the steady-state operating temperature of the transistor. Determine the temperature drop across the interface resistance. Suggest ways to eliminate the interface resistance.

1-30 The cross section of a typical home ceiling is shown in the figure. Draw the thermal circuit for a typical section of the ceiling. Estimate the heat-transfer rate through the ceiling per m² of area. Is more heat transferred through the studs or through insulation? Assume one-dimensional conduction.

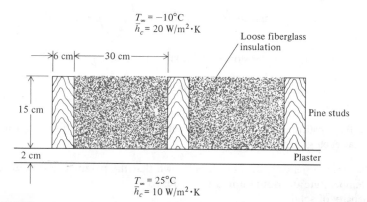

$T_\infty = -10°C$
$\bar{h}_c = 20$ W/m²·K

Loose fiberglass insulation

6 cm — 30 cm —

15 cm

Pine studs

2 cm

Plaster

$T_\infty = 25°C$
$\bar{h}_c = 10$ W/m²·K

1-31 A blackbody with a surface area of 0.1 m² radiates to a cold surrounding. Determine the heat-transfer rate by radiation from the body if it has a temperature of (a) 27°C, (b) 527°C, and (c) 1027°C.

1-32 The filament of a 100-W light bulb emits as a blackbody into a vacuum. The filament has a diameter of 0.13 mm and a length of 7 cm. Neglecting conduction from the filament, calculate the temperature of the filament.

1-33 A graybody with surface area of 10 cm² has an emittance of 0.3. Determine the rate at which it emits radiant energy if its temperature is 1000 K.

1-34 A graybody with area 1 m^2 and emittance of 0.5 is maintained at 700°C in a large black room whose temperature is 100°C. Determine the net rate of heat transfer between the gray body and room by the radiation mode of heat transfer.

1-35 Calculate the overall heat-transfer coefficient U for Problem 1-15. Base the value of U on the area A_D.

1-36 Calculate the overall heat-transfer coefficient U for Problem 1-26. Base the value of U on the total cross-sectional area of the wall.

1-37 Calculate the overall heat-transfer coefficient U for Problem 1-30. Base the value of U on a typical cross-sectional area of 1 m^2.

1-38 The physical parameters that govern the local temperature in a fin are known to be:

$\bar{h}_c$ = convective-heat-transfer coefficient for fluid surrounding fin
k = thermal conductivity of fin material
L = characteristic dimension of fin
T_∞ = ambient temperature of fluid surrounding fin
T_b = temperature of base of fin
x = location measured from base of fin

Using the Buckingham pi theorem, show that the dimensionless groups that can be used to describe the temperature distribution in the fin are

$$\text{dimensionless temperature: } \frac{T(x) - T_\infty}{T_b - T_\infty}$$

$$\text{dimensionless thermal resistance: } \frac{\bar{h}_c L}{k}$$

$$\text{dimensionless location: } \frac{x}{L}$$

1-39 By substituting dimensions for each physical quantity given in Table L-1, show that each group is dimensionless.

1-40 In transient conduction, it is known that the parameters that govern the local temperature distribution in a solid are:

ρ = density of solid
c_p = specific heat of solid
L = characteristic dimension of solid
k = thermal conductivity of solid
t = time
x = location within solid

Use the Buckingham pi theorem to show that the dimensionless temperature distribution can be expressed in terms of a dimensionless position

$$\frac{x}{L}$$

and a dimensionless group called the *Fourier number*,

$$\frac{kt}{\rho c_p L^2}$$

1-41 When a heated surface is placed in a stream of cooler fluid, the surface will lose heat by forced convection. For this case the parameters that govern the heat-transfer process are:

$\bar{h}_c =$ convective-heat-transfer coefficient
$L =$ characteristic dimension of surface
$\rho =$ density of fluid
$V =$ free-stream velocity of fluid
$k =$ thermal conductivity of fluid
$c_p =$ specific heat of fluid
$\mu =$ viscosity of fluid

Use the results of the Buckingham pi theorem to show that the three dimensionless groups that govern the heat-transfer process are:

Nusselt number: $\dfrac{\bar{h}_c L}{k}$

Reynolds number: $\dfrac{\rho V L}{\mu}$

Prandtl number: $\dfrac{\mu c_p}{k}$

Chapter 2

STEADY STATE CONDUCTION

2-1 INTRODUCTION

Conduction is that mode of heat transfer in which heat travels from a region of high temperature to a region of lower temperature because of direct contact between the molecules of the medium. The relationship between the heat-transfer rate by conduction and the temperature distribution in the medium is the Fourier law.

Conduction can occur in solids, liquids, and gases. However, in liquids and gases that are allowed to circulate, it is usually combined with convection. Therefore, pure conduction occurs primarily in opaque solids, where motion of the material is restricted. In this chapter we will consider the conducting medium to be a solid, but the principles developed can be applied to liquids and gases in which convective motion is restricted.

A discussion of heat conduction can be broken down into three major subject areas. The first involves steady conduction in which the temperature is a function of only one coordinate direction (see Sections 2-3 through 2-6). The second area concerns steady conduction in which the temperature is a function of two or three coordinate directions (see Section 2-7). The third area is transient or unsteady conduction. This subject will be dealt with in Chapter 3.

Digital computers and computer programming are important aspects of heat transfer, as we frequently encounter problems too difficult or time-consuming to solve by hand. In such cases we often turn to computers, either digital or programmable hand calculators, to provide solutions. To illustrate the types of programs that can be used to solve conduction problems, four computer programs are included here. The programs are written in a general form so that they can be applied to a broad range of problems.

The general conduction equation is derived in the next section. Much of the material that follows starts with a solution of this equation. It provides the temperature distribution in the material, and once the temperature distribution is known, the heat-transfer rate by the conduction mode can be evaluated by applying the Fourier law.

2-2 CONDUCTION EQUATION

The *conduction equation* is a mathematical expression of the conservation of energy in a solid substance. It is derived by performing an energy balance on an elemental volume of material in which heat is being transferred by conduction. Heat transfer by convection and radiation is assumed to be negligible within the solid. Heat-transfer rates by conduction are related to the temperature distribution in the solid by the Fourier law (Eq. 1-2).

The energy balance accounts for the fact that energy can be generated inside the material. Typical examples of internal energy generation in a solid would be heat generated by chemical reactions, heat generated as a result of electric currents passing through a resistance, and heat generated by nuclear reactions.

The general form of the conduction equation accounts for storage of energy in the material. We know from thermodynamic considerations that the internal energy of a material will increase if the temperature of the material increases. A solid material can therefore experience a net increase in stored energy when its temperature increases with time and a net decrease in stored energy when its temperature decreases with time. If the material temperature remains constant, no energy can be stored and steady conditions are said to exist.

Heat-transfer problems are classified in broad categories according to the variables that the temperature depends on. If the temperature is independent of time, the problem is called a *steady* or *steady-state problem*. If the temperature is a function of time, the problem is classified as *unsteady* or *transient*. Problems are also classified by the number of coordinate dimensions the temperature depends on. If the temperature is a function of a single coordinate, the problem is said to be *one-dimensional*. If the temperature is a function of two or three coordinate dimensions, it is said to be a *two-* or *three-dimensional problem*, respectively. If the temperature is a function of time and the x direction in rectangular coordinates, or

$T = T(x,t)$, the problem is classified as *one-dimensional and transient*. If $T = T(r,\theta)$ in cylindrical coordinates, the problem is classified as *two-dimensional and steady*.

Rectangular Coordinates

To simplify the derivation of the conduction equation, we will consider a one-dimensional rectangular coordinate system as shown in Fig. 2-1 and assume that the temperature in the material is a function of only the x coordinate and time, or $T = T(x,t)$. We will also assume that the conductivity k, density ρ, and specific heat c of the solid are all constant. The effects of variable conductivity will be discussed in Section 2-4.

A statement of the conservation of energy applied to the control volume of Fig. 2-1 is as follows:

$$\begin{bmatrix} \text{rate of energy} \\ \text{conducted into} \\ \text{control volume} \end{bmatrix} + \begin{bmatrix} \text{rate of energy} \\ \text{generated inside} \\ \text{control volume} \end{bmatrix}$$

$$= \begin{bmatrix} \text{rate of energy} \\ \text{conducted} \quad \text{out} \\ \text{of control} \\ \text{volume} \end{bmatrix} + \begin{bmatrix} \text{rate of energy} \\ \text{stored inside} \\ \text{control volume} \end{bmatrix}$$

$$(2\text{-}1)$$

Using the Fourier law to express the two conduction terms and defining the symbol q_G''' as the rate of energy generated inside the control volume

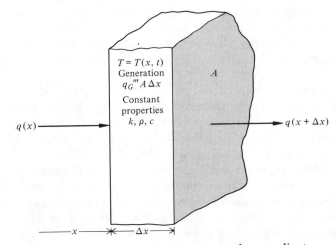

Figure 2-1 Control volume in rectangular coordinates.

per unit volume, Eq. 2-1 may be expressed in the form

$$-kA\frac{\partial T}{\partial x}(x)+q_G''' A\,\Delta x=-kA\frac{\partial T}{\partial x}(x+\Delta x)+\rho A\,\Delta x\,c\frac{\partial T}{\partial t}\quad\text{(2-2)}$$

Dividing by the volume of the control volume, $A\,\Delta x$, and rearranging produces

$$k\frac{\dfrac{\partial T}{\partial x}(x+\Delta x)-\dfrac{\partial T}{\partial x}(x)}{\Delta x}+q_G'''=\rho c\frac{\partial T}{\partial t}\quad\text{(2-3)}$$

When the limit is taken as $\Delta x\to0$, the first term on the left-hand side of Eq. 2-3 becomes the definition of the second derivative of the temperature with respect to the x coordinate:

$$k\frac{\partial^2 T}{\partial x^2}+q_G'''=\rho c\frac{\partial T}{\partial t}\qquad\text{ONE DIMENSIONAL}\quad\text{(2-4)}$$

Equation 2-4 is not a general equation because it was derived on the assumption that the temperature distribution was one-dimensional. If this restriction is now removed, and the temperature assumed to be a function of all three coordinates as well as time, or $T=T(x,y,z,t)$, terms like the first term in Eq. 2-4 representing the net conduction in the y and z directions will appear. The three-dimensional form of the conduction equation then becomes

$$k\left(\frac{\partial^2 T}{\partial x^2}+\frac{\partial^2 T}{\partial y^2}+\frac{\partial^2 T}{\partial z^2}\right)+q_G'''=\rho c\frac{\partial T}{\partial t}\quad\text{(2-5)}$$

It is important to understand the physical significance of each term in Eq. 2-5. The first three terms on the left-hand side of the equation represent the net rate of heat conducted into the control volume per unit volume. The last term on the left-hand side is the rate of energy generated per unit volume inside the control volume. The right-hand side of Eq. 2-5 represents the rate of increase in internal energy inside the control volume per unit volume. Each term has dimensions of energy per unit time and volume. Each term has the units $[\text{W}/\text{m}^3]$ in the SI system of units.

Equation 2-5 is often used in the form

$$\frac{\partial^2 T}{\partial x^2}+\frac{\partial^2 T}{\partial y^2}+\frac{\partial^2 T}{\partial z^2}+\frac{q_G'''}{k}=\frac{1}{\alpha}\frac{\partial T}{\partial t}\qquad\text{3 Dim.}\quad\text{(2-6)}$$

where the thermal diffusivity, α, is a group of material properties defined as

$$\alpha=\frac{k}{\rho c}\quad\text{(2-7)}$$

The thermal diffusivity has units of $[\text{m}^2/\text{s}]$. Numerical values of the thermal conductivity, density, specific heat, and thermal diffusivity for numerous engineering materials are given in the Appendixes.

We are seldom required to determine the temperature distribution in a solid material by solving the conduction equation in the form of Eq. 2-6. A solution to Eq. 2-6 would involve solving a partial differential equation. In most problems we can make simplifying assumptions that will eliminate terms from the conduction equation, and we can often reduce the complexity of the problem and resulting solution. A few special cases of the conduction equation will now be discussed and examples of these special cases will be considered in later sections of this chapter.

— If the temperature of a material is not a function of time, the problem is referred to as *steady* and the material is unable to store energy. The steady form of a three-dimensional conduction equation in rectangular coordinates then becomes

SPECIAL CASES

STEADY
STATE
$\Delta T = 0$

$$\frac{\partial^2 T}{\partial x^2} + \frac{\partial^2 T}{\partial y^2} + \frac{\partial^2 T}{\partial z^2} + \frac{q_G'''}{k} = 0 \qquad (2\text{-}8)$$

If in addition to being a steady problem, there is no internal energy generation in the material, the conduction equation can be further simplified to

$\Delta T = 0$
$q_G''' = 0$

$$\frac{\partial^2 T}{\partial x^2} + \frac{\partial^2 T}{\partial y^2} + \frac{\partial^2 T}{\partial z^2} = 0 \qquad (2\text{-}9)$$

Equation 2-9 is the *Laplace equation*, which occurs in several scientific disciplines.

For steady conduction with no generation and a purely one-dimensional problem such that the temperature is only a function of the x coordinate, the conduction equation may be reduced further to

$\Delta T = 0$
$q_G''' = 0.$
ONE DIMENSIONAL

$$\frac{d^2 T}{dx^2} = 0 \qquad (2\text{-}10)$$

UPON INTEGRATING TWICE : $T = C_1 x + C_2$

Dimensionless Form

The conduction equation written in the form of Eq. 2-6 is a dimensional equation. It is often convenient to rewrite this equation so that each term is dimensionless. In doing so we will identify the dimensionless groups that govern the heat-conduction process. We will nondimensionalize the one-dimensional form of the conduction equation (Eq. 2-4) by defining a dimensionless temperature as

$$\theta = \frac{T}{T_r} \qquad (2\text{-}11)$$

a dimensionless coordinate as

$$\xi = \frac{x}{L_r} \qquad (2\text{-}12)$$

and a dimensionless time as

$$\tau = \frac{t}{t_r} \tag{2-13}$$

The symbols T_r, L_r, and t_r represent a reference temperature, length, and time, respectively. The choice of reference quantities is arbitrary, although meaningful values should be selected once the problem is completely specified. A dimensionless ratio of temperature differences is often preferable to a ratio of temperatures; and the choice of dimensionless groups may vary from problem to problem. The form of the dimensionless groups are often selected so that they limit the dimensionless variables between convenient extremes such as zero and 1. The value for L_r is usually selected as the maximum x dimension of the system for which the temperature distribution is being determined.

When the definitions of the dimensionless temperature, coordinate, and time are substituted into Eq. 2-4, the conduction equation written in a nondimensional form becomes

$$\frac{\partial^2 \theta}{\partial \xi^2} + \frac{q_G''' L_r^2}{k T_r} = \frac{L_r^2}{\alpha t_r} \frac{\partial \theta}{\partial \tau} \tag{2-14}$$

The quantity $\alpha t_r / L_r^2$ is a dimensionless group called the *Fourier number*, which is designated by the symbol Fo:

$$\text{Fo} = \frac{\alpha t_r}{L_r^2} \tag{2-15}$$

The choice of reference time and length used in the Fourier number may vary from problem to problem, but the basic form remains unchanged. The Fourier number will always be a thermal diffusivity multiplied by time divided by the square of a characteristic length.

The Fourier number is the rate of heat transfer by conduction divided by the rate of energy stored in a material. The Fourier number is an important dimensionless group used in transient conduction problems and it will appear frequently in the work that follows.

The other dimensionless group appearing in Eq. 2-14 is a term involving heat generation. We will use the symbol $\bar{q}_G$ to represent the dimensionless generation:

$$\bar{q}_G = \frac{q_G''' L_r^2}{k T_r} \tag{2-16}$$

This term is a ratio of internal heat generated per unit time to heat conducted through the volume per unit time.

The one-dimensional form of the conduction equation expressed in dimensionless form now becomes

$$\frac{\partial^2\theta}{\partial\xi^2} + \bar{q}_G = \frac{1}{\text{Fo}}\frac{\partial\theta}{\partial\tau} \tag{2-17}$$

Example 2-1. Determine the simplified form of the general conduction equation that applies to steady, one-dimensional conduction in a rectangular solid with constant properties and no generation. Solve the resulting conduction equation for the temperature distribution and heat-transfer rate in the solid in terms of constants of integration.

Solution: The general form of the conduction equation is

$$k\left(\frac{\partial^2 T}{\partial x^2} + \frac{\partial^2 T}{\partial y^2} + \frac{\partial^2 T}{\partial z^2}\right) + q_G''' = \rho c\frac{\partial T}{\partial t}$$

Since the problem is steady and one-dimensional, the temperature is independent of time and is only a function of one of the three coordinate directions. If we assume that T is only a function of the x coordinate,

$$\frac{\partial T}{\partial t} = 0 \qquad \frac{\partial^2 T}{\partial y^2} = 0 \qquad \frac{\partial^2 T}{\partial z^2} = 0$$

Also, since no generation is present,

$$q_G''' = 0$$

The simplified form of the conduction equation becomes

$$\frac{d^2 T}{dx^2} = 0$$

Integrating this second-order differential equation twice yields the temperature distribution in terms of two constants of integration, C_1 and C_2:

$$T = C_1 x + C_2$$

Values for the two constants of integration would be determined in a particular problem by specifying and applying two boundary conditions.

The heat-transfer rate conducted through the solid in the x direction is given by the Fourier law:

$$q = -kA\frac{dT}{dx} = -kAC_1$$

Notice that the heat-transfer rate is uniform in agreement with Examples 1-1 and 1-2 and it remains so at all locations in the solid under steady conditions.

Cylindrical Coordinates

The conduction equation written in the form of Eq. 2-6 applies only to a rectangular coordinate system. The generation and energy storage terms are independent of coordinate system, but the net conduction terms depend on geometry and therefore on the coordinate system. The dependence on the coordinate system used to formulate the problem can be removed from the analysis by replacing the net conduction terms with the Laplacian operator. The form of the Laplacian is different for each coordinate system. The Laplacian operation in rectangular, cylindrical, and spherical coordinate systems is given in Appendix B. The conduction equation written in terms of the Laplacian is

$$\nabla^2 T + \frac{q_G'''}{k} = \frac{1}{\alpha}\frac{\partial T}{\partial t} \tag{2-18}$$

For a general transient three-dimensional problem in cylindrical coordinates, $T = T(r, \phi, z, t)$. The coordinate symbols are shown in Fig. 2-2(a). If the Laplacian is substituted into Eq. 2-18, the general form of the conduction equation in cylindrical coordinates becomes

$$\frac{1}{r}\frac{\partial}{\partial r}\left(r\frac{\partial T}{\partial r}\right) + \frac{1}{r^2}\frac{\partial^2 T}{\partial \phi^2} + \frac{\partial^2 T}{\partial z^2} + \frac{q_G'''}{k} = \frac{1}{\alpha}\frac{\partial T}{\partial t} \tag{2-19}$$

If the transient temperature in a cylindrical shape is one-dimensional so that $T = T(r, t)$, the special case of the conduction equation becomes

$\mathcal{T} = $ TRANSIENT HEAT FLOW

$\Upsilon = $ ONE DIMENSIONAL, DIRECTION

$$\frac{1}{r}\frac{\partial}{\partial r}\left(r\frac{\partial T}{\partial r}\right) + \frac{q_G'''}{k} = \frac{1}{\alpha}\frac{\partial T}{\partial t} \tag{2-20}$$

Furthermore, if the temperature is steady and a function of only the radial

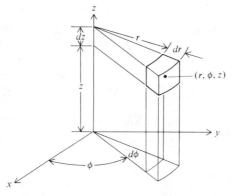

Figure 2-2 (a) Cylindrical coordinate system.

coordinate, the conduction equation can be reduced to

STEADY STATE
ONE DIMENSIONAL

$$\frac{1}{r}\frac{d}{dr}\left(r\frac{\partial T}{\partial r}\right) + \frac{q_G'''}{k} = 0 \qquad (2\text{-}21)$$

Notice in Eq. 2-21 that the temperature is now a function of only a single variable r and the equation can be written as an ordinary differential equation.

When no internal energy generation is present and the temperature is a function of the radius only, the steady form of the conduction equation for cylindrical coordinates is

STEADY STATE
ONE DIMENSIONAL
No HEAT GENERATION

$$\frac{d}{dr}\left(r\frac{dT}{dr}\right) = 0 \qquad (2\text{-}22)$$

Example 2-2. Determine the steady temperature distribution and heat-transfer rate in a cylinder with length l in terms of two constants of integration. The temperature is a function of the radius r only, and no internal generation is present in the cylinder.

Solution: The appropriate form of the conduction equation is Eq. 2-22:

$$\frac{d}{dr}\left(r\frac{dT}{dr}\right) = 0$$

Integrating once with respect to radius yields

$$r\frac{dT}{dr} = C_1$$

or

$$\frac{dT}{dr} = \frac{C_1}{r}$$

A second integration gives

$$T = C_1 \ln r + C_2$$

The constants of integration can be determined once two boundary conditions are specified. The heat-transfer rate across a cylindrical surface with arbitrary radius r is

$$q = -kA\frac{dT}{dr} = -k(2\pi rl)\frac{C_1}{r} = -2\pi klC_1$$

Notice that the heat-transfer rate across any cylindrical surface is constant for steady conditions.

Spherical Coordinates

For a spherical shape where the temperature is a function of the three coordinates and time, or $T = T(r, \theta, \phi, t)$, the general form of the conduc-

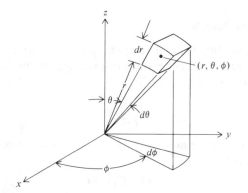

Figure 2-2 (b)   Spherical coordinate system.

tion equation which includes generation of energy is

$$\frac{1}{r^2}\frac{\partial}{\partial r}\left(r^2\frac{\partial T}{\partial r}\right) + \frac{1}{r^2\sin\theta}\frac{\partial}{\partial\theta}\left(\sin\theta\frac{\partial T}{\partial\theta}\right)$$

$$+ \frac{1}{r^2\sin^2\theta}\frac{\partial^2 T}{\partial\phi^2} + \frac{q_G'''}{k} = \frac{1}{\alpha}\frac{\partial T}{\partial t} \qquad (2\text{-}23)$$

NET HEAT CONDUCTION
PER UNIT VOLUME

The spherical coordinate system is illustrated in Fig. 2-2(b).

Special cases for one-dimensional, transient conduction and one-dimensional, steady conduction for spherical coordinates can be simplified from Eq. 2-23. The simplified forms are left as an exercise.

2-3 STEADY, ONE-DIMENSIONAL CONDUCTION WITHOUT GENERATION

We will now apply the conduction equation to problems in which the temperature is a function of a single coordinate only. In the rectangular coordinate system the temperature will be a function of the x coordinate only, and in both the cylindrical and spherical coordinate systems the temperature will be a function of the radial coordinate only. The thermal conductivity is assumed constant and no generation is considered.

The general procedure we will use consists basically of two steps. The first involves determining the temperature distribution by solving the appropriate simplified form of the conduction equation. This process consists of solving an ordinary, second-order differential equation. Once the differential equation is solved, two boundary conditions are imposed to determine the two constants of integration. The second step involves solving for the rate of heat transfer through the solid by applying the Fourier law.

Rectangular Coordinates

The steady, one-dimensional temperature distribution in a rectangular plane wall with no energy generation is governed by the simplified form of the conduction equation (Eq. 2-10),

$$\frac{d^2T}{dx^2} = 0$$

Solving this differential equation in terms of two constants of integration, C_1 and C_2, results in

$$T(x) = C_1 x + C_2$$

See Example 2-1.

The constants of integration can be determined once two boundary conditions are specified. Let us assume that the two boundary conditions are determined by specifying the temperatures at the two extreme surfaces of the wall as shown in Fig. 2-3:

$$T(0) = T_1$$
$$T(L) = T_2$$

Applying these two boundary conditions results in the dimensionless temperature in the wall of

$$\frac{T(x) - T_1}{T_2 - T_1} = \frac{x}{L} \qquad (2\text{-}24)$$

The temperature distribution is therefore linear with x. The heat-transfer rate through the wall as determined by the Fourier law is

$$q = -kA\frac{dT}{dx} = kA\frac{T_1 - T_2}{L} \qquad (2\text{-}25)$$

The heat-transfer rate per unit area through the wall is the heat flux and is

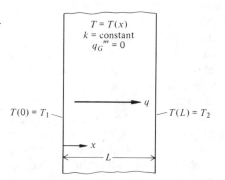

Figure 2-3. Rectangular geometry and boundary conditions.

denoted by q''. The double-prime superscript denotes that the quantity q is evaluated per unit area. For a plane wall

$$q'' = \frac{q}{A} = \frac{k(T_1 - T_2)}{L} = \text{WATTS}/\text{m}^2 \qquad \text{HEAT FLUX}$$

When Eq. 2-25 is written in the form of *Ohm's law*,

$$q = \frac{\Delta T}{R_t} = \frac{\Delta T}{L/kA} \tag{2-26}$$

the thermal resistance for a plane wall as noted in Chapter 1 becomes

$$R_t = \frac{L}{kA} \tag{2-27}$$

The flow of heat by conduction through a plane wall is a result of a temperature difference across the wall and is inhibited by thermal resistance, which is proportional to the wall thickness and inversely proportional to the thermal conductivity of the wall and its cross-sectional area.

If heat flows by conduction through several plane surfaces, the temperature distribution and heat-transfer rate can be determined by assuming that the heat flows through an equivalent thermal circuit in which the heat flows consecutively through a series of resistances, each corresponding to a separate wall material.

As an example of a series circuit, consider a plane wall, denoted by the subscript 1, covered with two different types of insulating materials, denoted by subscripts 2 and 3. The geometry is shown in Fig. 2-4. The same amount of heat flows consecutively through each resistance, and

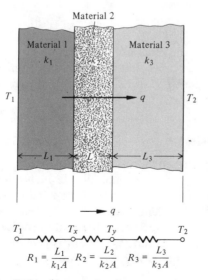

Figure 2-4 Series thermal circuit, rectangular coordinates.

therefore the thermal circuit is a series circuit. If the three material properties are known, the geometry is given and the two extreme surface temperatures are specified. The heat-transfer rate can then be determined from an expression similar to Ohm's law, as given by Eq. 1-14:

$$q = \left(\frac{\Delta T}{R_t}\right)_{total} = \frac{\cdot \Delta T_{total}}{\sum\limits_{i=1}^{3} R_i} = \frac{T_1 - T_2}{R_1 + R_2 + R_3} \tag{2-28}$$

Once the heat-transfer rate through the composite wall is known, interface temperatures among the three materials can be determined by applying Ohm's law to only one wall material. For example, the temperature T_x at the interface between materials 1 and 2 can be determined from

$$q = \frac{T_1 - T_x}{R_1} = \frac{T_1 - T_x}{L_1/k_1 A} \tag{2-29}$$

Frequently, plane walls are composed of composite materials subdivided so that the heat must flow through several materials simultaneously rather than consecutively. When this is the case, the thermal circuit becomes a parallel circuit. A typical example of a parallel circuit is shown in Fig. 1-6. The heat flow is determined by

$$q = \left(\frac{\Delta T}{R_t}\right)_{total} = \frac{T_1 - T_2}{R_1 + \left(\dfrac{R_2 R_3}{R_2 + R_3}\right) + R_4} \tag{2-30}$$

Individual resistances are determined by the equation

$$R_i = \frac{L_i}{k_i A_i} \qquad (i = 1, 2, 3, 4)$$

Intermediate temperatures such as T_x may be determined by Eq. 2-29.

The parallel circuit assumes that the heat flow is one-dimensional, and if the resistances R_2 and R_3 are significantly different, two-dimensional effects can become important (see Section 2-7).

Cylindrical Coordinates

The most common conduction problem involving a cylindrical geometry is one in which heat is conducted radially through a long, hollow cylinder as shown in Fig. 2-5. The temperature of the inside surface of the cylinder is known to be T_i and the temperature of the outside surface is known to be T_o. The steady-state temperature distribution in the constant property solid when no internal generation is present is given by the solution of Eq. 2-22 subjected to the two boundary conditions

$$T(r_i) = T_i$$

$$T(r_o) = T_o$$

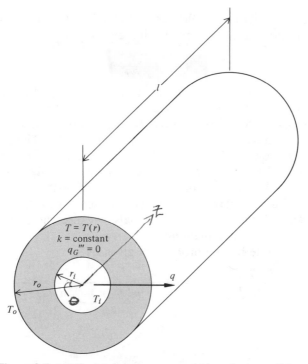

Figure 2-5 Cylindrical geometry and boundary conditions.

The solution for the local temperature, $T(r)$, is

$$T(r) = T_i + (T_o - T_i) \frac{\ln(r/r_i)}{\ln(r_o/r_i)} \qquad (2\text{-}31)$$

See Example 2-2. Equation 2-31 written in dimensionless form is

$$\frac{T(r) - T_i}{T_o - T_i} = \frac{\ln(r/r_i)}{\ln(r_o/r_i)} \qquad (2\text{-}32)$$

The temperature distribution in the cylinder is therefore logrithmic with radius.

Once the temperature distribution is known, the heat flow radially through the cylinder may be determined by using the Fourier law for cylindrical coordinates,

$$q = -kA(r)\frac{dT}{dr} = -k(2\pi rl)\frac{dT}{dr} \qquad (2\text{-}33)$$

where l is the length of the cylinder.

Differentiating the temperature distribution given in Eq. 2-31 and substituting the result into Eq. 2-33 yields

$$q = \frac{T_i - T_o}{\ln(r_o/r_i)/2\pi kl} \qquad (2\text{-}34)$$

Equation 2-34 is written in the form of Ohm's law and the denominator represents the thermal resistance of a hollow cylinder:

$$R_t = \frac{\ln(r_o/r_i)}{2\pi kl} \tag{2-35}$$

The principles of a series and parallel circuit developed for a plane-wall rectangular coordinate system can also be applied to a hollow-cylinder problem. For example, suppose that a fluid flows through a tube which is covered by an insulating material, as shown in Fig. 2-6. The average fluid temperature is known to be T_1 and the outside surface temperature of the insulation is T_2. The tube material is designated by subscript 1 and the insulation is number 2. The convective resistance of the fluid is given by Eq. 1-18. The fluid resistance is connected in series with the two conductive resistances of the two solid materials because the heat must flow consecutively through each material.

The heat-flow rate for the problem is given by

$$q = \left(\frac{\Delta T}{R_t}\right)_{\text{total}} = \frac{T_1 - T_2}{\dfrac{1}{\bar{h}_c 2\pi r_i l} + \dfrac{\ln(r_2/r_1)}{2\pi k_1 l} + \dfrac{\ln(r_3/r_2)}{2\pi k_2 l}} \tag{2-36}$$

AREA = $2\pi RL$

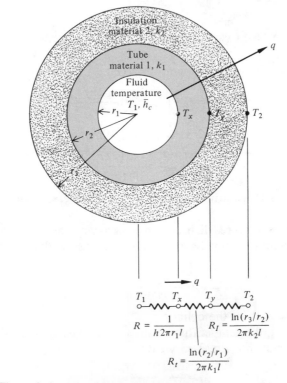

Figure 2-6 Series thermal circuit, cylindrical geometry.

The thermal resistance used in Eq. 2-36 must be the total resistance between the two known temperatures. If the two known temperatures had been T_x and T_2, the equivalent resistance would be only the sum of the conductive resistances of tube and insulation. The expression used to determine the temperature T_x once the heat-transfer rate is known is

$$q = \frac{T_x - T_2}{\ln(r_2/r_1)/2\pi k_1 l + \ln(r_3/r_2)/2\pi k_2 l} \tag{2-37}$$

Example 2-3. An aluminum pipe carries steam at 110°C. The pipe ($k = 185$ W/m·K) has an internal diameter (i.d.) of 10 cm and an outside diameter (o.d.) of 12 cm. The pipe is located in a room where the ambient air temperature is 30°C and the convective-heat-transfer coefficient between the pipe and air is 15 W/m²·K. Determine the heat-transfer rate per unit length of pipe if the pipe is uninsulated.

To reduce the heat loss from the pipe, it is covered with a 5-cm-thick layer of insulation ($k = 0.20$ W/m·K). Determine the heat-transfer rate per unit length from the insulated pipe. Assume that the convective resistance of the steam is negligible.

Solution: For the uninsulated pipe the only significant resistances to heat flow are the conductive resistance of the pipe and the convective resistance of the room air. Since the convective resistance of the steam is negligible, the inside surface temperature of the pipe is the same as the steam temperature. The heat-transfer rate per unit length in terms of symbols shown in the figure is

$$q' = \frac{q}{l} = \frac{T_s - T_\infty}{\ln(r_2/r_1)/2\pi k_p + 1/2\pi r_2 \bar{h}_{c0}}$$

$$= \frac{110 - 30}{\ln(6/5)/2\pi \times 185 + 1/2\pi \times 0.06 \times 15}$$

$$= \frac{80}{(1.57 \times 10^{-4}) + 0.177} = 452 \text{ W/m}$$

For the insulated pipe, the resistance of the insulation must be added and the expression for the heat loss becomes

$$q' = \frac{q}{l} = \frac{T_s - T_\infty}{\dfrac{\ln(r_2/r_1)}{2\pi k_p} + \dfrac{\ln(r_3/r_2)}{2\pi k_I} + \dfrac{1}{2\pi r_3 \bar{h}_{c0}}}$$

$$= \frac{110 - 30}{\dfrac{\ln(6/5)}{2\pi \times 185} + \dfrac{\ln(11/6)}{2\pi \times 0.2} + \dfrac{1}{2\pi \times 0.11 \times 15}}$$

$$= \frac{80}{(1.57 \times 10^{-4}) + 0.482 + 0.096} = 138 \text{ W/m}$$

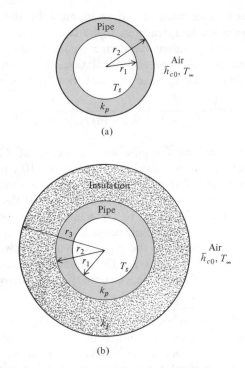

(a)

(b)

The presence of the insulation has reduced the heat loss from the steam by 70 percent. Notice that in both cases the resistance of the aluminum pipe can be neglected without loss in accuracy for the calculation of heat-transfer rate.

Spherical Coordinates

The temperature distribution and heat-transfer rate through a hollow sphere are determined in a manner similar to that outlined for the plane wall and hollow cylinder. The steady, one-dimensional temperature distribution with no generation present is determined by solving the simplified form of the conduction equation written in spherical coordinates. This equation is

$$\frac{1}{r^2}\frac{d}{dr}\left(r^2\frac{dT}{dr}\right) = \frac{1}{r}\frac{d^2(rT)}{dr^2} = 0$$

Assuming that the boundary conditions specify that the inner and outer surfaces of the sphere are at known temperatures as shown in Fig. 2-7,

$$T(r_i) = T_i$$
$$T(r_o) = T_o$$

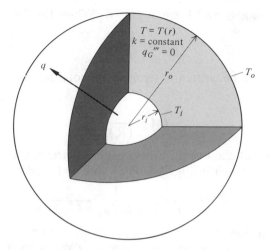

Figure 2-7 Spherical geometry and boundary conditions.

The temperature distribution in the hollow sphere is then

$$\frac{T(r) - T_i}{T_o - T_i} = \frac{r_o}{r_o - r_i}\left(1 - \frac{r_i}{r}\right) \tag{2-38}$$

The temperature in a hollow sphere therefore varies hyperbolically with radius.

The heat-transfer rate in the sphere is determined by applying the Fourier law to Eq. 2-38. The result is

$$q = \frac{T_i - T_o}{(r_o - r_i)/4\pi k r_o r_i} \tag{2-39}$$

The thermal resistance for a hollow sphere is therefore

$$R_t = \frac{r_o - r_i}{4\pi k r_o r_i} \tag{2-40}$$

Overall Heat-Transfer Coefficient

As shown in Chapter 1, when a heat-transfer problem involves several thermal resistances in series, parallel, or combinations of the two, it is convenient to define an overall heat-transfer coefficient or overall conductance. The symbol for the overall heat-transfer coefficient is U, and it is defined by the equation

$$q = UA(\Delta T)_{\text{total}} \tag{2-41}$$

The role of U is similar to that of the convective-heat-transfer coefficient.

The units of both U and h are W/m^2·K. When Eq. 2-41 is compared to

$$q = \left(\frac{\Delta T}{R_t}\right)_{total} \tag{2-42}$$

we see that U can be written in terms of the total thermal resistance of the circuit:

$$UA = \frac{1}{(R_t)_{total}} \tag{2-43}$$

As an example of the overall heat-transfer coefficient, consider the three plane materials shown in Fig. 2-4. The value of U for this example is

$$U = \frac{1}{L_1/k_1 + L_2/k_2 + L_3/k_3}$$

In this example, the cross-sectional areas of all three materials are equal, so there can be little confusion about what area should be used in Eq. 2-43. However, when the area in each resistance term varies, we must be consistent when selecting an area to be used in Eq. 2-43.

A problem in which the area does vary is the one involving a composite cylinder in which the resistances are connected in series. The value for UA for the circuit shown in Fig. 2-6 is

$$q = UA(\Delta T)_{total} = \frac{T_1 - T_2}{\dfrac{1}{\bar{h}_c(2\pi r_1 l)} + \dfrac{\ln(r_2/r_1)}{2\pi k_1 l} + \dfrac{\ln(r_3/r_2)}{2\pi k_2 l}}$$

or

$$UA = \frac{1}{\dfrac{1}{\bar{h}_c(2\pi r_1 l)} + \dfrac{\ln(r_2/r_1)}{2\pi k_1 l} + \dfrac{\ln(r_3/r_2)}{2\pi k_2 l}}$$

Notice that the product of UA is a constant, but the value of U varies depending upon the choice of the corresponding area. For example, suppose that we choose the inside pipe area, A_i, as our reference area, where

$$A_i = 2\pi r_1 l$$

Then the U value based on A_i would be

$$U_i = \frac{1}{\dfrac{1}{\bar{h}_c} + \dfrac{r_1 \ln(r_2/r_1)}{k_1} + \dfrac{r_1 \ln(r_3/r_2)}{k_2}}$$

If U is based on the outside pipe area, A_o, where

$$A_o = 2\pi r_3 l$$

then

$$U_o = \cfrac{1}{\cfrac{r_3}{r_1 \bar{h}_c} + \cfrac{r_3 \ln(r_2/r_1)}{k_1} + \cfrac{r_3 \ln(r_3/r_2)}{k_2}}$$

Even though the values for U_i and U_o are different, the UA product is always constant:

$$U_i A_i = U_o A_o$$

Example 2-4. A plastic ($k = 0.5$ W/m·K) pipe carries a fluid such that the convective heat-transfer coefficient is 300 W/m²·K. The average fluid temperature is 100°C. The pipe has an i.d. of 3 cm and an o.d. of 4 cm. If the heat-transfer rate through the pipe per unit length is 500 W/m, calculate the external pipe temperature. Also calculate the overall heat-transfer coefficient based on the pipe outside surface area.

Solution: A sketch of the pipe system is shown in the accompanying figure. The heat-transfer rate is given by

$$q = \cfrac{T_1 - T_2}{\cfrac{1}{\bar{h}_c(2\pi r_1 l)} + \cfrac{\ln(r_2/r_1)}{2\pi k_1 l}}$$

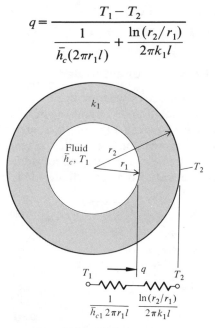

or the heat-transfer rate per unit length is

$$q' = \cfrac{q}{l} = \cfrac{T_1 - T_2}{\cfrac{1}{\bar{h}_c 2\pi r_1} + \cfrac{\ln(r_2/r_1)}{2\pi k_1}}$$

$$500 = \frac{100 - T_2}{\dfrac{1}{300 \times 2\pi \times 0.015} + \dfrac{\ln(2/1.5)}{2\pi \times 0.5}}$$

$$T_2 = 36.5°C$$

The overall heat-transfer coefficient based on A_o is

$$U_o A_o = \frac{1}{\dfrac{1}{\bar{h}_c 2\pi r_1 l} + \dfrac{\ln(r_2/r_1)}{2\pi k_1 l}}$$

$$U_o = \frac{1}{\dfrac{r_2}{r_1 \bar{h}_c} + \dfrac{r_2 \ln(r_2/r_1)}{k_1}}$$

$$= \frac{1}{\dfrac{2}{1.5 \times 300} + \dfrac{0.02 \times \ln(2/1.5)}{0.5}} = 62.69 \text{ W/m}^2\cdot\text{K}$$

As a check on the value for U, we can calculate the heat-transfer rate based on the calculated value of U_o:

$$q' = U_o A_o (T_1 - T_2) = 62.69 \times 2\pi \times 0.02(100 - 36.53) = 500 \text{ W/m}$$

Critical Insulation Thickness for a Cylinder

An interesting situation arises when a cylinder with low thermal resistance is covered by an insulation layer and the insulation is surrounded by a fluid. The geometry is shown in Fig. 2-8. Assume that the inner surface of the insulation has a known constant value of temperature equal to T_i. Suppose that we wish to determine the effect additional insulation will have on the heat-transfer rate from the cylinder. It is not obvious whether

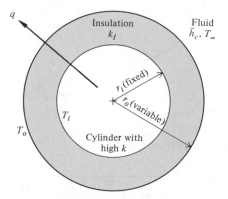

Figure 2-8 Critical radius of pipe insulation.

additional insulation will increase or decrease the heat-transfer rate. The heat transfer at steady state through the cylinder and insulation should be equal to the convection rate from the surface:

$$q = \bar{h}_c A_o (T_o - T_\infty)$$

where the symbols are defined in Fig. 2-8.

As insulation is added, A_o increases but T_o decreases. To determine which effect dominates, we can write the heat-transfer rate as

$$q = \frac{T_i - T_\infty}{\ln(r_o/r_i)/2\pi k_I l + 1/\bar{h}_c 2\pi r_o l}$$

To determine the effect of variable insulation thickness on the heat-transfer rate, we can take the derivative of q with respect to r_o and equate the result to zero to determine optimum conditions. The resulting condition for optimum heat flow is the condition

$$\frac{\bar{h}_c r_o}{k_I} = 1.0 \tag{2-44}$$

The quantity $\bar{h}_c r_o/k_I$ is a dimensionless group called the *Biot number*. It represents the ratio of conductive resistance in the solid insulation to the convective resistance in the fluid. The symbol used for the Biot number is Bi. The Biot number frequently occurs in problems involving the combined modes of conduction and convection.

The condition for optimum heat transfer from the cylinder is therefore

$$\text{Bi} = \frac{\bar{h}_c r_o}{k_I} = 1.0 \tag{2-45}$$

By plotting the heat-transfer rate as a function of the outside insulation radius r_o, we could show that the heat-transfer rate reaches a maximum when $\text{Bi} = 1.0$. If $\text{Bi} < 1.0$, addition of insulation will increase the heat-transfer rate. Once the Biot number exceeds unity, additional insulation will decrease the heat-transfer rate.

When the Biot number is unity, the outside radius of the insulation is termed the *critical radius* because of the heat-transfer rate from the cylinder is a maximum when

$$r_{\text{crit}} = \frac{k_I}{\bar{h}_c} \tag{2-46}$$

By examining the order of magnitude of both k_I and $\bar{h}_c$ that one can expect to encounter in typical problems we would see that the critical radius is on the order of a few millimeters. Therefore, we should be aware that additional insulation on small-diameter cylinders such as small-gauge electrical wires could actually increase the heat dissipation from the wire. On the other hand, we should expect that the addition of insulation to large-diameter pipes and ducts will always decrease the heat-transfer rate.

Example 2-5. A 1-mm-diameter electrical wire is covered with a 2-mm-thick layer of plastic insulation ($k=0.5$ W/m·K). The wire is surrounded by air with an ambient temperature of 25°C and $\bar{h}_c=10$ W/m²·K. The wire temperature is 100°C.

Determine the rate of heat dissipated from the wire per unit length with and without the insulation. Assume that the wire temperature is not affected by presence of the insulation.

Solution: First we calculate the Biot number:

$$\text{Bi} = \frac{\bar{h}_c r_o}{k_I} = \frac{10(2+0.5)\times 10^{-3}}{0.5} = 0.05$$

Since the Biot number is less than 1, the presence of insulation will increase the heat transfer from the wire. The heat-transfer rate per unit length with the insulation on the wire is

$$q' = \frac{T_i - T_\infty}{\dfrac{\ln(r_o/r_i)}{2\pi k_I} + \dfrac{1}{2\pi r_o \bar{h}_c}}$$

$$= \frac{100 - 25}{\dfrac{\ln(2.5/0.5)}{2\pi \times 0.5} + \dfrac{1}{2\pi \times (2.5\times 10^{-3})\times 10}} = 10.90 \text{ W/m}$$

Without insulation the heat-transfer rate is

$$q' = \bar{h}_c \frac{A_o}{l}(T_o - T_\infty) = 10\times 2\pi \times (0.5\times 10^{-3})(100-25) = 2.36 \text{ W/m}$$

The addition of the insulation increases the rate of heat transfer from the wire by a factor of 4.6.

2-4 EFFECT OF VARIABLE THERMAL CONDUCTIVITY

The thermal conductivity of most materials is not constant but varies with temperature. So far we have assumed that the thermal conductivity was constant. In this section, however, we will determine the effect a variable thermal conductivity would have on the heat flow and temperature distribution in a plane wall, a hollow cylinder, and a hollow sphere.

If the conduction equation for the rectangular coordinate system derived in Section 2-2 had been derived assuming the thermal conductivity was a variable, the form of Eq. 2-5 would become

$$\frac{\partial}{\partial x}\left(k\frac{\partial T}{\partial x}\right) + \frac{\partial}{\partial y}\left(k\frac{\partial T}{\partial y}\right) + \frac{\partial}{\partial z}\left(k\frac{\partial T}{\partial z}\right) + q_G''' = \rho c \frac{\partial T}{\partial t} \qquad (2\text{-}47)$$

If the temperature distribution in the rectangular solid is steady, one-dimensional, and does not involve internal energy generation, Eq. 2-47 can be simplified to

STEADY, STATE
ONE DIMENSIONAL

$$\frac{d}{dx}\left[k(T)\frac{dT}{dx}\right]=0 \tag{2-48}$$

$\dot{q}''_G=0$

Before we can solve this equation, we must know how the thermal conductivity varies with temperature, $k(T)$, over the range of temperatures encountered in the solid. For many materials, very little accuracy is lost by assuming the conductivity-temperature variation is linear:

$$k(T)=k_0(1+\beta T) \tag{2-49}$$

where β is a constant.

Integrating Eq. 2-48 once with respect to x gives

$$k(T)\frac{dT}{dx}=C_1 \tag{2-50}$$

The heat flow through the wall will be a constant for steady-state conditions. The Fourier law applied to the wall is

$$q''=-k(T)\frac{dT}{dx} \tag{2-51}$$

By comparing Eq. 2-50 with Eq. 2-51, we see that the heat flux is

$$q''=-C_1$$

Substituting Eq. 2-49 into Eq. 2-50 followed by integration with respect to x yields

$$k_0\left(T+\beta\frac{T^2}{2}\right)=C_1x+C_2$$

The values for the two constants of integration can be determined by specifying two boundary conditions. Assuming that the boundaries of the solid are at known temperatures such as those shown in Fig. 2-3, the boundary conditions are

$$T(0)=T_1$$
$$T(L)=T_2$$

We can determine the values of the constants of integration C_1 and C_2, which are

$$C_1=\frac{k_0}{L}\left[(T_2-T_1)+\frac{\beta}{2}(T_2^2-T_1^2)\right]$$

$$C_2=k_0\left(T_1+\frac{\beta T_1^2}{2}\right)$$

The dimensionless temperature distribution in the wall is then

$$\frac{T(x)-T_1}{T_2-T_1} = \frac{x}{L} + \frac{\beta}{2}\left[(T_2+T_1)\frac{x}{L}\right.$$

IF ~~TEMP~~ -
CONDUCTIVITY - TEMP.
VARIATION IS LINEAR

$$\left. + \frac{T_1^2 - T(x)^2}{T_2-T_1}\right] \tag{2-52}$$

The temperature profile in a plane wall with varying thermal conductivity is not linear, but we can see that Eq. 2-52 will reduce to the linear result (Eq. 2-24) when the thermal conductivity is constant, or when $\beta = 0$.
The heat flux through the wall is

$$q'' = -C_1 = -\frac{k_0}{L}\left[(T_2-T_1)+\frac{\beta}{2}(T_2^2-T_1^2)\right]$$

which can be rewritten in the form

$$q'' = k_0\left(1+\beta\frac{T_2+T_1}{2}\right)\frac{T_1-T_2}{L}$$

The quantity in parentheses is the thermal conductivity evaluated at the mean or average temperature, T_m, of the wall, where

$$T_m = \frac{T_2+T_1}{2}$$

The thermal conductivity evaluated at T_m is

$$k_m = k_0\left(1+\beta\frac{T_2+T_1}{2}\right)$$

The heat flux in terms of k_m becomes simply

$$q'' = \frac{k_m(T_1-T_2)}{L} \tag{2-53}$$

Equation 2-53 is written in a particularly convenient form. It shows that the heat flux through a wall with thermal conductivity that varies linearly with temperature can be calculated by using the form of the heat-flux equation developed for constant conductivity if the conductivity is evaluated at the average of the two wall-surface temperatures.

If a hollow cylinder or hollow sphere consists of a material for which the thermal conductivity varies linearly with temperature, a similar procedure will allow us to determine the temperature profile and heat flux through these materials. The details are left for problems at the end of the chapter.

The heat flux through a hollow cylinder with linearly varying conductivity and known surface temperatures is

$$q = \frac{T_i-T_0}{\ln(r_0/r_i)/2\pi k_m l}$$

and for a hollow sphere the heat flux is

$$q = \frac{T_i - T_o}{(r_0 - r_i)/4\pi k_m r_0 r_i}$$

where the values for k_m are determined by

$$k_m = k_0\left(1 + \beta\frac{T_i + T_o}{2}\right)$$

We can now see that the previous expressions for heat flux through a plane wall, a hollow cylinder, and a hollow sphere with constant thermal conductivity may still be used to determine the heat transfer by simply replacing the constant thermal conductivity with the thermal conductivity evaluated at the average solid temperature.

When the thermal conductivity does not vary linearly, the heat flux can be shown to be summarized by the Fourier law written in the form of Ohm's law:

$$q = \frac{\Delta T}{R_m}$$

where R_m represents the mean thermal resistance of the solid material. Regardless of geometry, the mean thermal resistance is based upon the mean thermal conductivity of the solid defined by

$$k_m = \frac{1}{T_2 - T_1}\int_{T_1}^{T_2} k(T)dT$$

where the temperatures T_1 and T_2 are the extreme temperatures across the surface, or $\Delta T = T_1 - T_2$. Then the mean thermal resistance for a plane wall is

$$R_m = \frac{L}{k_m A}$$

For a hollow cylinder R_m is

$$R_m = \frac{\ln(r_o/r_i)}{2\pi k_m l}$$

and for a hollow sphere

$$R_m = \frac{r_o - r_i}{4\pi k_m r_o r_i}$$

Example 2-6. A large plane wall is 0.35 m thick. One surface is maintained at a temperature of 35°C and the other surface is at 115°C. Only two values of thermal conductivity are available for the wall material. At 0°C $k = 26$ W/m·K and at 100°C $k = 32$ W/m·K. Determine the heat flux through the wall assuming the thermal conductivity varies linearly with temperature.

Solution: The mean temperature of the wall is

$$T_m = \frac{T_1 + T_2}{2} = \frac{35 + 115}{2} = 75°C$$

The mean thermal conductivity can be obtained by linearly interpolating between the two given conductivity values

$$\frac{32 - k_m}{32 - 26} = \frac{100 - 75}{100 - 0}$$

or

$$k_m = 30.5 \, \text{W/m} \cdot \text{K}$$

The heat flux through the wall becomes

$$q'' = \frac{q}{A} = \frac{\Delta T}{L/k_m} = \frac{115 - 35}{0.35/30.5} = 6970 \, \text{W/m}^2$$

2-5 STEADY, ONE-DIMENSIONAL CONDUCTION WITH GENERATION

Until now we have not considered a conduction problem that involves generation of heat inside the material. The approach to a problem involving internal generation is identical to that used in the previous sections. First, the appropriate form of the energy equation is solved for the temperature distribution in the material. The solution will result in two constants of integration that must be determined by two boundary conditions. Next, the Fourier law is used to determine the heat flux through the solid.

Heat can be generated internally in a number of ways. Chemical reactions, both endothermic or exothermic, can occur in a solid material. An exothermic reaction will generate heat, whereas an endothermic reaction will absorb heat from the material, causing a *negative source* or a *heat sink*. Electric current passing through a resistance generates heat in the conductor. Heat generation also occurs in fissionable materials as a result of the nuclear reaction that takes place within the material.

Rectangular Coordinates

As an example of a problem involving internal heat generation, consider a plane wall with a constant generation distributed uniformly throughout the entire volume. The generation rate per unit volume is denoted by the symbol q_G''', which in this example is a constant value. Suppose that the plane wall has one surface maintained at a known temperature T_1 and that the other surface is insulated. The geometry and boundary conditions given in the problem are illustrated in Fig. 2-9.

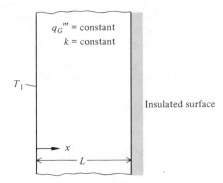

Figure 2-9   Conduction in a plane wall with uniform generation.

The appropriate form of the conduction equation is

$$\frac{d^2 T}{dx^2} + \frac{q_G'''}{k} = 0 \qquad (2\text{-}54) \quad \text{\Large *}$$

since the problem assumes a steady, one-dimensional temperature distribution. After integrating Eq. 2-54 twice, the temperature distribution is

$$T(x) = -\frac{q_G'''}{2k} x^2 + C_1 x + C_2 \qquad (2\text{-}55) \quad \text{\Large *}$$

where C_1 and C_2 are constants of integration that will be satisfied by the boundary conditions.

The first boundary condition for this problem is simply

$$T(0) = T_1 = C_2 \qquad (2\text{-}56)$$

The second boundary condition must specify that the surface at $x = L$ is an insulated or adiabatic boundary. Since the heat transferred to this boundary is conducted to the surface, the condition of an adiabatic surface would be

$$q''|_{x=L} = -k \frac{dT}{dx}\Big|_{x=L} = 0$$

$$\frac{dT}{dx}\Big|_{x=L} = \frac{-q_G'''}{k} L + C_1 = 0$$

or

$$\frac{dT}{dx}\Big|_{x=L} = 0 \qquad C_1 = \frac{q_G'''}{k} L \qquad (2\text{-}57)$$

An insulated boundary in a solid material is one for which the temperature gradient is zero at the boundary. Substituting both boundary conditions, Eqs. 2-56 and 2-57, into Eq. 2-55 results in the temperature distribution in the solid:

$$\frac{T(x) - T_1}{T_1} = \frac{q_G''' x L}{kT_1}\left(1 - \frac{x}{2L}\right) \qquad (2\text{-}58)$$

The temperature distribution is parabolic with x and its maximum value occurs at the insulated surface, $x = L$. The condition for a maximum temperature, $dT/dx = 0$, was satisfied by the insulated boundary condition at $x = L$. The maximum temperature in the wall is therefore

$$T(L) = T_{max} = T_1 + \frac{q_G''' L^2}{2k}$$

The equation above can be recast in terms of dimensionless parameters, as shown in Example 1-10:

$$\frac{T_{max}}{T_1} = 1 + \frac{q_G''' L^2}{2kT_1}$$

We should also notice that all the energy generated inside the wall must be conducted from the surface at $x = 0$. No heat may be transferred through the right-hand surface because it is insulated and no energy may be stored in the material because steady conditions have been assumed. Therefore, an energy balance on the wall at the surface $x = 0$ requires that

$$q|_{x=0} = -q_G''' V$$

or

$$-kA \frac{dT}{dx}\bigg|_{x=0} = -q_G''' AL$$

Differentiation of Eq. 2-58 will show that this condition is automatically satisfied.

Problems involving nonuniform heat generation, or ones with different boundary conditions, are approached in a manner similar to the procedure illustrated above.

Cylindrical Coordinates

A common problem in cylindrical coordinates involving energy generation is the case of a solid wire carrying an electric current such as that shown in Fig. 2-10. The current is I and the electrical resistance of the wire is R. The external surface temperature of the wire is a known value T_o. The energy generated per unit volume within the wire is

$$q_G''' = \frac{I^2 R}{V_{\text{OLUME}}}$$

If the current and electrical resistance are constants, the internal heat generation is also a constant.

The steady, one-dimensional form of the conduction equation in cylindrical coordinates which includes constant generation is Eq. 2-21:

$$\frac{1}{r} \frac{d}{dr}\left(r \frac{dT}{dr}\right) + \frac{q_G'''}{k} = 0 \tag{2-59}$$

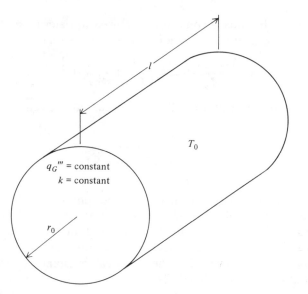

Figure 2-10   Conduction in a cylinder with uniform generation.

Integrating Eq. 2-59 twice yields the temperature distribution in the wire in terms of the two constants of integration, C_1 and C_2:

$$T(r) = C_1 \ln r - \frac{q_G''' r^2}{4k} + C_2 \tag{2-60}$$

To determine values for C_1 and C_2 we must have two boundary conditions. At first glance it appears that we have only one boundary condition, which is

$$T(r_o) = T_o$$

But we also know that all locations in the wire must have a finite temperature. If we try to determine the centerline temperature of the wire by evaluating Eq. 2-60 at $r = 0$, we would arrive at an infinite temperature as long as the $\ln r$ term remains. To prevent the unrealistic temperature at the centerline of the wire, we must set $C_1 = 0$.

Another way of visualizing the second boundary condition is to realize that the centerline of the wire as an insulated location:

$$\left. \frac{dT}{dr} \right|_{r=0} = 0$$

The centerline must be insulated because of the fact that it is a line of symmetry. This boundary condition provides the same result as before, $C_1 = 0$.

Temp is finite everywhere
so at $R = 0$ $\ln R = \infty \Rightarrow C_1 \ln R = T \Rightarrow \infty$

So $C_1 = 0$

When the two boundary conditions are used to determine values for C_1 and C_2, the temperature distribution in the wire becomes

$$\frac{T(r) - T_o}{T_o} = \frac{q_G''' r_o^2}{4kT_o}\left[1 - \left(\frac{r}{r_o}\right)^2\right]$$ (2-61)

The maximum temperature in the wire occurs at the center and is equal to

$$T(0) = T_{max} = \frac{q_G''' r_o^2}{4k} + T_o$$

Example 2-7. Determine the maximum current that a 1-mm-diameter bare aluminum ($k = 204$ $W/m\cdot K$) wire can carry without exceeding a temperature of 200°C. The wire is suspended in air with an ambient temperature of 25°C, and the convective heat-transfer coefficient between the wire and air is $10 W/m^2\cdot K$. The electrical resistance of this wire per unit length of conductor is 0.037 Ω/m.

Solution: This example is a slight variation of Example 2-6. In this problem the ambient air temperature is known rather than the surface temperature of the wire. The appropriate boundary condition is therefore one for which the heat conducted to the exterior surface of the wire is equal to the heat convected into the air. Mathematically, this boundary condition is expressed as

$$\bar{h}_c\left[T(r_0) - T_\infty\right] = -k\frac{dT}{dr}\bigg|_{r=r_o}$$ (2-62)

As before, the second boundary condition is

$$\frac{dT}{dr}\bigg|_{r=0} = 0$$ (2-63)

This boundary condition implies that the maximum wire temperature will occur at the center of the wire.

The appropriate form of the conduction equation is Eq. 2-59 and the solution for constant generation is given in Eq. 2-60. Substituting the two boundary conditions 2-62 and 2-63 results in the temperature distribution in the wire of

$$\frac{T(r) - T_\infty}{T_\infty} = \frac{q_G''' r_o}{2\bar{h}_c T_\infty}\left(1 + \frac{\bar{h}_c r_o}{2k} - \frac{\bar{h}_c r^2}{2r_o k}\right)$$ (2-64)

The maximum wire temperature is therefore

$$T(0) = T_{max} = T_\infty + \frac{q_G''' r_o}{2\bar{h}_c}\left(1 + \frac{\bar{h}_c r_o}{2k}\right)$$

The generation term expressed in terms of the current and resistance per unit length is

$$q_G''' = \frac{I^2 R}{V} = \frac{I^2}{A} \frac{R}{l} = \frac{I^2}{\pi r_o^2} \frac{R}{l}$$

so

$$T_{max} = T_\infty + \frac{I^2}{2\pi r_o \bar{h}_c} \frac{R}{l} \left(1 + \frac{\bar{h}_c r_o}{2k}\right)$$

$$200 = 25 + \frac{I^2}{2\pi(10^{-3}/2) \times 10} 0.037 \left[1 + \frac{10 \times (10^{-3}/2)}{2 \times 204}\right]$$

Solving for the current yields

$$I = 12.2 \text{ A}$$

At this point in our development of the principles of conduction, we should again recognize the occurrence of several dimensionless groups which recur throughout the chapter. Equation 2-64 is written in dimensionless form. Therefore, the groups

$$\frac{q_G''' r_o}{\bar{h}_c T_\infty} \quad \text{and} \quad \frac{\bar{h}_c r_o}{k}$$

are also dimensionless. The first term is actually a dimensionless generation and the product of the two is the dimensionless generation first identified in Eq. 2-16. The second dimensionless group is the Biot number, which appears in problems involving the combined conduction/convection modes of heat transfer.

In addition to recognizing the existence of the Biot number, we should also be aware of its effect on the heat-transfer process. The Biot number is the ratio of conductive resistance in the solid to convective resistance in the fluid. Therefore, the physical limits on the Biot number are

$$\text{Bi} \to 0 \quad \text{when} \quad R_{cond} \to 0$$

or when $k \to \infty$, and

$$\text{Bi} \to \infty \quad \text{when} \quad R_{conv} \to 0$$

or when $\bar{h}_c \to \infty$.

When the Biot number approaches zero, the solid is practically isothermal and the temperature varies most in the fluid. As the Biot number approaches infinity, the opposite is true. The resistance in the solid is much larger than that in the fluid, the fluid is nearly isothermal, and the temperature differences occur predominantly in the solid.

2-6 HEAT TRANSFER FROM FINS

Heat conducted through a solid substance is often removed from the solid purely by the convection mode. Since the convection rate is proportional to the surface area, the heat dissipated at the surface can be increased by merely extending the surface. The extended surface is called a *fin*.

A simple straight fin with constant cross-sectional area A is shown in Fig. 2-11. The heat is conducted through the solid material of the fin and it is removed from the surface to the surrounding fluid by convection. The temperature of the ambient fluid is T_∞ and the combined-heat-transfer coefficient is $\bar{h}_c$, both of which are assumed constant.

To determine the temperature distribution in the fin, and eventually the heat-transfer rate from the surface, we must first perform an energy balance on a differential volume of fin material. We cannot use the conduction equation developed in Section 2-2 because it accounts only for the conduction mode and does not consider convection from the surface.

For steady conditions the rate of heat conducted into the elemental volume at x shown in Fig. 2-11 is equal to the sum of the rate of heat conducted out of the volume at $x + \Delta x$ plus the rate of heat convected from the surface of the volume:

$$q_x = q_{x+\Delta x} + q_c$$

Substituting the Fourier law for the two conduction terms and Newton's law of cooling for the single convection term yields

$$-kA\frac{dT}{dx}\bigg|_x = -kA\frac{dT}{dx}\bigg|_{x+\Delta x} + \bar{h}_c P \Delta x \big[\, T(x) - T_\infty \big]$$

where P is the perimeter of the fin. Dividing all terms by Δx and taking the limit as $\Delta x \rightarrow 0$ gives a second-order differential equation for the tempera-

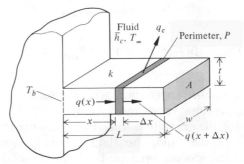

Figure 2-11   Fin with constant cross-sectional area.

ture distribution,

$$\frac{d^2T}{dx^2} - \frac{\bar{h}_c P}{kA}[T(x) - T_\infty] = 0 \tag{2-65}$$

Equation 2-65 may be nondimensionalized by defining a dimensionless temperature and coordinate as

$$\theta(x) = \frac{T(x) - T_\infty}{T_b - T_\infty}$$

and

$$\xi = \frac{x}{L}$$

where T_b is the base $(x=0)$ temperature of the fin. In terms of the new variables, Eq. 2-65 becomes

$$\frac{d^2\theta}{d\xi^2} - \frac{\bar{h}_c PL^2}{kA}\theta = 0 \tag{2-66}$$

The dimensionless group $(\bar{h}_c PL^2/kA)$ can be simplified to a form resembling the Biot number. The perimeter times the length of the fin is equal to the total surface area A_s of the fin:

$$A_s = PL$$

Then

$$\frac{PL^2}{A} = \frac{A_s L}{A} \tag{2-67}$$

where A is the cross-sectional area of the fin. Equation 2-67 has dimensions of length and it can therefore be considered to be the characteristic length of the fin l:

$$\frac{PL^2}{A} = l$$

The dimensionless group in Eq. 2-66 can now be expressed as

$$\frac{\bar{h}_c PL^2}{kA} = \frac{\bar{h}_c l}{k} \tag{2-68}$$

which is similar to the Biot number used in previous problems involving combined conduction and convection. The Biot number is then

$$\mathrm{Bi} = \frac{\bar{h}_c l}{k} = \frac{\bar{h}_c PL^2}{kA} = \frac{h_c A_s L}{kA} \tag{2-69}$$

We should have expected some form of the Biot number to appear in a fin problem which combines the conductive and convective modes of heat transfer.

The dimensionless form of the fin-energy equation (2-66) can now be written in terms of the Biot number:

$$\frac{d^2\theta}{d\xi^2} - (\text{Bi})\theta = 0 \tag{2-70}$$

The solution of Eq. 2-70 is

$$\theta(\xi) = C_1 e^{-(\text{Bi})^{1/2}\xi} + C_2 e^{(\text{Bi})^{1/2}\xi} \tag{2-71}$$

The values of the two constants of integration can be determined once two boundary conditions are specified. The most frequently known temperature along the length of the fin is the base temperature, T_b; written in the form of a boundary condition,

B.C.

$$T(0) = T_b \tag{2-72}$$

This equation will serve as the first boundary condition. The second boundary condition may take one of several different forms. Three of the most commonly used boundary conditions are considered in the following three cases.

Case I: A very long fin such that the tip temperature reaches the ambient temperature of the fluid:

$$T(L \to \infty) = T_\infty \tag{2-73}$$

or

$$\theta(1) = 0$$

Case II: A fin with an insulated tip at $x = L$:

$$\left.\frac{dT}{dx}\right|_L = 0 \tag{2-74}$$

or

$$\left.\frac{d\theta}{d\xi}\right|_{1.0} = 0$$

Case III: A fin with a convective heat loss from the tip surface area. This boundary condition becomes

$$-k\left.\frac{dT}{dx}\right|_L = \bar{h}_c\left[T(L) - T_\infty\right] \tag{2-75}$$

or

$$-\left.\frac{d\theta}{d\xi}\right|_{1.0} = \frac{\bar{h}_c L}{k}\theta(1)$$

The boundary condition 2-72, along with one of each of the three

$$\xi = \frac{x}{L}$$

boundary conditions 2-73, 2-74, or 2-75, will provide three different forms for the temperature distribution in a fin of constant cross-sectional area.

Once the temperature distribution in the fin is known, the heat dissipated from the fin can be determined. The easiest method of evaluating the heat-transfer rate from the fin involves determining the amount of heat conducted through the base of the fin:

$$q_f = -kA \frac{dT}{dx}\bigg|_{x=0} = -\frac{kA}{L}(T_b - T_\infty)\frac{d\theta}{d\xi}\bigg|_{\xi=0} \tag{2-76}$$

We can now determine the temperature distribution and heat-transfer rates from the fins that satisfy the three given sets of boundary conditions.

Case I: For an infinitely long fin, the dimensionless temperature distribution is

$$\theta(\xi) = \frac{T(\xi) - T_\infty}{T_b - T_\infty} = e^{-\sqrt{\text{Bi}}\,\xi^2}$$

But the length of the fin is indeterminate, so it is more convenient to express the temperature distribution in terms of x:

$$\theta(x) = \frac{T(x) - T_\infty}{T_b - T_\infty} = e^{-\sqrt{\bar{h}_c P x^2 / kA}} \tag{2-77}$$

The heat-transfer rate is

$$q_f = \sqrt{\bar{h}_c P k A}\,(T_b - T_\infty) = \sqrt{\text{Bi}}\,\frac{kA}{L}(T_b - T_\infty) \tag{2-78}$$

Case II: For a fin with an insulated tip, the dimensionless temperature distribution is

$$\theta(\xi) = \frac{T(\xi) - T_\infty}{T_b - T_\infty} = \frac{\cosh\left[(\text{Bi})^{1/2}(1 - \xi)\right]}{\cosh(\text{Bi})^{1/2}} \tag{2-79}$$

and the heat-transfer rate from the fin is

$$q_f = (\text{Bi})^{1/2}\frac{kA}{L}(T_b - T_\infty)\tanh(\text{Bi})^{1/2} \tag{2-80}$$

Case III: For a fin with convection from its tip, the temperature distribution is

$$\theta(\xi) = \frac{T(\xi) - T_\infty}{T_b - T_\infty}$$

$$= \frac{\cosh\left[(\text{Bi})^{1/2}(1 - \xi)\right] + (\text{Bi})^{1/2}(A/PL)\sinh\left[(\text{Bi})^{1/2}(1 - \xi)\right]}{\cosh(\text{Bi})^{1/2} + (\text{Bi})^{1/2}(A/PL)\sinh(\text{Bi})^{1/2}}$$

$$\tag{2-81}$$

and the heat-transfer rate is

$$q_f = (\text{Bi})^{1/2} \frac{kA}{L} (T_b - T_\infty) \left[\frac{\sinh(\text{Bi})^{1/2} + (\text{Bi})^{1/2}(A/PL)\cosh(\text{Bi})^{1/2}}{\cosh(\text{Bi})^{1/2} + (\text{Bi})^{1/2}(A/PL)\sinh(\text{Bi})^{1/2}} \right]$$

(2-82)

Example 2-8. A stainless steel ($k = 20$ W/m·K) fin has a circular cross-sectional area with a diameter of 2 cm and a length of 10 cm. The fin is attached to a wall that has a temperature of 300°C. The fluid surrounding the fin has an ambient temperature of 50°C and the heat-transfer coefficient is 10 W/m²·K. The end of the fin is insulated. Determine:

a. The rate of heat dissipated from the fin.
b. The temperature at the end of the fin.
c. The rate of heat transfer from the wall area covered by the fin if the fin is not used.
d. The heat-transfer rate from the same fin geometry if the stainless steel fin is replaced by a ficticious fin with infinite thermal conductivity.

Solution: The temperature distribution and heat-transfer rate from the fin are given by Eqs. 2-79 and 2-80, respectively. First, we will calculate the fin parameters:

$$A = \pi R^2 = \pi (0.01)^2 = \pi \times 10^{-4} \text{ m}^2$$

$$\text{Bi} = \frac{\bar{h}_c PL^2}{kA} = \frac{10 \times \pi(0.02)(0.1)^2}{20 \times \pi \times 10^{-4}} = 1.0$$

$$\frac{kA}{L} = \frac{20.0 \times \pi \times 10^{-4}}{0.1} = 0.06283 \text{ W/K}$$

a. The heat-transfer rate is

$$q_f = (\text{Bi})^{1/2} \frac{kA}{L} (T_b - T_\infty) \tanh(\text{Bi})^{1/2}$$

$$= (1.0)(0.06283)(300 - 50)\tanh(1.0) = 11.96 \text{ W}$$

b. The fin-tip temperature is the temperature at $\xi = 1$:

$$\theta(1) = \frac{\cosh 0}{\cosh(\text{Bi})^{1/2}} = \frac{1}{\cosh(1.0)} = \frac{1}{1.543} = 0.648$$

$$T(L) = T_\infty + 0.648(T_b - T_\infty) = 50 + 0.648(300 - 50)$$

$$= 212°C$$

c. If we assume that the heat-transfer coefficient over the surface of the wall is the same as that over the surface of the fin, the heat-transfer rate from the wall without a fin attached is

$$q = \bar{h}_c A (T_b - T_\infty) = 10 \times \pi \times 10^{-4}(300 - 50) = 0.785 \text{ W}$$

The presence of the fin has increased the heat dissipation from the surface area covered by the fin by a factor of $11.96/0.782 = 15.2$.

d. If the fin thermal conductivity approaches infinity, the Biot number would approach zero. The heat flow by conduction through the fin material would have no resistance and the entire length of the fin would become isothermal at the base temperature. The heat-transfer rate from this ideal fin would then become

$$q_{\text{ideal}} = \bar{h}_c A_s (T_b - T_\infty)$$
$$= 10\pi(0.02)(0.1)(300 - 50) = 15.71 \text{ W}$$

The ideal heat-transfer rate is the maximum possible amount of heat that can be transferred from a fin of equal size. The stainless steel fin dissipates

$$\frac{15.71 - 11.96}{15.71} = 24\%$$

less heat than the ideal fin.

Fin Efficiency

The previous analysis used to determine the temperature distribution and heat-transfer rate from fins only applies to fins that have constant cross-sectional areas. When the fin is tapered, the cross-sectional area varies resulting in a more complex equation for the temperature distribution. The temperature distribution and heat-transfer rates from tapered fins are expressed in terms of Bessel functions. A complete treatment on the subject of tapered fins can be found in References 1 and 2.

A convenient concept that can be used to provide a value for the heat-transfer rate from fins is the fin efficiency. The *fin efficiency* is defined as the ratio of the actual heat-transfer rate from a fin to the heat-transfer rate from an ideal fin:

$$\eta = \frac{q_{\text{actual}}}{q_{\text{ideal}}} \tag{2-83}$$

The ideal fin transfers the maximum amount of heat of any fin of equal size and base temperature. The ideal fin has an infinite thermal conductivity, and therefore its entire length is isothermal at the base temperature. The actual and ideal fins have the same geometry and the same base temperature. The heat-transfer rate from the ideal fin is

$$q_{\text{ideal}} = \bar{h}_c A_s (T_b - T_\infty)$$

where A_s is the entire surface area of the fin exposed to the fluid at a temperature of T_∞.

The heat-transfer rate from the actual fin will then be

$$q_{\text{actual}} = \eta \bar{h}_c A_s (T_b - T_\infty) \tag{2-84}$$

We are already in a position to determine expressions for the fin efficiency. For example, the fin efficiency for a fin with constant cross-sectional area and an insulated tip would be

$$\eta = \frac{q_{\text{actual}}}{q_{\text{ideal}}} = \frac{(\text{Bi})^{1/2}(kA/L)(T_b - T_\infty)\tanh(\text{Bi})^{1/2}}{\bar{h}_c PL(T_b - T_\infty)}$$

or

$$\eta = \frac{1}{(\text{Bi})^{1/2}}\tanh(\text{Bi})^{1/2} \tag{2-85}$$

A plot of Eq. 2-85 is shown in Fig. 2-12. The figure shows that the efficiency drops rapidly as the Biot number increases. A fin with a large value of Biot number dissipates less heat than one with a smaller Biot number. If the efficiency drops to a very low value, it is possible for the surface of the wall without the fin present to transfer more heat than from the wall with fin in place. We could have anticipated this situation. The Biot number represents the ratio of conductive to convective resistances. For a large value of Bi the conductive resistance is large compared to the convective resistance, and the temperature drop in the fin is significant.

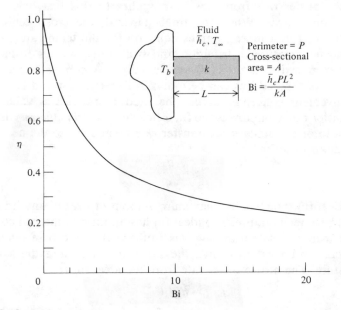

Figure 2-12   Efficiency for constant-cross-sectional-area fin with insulated tip.

When the Biot number is large, the poorly conducting fin occupies an area that can effectively transfer heat by convection, and the presence of the fin reduces the heat dissipation from the wall.

Fin materials should be selected that have high values of thermal conductivity: that is, metallic fins are superior to fins made of insulating materials. In situations when the convective-heat-transfer coefficient is large, the Biot number increases and the advantage of adding fins for the purpose of increasing the heat transfer rate is diminished. If the fluid changes phase by either boiling or condensing, the heat-transfer coefficient becomes quite large, as shown in Table 1-2. Therefore, when the fluid changes phase, it is possible for the fin to actually reduce the heat dissipated from the plane surface.

Figure 2-12 gives the fin efficiency for a fin with constant cross-sectional area if the fin has an insulated tip. The curve must be modified if it is expected to apply to a fin which loses heat from its end surface area. The heat transfer from the end surface may be compensated for by adding an imaginary length to the fin. The added length is such that the additional surface area will transfer the same amount of heat as from the tip area of the actual fin and the end surface of the extended fin will be insulated.

Jakob (Ref. 3) recommends that the added length be equal to the ratio of the fin cross-sectional area to perimeter. The corrected length of the fin,

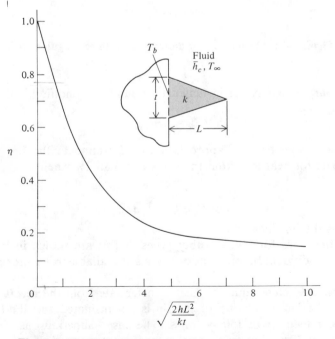

Figure 2-13 Efficiency for fin with triangular profile.

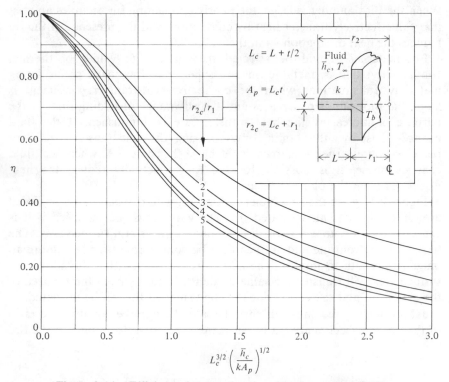

Figure 2-14 Efficiency for annular fin with rectangular profile.

L_c, necessary to satisfy the insulated tip boundary condition is then

$$L_c = L + \frac{A}{P}$$

The error involved in the approximation of adding to the fin length to compensate for heat loss from the tip is less than 1% when

$$\sqrt{\frac{\bar{h}_c t}{k}} \leqslant \frac{1}{4}$$

where t is the fin thickness.

Fin efficiencies for several other types of fins are shown in Figs. 2-13 and 2-14. Additional fin-efficiency curves are available in Reference 4.

Example 2-9. Determine the heat-transfer rate from the rectangular fin shown in the figure. The tip of the fin is not insulated and the fin has a thermal conductivity of 150 W/m·K. The base temperature is 100°C and the fluid is at 20°C. The heat-transfer coefficient between the fin and fluid is 30 W/m²·K.

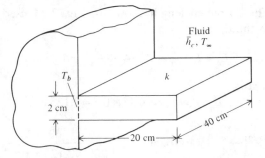

Solution: To account for heat loss from the tip area we determine the corrected length of the fin:

$$L_c = L + \frac{A}{P} = 20 + \frac{40 \times 2}{84} = 20.95 \text{ cm}$$

The Biot number based on the corrected length is then

$$\text{Bi} = \frac{\bar{h}_c P L_c^2}{kA} = \frac{30 \times 0.84 \times (0.2095)^2}{150 \times 0.008} = 0.922$$

and the surface area of the fin with length L_c is

$$A_s = L_c P = 0.2095 \times 0.84 = 0.176 \text{ m}^2$$

The efficiency from Fig. 2-12 is

$$\eta = 0.775$$

The heat-transfer rate is then

$$q = \eta \bar{h}_c A_s (T_b - T_\infty) = 0.775 \times 30 \times 0.176(100 - 20)$$
$$= 327 \text{ W}$$

Example 2-10. An aluminum ($k = 200$ W/m·K) annular fin is placed on a copper tube that carries a fluid. The tube is 8 cm o.d. The fluid is at 250°C. The fin is 0.5 cm thick and 16 cm o.d. The surrounding fluid is at 70°C and the convective-heat-transfer coefficient is 60 W/m²·K. Determine the heat-transfer rate from the fin.

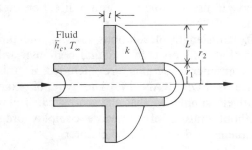

Solution: The corrected length shown in Fig. 2-14 used to account for heat loss from the tip is

$$L_c = L + \frac{t}{2} = (8-4) + \frac{0.5}{2} = 4.25 \text{ cm}$$

The profile area is

$$A_p = L_c t = 4.25 \times 0.5 = 2.125 \text{ cm}^2$$

$$L_c^{3/2} \left(\frac{\bar{h}_c}{kA_p} \right)^{1/2} = (0.0425)^{3/2} \left(\frac{60}{200 \times 2.125 \times 10^{-4}} \right)^{1/2}$$

$$= 0.33$$

Fig. 2-14

$$\frac{r_{2c}}{r_1} = \frac{r_1 + L_c}{r_1} = 1 + \frac{L_c}{r_1} = 1 + \frac{4.25}{4} = 2.06$$

The efficiency from Fig. 2-14 is

$$\eta = 0.89$$

and the heat-transfer rate from the fin is

$$q = \eta \bar{h}_c A_s (T_b - T_\infty) = \eta \bar{h}_c 2\pi (r_{2c}{}^2 - r_1{}^2)(T_b - T_\infty)$$

$$= 0.89 \times 60 \times 2 \times \pi (0.0825^2 - 0.04^2)(250 - 70)$$

$$= 314 \text{ W}$$

The base temperature of the fin is assumed to be the same as the fluid temperature inside the tube because the temperature drop across the copper tube will be small.

2-7 STEADY, TWO- AND THREE-DIMENSIONAL CONDUCTION

We have assumed so far that the temperature distribution in the solid was a function of only a single coordinate; that is, the situation involved only one-dimensional conduction. However, we now need to develop techniques that can be used to determine the heat-transfer rate and temperature distribution when the temperature is a function of two or three coordinate variables. The two- and three-dimensional solution will be more involved, and so we will have to use approximate and indirect or analog methods to provide a solution.

The complexity and length of solutions to two- and three-dimensional problems suggest that solution with a digital computer will be desirable. Therefore, two computer programs are included in this section. The program language is FORTRAN IV. The type of example programs selected are relatively simple, so that the reader can follow the program development without unusual effort. More complex programs are suggested in the problems at the end of the chapter.

Analytical Methods

The most obvious approach to determining the temperature distribution in a solid for which the temperature is a function of two or three coordinates would be to attempt an exact solution of the governing equation. For the case of steady conduction in a solid with constant thermal conductivity and no internal generation, the governing equation is the conduction equation derived in Section 2-2:

$$\nabla^2 T = 0$$

This equation is Laplace's equation. The form of Laplace's equation in the different coordinate systems is given in Appendix B.

Laplace's equation is a linear partial differential equation. Several standard techniques for solving it are available. One method, separation of variables, is particularly useful in heat-transfer work. Although this method is not covered here, the interested reader is referred to References 5, 6, and 7 for complete details on this and other methods of solving Laplace's equation.

Once the temperature distribution is determined, regardless of method, the heat flux is determined by the Fourier law. In two- and three-dimensional systems this law is most conveniently expressed in vector form as

$$\bar{q}'' = -k \nabla T \tag{2-86}$$

where ∇T is the gradient of the scalar temperature. The form of the gradient in rectangular, cylindrical and spherical coordinates is given in Appendix B.

The gradient of a scalar quantity such as the temperature results in a vector quantity which, according to the vector form of the Fourier law (Eq. 2-86), is the heat flux, $\bar{q}''$. Usually, we do not consider the heat flux to be a vector quantity since it has dimensions of energy per unit area, neither of which are vector quantities. However, it is convenient to imagine heat to be "flowing" in a certain direction; therefore, $\bar{q}''$ is often referred to as the *heat-flux vector*.

An important geometric property of the gradient is the fact that the heat-flux vector is directed perpendicular to an *isotherm*, a line of constant temperature, at each point in the solid. As an illustration of this property, Fig. 2-15 shows several isotherms and representative heat-flux vectors at points A, B, and C in a two-dimensional rectangular solid. The length of each of the three heat-flux vectors is proportional to the local temperature gradient. That is, where the isotherms are closely spaced, the gradient is large and the heat flux is also large. Where the isotherms are widely spaced, the heat flux is proportionally smaller. In Fig. 2-15 the heat flux at point A is greater than at point B, where the temperature gradient is smaller.

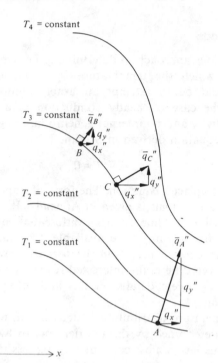

Figure 2-15 Heat-flux vector and its geometric relationship to isotherms.

Since we can visualize the heat flux as a vector, it should have properties like any other vector quantity. We should therefore be able to resolve the heat-flux vector into its components in the directions of the coordinate axes. Expressions for the vector components can be determined by expanding the form of the gradient. For a rectangular coordinate system, the heat-flux vector is

$$\vec{q}'' = -k\left(\frac{\partial T}{\partial x}\hat{i} + \frac{\partial T}{\partial y}\hat{j} + \frac{\partial T}{\partial z}\hat{k}\right)$$

Therefore, the heat flux in the x direction would be

$$q_x'' = -k\frac{\partial T}{\partial x}$$

Similar expressions can be written for the components of the heat-flux vector in the y and z directions.

The heat-transfer rate in the x direction across a plane area P which lies in the yz plane is then

$$q_x = -k\int_{A_P}\left(\frac{\partial T}{\partial x}\right)_P dA \tag{2-87}$$

The subscript P indicates that the derivative of the temperature must be evaluated at each point on the plane before integration over the area of the plane.

Graphical Methods

Exact, analytical solutions to the conduction equation for two and three dimensions are often impossible to achieve. For cases in which analytical solutions are difficult to obtain, approximate methods are valuable. The graphical method is a simple technique that can provide answers for the heat-transfer rate with surprising accuracy.

The graphical method is based on the geometrical requirement of the vector form of the Fourier law, which specifies that the isotherms and constant heat-flux lines are always perpendicular at points where the two lines intersect. We can therefore sketch the isotherms and constant flux lines and continue to revise them until they satisfy the perpendicular condition.

The accuracy of the sketched temperature distribution will be directly related to the care taken in the construction of the lines. With a little experience we can obtain reasonably accurate results in a short amount of time.

The steps used in the graphical method can be outlined as follows:

Step 1: Draw an accurate scale model of the material in which the temperature distribution and heat-transfer rate are desired.

Step 2: Sketch the heat-flux lines and isotherms on the model. The flux lines and isotherms form curvilinear squares. At all intersections of the heat flux and isotherms, the tangents to the curves will be perpendicular. The diagonals of curvilinear squares bisect each other and are perpendicular. Remember that adjacent isotherms and heat-flux lines cannot cross each other. Isotherms are perpendicular to adiabatic boundaries since an adiabatic boundary is a line of constant flux, that is, $q'' = 0$. Also, lines of symmetry are adiabatic boundaries.

Step 3: Continue to redraw the isotherms and flux lines by adjusting their location until they meet the conditions specified in Step 2.

Once you are satisfied with the accuracy of your drawing, the temperature distribution is known and the heat flux is determined, as usual, by applying the Fourier law. To illustrate this procedure, consider the problem of determining the heat-transfer rate through a structural I beam used in a furnace wall. The beam is surrounded on either side by insulation placed in the wall of the furnace, as shown in Fig. 2-16(a). The surface of

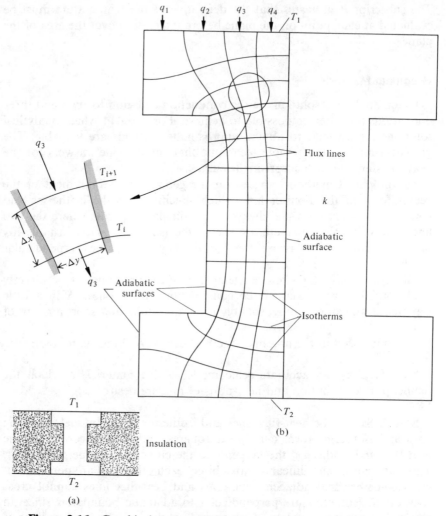

Figure 2-16 Graphical method applied to an I beam: (a) physical model; (b) scale drawing of beam and sketch of flux lines and isotherms.

the beam on the inside of the furnace has a temperature of T_1 and the surface of the beam near the exterior wall is at T_2. The thermal conductivity of the beam is k.

The centerline of the beam is a line of symmetry and it is therefore an adiabatic surface. Constant-flux lines and isotherms sketched on a scale drawing of the beam are shown in Fig. 2-16(b).

The heat is restricted to flow in four channels limited on each side by flux lines. The total heat transfer rate through one-half of the beam is therefore

$$q_{\text{total}} = \sum_{i=1}^{4} q_i$$

The inset in the figure shows a typical curvilinear square through which the heat-transfer rate is q_3. The Fourier law applied to the single curvilinear square per unit depth of the figure is

$$q_3' \simeq k \Delta y \frac{T_{i+1} - T_i}{\Delta x} \tag{2-88}$$

If each curvilinear square is sketched such that it meets the condition $\Delta x = \Delta y$, each temperature subdivision is equal. The temperature difference between two adjacent isotherms can then be expressed in terms of the overall temperature difference across the entire surface and, M, the number of equal temperature subdivisions in the figure:

$$T_{i+1} - T_i = \frac{(\Delta T)_{\text{overall}}}{M} = \frac{T_1 - T_2}{M} \tag{2-89}$$

If the flux lines have been divided into N equal subdivisions, the heat transfer through each of the channels formed by adjacent heat-flux lines is equal, and the total heat transfer through the beam is

$$q_{\text{total}} = N q_i \tag{2-90}$$

Substituting Eqs. 2-88 and 2-89 into 2-90 gives an expression for the total heat-transfer rate of

$$q_{\text{total}}' = \frac{N}{M} k (\Delta T)_{\text{overall}} \tag{2-91}$$

when the grid is square, that is, when $\Delta x = \Delta y$.

The heat-transfer rate can therefore be determined by drawing a series of curvilinear squares and then counting the number of equal temperature subdivisions, M, and the number of equal heat-flux subdivisions, N.

Example 2-11. Determine the heat-transfer rate through the beam shown in Fig. 2-16 if $T_1 = 500°C$, $T_2 = 200°C$, and $k = 70$ W/m·K.

Solution: For this example

$$M = 13 \qquad N = 4 \qquad (\Delta T)_{\text{overall}} = 300°C$$

Then the heat-transfer rate through the beam per unit depth is given by Eq. 2-91:

$$q' = 2 \left[\frac{N}{M} k (\Delta T)_{\text{overall}} \right] = 12{,}923 \text{ W/m}$$

The factor of 2 is used because the value of N was shown for only one-half of the beam.

Table 2-1   Conduction Shape Factor for Several Geometries, $q = kS(T_1 - T_0)$

GEOMETRY	SYMBOLS	CONDUCTION SHAPE FACTOR
Eccentric circular cylinders, length L		$\dfrac{2\pi L}{\cosh^{-1}\left(\dfrac{r_0^2 + r_1^2 - e^2}{2 r_0 r_1}\right)}$
Circular cylinder in a square, length L		$\dfrac{2\pi L}{\ln(0.54 a / r_1)}$
Parallelepiped shell with equal wall thicknesses, internal surface area A_1, and external surface area A_2		$\left[\dfrac{A_1}{t} + 2.16(a+b+c) + 1.2t\right]$ when $a,b,c > \dfrac{t}{5}$ $\dfrac{0.79\sqrt{A_1 A_2}}{t}$ when $a,b,c < \dfrac{t}{5}$

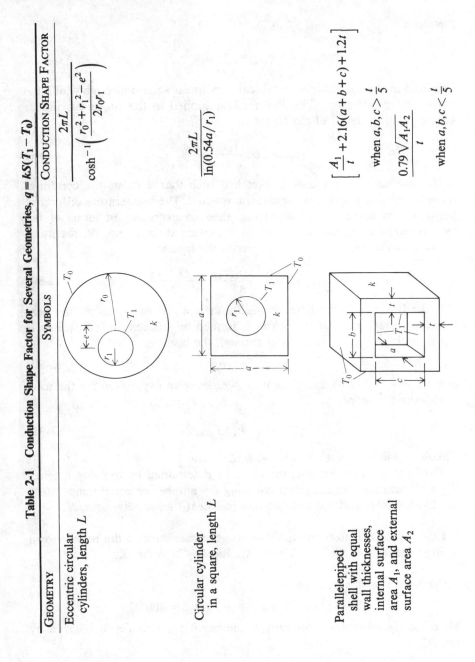

Thin, buried rectangular plate, length L

$$\frac{\pi a}{\ln(4a/L)} \quad (z=0)$$
$$\frac{2\pi a}{\ln(4a/L)} \quad (z \gg a)$$

Buried horizontal cylinder, length L

$$\frac{2\pi L}{\cosh^{-1}(z/r_1)} \quad (z \simeq 2r_1)$$
$$\frac{2\pi L}{\ln(2z/r_1)} \quad (z \gg 2r_1)$$

Buried sphere

$$\frac{4\pi r_1}{1-(r_1/2z)} \quad (z \gg r_1)$$
$$4\pi r_1 \quad (z \to \infty)$$

Two parallel cylinders of length L in an infinite medium

$$\frac{2\pi L}{\cosh^{-1}\left(\dfrac{a^2 - r_1^2 - r_0^2}{2r_1 r_0}\right)}$$

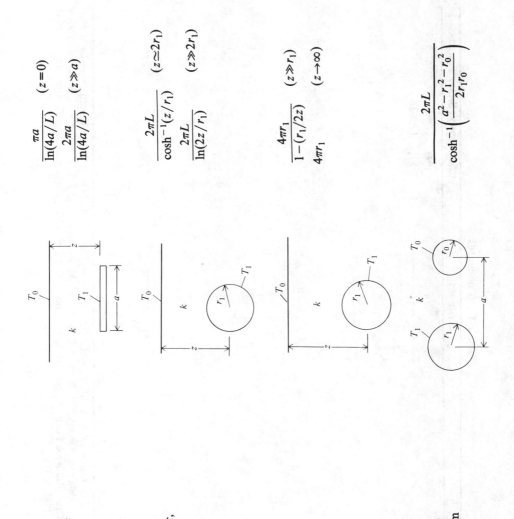

Table 2-1 (Continued)

GEOMETRY	SYMBOLS	CONDUCTION SHAPE FACTOR
Thin, buried horizontal disk		$4r_1 \quad (z=0)$ $8r_1 \quad (z \gg 2r_1)$
Vertical cylinder, length L, in semi-infinite solid		$\dfrac{2\pi L}{\ln(2L/r_1)}$
Hemisphere buried in a semi-infinite solid		$2\pi r_1$

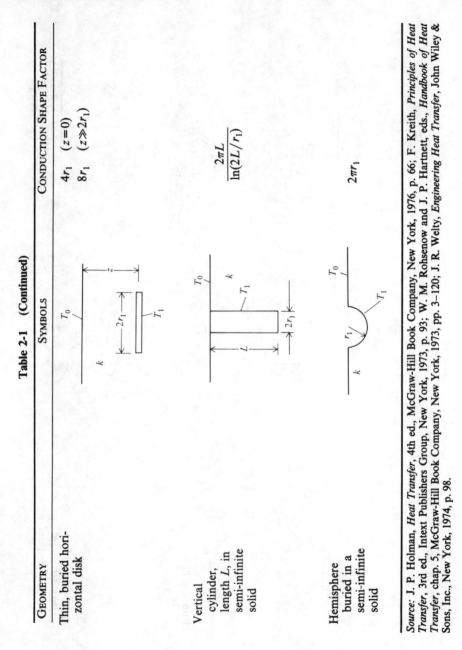

Source: J. P. Holman, *Heat Transfer*, 4th ed., McGraw-Hill Book Company, New York, 1976, p. 66; F. Kreith, *Principles of Heat Transfer*, 3rd ed., Intext Publishers Group, New York, 1973, p. 93; W. M. Rohsenow and J. P. Hartnett, eds., *Handbook of Heat Transfer*, chap. 5, McGraw-Hill Book Company, New York, 1973, pp. 3–120; J. R. Welty, *Engineering Heat Transfer*, John Wiley & Sons, Inc., New York, 1974, p. 98.

The factor N/M in Eq. 2-91 is called the *conduction shape factor*, S:

$$S = \frac{N}{M}$$

so that the heat flux written in terms of S is

$$q = kS(\Delta T)_{overall} \qquad (2\text{-}92)$$

The conduction shape factor for several shapes are cataloged in Table 2-1.

Expressions for the conduction shape factor are known for several simple geometries. For example, when Eq. 2-26 is arranged in the form of Eq. 2-92, the conduction shape factor for a plane wall is A/L. For a hollow cylinder of length l, the conduction shape factor is $2\pi l/\ln(r_o/r_i)$. Notice that the shape factor has dimensions of length.

Analog Methods

The steady electrical potential E in a material with constant resistivity and no internal sources of potential is governed by Laplace's equation, the same equation that governs the temperature distribution in a constant property solid with no internal energy generation. The rate equations for transport of heat (Fourier's law) and the transport of charge (Ohm's law) are also similar, as shown in the table. Owing to the similarity of the equations that govern the two phenomena, the transport of charge and heat are said to be analogous.

Analogous Equations for Thermal and Electrical Systems

	ELECTRICAL SYSTEM	THERMAL SYSTEM
Conservation equation	$\nabla^2 E = 0$	$\nabla^2 T = 0$
Rate equation	$I = \dfrac{\Delta E}{R}$	$q = \dfrac{\Delta T}{R_t}$

The dimensionless potential in the electrical system is analogous to the dimensionless temperature in a thermal system. It is to our advantage to use this analogy because voltages are easier to measure than temperatures. By measuring the location of the constant potential lines with a voltmeter, we can determine the locations of the isotherms.

The analogy is carried out as follows. A scale model of the thermal geometry is cut from a commercially available electrically conducting paper, and a battery is connected to the paper to provide the overall driving potential across the model. The electrical boundary conditions imposed on the paper must be analogous to the corresponding boundary conditions in the thermal problem. For isothermal boundaries the conducting paper must have a boundary of constant potential. This can easily be achieved by coating the boundary with a highly conducting paint and connecting it to a battery. The thermally insulated boundary can be simulated in the electrical system by an electrically insulated boundary, which is simply the edge of the paper.

Once the constant potential lines are located with a probe connected to a voltmeter, the boundary conditions may be switched and the orthogonal lines, or the lines of constant current, may be located. These lines correspond to lines of constant heat flux. By using this procedure we can accurately generate the complete set of curvilinear squares in the model, and a value for the conduction shape factor can be determined with greater accuracy than by the graphical technique.

The analog method has the advantage that it locates the isotherms and flux lines without the trial-and-error procedure of the graphical method, but it has the disadvantage that it requires the purchase of special equipment. The graphical method requires only pencil, paper, and patience. Both methods, however, are practically limited to two-dimensional geometries and simple boundary conditions such as isothermal and adiabatic boundaries. A more detailed discussion of the analog method can be found in References 8 and 9.

Numerical Methods

Numerical solutions are powerful and versatile techniques when applied to steady conduction problems. Numerical methods can be successfully applied to problems that cannot be solved conveniently by other techniques. For example, numerical methods can be used to solve problems involving radiative boundary conditions or internal energy generation. Graphical and analog techniques cannot be conveniently used to provide solutions to these two types of problems.

The finite-difference numerical method involves dividing the solid into a number of *nodes*. An energy balance is applied to each node, which results in an algebraic equation for the temperature of each node. A separate equation is derived for each node located on the boundary of the solid. The result of the finite-difference technique is n algebraic equations for the n nodes in the solid. The n algebraic equations replace the single partial differential equation and the applicable boundary conditions.

If the number of nodes in the solid is relatively small, we can use standard mathematical techniques to solve the resulting algebraic equations. As the number of nodes increases, the time required to achieve an exact solution becomes unreasonable. Approximate solutions become advantageous in these cases. We will first consider an approximate method, called *relaxation*.

As the number of equations grows large, the application of programmable calculators and digital computers becomes important. Two computer programs are included later in this section to illustrate the types of two-dimensional conduction problems that can be best solved by digital computers coupled with numerical techniques.

The finite-difference technique will be illustrated by considering a two-dimensional conduction problem. First, we divide the solid into a number of equal-size squares. The solid within each subdivision is imagined to be concentrated at the center of the square and the concentrated mass is the node. The interior region of a typical two-dimensional solid is shown in Fig. 2-17. Each subdivided square has a length in the x direction of Δx and a length in the y direction of Δy. The node designated by the subscript zero is shown surrounded by the four adjacent nodes. Each node is imagined to be connected to adjacent nodes by a small conducting rod. Heat can be conducted only along the imaginary rods. That is, conduction between node 0 and node 1, which actually occurs across an interface of height, Δy, in the continuous material is imagined to take place through the imaginary rod connecting nodes 0 and 1.

For steady conditions an energy balance applied to node 0 when there is no energy generation gives

$$\sum_{i=1}^{4} q_{i \to 0} = 0 \tag{2-93}$$

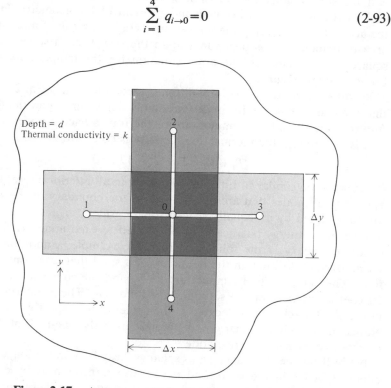

Figure 2-17   Arrangement of nodes for an interior section of a two-dimensional solid.

Next, we apply the Fourier law to each of these terms to express the equation in terms of nodal temperatures. The first term would be, for example,

$$q_{1\to0} = -kA\frac{\partial T}{\partial x} \simeq k(\Delta y\, d)\frac{T_1 - T_0}{\Delta x}$$

where the temperature gradient is evaluated at the midplane between the two nodes and d is the depth of the two-dimensional geometry measured into the plane of the figure. Similar expressions can be written for the three remaining terms:

$$q_{2\to0} \simeq k\Delta x\, d\frac{T_2 - T_0}{\Delta y}$$

$$q_{3\to0} \simeq k\Delta y\, d\frac{T_3 - T_0}{\Delta x}$$

$$q_{4\to0} \simeq k\Delta x\, d\frac{T_4 - T_0}{\Delta y}$$

If the subdivisions are drawn so that they are all square, $\Delta x = \Delta y$ and each heat-flow equation becomes independent of the geometry. However, the accuracy of replacing the temperature gradient by the finite difference of two temperatures is dependent upon the size of each square. As each square is made smaller, the approximation for the temperature gradient becomes more accurate.

By substituting the four finite-difference equations into Eq. 2-93 we see that the energy balance for node 0 is simply dependent upon the temperature of node 0, and the temperature of the four adjacent nodes when the grid is square and the thermal conductivity is constant:

$$T_1 + T_2 + T_3 + T_4 - 4T_0 = 0 \tag{2-94}$$

An equation similar to Eq. 2-94 will apply to all interior nodes; that is, it applies to all nodes that are not located on the boundary of the solid and are surrounded by an equally spaced square grid.

A separate energy balance must be applied to each node that is located on the boundary of the solid. Consider, for example, a node denoted by the subscript 0 located on the boundary of a solid that is in contact with a fluid. The ambient fluid temperature is T_∞ and the convective-heat-transfer coefficient between the solid and the fluid is $\bar{h}_c$. The geometry is shown in Fig. 2-18. Each node is located at the center of its respective subdivision. Notice that each boundary node represents only one-half of the mass represented by each interior node.

Node 0 on the boundary can exchange heat by conduction with three adjacent nodes in the solid, and it can also transfer heat by convection with the fluid. The energy balance applied to node 0 is therefore

$$q_{1\to0} + q_{2\to0} + q_{3\to0} + q_{\infty\to0} = 0$$

$$q_{IN} - q_{OUT} + q_G''' = q_{STORED}$$

$\Delta x = \Delta Y$

$q = KA \dfrac{\Delta T}{\Delta x}$

$= K \Delta x\, d\; \dfrac{T_1 - T_0}{\Delta x}$

$q = \bar{h}_c\, \dot{A}\; \Delta T$

$= \bar{h}_c\, \Delta x\, d\, (T_\infty - T_0)$

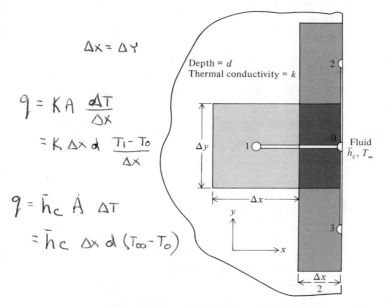

Figure 2-18   Arrangement of nodes for a two-dimensional
solid in contact with a fluid.

The first three terms represent conduction in the solid and the last term
represents the convection rate to node 0 from the ambient fluid designated
by the symbol ∞. Substituting finite-difference approximations for the
Fourier law for the first three terms and Newton's law for the last term
yields

$$k\,\Delta y\, d\frac{T_1 - T_0}{\Delta x} + k\frac{\Delta x}{2}\, d\frac{T_2 - T_0}{\Delta y} + k\frac{\Delta x}{2}\, d\frac{T_3 - T_0}{\Delta y}$$

$$+\, \bar{h}_c\, \Delta y\, d(T_\infty - T_0) = 0 \qquad (2\text{-}95)$$

Once again Eq. 2-95 can be simplified if we choose a square grid, or
$\Delta x = \Delta y$. Equation 2-95 may be rewritten in the form

$$\tfrac{1}{2}(T_2 + T_3) + T_1 + \left(\frac{\bar{h}_c\, \Delta x}{k}\right) T_\infty - \left[2 + \left(\frac{\bar{h}_c\, \Delta x}{k}\right)\right] T_0 = 0 \qquad (2\text{-}96)$$

The temperatures at the boundary are functions of the temperatures of
the neighboring nodes and the parameter $\bar{h}_c\, \Delta x/k$. We should recognize
this dimensionless group as the Biot number.

The finite-difference method is illustrated in the following example.

Example 2-12. Determine the steady temperature distribution and heat-
transfer rates from all four surfaces of the two-dimensional solid shown in
the figure. Two of the boundaries are isothermal, a third is insulated, and
the fourth transfers heat by convection.

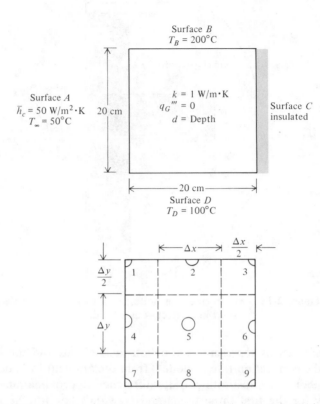

Solution: The solid is first subdivided into a square grid, as shown in the accompanying figure. The nodes are numbered from 1 to 9. The grid is square with $\Delta x = \Delta y = 10$ cm. The only nodes with unknown temperatures are nodes 4, 5, and 6. Node 5 is an interior node, so Eq. 2-94 applies.

$$T_4 + T_2 + T_6 + T_8 - 4T_5 = 0$$

Node 4 is on a boundary that transfers heat by convection, so Eq. 2-96 applies.

$$\tfrac{1}{2}(T_1 + T_7) + T_5 + (\text{Bi}) T_\infty - (2 + \text{Bi}) T_4 = 0$$

where

$$\text{Bi} = \frac{\bar{h}_c \Delta x}{k} = \frac{50 \times 0.10}{1} = 5$$

Node 6 is on an insulated boundary, so the appropriate energy balance is

$$q_{3 \to 6} + q_{5 \to 6} + q_{9 \to 6} = 0$$

or

$$k \frac{\Delta x}{2} d \frac{T_3 - T_6}{\Delta y} + k \Delta y\, d \frac{T_5 - T_6}{\Delta x} + k \frac{\Delta x}{2} d \frac{T_9 - T_6}{\Delta y} = 0$$

or

$$\tfrac{1}{2}(T_3 + T_9) + T_5 - 2T_6 = 0$$

The remaining six boundary nodes are maintained at known temperatures, so energy balances are not needed at these nodes. The six boundary temperatures are

$$T_1 = T_2 = T_3 = 200°C$$
$$T_7 = T_8 = T_9 = 100°C$$

Substituting these temperatures into the energy-balance equations for nodes 4, 5, and 6 yields

$$400 + T_5 - 7T_4 = 0 \qquad \text{(node 4)}$$
$$300 + T_4 + T_6 - 4T_5 = 0 \qquad \text{(node 5)}$$
$$150 + T_5 - 2T_6 = 0 \qquad \text{(node 6)}$$

Values for the temperatures T_4, T_5, and T_6 may be determined by simultaneously solving these three equations. The solutions are:

$$T_4 = 75.5°C$$
$$T_5 = 128.7°C$$
$$T_6 = 139.4°C$$

To determine the heat-transfer rates per unit depth at each surface, we will use the finite-difference form of the Fourier law when heat is transferred by conduction and Newton's law when heat is transferred by convection.

At surface A the heat-transfer rate per unit depth into the solid is

$$q_A' = q_{\infty \to 1}' + q_{\infty \to 4}' + q_{\infty \to 7}'$$
$$q_A' = \bar{h}_c \Delta y \left[\frac{T_\infty - T_1}{2} + (T_\infty - T_4) + \frac{T_\infty - T_7}{2} \right]$$
$$= -627.5 \text{ W/m}$$

That is, 627.5 W/m is convected away from the solid at the surface A. The negative sign indicates that the heat is removed from the solid.

At surface B the heat-transfer rate per unit depth into the solid is

$$q_B' = q_{1 \to 4}' + q_{2 \to 5}' + q_{3 \to 6}' + q_{1 \to \infty}'$$
$$= k \Delta x \left(\frac{1}{2} \frac{T_1 - T_4}{\Delta y} + \frac{T_2 - T_5}{\Delta y} + \frac{1}{2} \frac{T_3 - T_6}{\Delta y} \right)$$
$$+ \bar{h}_c \frac{\Delta y}{2} (T_1 - T_\infty)$$
$$= 538.8 \text{ W/m}$$

Surface C is insulated, so

$$q'_C = 0$$

At surface D

$$q'_D = k\,\Delta x\left(\frac{1}{2}\frac{T_9 - T_6}{\Delta y} + \frac{T_8 - T_5}{\Delta y} + \frac{1}{2}\frac{T_7 - T_4}{\Delta y}\right)$$
$$+ \bar{h}_c \frac{\Delta y}{2}(T_7 - T_\infty)$$
$$= 88.8 \text{ W/m}$$

As an overall check on our heat-transfer rates, we know that for steady-state conditions, the net heat-transfer rate into the solid must be zero:

$$q'_{net} = q'_A + q'_B + q'_C + q'_D = -627.5 + 538.8 + 0 + 88.8$$
$$= 0.1 \text{ W/m}$$

The size of the net heat flow into the solid gives an indication of the accuracy of the finite-difference method for this particular problem.

Relaxation Techniques

In Example 2-12 the solid material was subdivided into a grid in which three nodes had unknown temperatures. The solution resulted in three algebraic equations for the three unknown temperatures. If we had wanted to increase the accuracy of the solution by decreasing the grid spacing, we would have had more nodes with unknown temperatures and additional equations to solve. In general each node with an unknown temperature results in an algebraic equation that must be solved simultaneously with the other nodal equations.

When the number of nodes is relatively small, as it was in Example 2-12, we have little trouble solving the system of simultaneous equations. When the number of equations becomes large, however, it becomes necessary to use an approximate method to solve the equations. A technique occasionally used in heat transfer is the relaxation method. While the relaxation method has limited applications to practical heat transfer problems, it is a pedagogical tool that can illustrate how numerical techniques can be applied to simple problems. The conceptual scheme of the relaxation method will also help us understand the more practical numerical methods which follow later in the chapter.

The purpose of the relaxation method is to estimate the temperatures of each node in the solid such that the energy-balance equations are approximately satisfied. Instead of setting all energy-balance equations such as Eqs. 2-94 and 2-96 equal to zero, we could equate them to a term called a *residual*. The temperatures are then systematically changed until the residual is reduced to a small value. The size of the residual will indicate the accuracy to which the temperatures of all the nodes have been estimated. If residuals for all nodal equations are reduced exactly to zero, then the temperatures are exact solutions to the energy balance equations.

To illustrate the relaxation method, we can apply it to the three equations used in Example 2-12. The three nodal energy balances in the example were

$$400 + T_5 - 7T_4 = R_4$$

$$300 + T_4 + T_6 - 4T_5 = R_5$$

$$150 + T_5 - 2T_6 = R_6$$

The right-hand side of each equation has been replaced by a residual, R_i, where the subscript indicates the respective node. Our job now is to determine values of T_4, T_5, and T_6 so that the residuals are reduced to reasonably small values. The magnitude of the residuals will determine the accuracy of the approximation of the temperature. We notice, for example, that an error in the temperature of node 4 of one degree will produce a residual of seven degrees. The dimensions of the residuals are temperature. Once the nodal energy-balance equations have been derived, the relaxation technique proceeds according to the following four steps.

Step 1: The first step in the relaxation method is to assume values for all unknown nodal temperatures. We should use our knowledge of heat transfer to guess temperatures as close to the actual values as possible. In Example 2-12 we must guess values for T_4, T_5, and T_6. The extreme temperature limits in the problem are the 50°C fluid and the 200°C temperature on the upper surface of the solid. Therefore, the guesses for the unknown steady temperatures must lie between these limits. We would expect T_4 to be the lowest of the three temperatures and T_6 to be the highest because it is on an insulated boundary. Suppose we assume that the three initial values for the temperatures are

$$T_4 = 80°C$$

$$T_5 = 100°C$$

$$T_6 = 150°C$$

Step 2: The next step involves substituting the initial temperature guesses into the residual equations and calculating each residual. The residuals for this example are

$$R_4 = -60°C$$

$$R_5 = 130°C$$

$$R_6 = -50°C$$

Since the residuals are nonzero, we must continue to change the temperatures until each residual is reduced toward zero.

Step 3: To reduce the residuals we change the temperature corresponding to the absolute value of the largest residual until that residual is reduced to zero. The convergence to the correct set of temperatures is often helped by changing the nodal temperature so that the residual is not reduced exactly to zero but is changed to a small value with a sign opposite the sign of the residual prior to change in temperature. This process is called *overrelaxation*.

In our example the largest residual corresponds to node 5. The residual equation for node 5 shows that if we increase T_5 by 35°C, the new value for the residual R_5 will be reduced by 140°C, thereby changing it to a low value with an opposite sign. Notice that a change in T_5 will also affect the values for R_5 and R_6. A summary of the new values for the three temperatures and corresponding residuals are

$$T_4 = 80°C \qquad R_4 = -25°C$$
$$T_5 = 135°C \qquad R_5 = -10°C$$
$$T_6 = 150°C \qquad R_6 = -15°C$$

Step 4: The next step in the relaxation process is to repeat the previous step until the desired degree of accuracy is achieved. The largest residual is now R_4, so we change T_4 by an amount to change R_4 to a small positive value. Assume that we decrease T_4 by 4°C. The new temperature values and corresponding residuals are

$$T_4 = 76°C \qquad R_4 = 3°C$$
$$T_5 = 135°C \qquad R_5 = -14°C$$
$$T_6 = 150°C \qquad R_6 = -15°C$$

Repeating this step twice, first changing T_6, then T_5, results in the following values:

Decrease T_6 by 10°C:

$$T_4 = 76°C \qquad R_4 = 3°C$$
$$T_5 = 135°C \qquad R_5 = -24°C$$
$$T_6 = 140°C \qquad R_6 = 5°C$$

Decrease T_5 by 7°C:

$$T_4 = 76°C \qquad R_4 = -4°C$$
$$T_5 = 128°C \qquad R_5 = 4°C$$
$$T_6 = 140°C \qquad R_6 = -2°C$$

In four relaxation steps the temperatures are all within 1°C of the exact values determined in Example 2-12. The preceding steps are best organized in a table similar to the one shown in Table 2-2. By organizing the relaxation steps and recording the data in tabular form, the amount of repeated work is minimized.

Table 2-2 Summary of Temperatures and Residuals for Example 2-12

STEP	T_4	R_4	T_5	R_5	T_6	R_6
Initial guess	80		100		150	
Initial residuals		-60		130		-50
Increase T_5 by 35	80		135		150	
New residuals		-25		-10		-15
Decrease T_4 by 4	76		135		150	
New residuals		3		-14		-15
Decrease T_6 by 10	76		135		140	
New residuals		3		-24		5
Decrease T_5 by 7	76		128		140	
New residuals		-4		4		-2

The finite-difference approach using a relaxation method can be extended to cylindrical coordinates, and the resulting difference equations are described in Reference 5.

If internal generation is present in the solid, the relaxation technique can be used without any complication. Suppose that at a particular node the energy generation rate per unit volume is q_G'''. The energy balance for an interior node 0 in a two-dimensional system with four neighboring nodes as shown in Fig. 2-17 is

$$q_{1\to0} + q_{2\to0} + q_{3\to0} + q_{4\to0} + q_G = 0$$

Replacing each heat-flow term by the finite-difference form of the Fourier law gives

$$k\,\Delta y\,d\frac{T_1 - T_0}{\Delta x} + k\,\Delta x\,d\frac{T_2 - T_0}{\Delta y} + k\,\Delta y\,d\frac{T_3 - T_0}{\Delta x}$$

$$+ k\,\Delta x\,d\frac{T_4 - T_0}{\Delta y} + q_G'''\,\Delta x\,\Delta y\,d = 0 \qquad (2\text{-}97)$$

If the grid is square, Eq. 2-97 becomes

$$T_1 + T_2 + T_3 + T_4 - 4T_0 + q_G'''\frac{(\Delta x)^2}{k} = 0 \qquad (2\text{-}98)$$

Whenever a node is located on the boundary of a solid, the residual equation depends upon the type of boundary condition at the surface. For example, the residual equation for a surface node on a flat surface in contact with fluid is given by Eq. 2-96. Residual equations for other boundary conditions are summarized in Table 2-3. In each case the energy-balance equation is written for the node denoted by the subscript 0.

To this point we have considered problems in which the temperature is a function of two coordinates only. However, the techniques we have developed for two-dimensional problems can easily be extended to three-dimensional problems. For example, if we consider a typical node 0 in a constant property solid with no generation surrounded by six nodes as shown in

Table 2-3 Residual Equations for Boundary Nodes in Two-Dimensional Systems, Square Grids ($\Delta x = \Delta y$)

CONDITION	GEOMETRY	NODAL EQUATION
Flat surface, isothermal boundary		$q'' \dfrac{\Delta x}{k} + T_1 - T_0 = R_0$ ($T_0 = T_2 = T_3$, heat input at surface per unit area $= q''$)
Flat surface, insulated boundary		$\frac{1}{2}(T_2 + T_3) + T_1 - 2T_0 = R_0$
Flat surface in contact with fluid		$\frac{1}{2}(T_2 + T_3) + T_1 + (\text{Bi})T_\infty$ $\quad - (2 + \text{Bi})T_0 = R_0$ $(\text{Bi} = \bar{h}_c \Delta x / k)$
Exterior corner, both surfaces insulated		$\frac{1}{2}(T_1 + T_2) - T_0 = R_0$
Exterior corner, both surfaces in contact with a fluid		$\frac{1}{2}(T_1 + T_2) + (\text{Bi})T_\infty$ $\quad - (1 + \text{Bi})T_0 = R_0$ $(\text{Bi} = \bar{h}_c \Delta x / k)$

Interior corner, both surfaces insulated

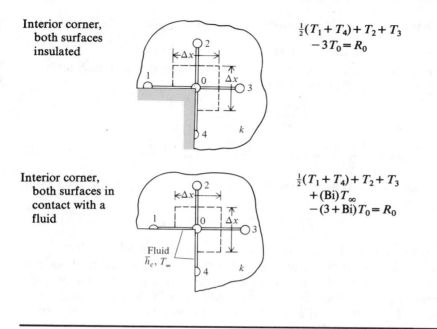

$$\tfrac{1}{2}(T_1 + T_4) + T_2 + T_3 - 3T_0 = R_0$$

Interior corner, both surfaces in contact with a fluid

$$\tfrac{1}{2}(T_1 + T_4) + T_2 + T_3 + (\mathrm{Bi})T_\infty - (3 + \mathrm{Bi})T_0 = R_0$$

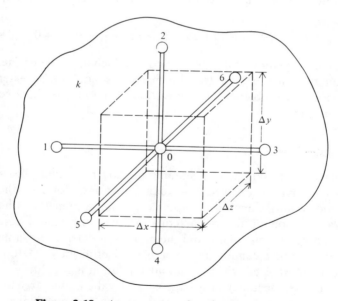

Figure 2-19 Arrangement of nodes for an interior section of a three-dimensional solid.

Fig. 2-19, an energy balance applied to node 0 will give

$$q_{1\to0}+q_{2\to0}+q_{3\to0}+q_{4\to0}+q_{5\to0}+q_{6\to0}=0 \qquad (2\text{-}99)$$

Equation 2-99 can be expressed in terms of temperatures of each node by replacing the heat-flow terms with Fourier's law:

$$k\Delta y\,\Delta z\frac{T_1-T_0}{\Delta x}+k\,\Delta x\,\Delta z\frac{T_2-T_0}{\Delta y}+k\,\Delta y\,\Delta z\frac{T_3-T_0}{\Delta x}$$

$$+k\,\Delta x\,\Delta z\frac{T_4-T_0}{\Delta y}+k\,\Delta x\,\Delta y\frac{T_5-T_0}{\Delta z}+k\,\Delta x\,\Delta y\frac{T_6-T_0}{\Delta z}=0 \qquad (2\text{-}100)$$

If the grid is drawn as a series of cubes, or $\Delta x=\Delta y=\Delta z$, Eq. 2-100 can be simplified to

3 - Dim.
$$T_1+T_2+T_3+T_4+T_5+T_6-6T_0=0 \qquad (2\text{-}101)$$

Therefore, the nodal equation in a three-dimensional problem, with no generation when the node is in the interior of a solid and each node is at the center of a cube, is simply the sum of the temperatures of each neighboring node minus six times the temperature of the central node. The form of Eq. 2-101 is similar to the nodal energy balance in a two-dimensional problem, Eq. 2-94.

If internal energy generation is present in a three-dimensional solid at a rate per unit volume of q_G''', the nodal equation for an interior node is

$$T_1+T_2+T_3+T_4+T_5+T_6-6T_0+\frac{q_G'''(\Delta x)^2}{k}=0 \qquad (2\text{-}102)$$

The residual equations listed in Table 2-3, which apply when the node is located on the surface of a two-dimensional solid, can be extended to three-dimensional problems. The derivation of the new residual equations is left as an exercise.

Matrix Techniques

The relaxation technique is a suitable method for solving heat-transfer problems that involve relatively few nodal equations. However, the relaxation method is not particularly adaptable to computer methods because it requires a selection of the nodal equation with the largest residual. A computer operates sequentially and the process of determining a maximum value from an array cannot be accomplished efficiently. Methods other than relaxation are used when a computer solution is sought.

When increased accuracy or a large geometry dictates a large number of nodes, the use of a computer becomes desirable. A convenient method of determining the temperature distribution in a two- or three-dimensional solid which is easily adaptable to a computer is the *matrix-inversion method*. The matrix method is based on representing the energy-balance equations for each node in the form of a matrix. If we subdivide the solid into n nodes, for example, each nodal equation can be expressed as

$$
\begin{aligned}
a_{11}T_1 + a_{12}T_2 + \cdots + a_{1n}T_n &= b_1 \\
a_{21}T_1 + a_{22}T_2 + \cdots + a_{2n}T_n &= b_2 \\
\vdots \qquad\qquad\qquad \vdots & \\
a_{n1}T_1 + a_{n2}T_2 + \cdots + a_{nn}T_n &= b_n
\end{aligned}
\tag{2-103}
$$

where the a_{ij}'s and b_i's are known constants and the T_i's are the unknown temperatures.

Equations 2-103 can be condensed and written in matrix notation as

$$
\text{AT} = \text{B} \tag{2-104}
$$

where A is a $n \times n$ coefficient matrix defined by

$$
A = \begin{Bmatrix}
a_{11} & a_{12} & \cdots & a_{1n} \\
a_{21} & a_{22} & \cdots & a_{2n} \\
\vdots & & & \vdots \\
a_{n1} & a_{n2} & \cdots & a_{nn}
\end{Bmatrix}
\tag{2-105a}
$$

while T and B are column matrices consisting of n elements each:

$$
T = \begin{Bmatrix} T_1 \\ T_2 \\ \vdots \\ T_n \end{Bmatrix}
\qquad
B = \begin{Bmatrix} b_1 \\ b_2 \\ \vdots \\ b_n \end{Bmatrix}
\tag{2-105b}
$$

To calculate the unknown temperatures, we must determine the inverse of the matrix, A^{-1}, which satisfies the equation

$$
T = A^{-1}B \tag{2-106}
$$

If the elements of the inverse of matrix A are given by

$$
C = A^{-1} = \begin{Bmatrix}
c_{11} & c_{12} & \cdots & c_{1n} \\
c_{21} & c_{22} & \cdots & c_{2n} \\
\vdots & & & \\
c_{n1} & c_{n2} & \cdots & c_{nn}
\end{Bmatrix}
\tag{2-107}
$$

the unknown nodal temperatures are given by the equations

$$
\begin{aligned}
c_{11}b_1 + c_{12}b_2 + c_{13}b_3 + \cdots + c_{1n}b_n &= T_1 \\
c_{21}b_1 + c_{22}b_2 + c_{23}b_3 + \cdots + c_{2n}b_n &= T_2 \\
\vdots \qquad\qquad\qquad \vdots & \\
c_{n1}b_1 + c_{n2}b_2 + c_{n3}b_3 + \cdots + c_{nn}b_n &= T_n
\end{aligned}
\tag{2-108}
$$

Since the values for all the b_i's are known, the problem of calculating the temperatures depends upon determining the inverse of matrix **A**. The inverse matrix can be calculated by using standard mathematical techniques (see Ref. 6, for example) if the number of nodes is reasonably small. When the number of nodes becomes large, the elements in the inverted matrix can be determined quickly with a computer by using standard library subroutines. The computer method is illustrated in the following example.

Example 2-13. Determine the steady temperature distribution in the two-dimensional solid shown in the figure. Use a square grid with $\Delta x = \Delta y = 5$ cm. Two of the boundaries are isothermal, a third is insulated, and the fourth transfers heat by convection. Compare your answers with those in Example 2-12.

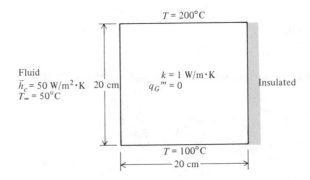

Solution: The solid is subdivided into a square grid as shown in the figure. Each node with an unknown temperature is assigned a number. The nodes on the upper and lower surfaces have known temperatures, leaving 15 nodal temperatures to be determined. Fifteen equations require significant time to solve by the relaxation technique, so a computer program using a matrix inversion subprogram becomes a reasonable alternative.

The first step in the matrix method is the determination of the values for the elements in the matrices **A** and **B** in Eq. 2-103. All interior nodes have nodal equations similar to Eq. 2-94. The boundary node equations can be determined by referring to Table 2-3. Constant values that are used in the equations are

$$\text{Bi} = \frac{\bar{h}_c \Delta x}{k} = \frac{50 \times 0.05}{1} = 2.5$$

$$(\text{Bi}) T_\infty = 2.5 \times 50 = 125°\text{C}$$

$$2 + \text{Bi} = 4.5$$

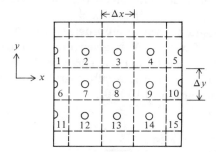

The 15 nodal equations written in the matrix form of Eq. 2-103 are then

Node 1: $\quad -4.5T_1 + T_2 + \frac{1}{2}T_6 = -225$

Node 2: $\quad T_1 - 4T_2 + T_3 + T_7 = -200$

Node 3: $\quad T_2 - 4T_3 + T_4 + T_8 = -200$

Node 4: $\quad T_3 - 4T_4 + T_5 + T_9 = -200$

Node 5: $\quad T_4 - 2T_5 + \frac{1}{2}T_{10} = -100$

Node 6: $\quad \frac{1}{2}T_1 - 4.5T_6 + T_7 + \frac{1}{2}T_{11} = -125$

Node 7: $\quad T_2 + T_6 - 4T_7 + T_8 + T_{12} = 0$

Node 8: $\quad T_3 + T_7 - 4T_8 + T_9 + T_{13} = 0$

Node 9: $\quad T_4 + T_8 - 4T_9 + T_{10} + T_{14} = 0$

Node 10: $\quad \frac{1}{2}T_5 + T_9 - 2T_{10} + \frac{1}{2}T_{15} = 0$

Node 11: $\quad \frac{1}{2}T_6 - 4.5T_{11} + T_{12} = -175$

Node 12: $\quad T_7 + T_{11} - 4T_{12} + T_{13} = -100$

Node 13: $\quad T_8 + T_{12} - 4T_{13} + T_{14} = -100$

Node 14: $\quad T_9 + T_{13} - 4T_{14} + T_{15} = -100$

Node 15: $\quad \frac{1}{2}T_{10} + T_{14} - 2T_{15} = -50$

The elements of the two matrices A and B can now be determined by examining the coefficients in the 15 nodal equations. The constants on the right side of each equation are the elements that make up the column matrix B. Many of the elements in matrix A are zero and the nonzero terms are gathered close to the diagonal.

The next step is inversion of matrix A. The elements of the inverted matrix are the elements c_{ij} in Eq. 2-107. To invert a matrix of this size by hand is quite time-consuming. Fortunately, standard computer library subroutines are available and they can be employed in a short program to solve for the 15 unknown temperatures.

The following program is an example of a FORTRAN program that will solve for the nodal temperatures. It is written in a general form so that an interested user can copy it and apply it to an individual problem which may be different from the example problem considered here. Input values are N, the number of nodes with unknown temperatures, the N^2 values of matrix A, and the N elements of matrix B.

The example program uses a subprogram called MATINV to invert matrix A. A listing of MATINV is given in Appendix M. Other subprograms are available to invert a matrix. The user should check with his own computer center to determine if a subprogram can be accessed without duplicating the subprogram MATINV.

A listing of the program is:

Program Listing for Example 2-13

```
      DIMENSION A(50,50),B(50),C(50,50),T(50)          N = # of Nodes
      READ , N
      READ , ((A(I,J),J=1,N),I=1,N),(B(I),I=1,N)
      CALL MATINV (A,N,C)      ---> PAGE 538
      DO 20 I=1,N
      SUM=0.0
      DO 10 J=1,N
   10 SUM=SUM+C(I,J)*B(J)
   20 T(I)=SUM
      WRITE(6,40)
      WRITE (6,50) (I,T(I), I=1,N)
      STOP
   40 FORMAT (1H ,'*** STEADY TEMPERATURE DISTRIBUTION IN DEGREES',/,
     1'CELSIUS DETERMINED BY A MATRIX INVERSION TECHNIQUE ***',//)
   50 FORMAT (4('T(',I2,')=',F8.2,2X))
      END
```

The program input is shown below. The first line contains the single value for N, which is the number of nodes for this particular problem. The next N lines list the elements a_{ij} of matrix A. Each of the 15 lines of data contain 15 values, and each line represents a fixed value of the subscript i. The last two lines of data are the 15 values b_i of the column matrix B.

Program Input for Example 2-13

```
15
 -4.5, 1., 0., 0., 0., 0.5, 0., 0., 0., 0., 0., 0., 0., 0., 0.,
 1., -4., 1., 0., 0., 0., 1., 0., 0., 0., 0., 0., 0., 0., 0.,
 0., 1., -4., 1., 0., 0., 0., 1., 0., 0., 0., 0., 0., 0., 0.,
 0., 0., 1., -4., 1., 0., 0., 0., 1., 0., 0., 0., 0., 0., 0.,
 0., 0., 0., 1., -2., 0., 0., 0., 0., 0.5, 0., 0., 0., 0., 0.,
 0.5, 0., 0., 0., 0., -4.5, 1., 0., 0., 0., 0.5, 0., 0., 0., 0.,
 0., 1., 0., 0., 0., 1., -4., 1., 0., 0., 0., 1., 0., 0., 0.,
 0., 0., 1., 0., 0., 0., 1., -4., 1., 0., 0., 0., 1., 0., 0.,
 0., 0., 0., 1., 0., 0., 0., 1., -4., 1., 0., 0., 0., 1., 0.,
 0., 0., 0., 0., 0.5, 0., 0., 0., 1., -2., 0., 0., 0., 0., 0.5,
 0., 0., 0., 0., 0., 0.5, 0., 0., 0., 0., -4.5, 1., 0., 0., 0.,
 0., 0., 0., 0., 0., 0., 1., 0., 0., 0., 1., -4., 1., 0., 0.,
```

0., 0., 0., 0., 0., 0., 0., 1., 0., 0., 0., 1., −4., 1., 0.,
0., 0., 0., 0., 0., 0., 0., 0., 1., 0., 0., 0., 1., −4., 1.,
0., 0., 0., 0., 0., 0., 0., 0., 0.5, 0., 0., 0., 1., −2.,
−225., −200., −200., −200., −100., −125., 0., 0., 0., 0., −175.,
−100., −100., −100., −50.

The program output is shown below.

Program Output for Example 2-13

***** STEADY TEMPERATURE DISTRIBUTION IN DEGREES
CELSIUS DETERMINED BY A MATRIX INVERSION TECHNIQUE *****

T(1)= 88.57	T(2)=138.91	T(3)=158.56	T(4)=166.30
T(5)=168.38	T(6)= 69.30	T(7)=108.51	T(8)=129.01
T(9)=138.28	T(10)=140.91	T(11)= 68.11	T(12)= 96.83
T(13)=110.69	T(14)=116.91	T(15)=118.68	

This example is identical to Example 2-12 except that the grid is smaller. A comparison of the equivalent nodes in Table 2-4 shows the influence that the grid size has on the values for the local temperatures.

**Table 2-4 Comparison of Temperatures
Resulting from Coarse and Fine Numerical Grids**

Temperature, 10×10 cm grid (Example 2-12)	Temperature, 5×5 cm grid (Example 2-13)
$T_4 = 75.5°C$	$T_6 = 69.3°C$
$T_5 = 128.7°C$	$T_8 = 129.0°C$
$T_6 = 139.4°C$	$T_{10} = 140.9°C$

The program written for Example 2-13 is completely general. It can be simply copied exactly and applied to a large number of steady conduction problems. It can solve for the temperature distribution in a one-, two-, or three-dimensional problem that contains internal energy generation under a variety of boundary conditions. The user must first derive all the nodal equations and determine the elements of matrices A and B. These elements along with the number of nodes N must be organized as input data exactly as illustrated in the example program.

The matrix-inversion program should not be used when the number of nodes exceeds 50. When a problem involves more than 50 nodes, a more sophisticated numerical method such as Cholesky's method (Ref. 10) should be used to solve the set of simultaneous equations. Cholesky's method is more economical of computer time than the program MATINV, because it eliminates nonzero elements in the matrix and thereby minimizes storage requirements.

To help in the use of the program, the table below lists the symbols used in program along with their definitions and typical units.

Symbols Used in Program for Example 2-13

SYMBOL	DEFINITION	UNIT
A(I,J)	Two-dimensional array of the elements in matrix A, defined in Eq. 2-105a	—
B(I)	One-dimensional array of the elements in matrix B, defined in Eq. 2-105b	—
C(I,J)	Two-dimensional array of the elements in the matrix $C = A^{-1}$, defined in Eq. 2-107	—
N	Integer value equal to the number of nodes	—
T(I)	One-dimensional array of elements in matrix T, defined in Eq. 2-105b	°C

Iteration Techniques

A numerical method that is particularly well-suited for a computer solution is an *iteration method* based on solving each nodal equation explicitly for the temperature of that node. For example, if we consider an energy-balance equation for an interior node in a two-dimensional solid, the result is given in Eq. 2-94 as

$$T_1 + T_2 + T_3 + T_4 - 4T_0 = 0$$

If we solve for the temperature at node zero, we get

$$T_0 = \tfrac{1}{4}(T_1 + T_2 + T_3 + T_4)$$

This equation is typical of an interior node in a solid with constant properties subdivided into a square grid with no energy generation. The procedure of solving for the temperature when the node is located on the boundary will result in a similar equation. For example, if the node zero is located on a boundary that transfers heat to a fluid by convection, the equation for the temperature T_0 can be determined from Eq. 2-96 to be

$$T_0 = \frac{\tfrac{1}{2}(T_2 + T_3) + T_1 + (\mathrm{Bi})\,T_\infty}{2 + \mathrm{Bi}}$$

Equations for other boundary conditions may easily be determined by rearranging the nodal equations in Table 2-3.

An equation for the temperature of each node may be written in terms of the temperatures of the neighboring nodes. The number of equations equals the number of nodes with unknown temperatures.

The iteration procedure can be divided into four steps.

Step 1: First derive nodal equations by performing an energy balance for each node that has an unknown temperature. Solve each of the equations explicitly in terms of the temperature at the node at which the energy balance is being performed. All interior nodes will have similar equations. Boundary node equations will vary depending upon the type of boundary condition applicable to the particular problem. Table 2-3 should help in determining the boundary node equations.

Step 2: Next, assume a set of values for all the nodal temperatures. If the problem is being solved by hand, an intelligent guess for all the temperatures will help reduce the amount of time required to arrive at the correct nodal temperatures. If a computer solution is used, it is convenient to set all initial temperatures to zero.

Step 3: Next calculate new values for the temperatures using the nodal equations derived in Step 1. Once a new temperature is determined, immediately replace the old value with the new value so that new values for the nodal temperatures are always calculated using the most recent values. By using the most recent temperatures, the time required to converge to the eventual steady temperatures will be decreased. This particular type of iteration technique is often called the *Gauss-Siedel method*.

Step 4: Repeat Step 3 until the difference between the old and new values for all temperatures are within a tolerance level specified by the user.

The iteration technique is illustrated in the following two examples. The first is solved by hand using only three equations. The second involves 15 equations, which is large enough to justify using a computer for the solutions. Both examples have been worked previously using different methods, so we will have a check on the accuracy of the iteration technique.

Example 2-14. Determine the temperatures of nodes 4, 5, and 6 in Example 2-12 using the iteration method. Compare your results with the exact results given in Example 2-12 and the results in Table 2-4 obtained by the matrix-inversion method.

Solution: In Example 2-12 we were asked to determine the temperatures of nodes 4, 5, and 6. Energy-balance equations for these three nodes derived in the example were:

Node 4: $400 + T_5 - 7T_4 = 0$
Node 5: $300 + T_4 + T_6 - 4T_5 = 0$
Node 6: $150 + T_5 - 2T_6 = 0$

Now we complete the four steps outlined for the iteration method.

Step 1: Solve explicitly for the temperature of each node.

Node 4: $T_4 = \dfrac{400}{7} + \dfrac{T_5}{7}$

Node 5: $T_5 = \dfrac{300}{4} + \dfrac{T_4}{4} + \dfrac{T_6}{4}$

Node 6: $T_6 = \dfrac{150}{2} + \dfrac{T_5}{2}$

Step 2: Assume initial values for the nodal temperatures. An intelligent guess for the initial values for the three temperatures would be

$$T_4 = 80°C$$
$$T_5 = 100°C$$
$$T_6 = 150°C$$

These are the same initial values we used in the relaxation-method solution of the three equations.

Step 3: Calculate new values for the temperatures using the explicit form of nodal equations. As soon as a new temperature is calculated, use it in successive steps.

$$T_4 = \frac{400}{7} + \frac{100}{7} = 71.43°C$$

$$T_5 = \frac{300}{4} + \frac{71.43}{4} + \frac{150}{4} = 130.36°C$$

$$T_6 = \frac{150}{2} + \frac{130.36}{2} = 140.18°C$$

Step 4: Repeat Step 3 until successive temperatures converge to within a specified tolerance level. Assuming that we wish the temperatures for all nodes to differ by less than 0.1°C for two successive iteration steps, we repeat Step 3. A table summarizing the results is shown below.

ITERATION STEP	T_4 (°C)	T_5 (°C)	T_6 (°C)
Initial guess	80.	100.	150.
1	71.43	130.36	140.18
2	75.77	128.99	139.50
3	75.57	128.77	139.39
4	75.54	128.73	139.37

In four iteration steps, all the temperatures were within 0.1°C of the values in Step 3, so the process is stopped. Rounded to the nearest 0.1°C, these values are identical to the exact solutions for these same equations given in

Example 2-12. We should also notice that for this particular example the relaxation technique was able to converge to within only 1°C of the exact temperature values in four iteration steps, starting with the same initial values for the temperatures. The relaxation results are given in Table 2-2. The exact solution of the nodal equations, the relaxation solution after four steps, and the iteration solution after four iteration steps provide temperatures for the three nodes that differ by less than 1°C.

Example 2-15. Determine temperatures for the 15 nodes in Example 2-13 using the iteration method. Compare your results with those obtained by the matrix-inversion method.

Solution: The solution of 15 equations by hand using the iteration method is somewhat time-consuming, making a computer solution reasonable. The four steps in the iteration method are as follows.

Step 1: The 15 nodal equations are simply rearranged forms of the nodal equations derived in Example 2-13. They are:

$$T_1 = 50 + 0.222\,T_2 + 0.111\,T_6$$

$$T_2 = 50 + \tfrac{1}{4}(T_1 + T_3 + T_7)$$

$$T_3 = 50 + \tfrac{1}{4}(T_2 + T_4 + T_8)$$

$$T_4 = 50 + \tfrac{1}{4}(T_3 + T_5 + T_9)$$

$$T_5 = 50 + \tfrac{1}{2}T_4 + \tfrac{1}{4}T_{10}$$

$$T_6 = 27.778 + 0.111\,T_1 + 0.222\,T_7 + 0.111\,T_{11}$$

$$T_7 = \tfrac{1}{4}(T_2 + T_6 + T_8 + T_{12})$$

$$T_8 = \tfrac{1}{4}(T_3 + T_7 + T_9 + T_{13})$$

$$T_9 = \tfrac{1}{4}(T_4 + T_8 + T_{10} + T_{14})$$

$$T_{10} = \tfrac{1}{4}T_5 + \tfrac{1}{2}T_9 + \tfrac{1}{4}T_{15}$$

$$T_{11} = 38.889 + 0.111\,T_6 + 0.222\,T_{12}$$

$$T_{12} = 25 + \tfrac{1}{4}(T_7 + T_{11} + T_{13})$$

$$T_{13} = 25 + \tfrac{1}{4}(T_8 + T_{12} + T_{14})$$

$$T_{14} = 25 + \tfrac{1}{4}(T_9 + T_{13} + T_{15})$$

$$T_{15} = 25 + \tfrac{1}{4}T_{10} + \tfrac{1}{2}T_{14}$$

These equations appear between statements 21 and 22 in the program listing shown below.

Step 2: The initial nodal temperatures are all set equal to zero. We are aware that these are not intelligent initial guesses for the temperatures considering the given boundary conditions. The time required to reach the eventual solution will undoubtedly be increased because of this guess. But zero is a convenient value to set in a program and the added time required to obtain a solution is of little consequence when using a computer solution. The initial temperatures are set to zero in statement 15 in the program listing.

Step 3: The steps required to calculate the new temperatures of the 15 nodes are achieved in the steps between statements 21 and 22 in the program listing.

Step 4: The new values of each nodal temperature are compared with the old value and if the difference between the two is less than a tolerance level specified upon input, the iteration process is terminated and the temperature values are printed. The current values of the temperatures are stored in the array T(I) and the temperatures from the previous iteration step are stored in the array TT(I). The tolerance check is performed by the logical IF statement.

The program listing is:

Program Listing for Example 2-15   *Account # 20129539*

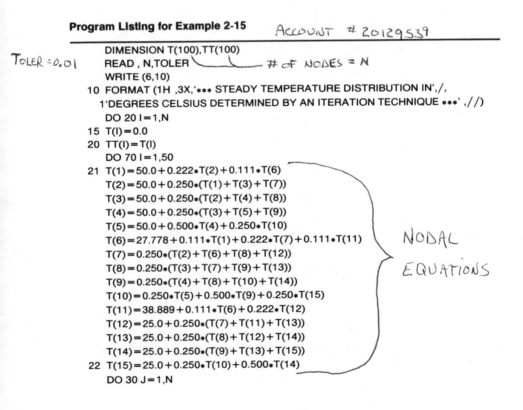

```
            DIMENSION T(100),TT(100)
            READ , N,TOLER          — # OF NODES = N
            WRITE (6,10)
        10  FORMAT (1H ,3X,'*** STEADY TEMPERATURE DISTRIBUTION IN',/,
            1'DEGREES CELSIUS DETERMINED BY AN ITERATION TECHNIQUE ***' ,//)
            DO 20 I=1,N
        15  T(I)=0.0
        20  TT(I)=T(I)
            DO 70 I=1,50
        21  T(1)=50.0+0.222*T(2)+0.111*T(6)
            T(2)=50.0+0.250*(T(1)+T(3)+T(7))
            T(3)=50.0+0.250*(T(2)+T(4)+T(8))
            T(4)=50.0+0.250*(T(3)+T(5)+T(9))
            T(5)=50.0+0.500*T(4)+0.250*T(10)
            T(6)=27.778+0.111*T(1)+0.222*T(7)+0.111*T(11)
            T(7)=0.250*(T(2)+T(6)+T(8)+T(12))
            T(8)=0.250*(T(3)+T(7)+T(9)+T(13))
            T(9)=0.250*(T(4)+T(8)+T(10)+T(14))
            T(10)=0.250*T(5)+0.500*T(9)+0.250*T(15)
            T(11)=38.889+0.111*T(6)+0.222*T(12)
            T(12)=25.0+0.250*(T(7)+T(11)+T(13))
            T(13)=25.0+0.250*(T(8)+T(12)+T(14))
            T(14)=25.0+0.250*(T(9)+T(13)+T(15))
        22  T(15)=25.0+0.250*T(10)+0.500*T(14)
            DO 30 J=1,N
```

TOLER =0.01

NODAL EQUATIONS

```
     IF (ABS(TT(J)-T(J)).GT. TOLER) GO TO 50
 30  CONTINUE
     WRITE (6,40) (K,T(K),K=1,N)
 40  FORMAT (1H ,'T(',I2,')=',F,8.2)
     STOP
 50  DO 60 J=1,N
 60  TT(J)=T(J)
 70  CONTINUE
     WRITE (6,80) TOLER
 80  FORMAT (1H1,'TEMPERATURES DO NOT CONVERGE TO WITHIN' ,F6.3,/,
    1'DEGREES IN 50 ITERATION STEPS')
     STOP
     END
```

The input consists of two values on a single line, which should appear in the following order:

$\qquad$ N = the number of nodes

TOLER = the tolerance level in °C between two different iterations steps. If the differences in the temperatures of each individual node in successive iteration steps are all less than TOLER, the iteration procedure stops and the answers for the temperatures are printed. If the difference in any nodal temperature for successive iteration steps is greater than TOLER, the iteration procedure continues. $\text{ToLER} = T(I) - TT(I)$

The input to this particular program is:

Program Input for Example 2-15

15, 0.1

The program output is shown below.

Program Output for Example 2-15

STEADY TEMPERATURE DISTRIBUTION IN DEGREES CELSIUS DETERMINED BY AN ITERATION TECHNIQUE

T(1)=	88.48	T(9)=	138.12
T(2)=	138.78	T(10)=	140.76
T(3)=	158.41	T(11)=	68.04
T(4)=	166.16	T(12)=	96.73
T(5)=	168.25	T(13)=	110.59
T(6)=	69.19	T(14)=	116.82
T(7)=	108.36	T(15)=	118.60
T(8)=	128.84		

Example 2-15 is written in a general form so that it can be copied and applied to a wide variety of steady conduction problems. The user must specify only two input values. The first is the number of nodes, N, for which the temperature is unknown, and the second is the tolerance, TOLER, to which the calculation will be evaluated.

The program is limited to 100 nodes by the number of storage locations allocated, although the DIMENSION statement can be changed to provide for more storage. If the number of iteration steps exceeds 50 without the temperatures converging, the program will print a diagnostic statement indicating a lack of convergence.

Whenever the program is applied to a different problem, new nodal equations must be derived and solved explicitly for the individual nodal temperatures. These N equations must be inserted between statements 21 and 22 in the program.

REFERENCES

1. P. J. Schneider, *Conduction Heat Transfer*, Addison-Wesley Publishing Co., Inc., Reading, Mass., 1955.
2. D. Q. Kern, and A. D. Kraus, *Extended Surface Heat Transfer*, McGraw-Hill Book Company, New York, 1972.
3. M. Jakob, *Heat Transfer*, vol. 1, John Wiley & Sons, Inc., New York, 1949.
4. K. A. Gardner, "Efficiency of Extended Surfaces," *Trans. ASME*, vol. 67, pp. 621–631, 1945.
5. V. Arpaci, *Conduction Heat Transfer*, Addison-Wesley Publishing Co., Inc., Reading, Mass., 1966.
6. M. N. Ozisik, *Boundary Value Problems of Heat Conduction*, Intext Publishers Group, New York, 1968.
7. G. E. Myers, *Analytical Methods in Conduction Heat Transfer*, McGraw-Hill Book Company, New York, 1971.
8. C. F. Kazan, "An Electrical Geometrical Analogue for Complex Heat Flow," *Trans. ASME*, vol. 67, pp. 113–716, 1945.
9. C. F. Kazan, "Heat Transfer Temperature Patterns of a Multicomponent Structure by Comparative Methods," *Trans. ASME*, vol. 71, pp. 9–16, 1949.
10. M. L. James, G. M. Smith, and J. C. Wolford, *Applied Numerical Methods for Digital Computation with FORTRAN and CSMP*, 2nd ed., Crowell, New York, 1977.

PROBLEMS

The problems in this chapter are organized in the manner shown in the table. Four problems suggest computer solutions. They are Problems 2-70, 2-71, 2-72, and 2-73. No original programs need to be written. All programs needed for solutions are developed in the examples within the chapter.

PROBLEM NUMBERS - SECTIONS		SUBJECT
2-1 to 2-23	2-1 to 2-3	Steady, one-dimensional conduction without generation
2-24 to 2-27	2-4	Effect of variable thermal conductivity
2-28 to 2-34	2-5	Steady, one-dimensional conduction with generation
2-35 to 2-46	2-6	Heat transfer from fins
2-47 and 2-48	2-7	Steady, two- and three-dimensional conduction, analytical methods
2-49 to 2-58	2-7	Steady, two- and three-dimensional conduction, graphical methods
2-59 to 2-73	2-7	Steady, two- and three-dimensional conduction, numerical methods

2-1 Compute the heat flux through a plane brick wall ($k = 0.65$ W/m·K) that is 15 cm thick and has an exterior temperature of 35°C and interior surface temperature of 25°C. Calculate the thermal resistance of the wall per unit area. Calculate the centerline temperature of the brick wall.

2-2 The heat flux through a 10-cm-thick aluminum plate is 5×10^4 W/m². One surface of the plate is at 150°C. Determine the temperature of the other surface of the plate.

2-3 Suppose that the aluminum plate in Problem 2-2 is replaced by a type 304 stainless steel plate of the same thickness. Assume that the heat flux and one surface temperature are the same as in Problem 2-2. Determine the other surface temperature. Compare your answer with the temperature in Problem 2-2. Explain the reason for the difference between the two temperatures.

2-4   A building wall has a surface area of 500 m² and a thermal conductivity of 0.7 W/m·K. The wall thickness is 20 cm. The outside wall temperature is 0°C during the winter and the inside wall surface is 20°C. Determine the capacity of the heating plant in watts necessary to make up for heat lost through the wall. Calculate the heat flux through the wall.

2-5   A single pane of glass is 4 mm thick and has an area of 2 m². Determine the heat-transfer rate through the glass if one side is 0°C and the other side is 20°C. Calculate the heat flux through the glass.

2-6   Compare the heat-flux values for Problems 2-4 and 2-5. Is more heat transferred through the wall or through the glass for similar temperature differences? Suggest ways to reduce the heat loss through the glass.

2-7   A cylinder with diameter of 20 cm and length 50 cm is insulated around its perimeter. One end of the cylinder is at a temperature of 300°C and a plane 25 cm from this end of the cylinder is 100°C. The thermal conductivity is 2 W/m·K. Determine the heat-transfer rate along the axis of the cylinder and calculate the end temperature of the cylinder.

2-8   A long aluminum wire with diameter 1 cm carries a current of 1000 A. The wire is covered with a 3-mm-thick layer of rubber ($k=0.15$ W/m·K) insulation. The temperature of the outside surface of the insulation is 30°C. Determine the inside insulation temperature. The electrical resistance of the wire per unit length is 3.7×10^{-4} Ω/m.

2-9   A steam pipe with exterior temperature 120°C and outside diameter of 10 cm is covered with a 5-cm layer of asbestos ($k=0.15$ W/m·K) insulation. If the outside surface of the insulation is at 35°C, calculate the heat-transfer rate from the steam pipe per unit length of pipe. Calculate the thermal resistance of the insulation for a unit length of pipe.

2-10   A circular duct carries hot combustion gases from a furnace. The duct has a temperature of 500°C and an outside diameter of 0.5 m. Determine the thickness of insulation ($k=0.2$ W/m·K) that is necessary to reduce the outside surface of the insulation to a level that would not injure anyone who touches the surface. The gases have a specific heat of 1000 J/kg·K, a flow rate of 1.0 kg/s and experience a temperature drop of 10°C over a duct length of 40 m. Assume that the thermal resistance of the duct is small compared to that of the insulation and that the highest temperature that one could be subjected to without injury is 65°C.

2-11   Determine the thermal resistance and heat-transfer rate through a hollow sphere with an internal diameter of 5 cm, an external diameter of 10 cm, and a thermal conductivity of 20 W/m·K. The internal and external surface temperatures are 100°C and 50°C, respectively.

$q_f = 628.5 \text{ w}$

$R_T = .0796$

$$\text{WATTS} = \frac{\text{Joule}}{\text{sec}} \Rightarrow q'' = Q/\text{Time}$$

2-12 The wall of a large furnace is made of cast iron and has a thickness of 1.5 cm. The ambient combustion gas temperature is 1100°C and the convective-heat-transfer coefficient on the inside surface of the furnace wall is 250 W/m². K. The exterior surface of the furnace is surrounded by air ($\bar{h}_c = 20$ W/m². K) that has an ambient temperature of 30°C. Draw the thermal circuit and identify all resistances. Calculate values for all thermal resistances per unit area. Determine the heat flux through the furnace wall. Calculate the inner and outer furnace wall temperatures.

2-13 Determine the thickness of insulation ($k = 0.5$ W/m·K) which when placed on the furnace wall given in Problem 2-12 will cut the heat-transfer rate in half. Assume that the presence of the insulation does not affect the outside heat-transfer coefficient. Calculate the temperatures of both surfaces of the insulation.

2-14 A styrofoam ($k = 0.035$ W/m·K) ice chest is made in the form of a hollow cylinder. The inside diameter is 0.50 m, the outside diameter is 0.60 m, and the length is 1.0 m. The ice chest is completely filled with ice at 0°C. The ambient air around the styrofoam is 30°C and $\bar{h}_c = 10$ W/m². K between the air and styrofoam. Determine the time required for the ice to completely melt, and the temperature of the outside surface of the styrofoam. Calculate the thermal resistance of the styrofoam and the air. Neglect any heat gain through the ends of the ice chest. The latent heat of fusion for water is 3.35×10^5 J/kg.

$$\text{USE} \quad K_{\text{WATER}} @ 0°C$$

2-15 A manufacturer builds two models of a hot water heater. Both models are cylindrically shaped, with double-walled metal construction. The inexpensive model has no insulation between the two walls and the convective heat transfer coefficient on both surfaces between the walls is 20 W/m². K. In the expensive model the gap between the walls is filled with fiberglass insulation ($k = 0.05$ W/m·K). Both models have an inner wall diameter of 0.60 m, an outer wall diameter of 0.70 m, and a height of 3.0 m. The inner wall has a temperature of 60°C and the ambient air temperature on the outside of the hot water heater is 25°C with $\bar{h}_c = 15$ W/m². K. Assume one-dimensional radial heat flow and neglect the thermal resistance of the metal walls. Determine the steady energy requirements of both models if no water is removed from the heater. Determine the percent energy savings that can be realized by buying the expensive model.

2-16 A long, cylindrical rod 4 cm in diameter has a surface temperature of 200°C. The rod is covered with two types of insulation as shown in the figure. The plane that separates the two materials is perfectly insulated. The thickness of both insulators is 5 cm, $k_A = 5$ W/m·K, and $k_B = 10$ W/m·K. The exterior surface of the insulators is surrounded by 20°C air for which $\bar{h}_c = 15$ W/m². K. Consider only radial steady conduction:
   (a) Draw the thermal circuit and label all resistances.
   (b) Determine the total heat-transfer rate from the rod per unit length.
   (c) Determine the exterior surface temperature of both insulators.

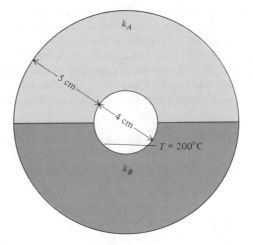

2-17 For the composite wall shown in the figure, determine the thermal conductivity, k_x. Also determine the interface temperatures T_x and T_y.

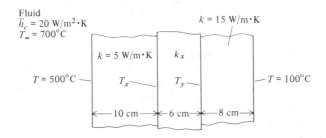

2-18 Air at 120°C flows over the top of a large horizontal 5-cm-thick stainless steel plate whose upper surface is maintained at 250°C. The convective-heat-transfer coefficient is 30 W/m²·K. The upper surface of the plate loses 700 W/m² by radiation to the air. Determine the steady temperature of the lower surface of the plate.

2-19 A hockey arena in a multifunction building has a surface area of 1600 m². The ice is maintained by cooling coils that circulate through the ice as illustrated in the figure. A basketball floor is laid over the ice, leaving a 1.5-cm air gap between the ice and the floor. The effective convective-heat-transfer coefficient in the air gap is 5 W/m²·K. The surface of the basketball floor is in contact with 25°C air and the convective-heat-transfer coefficient is 3 W/m²·K. The net radiant energy gain from the lights is 250 W/m². The wood has a thermal conductivity of 0.2 W/m·K and for ice $k=2.6$ W/m·K. The lower surface may be assumed to be adiabatic. If the basketball floor is left in place for 24 h, estimate the minimum amount of energy that must be removed from the ice to keep the ice from melting.

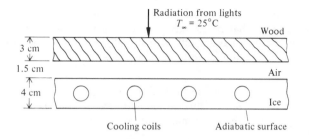

Estimate the cost to operate the cooling system for 24 h. Assume that the refrigeration unit has a coefficient of performance of 2.0 and electricity costs $0.06/kwh.

2-20   A large flat wall is exposed to a fluid that has a temperature of 200°C. The wall is covered with a 5-cm-thick layer of insulation with $k = 0.5$ W/m·K. The temperature of the interface between the insulation and wall is 100°C. Determine the value of convective-heat-transfer coefficient which must be maintained on the exposed surface of the insulation so that the surface will not exceed a temperature of 150°C.

2-21   A tank consists of cylindrical center section with two hemispherical end sections. The tank contains a heated fluid that maintains the inside surface temperature of the tank at 350°C. The tank is stainless steel with a constant thickness of 2.5 cm. The outside diameter of the cylindrical section is 2 m and it is 2 m in length. The air surrounding the tank has an ambient temperature of 25°C. The convective-heat-transfer coefficient between the air and tank is 7 W/m²·K. Determine the amount of heat that must be added to the fluid in the tank to maintain its temperature. Assume only radial conduction through the tank.

2-22   A 2-mm-diameter copper wire covered with a 1-mm-thick insulation $(k = 0.18$ W/m·K) has a tendency to overheat when used in a laboratory device. The heat-transfer coefficient between the insulation and air is 34 W/m²·K. Would the addition of thicker insulation increase the heat dissipation from the wire? If so, what is the maximum percentage increase in energy dissipation that can be realized by adding insulation? Determine the critical thickness of insulation that maximizes the heat-transfer rate from the wire.

2-23 Using the procedure outlined in Section 2-3 for determining the critical insulation thickness on a cylinder, show that the critical insulation thickness on a sphere is given by

$$\text{Bi} = \frac{\bar{h}_c r_0}{k_I} = 2.0$$

or

$$r_{\text{crit}} = \frac{2k_I}{\bar{h}_c}$$

2-24 Determine an expression for the temperature distribution and heat flux through a plane wall that has a variation in thermal conductivity of $k(T) = k_0 \times (1 + bT + cT^2)$. The wall has a width L and the boundary conditions are $T(0) = T_1$ and $T(L) = T_2$. The symbols b and c represent constants.

2-25 Determine an expression for the steady temperature distribution in a long, hollow cylinder with specified surface temperatures if the thermal conductivity of the cylinder varies linearly with temperature or $k(T) = k_0(1 + \beta T)$.

2-26 The steady temperature distribution for four plane surfaces in contact is shown in the figure. List the four thermal conductivities in order of lowest k first and the highest k last. Explain the reason for your decision.

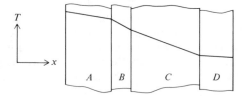

2-27 The steady temperature distribution in a plane surface is shown in the figure. Does the thermal conductivity of the material increase or decrease with temperature? Explain the reason for your decision.

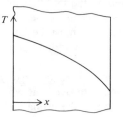

2-28 A current of 200 A is passed through a stainless steel wire which has a 2-mm diameter and a 1-m length. The electrical resistance of the wire is 0.125 Ω and its thermal conductivity is 17 W/m·K. The temperature of the outer surface of the wire is measured at 150°C:

(a) State the governing equation for the steady-state temperature, $T(r)$, in the wire.

(b) State the governing boundary conditions.

(c) Solve the differential equation.

(d) Calculate the wire centerline temperature.

(e) Assume that you wish to add insulation ($k = 0.15$ W/m·K to the surface of the wire and that the convective-heat-transfer coefficient on the insulation is 60 W/m²·K. Could the current in the wire be increased, or would it have to be decreased, assuming that the outer surface of the wire is at a constant temperature of 150°C?

2-29 Show that the steady temperature distribution in a constant property hollow cylinder with uniform generation q_G''' per unit volume is

$$\frac{T(r)-T_o}{T_i-T_o} = \frac{q_G''' r_o^2}{4k(T_i-T_o)} \left\{ \left[1-\left(\frac{r_i}{r_o}\right)^2\right] \frac{\ln(r/r_o)}{\ln(r_o/r_i)} + \left[1-\left(\frac{r}{r_o}\right)^2\right] \right\} - \frac{\ln(r/r_o)}{\ln(r_o/r_i)}$$

The boundary conditions are $T(r_i)= T_i$ and $T(r_o)= T_o$.

2-30 The wall shown in the figure has a constant thermal conductivity. The surface at $x=0$ is insulated and the surface at $x=L$ is maintained at temperature T_1 by a fluid. A constant heat-generation rate per unit volume q_G''' is distributed uniformly throughout the wall. Determine the maximum amount of heat that can be generated inside the wall per unit volume such that the steady temperature T_1 does not exceed 120°C when $L=0.1$ m, $k=15$ W/m·K. Calculate the temperature of the insulated surface of the wall.

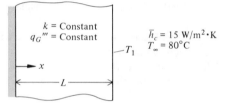

2-31 A plane wall of width L has a constant thermal conductivity k. The surface temperatures are $T(0)= T_1$ and $T(L)= T_2$. The heat generation per unit volume in the wall varies according to the expression $q_G''' = Bx^2$. Determine the following quantities in terms of k, B, T_1, T_2, and L:

    (a) The steady temperature distribution, $T(x)$.
    (b) The location of the plane of maximum temperature.
    (c) The heat flux leaving the wall at the surface $x=L$.

2-32 A plane wall is 1.0 m thick and it has one surface ($x=0$) insulated while the other surface ($x=L$) is maintained at a constant temperature of 350°C. The thermal conductivity of the wall is 25 W/m·K and a uniform generation per unit volume of 500 W/m³ exists throughout the wall. Determine a dimensionless expression for the temperature distribution in the wall. Determine the maximum temperature in the wall and the location of the plane where the maximum temperature occurs.

2-33 A long, solid cylinder has a constant thermal conductivity k and outside radius r_o. Generation exists within the cylinder at a rate per unit volume of $q_G''' = Ar$. If the exterior surface of the cylinder is maintained at a constant temperature T_1, derive an expression for the steady temperature distribution in the cylinder in terms of k, r_o, T_1, and A.

2-34 A long, hollow cylinder has a constant current I passing through it in an axial direction. The length of the cylinder is L. The inside surface of the cylinder,

$r = r_i$, is insulated and the outside surface, $r = r_o$, is maintained at T_o. The electrical resistance of the cylinder is R. Derive an expression for the steady temperature distribution in the cylinder. Your answer can be expressed in terms of I, R, k, L, r_i, r_o, and T_o.

2-35 A stainless steel pan is placed on the stove. The pan has a diameter of 30 cm and a thickness of 2 mm. The pan is used to boil water and the distance between the water and the rim of the pan is 15 cm. The heat-transfer coefficient between the air and the pan is 300 W/m·K. The average air temperature inside and outside the pan is 50°C. Estimate the temperature of the rim of the pan.

2-36 A fin with circular cross-sectional area of diameter d and length L is exposed to a fluid with ambient temperature T_∞. The convective-heat-transfer coefficient between the fluid and fin surface is $\bar{h}_c$ and the thermal conductivity of the fin is k. Plot the dimensionless temperature $\theta = (T - T_\infty)/(T_b - T_\infty)$ as a function of dimensionless distance down the fin axis $\xi = x/L$ for three values of the Biot number $\text{Bi} = \bar{h}_c PL^2/kA = 0.1, 1, 10$, assuming an insulated tip area.

2-37 A fin with triangular profile is attached to a surface with a temperature of 200°C. The fin material is aluminum and it has a length of 10 cm and a base thickness of 4 cm. The surrounding fluid is at 100°C and the heat-transfer coefficient is 35 W/m²·K. Determine the heat-transfer rate from the fin per unit width of fin surface.

2-38 A stainless steel fin with thickness 5 mm and outside diameter 3 cm surrounds a 1-cm-outside-diameter tube. The ambient fluid temperature is 50°C with a convective-heat-transfer coefficient of 40 W/m²·K. The tube-wall temperature is 150°C. Calculate the heat-transfer rate from the fin.

2-39 A copper fin with circular cross section with an area of 0.25 cm² and length of 2.5 cm is attached to a wall with temperature of 175°C. The ambient fluid temperature is 20°C, with $\bar{h}_c = 35$ W/m²·K. Calculate the heat-transfer rate and tip temperature for two cases:
  (a) the fin has an insulated tip, and
  (b) heat is convected from the tip surface area.

2-40 Two very long aluminum wires with diameter of 1 cm are to be soldered together. The wires are located in air ($\bar{h}_c = 20$ W/m²·K) with a temperature of 25°C. If the solder has a melting temperature of 250°C, determine the amount of heat input required at the interface of the two wires when the temperature distribution in the wire is steady.

2-41 A solid circular rod of length $2L$, diameter D, and thermal conductivity k extends between two walls that are both maintained at a temperature of T_b. The ambient temperature surrounding the rod is T_∞ and the convective-heat-transfer coefficient is $\bar{h}_c$. Determine an expression for the dimensionless steady temperature distribution in the rod and an expression for the total heat-transfer rate from the rod. Your answers may be expressed in terms of L, D, $\bar{h}_c$, T_∞, and T_b.

2-42 An engineer wishes to make accurate measurements of the heat flux leaving a test specimen. The test specimen is heated to 300°C and its temperature is monitored by a thermocouple consisting of two wires with diameter of 1 mm and thermal conductivity 75 W/m·K. The ambient laboratory temperature is 20°C and the convective-heat-transfer coefficient between the air and thermocouple wires is 25 W/m²·K. If the engineer monitors the heat flux from the specimen, determine the possible error in the heat-transfer measurement due to conduction through the thermocouple leads.

2-43 A single-cylinder air-cooled lawn mower engine operates under steady conditions. The cylinder temperature cannot exceed 300°C. To cool the engine, annular fins are placed around the cylinder. The fins are 0.3 cm thick and are 2 cm long from base to tip. The fins are cast iron. The outside diameter of the engine cylinder at the base of the fin is 0.3 m. Assume that the engine operates in 30°C air and that the heat-transfer coefficient on the sides and tip of the fin is 12 W/m²·K. Estimate the heat-transfer rate from a single fin. Determine the number of fins needed to cool a 3-kW engine to the given temperature if the engine has an efficiency of 30% and 50% of the total heat given off by the engine actually is dissipated through the fins.

2-44 One end of a circular poker is placed in a fire. The poker is made of steel, $k = 55$ W/m·K, and it has a diameter of 1.0 cm. The end of the poker in the fire is 350°C, the air around the poker ($\bar{h}_c = 25$ W/m²·K) is at a temperature of 80°C, and the length of the poker between the fire and handle is 0.6 m. Estimate the temperature of the handle of the poker.

2-45 A fin with constant cross-sectional area A, perimeter P, and length L has a uniform heat-generation rate per unit volume of q_G''' throughout its volume. The fin is surrounded by a fluid with constant ambient temperature T_∞ and constant-heat-transfer coefficient $\bar{h}_c$. Show that the dimensionless temperature distribution in the fin is given by

$$\theta(\xi) = (1 - Q_G)\left(\frac{e^{\sqrt{Bi}\,\xi}}{1 + e^{2\sqrt{Bi}}} + \frac{e^{-\sqrt{Bi}\,\xi}}{1 + e^{-2\sqrt{Bi}}}\right) + Q_G$$

where

$$\theta(\xi) = \frac{T(\xi) - T_\infty}{T_b - T_\infty}$$

$$Q_G = \frac{q_G''' A}{\bar{h}_c P(T_b - T_\infty)} \qquad \text{(dimensionless generation)}$$

$$Bi = \frac{\bar{h}_c P L^2}{kA}$$

The boundary conditions are

$$\theta(0) = 1.0$$

$$\left.\frac{d\theta}{d\xi}\right|_{\xi=1} = 0 \qquad \text{(insulated tip)}$$

Show that the temperature distribution reduces to the one derived in Section 2-6 for a fin with insulated tip when the generation is reduced to zero ($Q_G=0$).

2-46 A popular style of soldering gun has a power rating of 50 W. The heating element of the gun is shown in the sketch. The element is approximately 10 cm long and it has a 3-mm-square cross section with thermal conductivity of 55 W/m·K. The ambient air temperature is 25°C, with $\bar{h}_c=15$ W/m²·K. The base temperature of the heater element is 400°C. Estimate the tip temperature that could be reached after steady use. Do your calculations suggest that the soldering gun not be used continuously? Use the results of Problem 2-45 for the temperature distribution.

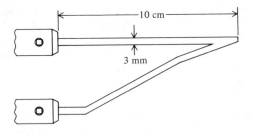

2-47 The temperature distribution in a solid with $k=2.5$ W/m·K is given by $T(x,y)=ax^2+by^2+cx+d$, where the units of T are K and both x and y are in centimeters. Values for the constants are, $a=2.0$ K/cm²; $b=1.5$ K/cm²; $c=1.0$ K/cm; $d=300$ K. Determine the direction and magnitude of the heat-flux vector at (x,y) locations of $(0,0)$, $(1,1)$, and $(3,0)$.

2-48 For the temperature distribution given in Problem 2-47, determine the heat-transfer rate per unit depth across the surfaces $x=0$ from $y=0$ to $y=3$ cm and $y=0$ from $x=0$ to $x=5$ cm.

2-49 Two steam pipes are placed in contact, as shown in the figure, and surrounded by asbestos insulation to reduce the heat loss from the pipes. The pipes have an exterior diameter of 10 cm and the outside diameter of the insulation is 30

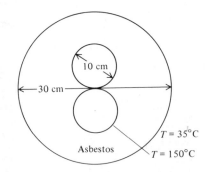

cm. The pipes carry steam with a temperature of 150°C at a flow rate of 0.08 kg/s. The exterior surface of the insulation averages 35°C. Estimate the heat-transfer rate through the insulation per unit length of pipe. Estimate the conduction shape factor for the insulation. Estimate the length of pipe in which the steam temperature will drop 5°C.

2-50 Hot water pipes are located on 0.5-m centers in a concrete $(k=1.5$ W/m·K) slab as shown in the figure. If the outside surfaces of the concrete are at 30°C and the water has an average temperature of 90°C, estimate the heat-transfer rate from each pipe per unit depth.

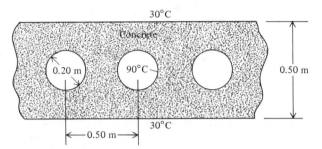

2-51 The cross section of a tall chimney shown in the figure has an inside surface temperature of 170°C and an exterior temperature of 50°C. Estimate the heat-transfer rate through the chimney per unit length if $k=2.0$ W/m·K.

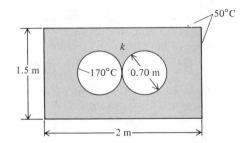

2-52 The long steel $(k=43$ W/m·K) angle shown in the figure has one surface at 100°C and the other surface maintained at 200°C. Estimate the heat-transfer rate between the two surfaces per unit of length.

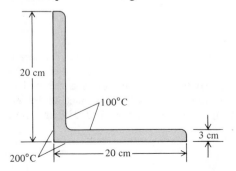

2-53   A square duct carries hot gases such that its surface temperature is 300°C. The duct passes through a long layer of circular asbestos ($k = 0.25$ W/m·K) insulation as shown in the figure. The outside surface temperature of the asbestos is 45°C. Estimate the heat-transfer rate from the gases per unit length of duct.

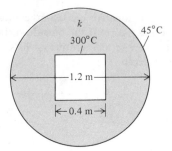

2-54   Radioactive wastes are sealed in a spherical container and buried in the earth. A 1-m-diameter sphere is buried at a depth of 25 m, where the soil thermal conductivity is 2.0 W/m·K. The surface of the earth has a constant temperature of 15°C. The waste material generates heat at a rate of 1000 W. What temperature does the container have to be designed to withstand?

2-55   A buried pipeline transports oil with an average temperature of 15°C. The pipe has an outside diameter of 0.5 m, inside diameter of 0.45 m, and it is buried at a depth of 5 m. If the surface of the earth is 5°C, estimate the steady heat-transfer rate from the oil per unit length of pipe. How far can the oil be transported until its average temperature drops to 12°C if the specific heat of the oil is 2000 W·s/kg·K and its flow rate is 50 kg/s? The thermal conductivity of the soil is 1.0 W/m·K.

2-56   An electrical power cable is buried in the earth at a depth of 1.5 m. The cable diameter is 10 cm. The thermal conductivity of the soil is 1.5 W/m·K and the temperature of the surface of the earth is 20°C. The electrical resistance of the cable per unit length is 10^{-4} Ω/m. If the insulation on the wire is limited to a temperature of 120°C, estimate the maximum current that can be carried by the cable.

2-57   A small electrical furnace is in the shape of a parallelepiped. The furnace wall is constructed of asbestos insulation that is 10 cm thick on all surfaces of the furnace. The internal cavity of the furnace is a cubic shape 0.50 m on a side. Determine the power consumption of the furnace under steady operating conditions if the internal surface of the asbestos is 220°C and the outside surface is at a temperature of 45°C.

2-58 Oil is transported in a pipe buried with its centerline 6 m below the surface of the earth. The metallic pipe has a 1-m o.d. and 0.95-m i.d. The thermal conductivity of the soil is 1.5 W/m·K and the soil air interface temperature is $-20°C$. Determine the heat input to the oil per kilometer of pipe length necessary to maintain the average oil temperature at 20°C if the oil flow rate is 900 kg/s.

2-59 The fin shown in the figure has a base temperature of 200°C. A source of energy is incident on the tip of the fin with an energy flux of 5000 W/m². The exterior surface of the fin is insulated. Use the relaxation method to estimate the temperatures at nodes 1 through 5.

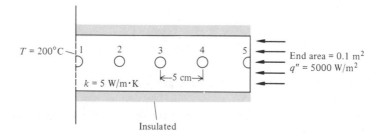

2-60 If the temperatures of nodes 1 through 8 in the figure in °C are assumed to be 50, 51, 52, 53, 54, 55, 56 and 57, calculate the residuals at nodes 1, 2, 4, and 5. Numerical values for the symbols in the figure are $\bar{h}_{c1}=5$ W/m²·K, $\bar{h}_{c2}=10$ W/m²·K, $T_\infty=40°C$, $L=5$ cm, $k=20$ W/m·K. By changing the temperature corresponding to the node with largest residual, determine the temperature of that node which will reduce the residual to zero.

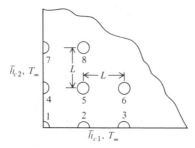

2-61 The figure shows a solid wall that has a thermal conductivity k. The surface is exposed to a fluid with ambient temperature T_∞. Derive the residual equations for nodes 1, 2, and 3 in terms of the temperatures of nodes 4 through 9 and $Bi = \bar{h}_c L / k$.

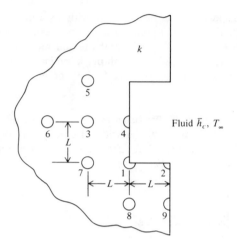

2-62 Verify all the relaxation equations given in Table 2-3.

2-63 Use the relaxation method to solve the five equations. The values for all the unknowns are all between 0 and 10.

$$x + 5y - 3z = 22$$
$$6y - 2z + u = 35$$
$$2z - 5w - 14u = -9$$
$$3x - 4z + 3w = -11$$
$$2x + 5w - 20u = 9$$

2-64 Calculate the two-dimensional steady temperatures of nodal points 1 through 4 in the figure. Surface temperatures are shown and the solid has a thermal conductivity of 2 W/m·K. Calculate the temperatures under two conditions:

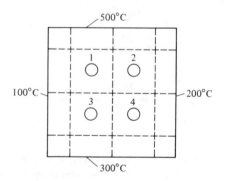

(a) No internal energy generation is present.
(b) A constant energy generation per unit volume exists throughout the solid equal to 1000 W/m³.

2-65 Repeat Problem 2-64 using an iteration technique.

2-66 Calculate the temperatures of the nine nodes shown in the figure. Use the relaxation technique. The thermal conductivity of the solid is 30 W/m·K.

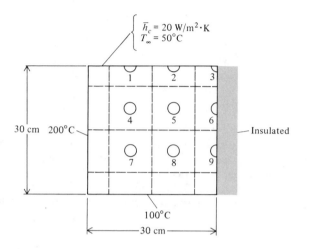

2-67 Repeat Problem 2-66 using an iteration technique.

2-68 A rectangular liquid nitrogen dewar is supported by two stainless steel legs as shown in the figure. The legs maintain the spacing between inner and outer walls of the dewar. The region separating the walls is evacuated and filled with an insulating material. Estimate the boil-off rate of the liquid nitrogen due to heat leak through the legs by using two different methods:

    (a) Nodal equation using the relaxation method.

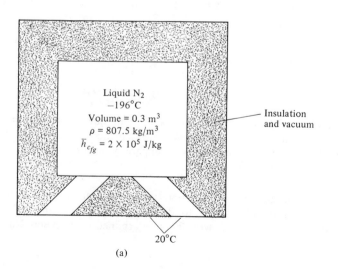

(a)

(b) Graphical method.

Support detail:

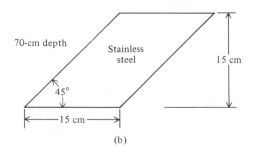

70-cm depth

Stainless steel

15 cm

45°

15 cm

(b)

2-69 The long, rectangular bar shown in the figure has two adjacent surfaces that are maintained at a constant temperature and two adjacent surfaces that transfer heat by convection. Numerical values are $k = 30$ W/m·K, $\bar{h}_{c1} = 50$ W/m². K, $\bar{h}_{c2} = 70$ W/m²·K, and $T_\infty = 100°C$. Using a nodal subdivision as shown in the figure, determine the steady temperatures of the 15 nodes by a relaxation method.

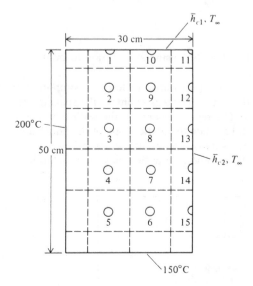

$\bar{h}_{c1}, T_\infty$

30 cm

200°C

50 cm

$\bar{h}_{c2}, T_\infty$

150°C

2-70 Work Problem 2-69 by a matrix-inversion technique. Derive the energy-balance equations for each node and determine the elements of matrices A and B. Use the program given in Example 2-13. Compare your answers with those determined in Problem 2-69.

2-71 Work Problem 2-69 by an iteration technique. Modify the computer program in Example 2-15 and use it to determine values for the 15 nodal temperatures. Compare your answers with those determined in Problems 2-69 and 2-70.

2-72 A long steel channel ($k = 45$ W/m·K) is shown in the figure. The top surface is isothermal at a temperature of 100°C while the bottom surface is at 300°C. One side is exposed to air ($\bar{h}_c = 100$ W/m²·K) at 40°C and the other side is insulated. The internal portion of the channel is in contact with a fluid with a temperature of 200°C and $\bar{h}_c = 25$ W/m²·K. Determine the steady temperatures of the 20 nodes with unknown temperatures. Use the matrix technique and the computer program in Example 2-13.

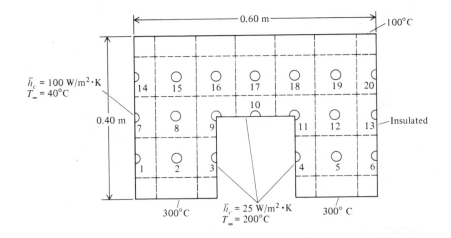

2-73 Work Problem 2-72 with an iteration technique. Use the computer program in Example 2-15.

Chapter 3

TRANSIENT CONDUCTION

3-1 INTRODUCTION

In Chapter 2 we considered only steady heat-transfer problems. Since most engineering situations involve changes with respect to time, we should consider techniques to solve for temperatures and heat-transfer rates in physical systems that involve unsteady, or transient conditions.

A heat-transfer problem is *transient* whenever the temperature within the system being considered changes with respect to time. There are many practical instances involving transient heat transfer. For example, many manufacturing processes require that the product is heated or cooled to convert it into a useful product. Furnaces and ovens are periodically cycled, and these processes produce temperature transients in the contents and the oven walls. Buildings undergo daily and seasonal transients. Metals are often heated and cooled to achieve desirable physical properties. Engines experience startup transients, in addition to shorter duration cyclic transients, during each portion of the thermodynamic cycle.

In general the effort required to solve a transient problem is greater than that to solve a similar problem that is steady. The next section develops equations for systems for which the spacial variation in temperature is negligible so that the governing conduction equation reduces to an ordinary differential equation. In later sections we will treat more complex situations. In these sections we will consider analytical solutions and numerical techniques that can be used to determine heat-transfer rates in solids that have both spacial and time variations.

3-2 TRANSIENT CONDUCTION WITH NEGLIGIBLE INTERNAL RESISTANCE

To determine the transient temperature distribution and ultimately the heat-transfer rate, we need to solve the general conduction equation, which for the first time includes the energy-storage term. The form of the conduction equation that must be solved is

$$\nabla^2 T + \frac{q_G'''}{k} = \frac{1}{\alpha} \frac{\partial T}{\partial t} \tag{3-1}$$

Equation 3-1 is a partial differential equation, and a general solution to it requires advanced mathematical techniques. Several excellent references are available (Refs. 1 through 4), and they should be referred to for specific solutions to Eq. 3-1.

One way of simplifying our approach to transient conduction problems would be to consider a class of problems for which the temperature in a solid body varies with time but at any instant the temperature does not vary with position. That is, the temperature at all locations inside the solid varies uniformly with time.

If we assume that the energy transferred from the solid is removed by convection to a fluid, then the condition for a uniformly varying temperature within the solid would be satisfied when the resistance to conduction is much less than the resistance to convection from the surface. Such a condition is referred to as a system with negligible internal resistance, although some texts refer to it as a lumped capacitance system.

When a body has negligible internal resistance, the temperature gradients inside the body are much smaller than those occurring in the surrounding fluid. To determine if a body surrounded by a fluid has negligible internal resistance, we must first check the magnitudes of the two resistances. The magnitude of the Biot number provides such a check because it is the dimensionless group defined as the ratio of conductive resistance to convective resistance. Therefore, if

INTERNAL RESISTANCE
IS NEGLIGABLE IF :

$$\text{Bi} = \frac{\bar{h}_c L}{k} \ll 1.0 \tag{3-2}$$

we can be assured that the internal resistance is negligible compared to the external or fluid resistance. The symbol L in Eq. 3-2 represents a characteristic length of the solid. As we proceed we will see that the characteristic length will change as we consider heat flow from solids with different geometries. For an irregularly shaped body the characteristic length is frequently defined as the volume of the body divided by its surface area.

$L = \frac{V}{A_S}$

Whenever the Biot number is much less than 1, we can easily obtain an equation for the transient temperature of the solid and heat-transfer rate from the surface to the fluid. Consider a solid of arbitrary shape as shown in Fig. 3-1. An energy balance applied to the solid indicates that the

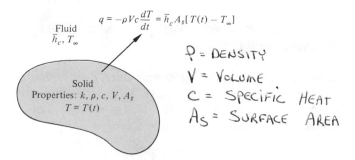

Figure 3-1 Energy balance for a solid with negligible internal resistance.

decrease in stored energy in the solid must be equal to the heat-transfer rate from the surface by the convective mode:

ENERGY
BALANCE

$$-\rho V c \frac{dT(t)}{dt} = \bar{h}_c A_s \left[T(t) - T_\infty \right] \qquad (3\text{-}3)$$

The symbol ρ represents the density of the solid, V its volume, c its specific heat, and A_s its surface area. Equation 3-3 is an ordinary differential equation with time as the only independent variable. The solution for the instantaneous temperature, $T(t)$ will specify the temperature for all points within the solid, including the surface, because we have assumed negligible internal resistance.

Equation 3-3 can be simplified slightly by defining a new dependent variable as

$$\theta(t) = T(t) - T_\infty \qquad (3\text{-}4)$$

To solve Eq. 3-3 we must specify the temperature of the body at a particular time. If we assume that the temperature of the body at an initial time, $t = 0$, is known to be T_0, we say that the initial condition for Eq. 3-3 is

INITIAL
CONDITION

$$\theta_0 = T_0 - T_\infty \qquad \text{at} \quad t = 0 \qquad (3\text{-}5)$$

The solution to Eq. 3-3 subject to the initial condition given in Eq. 3-5 is

SOLUTION

$$\frac{\theta(t)}{\theta_0} = e^{-(\bar{h}_c A_s / \rho V c)t} \qquad (3\text{-}6)$$

Equation 3-6 is written in dimensionless form, and therefore the term $\bar{h}_c A_s t / \rho V c$ must also be dimensionless.

This term is actually the product of two dimensionless groups that have been used previously. The first dimensionless group is the Biot number, defined in Eq. 3-2, and the second is the Fourier number, defined in Eq. 2-15. The product of the Biot and Fourier numbers is

$$(\text{Bi})(\text{Fo}) = \left(\frac{\bar{h}_c L}{k} \right) \left(\frac{\alpha t}{L^2} \right) = \frac{\bar{h}_c A_s t}{\rho V c} \qquad (3\text{-}7)$$

The characteristic length L for the arbitrary geometry considered here is defined as the ratio of volume to surface area of the solid.

The Fourier number has not appeared in previous problems involving steady-state conditions. The Fourier number is an important dimensionless group in transient heat-transfer problems. Equation 3-6 written in terms of the Biot and Fourier numbers is

SOLUTION

$$\frac{\theta(t)}{\theta_0} = e^{-(\text{Bi})(\text{Fo})} = \frac{T(t) - T_\infty}{T_0 - T_\infty} \qquad (3\text{-}8)$$

IF $Bi \ll 1.0$

We should remember that Eq. 3-8 only accurately predicts the time-temperature history of a solid when the condition of negligible internal resistance is met. The first step in any transient problem is to determine the magnitude of the Biot number. If Bi is less than 0.10, the error in the temperature history given by Eq. 3-8 is known to be less than 5 percent. As the Biot number becomes smaller, the accuracy is increased. If the Biot number is larger than 0.10, we must accept larger errors or use a different method to solve the problem. Methods that account for spacial temperature variations in the solid are developed in later sections.

Once the temperature history of the solid is known, the total heat-transfer and instantaneous heat-transfer rates from the surface of the solid can be calculated by determining the amount of heat that leaves the surface. The instantaneous heat-transfer rate at a particular time t would be

$$q''(t) = \bar{h}_c A_s \left[T(t) - T_\infty \right]$$

Substituting the instantaneous temperature from Eq. 3-8, the instantaneous heat-transfer rate from a solid with negligible internal resistance can be written in nondimensional form as

$$\frac{q''(t)}{\bar{h}_c A_s (T_0 - T_\infty)} = e^{-(\text{Bi})(\text{Fo})} \qquad (3\text{-}9)$$

$q''(t) = WATTS$

The total amount of heat transferred from the solid between time $t=0$ and an arbitrary time t can be determined by integrating Eq. 3-9 between the two time limits. The result is

$$Q(t) = \int_{t=0}^{t} q(t)\,dt = \bar{h}_c A_s (T_0 - T_\infty) \int_{t=0}^{t} e^{-(\text{Bi})(\text{Fo})}\,dt$$

or in the dimensionless form is

$Q(t) = W \cdot SEC = J$

TOTAL HEAT TRANSFERRED

$$\frac{Q(t)}{\bar{h}_c A_s (T_0 - T_\infty) t} = \left[1 - e^{-(\text{Bi})(\text{Fo})} \right] \frac{1}{(\text{Bi})(\text{Fo})} \qquad (3\text{-}10)$$

The symbol $q(t)$ is the instantaneous heat-transfer rate from the solid and it is usually expressed in units of watts. The symbol $Q(t)$ represents the total heat transfer from the body and it has units of watt-seconds, or joules. Both equations for $q(t)$ and $Q(t)$ are only accurate for a system with negligible internal resistance, that is, $Bi \ll 1.0$.

The equations for temperature history and heat-transfer rates for negligible internal resistance apply to another important class of problems. If the solid material is replaced by a fluid which is continually stirred so that the temperature differences at any instant of time in the fluid are not allowed to exist, the fluid will have a very small internal resistance compared to the external convective resistance. The *well-stirred fluid*, as it is called, will then be a type of problem for which the instantaneous fluid temperature will be given by Eq. 3-8; the instantaneous heat-transfer rate is given by Eq. 3-9; and the total heat transfer is given by Eq. 3-10.

The analogy between the flow of heat and the flow of an electrical current was described for a steady system in Section 2-7. A similar analogy exists for the transient heat flow from a body with negligible internal resistance. The two analogous systems are illustrated in Fig. 3-2, where the analogous conservation equations and solutions for the temperature in the thermal system and the electrical potential in the electrical system are compared.

(A) THERMAL SYSTEM	(B) ELECTRICAL SYSTEM
Physical System: 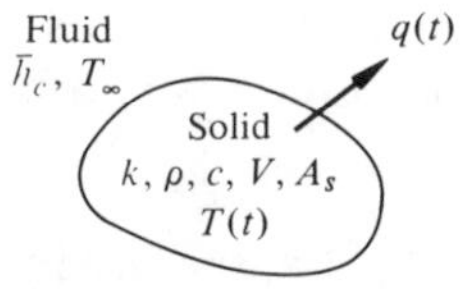	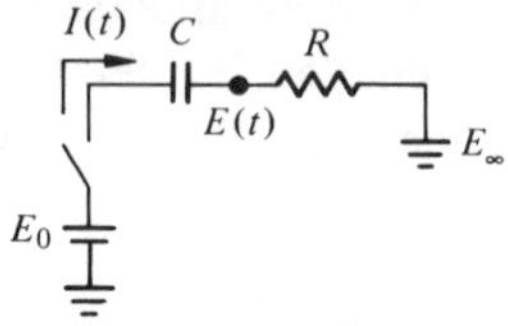
Fluid $\bar{h}_c$, T_∞ $q(t)$ Solid k, ρ, c, V, A_s $T(t)$	$I(t)$ C R $E(t)$ E_∞ E_0
Conservation equation:	
$q(t) = -\rho V c \dfrac{\partial \theta}{\partial t} = \dfrac{\theta}{1/\bar{h}_c A_s}$	$I(t) = -C \dfrac{\partial \epsilon}{\partial t} = \dfrac{\epsilon}{R}$
$\theta = T(t) - T_\infty$	$\epsilon = E(t) - E_\infty$
Initial condition:	
$\theta(0) = T_0 - T_\infty = \theta_0$ *Solution for potential*:	$\epsilon(0) = E_0 - E_\infty = \epsilon_0$
$\dfrac{\theta(t)}{\theta_0} = e^{-(\text{Bi})(\text{Fo})}$	$\dfrac{\epsilon(t)}{\epsilon_0} = e^{-t/RC}$

Figure 3-2 Analogy between transient electrical and thermal systems.

If the conservation equations for both systems are compared in Fig. 3-2, we see that the thermal resistance of the convection layer is $1/\bar{h}_c A_s$. Also, the thermal capacitance of the solid is

$$C_t = \rho V c$$

The thermal capacitance is directly proportional to the product of mass and specific heat of the solid. Therefore, an object with a high value of specific heat will respond more slowly to an external temperature change than will an object of equal mass, but with a lower value of specific heat. The slowly responding object is said to have a large value of thermal capacitance.

Example 3-1. Chromium steel ball bearings ($k = 50$ W/m·K, $\alpha = 1.3 \times 10^{-5}$ m^2/s) are to be heat-treated. They are heated to a temperature of 650°C and then quenched in a vat of oil that has a temperature of 55°C. The ball bearings have a diameter of 4.0 cm. The convective-heat-transfer coefficient between the ball bearings and oil is 300 W/m^2·K. Determine (a) the length of time that the bearings must remain in the oil before their temperature drops to 200°C, (b) the total amount of heat removed from each bearing during this time interval, and (c) the instantaneous heat-transfer rate from the bearings when they are first placed in the oil and when they reach 200°C.

Solution: To determine if the bearings have negligible internal resistance, we first check the magnitude of the Biot number:

$$\text{Bi} = \frac{\bar{h}_c L}{k} = \frac{\bar{h}_c (V/A_s)}{k} = \frac{\bar{h}_c (r/3)}{k} = \frac{300(0.02/3)}{50} = 0.04$$

Since the Biot number is less than 0.1, the equations developed in this section can be used with little error. We can calculate the Fourier number in terms of the unknown time:

$$\text{Fo} = \frac{\alpha t}{L^2} = \frac{\alpha t}{(r/3)^2} = \frac{1.3 \times 10^{-5}}{(0.02/3)^2} t$$

a. The time required for the ball bearings to reach 200°C is given by Eq. 3-8, or

$$\frac{200 - 55}{650 - 55} = e^{-(0.04)(0.293 t)} \implies \ln \frac{145}{595} = -(.04)(.293) t$$

Thus, $t = 120.5$ s, which corresponds to a Fourier number of 35.31.

b. The total amount of heat transferred from each bearing in the first 120.5 s is

$$Q = \bar{h}_c A_s (T_0 - T_\infty) \left[1 - e^{-(\text{Bi})(\text{Fo})} \right] \frac{t}{(\text{Bi})(\text{Fo})}$$

$$= 300 \times 4\pi (0.02)^2 (650 - 55) \left[1 - e^{-(0.04)(35.31)} \right] \frac{120.5}{(0.04)(35.31)}$$

$$= 5.79 \times 10^4 \text{ W·s}$$

c. The instantaneous heat-transfer rate at $t=0$ (or Fo$=0$) is

$$q=\bar{h}_c A_s (T_0 - T_\infty)=300\times 4\pi(0.02)^2(650-55)=897 \text{ W}$$

and at $t=120.5$ s (Fo$=35.31$) the instantaneous heat-transfer rate is

$$q=\bar{h}_c A_s (T_0 - T_\infty)e^{-(\text{Bi})(\text{Fo})}$$

$$=300\times 4\pi(0.02)^2 \times (650-55)e^{-(0.04)(35.31)}$$

$$=218 \text{ W}$$

3-3 TRANSIENT CONDUCTION IN A SEMI-INFINITE SOLID

A semi-infinite solid is a large body with one plane surface. A good example of a semi-infinite solid is the earth. If the temperature of the surface of the earth is changed, heat is conducted into the earth and due to its infinite extent, the temperature is a function of distance from the surface of the earth, x, and time, t, or expressed mathematically $T=T(x,t)$. The conduction equation (Eq. 2-6) simplified for transient conduction in a semi-infinite solid is

ASSUMING:
1 DIMENSIONAL FLOW
$\dot{q}_G''' = 0$

$$\frac{\partial^2 T}{\partial x^2}=\frac{1}{\alpha}\frac{\partial T}{\partial t} \tag{3-11}$$

where x is measured from the surface as shown in Fig. 3-3.

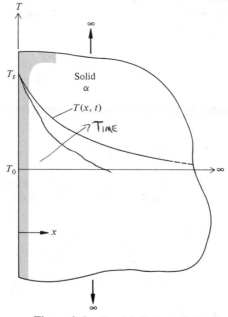

Figure 3-3 Semi-infinite solid.

Before solving Eq. 3-11, we must specify two boundary conditions and a single initial condition. The initial condition is

$$T(x,0)=T_0 \qquad\qquad (3\text{-}12)$$

That is, the entire semi-infinite solid is at a uniform temperature, T_0, at the initial time of $t=0$.

One of the boundary conditions requires that the material at a large distance from the surface never changes in temperature or

$$T(\infty,t)=T_0 \qquad\qquad (3\text{-}13)$$

Solutions are possible for several different choices for the second boundary condition. We will consider two possible cases.

Case I: Isothermal Boundary Condition

One possible boundary condition that is relatively easy to achieve physically is suddenly to change the surface ($x=0$) temperature to a value equal to T_s and maintain the value constant. The isothermal boundary condition can be mathematically specified by

$$T(0,t)=T_s \qquad\qquad (3\text{-}14)$$

The solution to Eq. 3-11 subject to the two boundary conditions 3-13 and 3-14 and the initial condition is

$$\frac{T(x,t)-T_s}{T_0-T_s}=\mathrm{erf}\left(\frac{x}{2\sqrt{\alpha t}}\right) \qquad\qquad (3\text{-}15)$$

A derivation of this equation is given in Reference 5. The *Gauss error function*, erf, is a function which occurs frequently in engineering work and it is defined as

$$\mathrm{erf}\left(\frac{x}{2\sqrt{\alpha t}}\right)=\frac{2}{\sqrt{\pi}}\int_0^{x/2\sqrt{\alpha t}}e^{-\eta^2}\,d\eta$$

η IS INDEPENDENT OF x

Values of the Gauss error function are tabulated in Appendix D and plotted in Fig. 3-4.

The heat flux conducted into the semi-infinite solid can be determined from the Fourier law evaluated at the surface or

$$q''(t)=-k\frac{\partial T}{\partial x}\bigg|_{x=0}=-k(T_0-T_s)\frac{\partial}{\partial x}\left[\mathrm{erf}\left(\frac{x}{2\sqrt{\alpha t}}\right)\right]\bigg|_{x=0}$$

or

$$q''(t)=\frac{k(T_s-T_0)}{\sqrt{\pi\alpha t}} \qquad\qquad (3\text{-}16)$$

The total amount of heat conducted into the solid during the time interval $t=0$ to $t=t$ is

$$Q(t)=\int_{t=0}^{t}q(t)\,dt=\frac{k(T_s-T_0)}{\sqrt{\pi\alpha}}\int_0^t t^{-1/2}\,dt$$

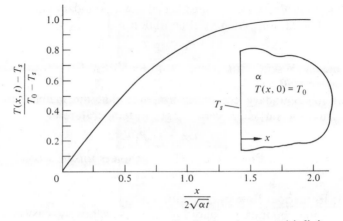

Figure 3-4 Temperature distribution in a semi-infinite solid subject to sudden change in surface temperature.

or

$$Q(t) = 2k(T_s - T_0)\sqrt{\frac{t}{\pi\alpha}} \qquad (3\text{-}17)\checkmark$$

Case II: Convection Boundary Condition

Instead of changing the surface temperature of the infinite solid, we could subject the surface to a fluid with an ambient temperature of T_∞ and an average convective-heat-transfer coefficient of $\bar{h}_c$. The heat transferred into the solid must then be convected through the fluid and then conducted into the solid. The appropriate boundary condition for this type of problem is then

$$\bar{h}_c[T_\infty - T(0,t)] = -k\left(\frac{\partial T}{\partial x}\right)\bigg|_{x=0} \qquad (3\text{-}18)$$

The solution to Eq. 3-11 subject to the initial condition (3-12) and the two boundary conditions (3-13) and (3-18) is given in Schneider (Ref. 5) as

$$\frac{T(x,t) - T_0}{T_\infty - T_0} = 1 - \operatorname{erf}\xi - \left\{ \exp\left[(\mathrm{Bi}) + \eta\right]\left[1 - \operatorname{erf}(\xi + \sqrt{\eta})\right]\right\} \qquad (3\text{-}19)$$

where

$$\xi = \sqrt{\frac{x^2}{4\alpha t}} = \frac{\mathrm{Fo}^{-1/2}}{2}$$

$$\mathrm{Fo} = \frac{\alpha t}{x^2}$$

$$\mathrm{Bi} = \frac{\bar{h}_c x}{k}$$

$$\eta = \frac{\bar{h}_c^2 \alpha t}{k^2} = (\mathrm{Bi})^2(\mathrm{Fo})$$

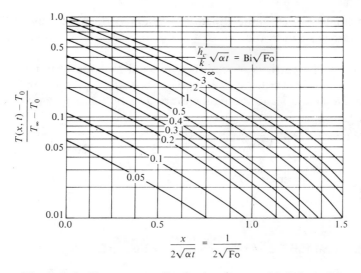

Figure 3-5 Temperature distribution in a semi-infinite solid
subject to convection from its surface to a fluid at T_∞.

Therefore, the dimensionless temperature distribution in a semi-infinite
solid with uniform initial temperature subjected to a fluid with temperature
T_∞ at time $t=0$ is only a function of the Fourier and Biot numbers. For
convenience, the solution to Eq. 3-19 is plotted in Fig. 3-5.

The equations for the temperature distribution and heat flux derived in
this section are valid specifically for a semi-infinite geometry. One logical
question at this point would be: How large does a solid have to be to
qualify as a semi-infinite solid? Certainly, if the times for which we wish a
solution are so short or if the thermal conductivity is so low that the
temperature at a given depth in a finite solid does not change from its
initial value, the geometry might also be considered a semi-infinite solid.
Kreith (Ref. 6) has stated that a large plate with a width L can be
considered to be a semi-infinite plate if   USE EQ. 3-15 OR 3-19 IF;

$$Fo = \frac{\alpha t}{L^2} < 1.0 \qquad (3\text{-}20)$$

Therefore, a plate with finite width that satisfies the condition in Eq. 3-20
will have its temperature distribution given by either Eq. 3-15 or 3-19,
depending upon whether the boundary condition satisfies Case I or Case
II, respectively.

Example 3-2. A large, flat heater with a surface temperature of 100°C is
placed in direct contact with the earth ($k=2.0$ W/m·K, $\alpha=5\times10^{-7}$
m^2/s). If the soil was originally at a uniform temperature of 15°C,
determine the temperature of the plane 5 cm below the heater 2 h after the

heat is applied. Determine the total amount of heat per unit area conducted into the soil in the first 2 h.

Solution: The earth is a semi-infinite solid and the heater is in direct contact with the earth. The boundary condition is therefore the one described by Case I and the temperature distribution is given by Eq. 3-15.

$$\frac{T(x,t)-T_s}{T_0-T_s}=\text{erf}\left(\frac{x}{2\sqrt{\alpha t}}\right)=\text{erf}\left[\frac{0.05}{2\sqrt{5\times10^{-7}\times7200}}\right]=\text{erf}(0.417)$$

$$\frac{T(x,t)-100}{15-100}=\text{erf}(0.417)=0.445$$

or

$$T(x,t)=62.2°C$$

The total heat flux during 2 h is given by Eq. 3-17:

$$Q''(t)=2k(T_s-T_0)\sqrt{\frac{t}{\pi\alpha}}=2(2.0)(100-15)\sqrt{\frac{7200}{\pi\times(5\times10^{-7})}}$$

$$=2.3\times10^7\text{ W}\cdot\text{s/m}^2=63.9\text{ kWh/m}^2$$

Example 3-3. A large steel plate is initially at a uniform temperature of 300°C. The plate is to be cooled by directing 50°C air over one surface of the plate. The plate is 10 cm thick, it has a thermal diffusivity of 10^{-5} m²/s, thermal conductivity of 40 W/m·K, and the convective-heat-transfer coefficient between the air and plate is 400 W/m²·K. Determine the length of time that the air must be directed over the surface of the plate before the surface temperature is reduced to 200°C. At the time the surface reaches 200°C, calculate the temperature at planes 1 cm and 10 cm from the surface.

Solution: First, we should check the Biot number to see if the system has negligible internal resistance.

$$\text{Bi}=\frac{\bar{h}_c L}{k}=\frac{400(0.10)}{40}=1.0$$

The system does have significant internal resistance, so the equations developed in Section 3-2 cannot be used.

Next, we should check the Fourier number to see if the plate can be considered to be a semi-infinite body.

$$\text{Fo}=\frac{\alpha t}{L^2}=\frac{10^{-5}t}{(0.10)^2}=10^{-3}t$$

Therefore, according to the condition of Eq. 3-20, the plate may be considered to be semi-infinite as long as we do not exceed a time of 1000 s ($16\frac{2}{3}$ min) in our solution.

To determine the time required for the surface temperature to reach 200°C, we could use either Eq. 3-19 or the values in Fig. 3-5. Using the values in Fig. 3-5, we note that the abscissa evaluated at the surface, $x = 0$, is

$$\frac{x}{2\sqrt{\alpha t}} = 0$$

and the dimensionless temperature at the surface at the unknown time is

$$\frac{T(0,t) - T_0}{T_\infty - T_0} = \frac{200 - 300}{50 - 300} = 0.40$$

From Fig. 3-5 we find that

$$(\text{Bi})\sqrt{\text{Fo}} = \frac{\bar{h}_c\sqrt{\alpha t}}{k} = 0.50$$

or

$$t = \frac{0.25k^2}{\bar{h}_c^2\alpha} = \frac{0.25(40)^2}{(400)^2(10^{-5})} = 250 \text{ s}$$

Since the surface reaches 200°C before 1000 s have elapsed, the plate can be considered to be a semi-infinite solid.

To determine local temperatures at 250 s at the planes located at $x = 1$ cm and 10 cm, we can organize our calculations in the following table:

LOCATION	$x/2\sqrt{\alpha t}$	$\bar{h}_c\sqrt{\alpha t}/k$	$\theta(x,t)/\theta_\infty$	$T(x,t)$
$x = 1$ cm	0.1	0.50	0.325	219°C
$x = 10$ cm	1.0	0.50	0.039	290°C

The temperature of the plane located at $x = 10$ cm has decreased only 10°C in the first 250 s, which is small enough to justify the use of semi-infinite equations for a plate of finite width.

3-4 CHART SOLUTIONS TO TRANSIENT CONDUCTION PROBLEMS

One-dimensional Solutions

Analytical solutions to the transient form of the general conduction equation have been obtained for simple geometries that are commonly encountered in engineering practice. The three geometries that are most practical are:

1. An infinite plate with width $2L$ for which $T = T(x,t)$, where x is measured from the plate centerline.
2. An infinitely long solid cylinder with outside radius r_0 for which $T = T(r,t)$.
3. A solid sphere with radius r_0 for which $T = T(r,t)$.

The boundary conditions for all three geometries are similar. The first boundary condition specifies an insulated location at the midplane of the infinite plate, the axis of the cylinder, and the center of the sphere.

The second boundary condition requires that the heat transferred from the exterior surface of the solid is removed by a fluid with an ambient temperature of T_∞ and a heat-transfer coefficient equal to $\bar{h}_c$. This boundary condition can be expressed mathematically as

$$\bar{h}_c[\, T_s - T_\infty\,] = -k \frac{\partial T}{\partial n}\bigg|_s$$

where the subscript s refers to the quantity evaluated at the surface of the solid and n refers to the coordinate direction normal to the surface of the solid.

There is a second possible boundary condition if we consider the limiting case of a very large value of the convective-heat-transfer coefficient. If $\bar{h}_c \to \infty$, the thermal resistance of the convection layer is negligible and the surface temperature of the solid is identical to the fluid temperature. Therefore, even though the results that follow consider heat loss from the surface of the solid by convection, the solution for the isothermal boundary condition is available by determining the results for the limiting case of $\bar{h}_c \to \infty$ or Bi$\to \infty$.

The initial condition for all these geometries are identical. The solid is uniformly isothermal at a temperature of T_0 at the time $t = 0$. At $t = 0$ the exterior surface of the solid is immersed in the fluid, which has temperature T_∞, and the transient-heat-transfer process is initiated.

The solutions for the temperature distribution and heat transfer from the three geometries are usually presented in graphical form with the variables expressed in dimensionless form. We have seen that transient conduction problems that involve convective boundary conditions are governed by the Biot and Fourier numbers. The three cases considered here are no exception. The local temperatures are functions of the dimensionless position within the solid, the Biot number, and Fourier number. A summary of the form of each dimensionless group is given in Table 3-1.

The dimensionless temperatures are plotted in Figs. 3-6(a), 3-7(a), and 3-8(a). Figure 3-6(a) gives values for the infinite plate of width $2L$; Fig. 3-7(a) gives values for an infinitely long cylinder; Fig. 3-8(a) gives values for a sphere. Each figure consists of two sets of curves; the first set plots the dimensionless centerline temperature:

$$\frac{\theta(0,t)}{\theta_0} = \frac{T(0,t) - T_\infty}{T_0 - T_\infty} \tag{3-21}$$

To determine a local temperature for a position not on the centerline, the second set of curves in each figure must be used. The second set of curves plots the dimensionless local temperature as a function of the centerline temperature, which for the case of the infinite plate is

$$\frac{\theta(x,t)}{\theta(0,t)} = \frac{T(x,t) - T_\infty}{T(0,t) - T_\infty} \tag{3-22}$$

The expression for an infinite cylinder and sphere is similar. To determine the value for a local temperature we can form the product of Eqs. 3-21 and 3-22:

$$\frac{T(x,t) - T_\infty}{T_0 - T_\infty} = \frac{\theta(0,t)}{\theta_0} \frac{\theta(x,t)}{\theta(0,t)}$$

Once the temperature distribution is known, the heat transferred from the surface of the solid can be evaluated using the Fourier law evaluated at the solid/fluid interface. Graphs that plot the dimensionless heat transfer from the surface are given in Fig. 3-6(c) for the infinite plate, Fig. 3-7(c) for the infinite cylinder, and Fig. 3-8(c) for the sphere. Each heat-transfer value $Q(t)$ is the total amount of heat that is transferred from the surface to the fluid during the time from $t=0$ to $t=t$. The normalizing factor, Q_0, is the initial amount of energy in the solid at $t=0$ when the reference temperature for zero energy is T_∞. The values for Q_0 for each of the three geometries are listed in Table 3-1 for convenience. Since the volume of the plate is infinite, the dimensionless heat-transfer for this geometry, given on a per unit surface area basis, is designated by the ratio $Q''(t)/Q_0''$. The volume of an infinitely long cylinder is also infinite, so the dimensionless heat transfer ratio is written on a per unit length basis $Q'(t)/Q_0'$. The sphere has a finite volume, so the heat-transfer ratio is simply $Q(t)/Q_0$ for that geometry. If the value for $Q(t)$ is positive, heat flows from the solid into the fluid; that is, the body is cooled. If it is negative, the solid is heated by the fluid.

Two general classes of transient problems can be solved by using the charts. One class of problem involves knowing the time while the local temperature at that time is unknown. In the other type of problem, the local temperature is the known quantity and the time required to reach that temperature is the unknown. The first class of problems can be solved in a straightforward fashion by the use of the charts. The second class of problem occasionally involves a trial-and-error procedure. Both types of solutions will be illustrated by the following examples.

Table 3-1 Summary of Dimensionless Groups To Be Used in Figures 3-6, 3-7, and 3-8

Situation	Infinite Plate Width $2L$	Infinitely Long Cylinder, Radius r_0	Sphere Radius r_0
Geometry	 Fluid $\bar{h}_c, T_\infty$; k, α; x; $2L$	Fluid $\bar{h}_c, T_\infty$; k, α; r; r_0	Fluid $\bar{h}_c, T_\infty$; k, α; r; r_0
Dimensionless position	$\dfrac{x}{L}$	$\dfrac{r}{r_0}$	$\dfrac{r}{r_0}$
Biot number	$\dfrac{\bar{h}_c L}{k}$	$\dfrac{\bar{h}_c r_0}{k}$	$\dfrac{\bar{h}_c r_0}{k}$
Fourier number	$\dfrac{\alpha t}{L^2}$	$\dfrac{\alpha t}{r_0^2}$	$\dfrac{\alpha t}{r_0^2}$

Dimensionless centerline temperature $\dfrac{\theta(0,t)}{\theta_0}$	Fig. 3-6(a)	Fig. 3-7(a)	Fig. 3-8(a)
Dimensionless local temperature $\dfrac{\theta(x,t)}{\theta(0,t)}$ or $\dfrac{\theta(r,t)}{\theta(0,t)}$	Fig. 3-6(b)	Fig. 3-7(b)	Fig. 3-8(b)
Dimensionless heat transfer $\dfrac{Q(t)}{Q_0}, \dfrac{Q'(t)}{Q'_0}, \dfrac{Q''(t)}{Q''_0}$	Fig. 3-6(c)	Fig. 3-7(c)	Fig. 3-8(c)
	$Q''_0 = \rho c L(T_0 - T_\infty)$	$Q'_0 = \rho c \pi r_0^2 (T_0 - T_\infty)$	$Q_0 = \rho c \dfrac{4}{3}\pi r_0^3 (T_0 - T_\infty)$

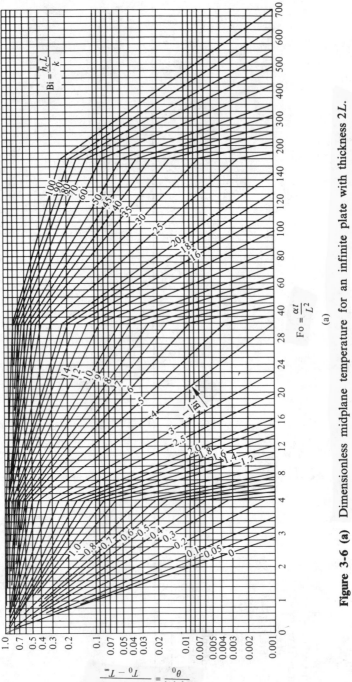

Figure 3-6 (a) Dimensionless midplane temperature for an infinite plate with thickness $2L$.

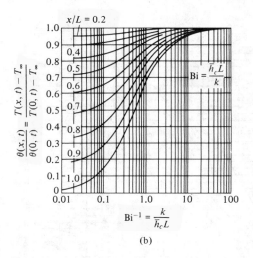

(b)

Figure 3-6 (b)   Dimensionless local temperature for an infinite plate with thickness $2L$ as a function of midplane temperature.

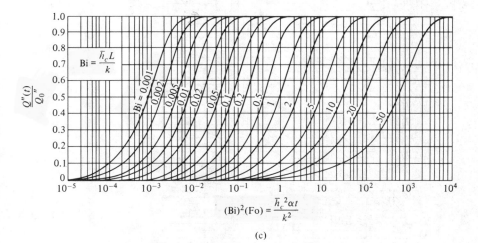

(c)

Figure 3-6 (c)   Dimensionless heat transfer from one surface of an infinite plate with thickness $2L$.

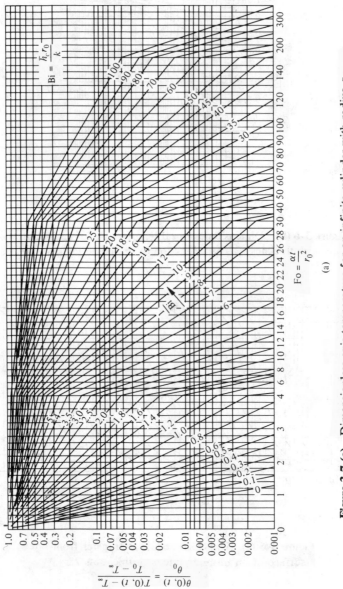

Figure 3-7 (a) Dimensionless axis temperature for an infinite cylinder with radius r_0.

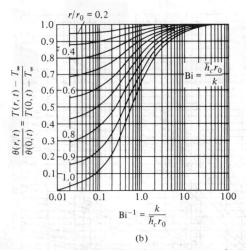

Figure 3-7 (b) Dimensionless local temperature for an infinite cylinder with radius r_0 as a function of axis temperature.

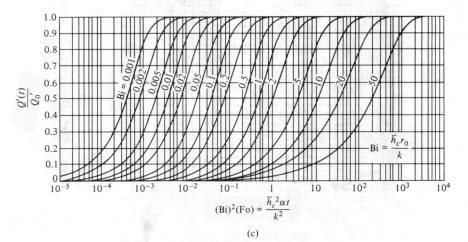

Figure 3-7 (c) Dimensionless heat transfer from an infinite cylinder with radius r_0.

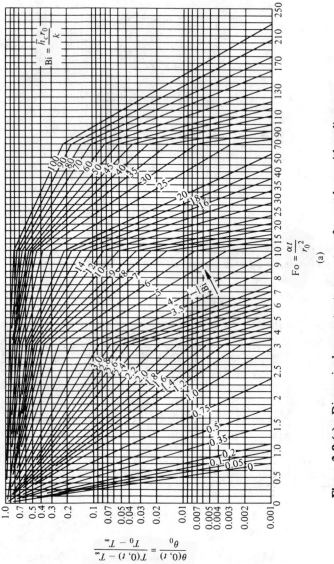

Figure 3-8 (a) Dimensionless center temperature for a sphere with radius r_0.

156

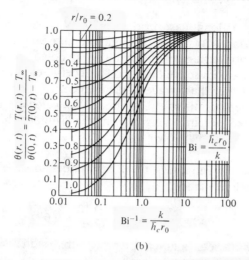

(b)

Figure 3-8 (b) Dimensionless local temperature for a sphere with radius r_0 as a function of center temperature.

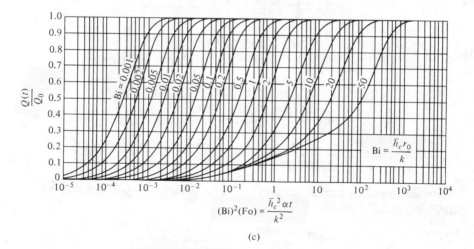

(c)

Figure 3-8 (c) Dimensionless heat transfer from a sphere with radius r_0.

Example 3-4. A long cast iron $(k=70$ W/m·K,$\alpha=2\times10^{-5}$ m²/s) cylinder with diameter of 20 cm is originally at a uniform temperature of 400°C. The exterior surface of the cylinder is cooled by air with temperature of 50°C and a convective-heat-transfer coefficient of 420 W/m²·K. If the air is directed over the surface of the cylinder for 20 min, determine the surface and axis temperatures at that time. Also, calculate the total amount of heat transferred per unit length from the cylinder during the 20-min period.

Solution: First, we will check the Biot number to see if the system has negligible internal resistance:

$$\text{Bi}=\frac{\bar{h}_c r_0}{k}=\frac{420\times0.10}{70}=0.60$$

The system does possess significant temperature gradients within the solid, so the chart solutions should be used.

Next, we will calculate the appropriate parameters indicated in Table 3-1:

$$\text{Fo}=\frac{\alpha t}{r_0^2}=\frac{2\times10^{-5}(20\times60)}{(0.10)^2}=2.40$$

$$(\text{Bi})^2(\text{Fo})=\frac{\bar{h}_c^2\alpha t}{k^2}=\frac{(420)^2(2\times10^{-5})(20\times60)}{(70)^2}=0.86$$

$$\theta_0=T_0-T_\infty=400-50=350°C$$

$$Q_0'=\rho c\pi r_0^2(T_0-T_\infty)=\frac{k}{\alpha}\pi r_0^2(T_0-T_\infty)$$

$$=\frac{70}{2\times10^{-5}}\pi(0.10)^2(350)=3.85\times10^7\text{W}\cdot\text{s/m}$$

The dimensionless centerline temperature is given in Fig. 3-7(a), evaluated at Fo=2.4, Bi^{-1}=1.66:

$$\frac{\theta(0,t)}{\theta_0}=\frac{T(0,t)-T_\infty}{T_0-T_\infty}=0.09$$

or

$$T(0,t)=0.09(T_0-T_\infty)+T_\infty=81.5°C$$

The surface temperature in terms of the axis temperature is given in Fig. 3-7(b), evaluated at $r/r_0=1.0$:

$$\frac{\theta(r_0,t)}{\theta(0,t)}=0.75$$

so the surface temperature is

$$\frac{\theta(r_0,t)}{\theta_0} = \frac{\theta(r_0,t)}{\theta(0,t)} \frac{\theta(0,t)}{\theta_0} = 0.75 \cdot 0.09 = 0.0675$$

$$T(r_0,t) = 0.0675 \times 350 + 50 = 73.6°C$$

the amount of heat transferred from the surface can be determined by using Fig. 3-7(c):

$$\frac{Q'(t)}{Q'_0} = 0.88$$

or

$$Q'(t) = 0.88 \times 3.85 \times 10^7 = 3.39 \times 10^7 \ W \cdot s/m = 9.41 \ kWh/m$$

Example 3-5. A large alloy steel plate $(k = 30 \ W/m \cdot K, \alpha = 1.5 \times 10^{-5}$ $m^2/s)$ comes from a rolling mill at a uniform temperature of 800°C. The plate has a thickness of 30 cm. To cool the plate, both surfaces are exposed to high-velocity air jets. The air has a temperature of 30°C and the heat-transfer coefficient between the air and plate surfaces is 500 W/m²·K. A layer of plastic insulating material is to be applied to the surface of the plate, but the surface temperature of the steel plate must be below 200°C before a layer is applied. Determine the minimum amount of time that the plate must be cooled before the insulating layer can be applied.

Solution: The Biot number as specified in Table 3-1 is

$$Bi = \frac{\bar{h}_c L}{k} = \frac{500 \times 0.15}{30} = 2.50$$

Therefore, chart solutions must be used. Figure 3-6(a) cannot be used directly because it contains the midplane temperature and the time, both of which are unknown. We must first use Fig. 3-6(b) to determine the midplane temperature.

$$\frac{\theta(L,t)}{\theta(0,t)} = \frac{200 - 30}{T(0,t) - 30} = \frac{170}{T(0,t) - 30} = 0.41$$

or

$$T(0,t) = 444.6°C$$

The dimensionless midplane temperature is

$$\frac{\theta(0,t)}{\theta_0} = \frac{444.6 - 30}{800 - 30} = 0.538$$

Now using Fig. 3-6(a) with $Bi^{-1} = 0.4$ and a dimensionless midplane temperature of 0.538, the Fourier number is

$$Fo = \frac{\alpha t}{L^2} = 0.60$$

or

$$t = 900 \text{ s} = 15 \text{ min}$$

Two- and Three-dimensional Solutions

The use of the one-dimensional transient charts can be extended to two- and three-dimensional problems. The method involves using the product of multiple values from the one-dimensional charts: Figs. 3-6, 3-7, and 3-8. The basis for obtaining two- and three-dimensional solutions from one-dimensional charts is the manner in which partial differential equations can be separated into the product of two or three ordinary differential equations. A proof of the method can be found in Arpaci (Ref. 3, sec. 5-2).

The product solution method can best be illustrated by an example. Suppose that we wish to determine the transient temperature at point P in a cylinder of finite length, as shown in Fig. 3-9. The point P is located by the two coordinates (x, r), where x is the axial location measured from the center of the cylinder and r is the radial position. The initial condition and boundary conditions are the same as those that apply to the transient one-dimensional charts. The cylinder is initially at a uniform temperature, T_0. At time $t = 0$ the entire surface is subjected to a fluid with constant ambient temperature T_∞, and the convective-heat-transfer coefficient between the cylinder surface area and fluid is a constant value, h.

The radial temperature distribution for an infinitely long cylinder is given in Fig. 3-7. For a cylinder with finite length the radial and axial

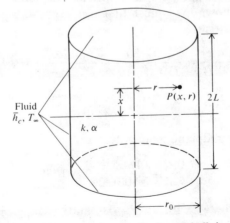

Figure 3-9 Geometry for a cylinder with finite length.

temperature distribution is given by the product solution of an infinitely long cylinder and infinite plate:

$$\frac{\theta_p(r,x)}{\theta_0} = C(r)P(x)$$

where the symbols $C(r)$ and $P(x)$ are the dimensionless temperatures for the infinite cylinder and infinite plate, respectively:

$$C(r) = \frac{\theta(r,t)}{\theta_0}$$

$$P(x) = \frac{\theta(x,t)}{\theta_0}$$

The solution for $C(r)$ is obtained from Fig. 3-7(a) and (b) while the value for $P(x)$ is obtained from Fig. 3-6(a) and (b).

Solutions for other two- and three-dimensional geometries may be obtained in a procedure similar to the one illustrated for the finite cylinder. Three-dimensional problems involve the product of three solutions while two-dimensional problems can be solved by taking the product of two solutions.

A summary of two-dimensional geometries that have chart solutions are given in Table 3-2. Three-dimensional solutions are outlined in Table 3-3. The symbols used in the two tables represent the following solutions:

$$S(x) = \frac{\theta(x,t)}{\theta_0} \qquad \text{for a semi-infinite solid, Fig. 3-4 or 3-5}$$

$$P(x) = \frac{\theta(x,t)}{\theta_0} \qquad \text{for an infinite plate, Fig. 3-6(a) and (b)}$$

$$C(r) = \frac{\theta(r,t)}{\theta_0} \qquad \text{for an infinite cylinder, Fig. 3-7(a) and (b)}$$

The extension of the one-dimensional charts to two- and three-dimensional geometries allows us to solve a surprisingly large variety of transient conduction problems.

Example 3-6. A 10-cm-diameter 16-cm-long cylinder with properties $k = 0.5$ W/m·K and $\alpha = 5 \times 10^{-7}$ m^2/s is initially at a uniform temperature of 20°C. The cylinder is placed in an oven where the ambient air temperature is 500°C and $\bar{h}_c = 30$ W/m^2·K. Determine the minimum and maximum temperatures in the cylinder 30 min after it has been placed in the oven.

Solution: The Biot number based on the cylinder radius is

$$\text{Bi} = \frac{\bar{h}_c r_0}{k} = \frac{30 \times 0.05}{0.5} = 3.0$$

Table 3-2 Product Solutions to Transient Conduction Problems Using Chart Information, Two-dimensional Systems

SITUATION	GEOMETRY	DIMENSIONLESS TEMPERATURE AT POINT P
Semi-infinite Plate		$\dfrac{\theta_p(x_1,x_2)}{\theta_0} = P(x_1)S(x_2)$
Infinite rectangular bar		$\dfrac{\theta_p(x_1,x_2)}{\theta_0} = P(x_1)P(x_2)$
One-quarter infinite solid		$\dfrac{\theta_p(x_1,x_2)}{\theta_0} = S(x_1)S(x_2)$
Semi-infinite cylinder		$\dfrac{\theta_p(x,r)}{\theta_0} = S(x)C(r)$
Finite cylinder		$\dfrac{\theta_p(x,r)}{\theta_0} = P(x)C(r)$

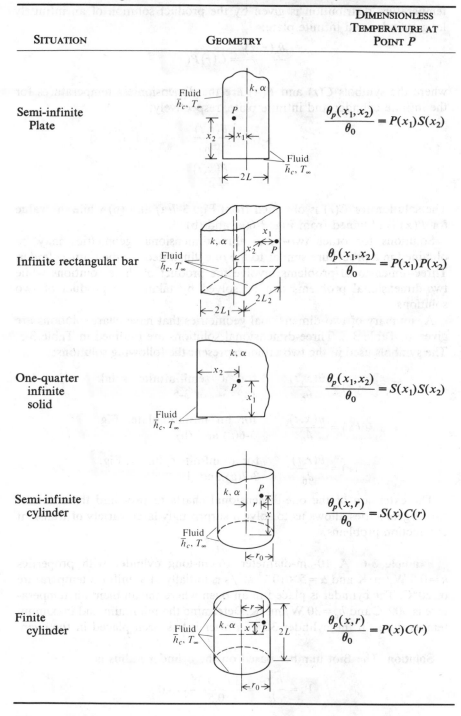

**Table 3-3   Product Solutions to Transient Conduction Problems
Using Chart Information, Three-dimensional System**

SITUATION	GEOMETRY	DIMENSIONLESS TEMPERATURE AT POINT P
Semi-infinite rectangular bar		$\dfrac{\theta_p(x_1,x_2,x_3)}{\theta_0} = S(x_1)P(x_2)P(x_3)$
Rectangular parallelepiped		$\dfrac{\theta_p(x_1,x_2,x_3)}{\theta_0} = P(x_1)P(x_2)P(x_3)$
One-quarter infinite plate		$\dfrac{\theta_p(x_1,x_2,x_3)}{\theta_0} = S(x_1)S(x_2)P(x_3)$
One-eighth infinite plate		$\dfrac{\theta_p(x_1,x_2,x_3)}{\theta_0} = S(x_1)S(x_2)S(x_3)$

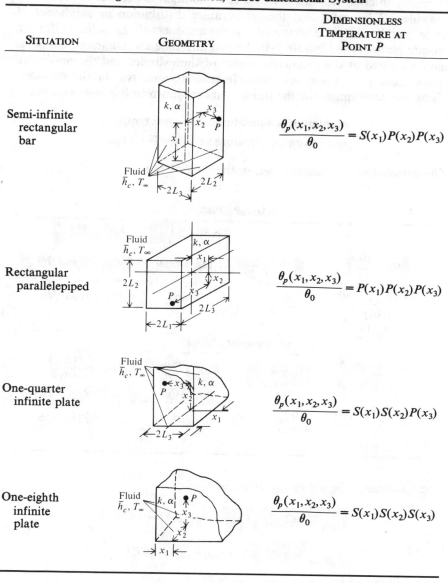

The problem cannot be solved using the simplified approach assuming negligible internal resistance, and a chart solution is necessary.

Table 3-2 indicates that the temperature distribution in a cylinder of finite length can be determined by the product of the solution for an infinite plate and an infinte cylinder. At any time the minimum temperature is located at the geometric center of the cylinder and the maximum temperature is at the outer circumference at each end of the cylinder. Using the coordinates for the finite cylinder shown in Fig. 3-9, we have

$$\text{minimum temperature at:} \quad x=0, r=0$$
$$\text{maximum temperature at:} \quad x=L, r=r_0$$

The calculations are summarized in the tables below.

Infinite Plate

$\text{Fo} = \dfrac{\alpha t}{L^2}$	$\text{Bi}^{-1} = \dfrac{k}{\bar{h}_c L}$	$P(0) = \dfrac{\theta(0,t)}{\theta_0}$ [Fig. 3-6(a)]	$P(L) = \dfrac{\theta(L,t)}{\theta_0}$ [Fig. 3-6(a) and (b)]
$\dfrac{(5 \times 10^{-7}) \times 1800}{(0.08)^2}$ $= 0.14$	$\dfrac{0.5}{30 \times 0.08} = 0.21$	0.90	$0.90 \times 0.27 = 0.249$

Infinite Cylinder

$\text{Fo} = \dfrac{\alpha t}{r_0^2}$	$\text{Bi}^{-1} = \dfrac{k}{\bar{h}_c r_0}$	$C(0) = \dfrac{\theta(0,t)}{\theta_0}$ [Fig. 3-7(a)]	$C(r_0) = \dfrac{\theta(r_0,t)}{\theta_0}$ [Fig. 3-7(a) and (b)]
$\dfrac{(5 \times 10^{-7}) \times 1800}{(0.05)^2}$ $= 0.36$	$\dfrac{0.5}{30 \times 0.05} = 0.33$	0.47	$0.47 \times 0.33 = 0.155$

The minimum cylinder temperature is

$$\frac{\theta_{\min}}{\theta_0} = P(0) \cdot C(0) = 0.90 \times 0.47 = 0.423$$

$$T_{\min} = 0.423(20 - 500) + 500 = 297°\text{C}$$

The maximum cylinder temperature is

$$\frac{\theta_{\max}}{\theta_0} = P(L)C(r_0) = 0.249 \times 0.155 = 0.039$$

$$T_{\max} = 0.039(20 - 500) + 500 = 481°\text{C}$$

3-5 NUMERICAL SOLUTIONS TO TRANSIENT CONDUCTION PROBLEMS

CONDITIONS:
~~1. ONE DIMENSIONAL~~

Explicit Method

2. NO HEAT GENERATION
3. CONSTANT PROPERTIES

The numerical approach to transient conduction problems is similar to the one used in Section 2-7 for steady conduction. The solid is first subdivided into a number of sections. A fictitious node is placed at the center of each section. An energy balance performed on each node results in an algebraic equation for the temperature of each node in terms of neighboring nodal temperatures, geometry, and thermal properties of the solid. The additional factor that must be considered in a transient problem is the energy stored in the material represented by each node. The stored energy is reflected as an increase in internal energy of the node, and the thermodynamic property that regulates the stored energy is the specific heat, c.

Let us first consider a one-dimensional problem involving an interior node as shown in Fig. 3-10. We can extend the analysis to two- and three-dimensional problems later. A verbal expression of the conservation of energy for node 0 surrounded by nodes 1 and 2 without internal energy generation is

$$\sum_{i=1}^{2} q_{i \to 0} = \left(\begin{array}{l} \text{time rate of change in internal} \\ \text{energy of node 0} \end{array} \right) \qquad (3\text{-}23)$$

Equation 3-23 is similar to Eq. 2-93 except that rate of change in the internal energy has been added for the transient problem. For a steady problem the increase in internal energy is zero.

When symbols are substituted into Eq. 3-23, we get

$$q_{1 \to 0} + q_{2 \to 0} = \frac{\partial U_0}{\partial t} \qquad (3\text{-}24)$$

where U_0 is the internal energy of node 0.

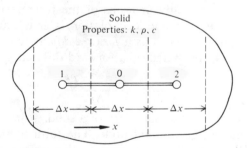

Figure 3-10 Arrangement of nodes for an interior section of a one-dimensional solid.

The conduction terms in Eq. 3-24 can be approximated by the finite-difference form of the Fourier law:

$$q_{1 \to 0} \simeq kA \frac{T_1^t - T_0^t}{\Delta x}$$

$$q_{2 \to 0} \simeq kA \frac{T_2^t - T_0^t}{\Delta x} \tag{3-25}$$

The superscript t in these terms refers to the fact that the temperatures are to be evaluated at the time t. The subscripts refer to the location of the nodes. Therefore, the subscripts specify spatial or x variation, while the superscripts specify time or t variation in the temperatures.

The increase in the internal energy of node 0 assuming constant density and specific heat for the material is

$$\frac{\partial U_0}{\partial t} \simeq mc \frac{\Delta T_0}{\Delta t} = \rho A \, \Delta x c_{\rho} \frac{T_0^{t+\Delta t} - T_0^t}{\Delta t} \tag{3-26}$$

Substituting Eqs. 3-25 and 3-26 into Eq. 3-24 yields the energy-balance equation for node 0:

$$kA \frac{T_1^t - T_0^t}{\Delta x} + kA \frac{T_2^t - T_0^t}{\Delta x} = \rho A \, \Delta x c \frac{T_0^{t+\Delta t} - T_0^t}{\Delta t}$$

Solving for $T_0^{t+\Delta t}$ gives

$$T_0^{t+\Delta t} = (\text{Fo})(T_1^t + T_2^t) + [1 - 2(\text{Fo})] T_0^t \tag{3-27}$$

where the Fourier number is defined as

$$\text{Eq (2-7)} \quad \alpha = \frac{K}{\rho c_p} \qquad \text{Fo} = \frac{\alpha(\Delta t)}{(\Delta x)^2} = \frac{K \, (\Delta t)}{\rho c_p \, (\Delta x)^2}$$

Equation 3-27 shows that the future temperature at time $t + \Delta t$ at an arbitrary interior node 0 can be evaluated by knowing the current temperatures at time t at the node 0 and its neighboring nodes. An equation similar to Eq. 3-27 can be written for all interior nodes, resulting in algebraic equations for the temperatures of the n interior nodes. Each of the nodal equations is written explicitly in terms of the future temperature of that node; hence this type of numerical solution is called an *explicit method*. The method is also referred to as a *forward-difference method* because the time derivative is approximated by a forward difference in time:

$$\frac{\partial T_i}{\partial t} \simeq \frac{T_i^{t+\Delta t} - T_i^t}{\Delta t}$$

If a node is located on the boundary of a solid, a special energy-balance equation must be written for that node. The form of the energy balance depends upon the boundary condition at the surface. One of the most commonly encountered boundary conditions involves convection from the

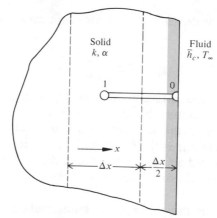

Figure 3-11 Arrangement of nodes for a one-
dimensional solid in contact with a fluid.

surface. We again consider a one-dimensional problem for which the node
0 resides on the surface as shown in Fig. 3-11, the energy balance for node
0 is

$$q_{1\to 0} + q_{\infty\to 0} = \frac{\partial U_0}{\partial t}$$

or

$$kA\frac{T_1{}^t - T_0{}^t}{\Delta x} + \bar{h}_c A(T_\infty{}^t - T_0{}^t) = \frac{\rho A \Delta x c}{2}\frac{T_0{}^{t+\Delta t} - T_0{}^t}{\Delta t} \qquad (3\text{-}28)$$

Notice that the volume of the surface node is $(\Delta x)A/2$ because the node 0
has only one-half the width of an interior node. All interior nodes have a
width of Δx while the boundary nodes are $\Delta x/2$ wide. CONVECTIVE

Solving Eq. 3-28 for the future temperature of the surface node, we get

$$T_0{}^{t+\Delta t} = 2(\text{Fo})\left[T_1{}^t + (\text{Bi})T_\infty{}^t\right]$$

$$+\left[1 - 2(\text{Fo}) - 2(\text{Fo})(\text{Bi})\right]T_0{}^t \qquad (3\text{-}29)$$

where

$$\text{Fo} = \frac{\alpha(\Delta t)}{(\Delta x)^2}$$

$$\text{Bi} = \frac{\bar{h}_c(\Delta x)}{k}$$

If the nodes are placed close together, the amount of mass that the surface
node represents is small and it becomes possible to neglect the energy
stored in the surface node; that is, the heat capacity of the surface node

can be neglected. For the case of negligible heat capacity, $T_0'^{t+\Delta t} = T_0'$ and Eq. 3-29 reduces to

$$T_0' = \frac{T_1' + (\text{Bi})\,T_\infty'}{1 + \text{Bi}} \qquad (3\text{-}30)$$

If the capacitance of the surface node is not neglected and Eq. 3-29 is used to calculate the temperature history of the surface, a knowledge of the current temperatures in the solid at the surface and one node removed from the surface determines the future surface temperature. If the surface capacitance is neglected and Eq. 3-30 is used, the current values of T_1 and T_∞ are sufficient to determine the current value of the surface temperature, T_0'.

A knowledge of the initial temperature distribution in a solid is necessary before a transient conduction problem can be solved by a numerical technique. The usual form of the initial condition is the specification of the initial temperatures in the solid. Often the solid is originally isothermal, which can easily be satisfied by setting all initial nodal temperatures equal to the known initial temperature. The numerical solution then proceeds by calculating the temperatures at time Δt by using Eq. 3-27 for all interior nodes and Eq. 3-29 for the surface node if the boundary transfers heat to a fluid at T_∞. With the temperatures known at Δt, the process is repeated to calculate the complete temperature distribution at time $2\Delta t$. The procedure is repeated until a time is reached at which the temperature distribution is needed.

At this point the choices for the spacing Δx between the nodal points and the time interval Δt appear to be completely at the discretion of the individual. However, certain values of Δx and Δt will lead to results that violate basic thermodynamic principles. For example, let us assume that the temperatures at an arbitrary time for three neighboring interior nodes in Fig. 3-10 are $T_1 = 100°C$, $T_2 = 100°C$, and $T_0 = 50°C$. We wish to calculate the temperature at node zero at the next time interval using Eq. 3-27. Furthermore, assume that we select values for Δx and Δt which, along with the given thermal diffusivity of the material, are such that the Fourier number is

$$\text{Fo} = \frac{\alpha\,\Delta t}{(\Delta x)^2} = 1.0$$

Substituting this value for Fourier number into Eq. 3-27 results in a future temperature for node zero of

$$T_0'^{t+\Delta t} = T_1' + T_2' - T_0'$$

or

$$T_0'^{t+\Delta t} = 100 + 100 - 50 = 150°C$$

But we know that the node zero cannot have a temperature in excess of

100°C because the heat supplied to node zero came from regions at 100°C. The fact that the numerical technique indicates that the temperature of node zero is above 100°C represents a violation of the second law of thermodynamics. If we were to try other values of the Fourier number to see what effect their values have on the future temperature of node zero, we would determine that the results predicted by Eq. 3-27 would not violate the second law of thermodynamics if

$$\text{INTERIOR NODE} \qquad \text{Fo} \leqslant \tfrac{1}{2} \qquad\qquad (3\text{-}31)$$

or

$$1 - 2(\text{Fo}) > 0$$

The restriction on the size of the Fourier number is often referred to as a *stability limit*. If the Fourier number exceeds one-half, the solution for the temperatures is said to be *unstable*. We notice that the stability criteria requires that the coefficient on the T_0' term in Eq. 3-27 be positive. This limit is a general result that can be proven mathematically (see Ref. 7).

The stability limit for the boundary nodes are different from the stability limits for the interior nodes. If we wish to determine the stability limit for a surface node that transfers heat to a fluid by convection, we can require the coefficient of the T_0' term in Eq. 3-29 be positive:

$$1 - 2(\text{Fo}) - 2(\text{Fo})(\text{Bi}) \geqslant 0$$

or BOUNDARY

$$\text{NODE} \qquad \text{Fo}(1 + \text{Bi}) \leqslant \tfrac{1}{2} \qquad\qquad (3\text{-}32)$$

For a particular conduction problem, the nodal equations for both the interior and boundary nodes must be stable. For example, if we selected $\text{Fo} = \tfrac{1}{4}$ to make the temperatures for the interior nodes stable according to the restriction of Eq. 3-31, the stability requirement for the boundary node would be $\text{Bi} \leqslant 1.0$, as given by Eq. 3-32. If the Fourier had been set equal to $\tfrac{1}{2}$, it would have been impossible to meet the stability condition for the boundary node because Eq. 3-32 would have required a negative Biot number.

Energy-balance equations for interior nodes in two- and three-dimensional problems can be easily derived by using a procedure similar to the one used to develop Eq. 3-27. Energy-balance equations can also be derived for two- and three-dimensional surface nodes under various boundary conditions.

A summary of interior and surface energy-balance equations is given in Table 3-4. A stability criterion is given with each equation. The stability criterion must be met for each equation to provide stable temperature results. Each nodal equation is written as though the node under consideration is called node 0 and the surrounding nodes are numbered as indicated in the individual figures shown in the table.

Table 3-4 Explicit Nodal Equations and Stability Requirements for Selected Geometries

SITUATION	GEOMETRY	NODAL EQUATION	STABILITY CRITERIA
One-dimensional geometry, interior node		$T_0^{t+\Delta t} = \mathrm{Fo}(T_1' + T_2') + [1 - 2(\mathrm{Fo})]T_0'$	$\mathrm{Fo} \leq \frac{1}{2}$
Two-dimensional geometry, interior node, square grid		$T_0^{t+\Delta t} = \mathrm{Fo}(T_1' + T_2' + T_3' + T_4')$ $+ [1 - 4(\mathrm{Fo})]T_0'$	$\mathrm{Fo} \leq \frac{1}{4}$
Three-dimensional geometry, interior node, cubic grid		$T_0^{t+\Delta t} = \mathrm{Fo}(T_1' + T_2' + T_3' + T_4' + T_5' + T_6')$ $+ [1 - 6(\mathrm{Fo})]T_0'$	$\mathrm{Fo} \leq \frac{1}{6}$
One-dimensional geometry, surface node, convection at boundary		$T_0^{t+\Delta t} = 2(\mathrm{Fo})[T_1' + (\mathrm{Bi})T_\infty']$ $+ [1 - 2(\mathrm{Fo}) - 2(\mathrm{Fo})(\mathrm{Bi})]T_0'$	$\mathrm{Fo}(1 + \mathrm{Bi}) < \frac{1}{2}$

Two-dimensional geometry, surface node, convection at boundary

$$T_0'^{t+\Delta t} = 2(Fo)\left[T_1' + \frac{T_2'}{2} + \frac{T_3'}{2} + (Bi)T_\infty'\right]$$
$$+ [1 - 4(Fo) - 2(Fo)(Bi)]T_0'$$

$$Fo(2+Bi) \le \tfrac{1}{2}$$

Two-dimensional geometry, surface node, exterior corner, convection at boundary

$$T_0'^{t+\Delta t} = 2(Fo)[T_1' + T_2' + 2(Bi)T_\infty']$$
$$+ [1 - 4(Fo) - 4(Fo)(Bi)]T_0'$$

$$Fo(1+Bi) \le \tfrac{1}{4}$$

Two-dimensional geometry, surface node, interior corner, convection at boundary

$$T_0'^{t+\Delta t} = \tfrac{4}{3}(Fo)\left[T_2' + T_3' + \frac{T_1'}{2} + \frac{T_4'}{2} + (Bi)T_\infty'\right]$$
$$+ [1 - 4(Fo) - \tfrac{4}{3}(Fo)(Bi)]T_0'$$

$$Fo(3+Bi) \le \tfrac{3}{4}$$

Example 3-7. A large, thick plate has a uniform temperature of 200°C. The surface of the plate is suddenly submerged in a fluid with an ambient temperature of 100°C and the convective-heat-transfer coefficient between the fluid and surface of the plate is 500 W/m²·K. The plate properties are $k=40$ W/m·K, $\alpha=3\times10^{-5}$ m²/s. Determine the temperature history of the plate for the first minute after the plate is submerged in the fluid.

Do 5
ITERATIONS

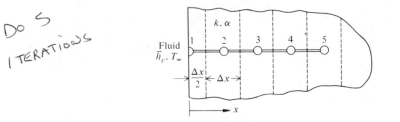

Solution: The plate is subdivided into nodes as shown in the figure and the nodes are numbered starting with one at the surface. If the surface capacitance is included, the future temperature of the surface node is given by Eq. 3-29:

$$T_1^{t+\Delta t}=2(\text{Fo})\left[T_2^t+(\text{Bi})T_\infty\right]+\left[1-2(\text{Fo})-2(\text{Fo})(\text{Bi})\right]T_1^t$$

All other nodes are interior nodes, so Eq. 3-27 applies for an arbitrary node designated by the subscript i:

$$T_i^{t+\Delta t}=(\text{Fo})\left[T_{i-1}^t+T_{i+1}^t\right]+\left[1-2(\text{Fo})\right]T_i^t$$

The stability limit for the surface node is

$$\text{Fo}(1+\text{Bi})\leqslant\tfrac{1}{2}$$

and for the internal nodes it is

$$\text{Fo}\leqslant\tfrac{1}{2}$$

Suppose that we select

$$\text{Fo}=\frac{\alpha\Delta t}{(\Delta x)^2}=\tfrac{1}{4}$$

$$\text{Bi}=\frac{\bar{h}_c\Delta x}{k}=\tfrac{1}{2}$$

to satisfy these two requirements. Once a value for the Biot number is selected, the spatial interval Δx is known:

$$\Delta x=\frac{k}{2\bar{h}_c}=\frac{40}{2\times500}=0.04 \text{ m}=4 \text{ cm}$$

The time interval Δt is restricted by the value of the Fourier number:

$$\Delta t = \frac{(\Delta x)^2}{4\alpha} = \frac{(0.04)^2}{4 \times (3 \times 10^{-5})} = 13.33 \text{ s}$$

The choices of the Fourier and Biot numbers were arbitrary within the limits of the stability criteria. We could have approached the problem differently. We could have selected values for Δx and Δt and then calculated values for Fo and Bi. If the calculated values for Fo and Bi had not met the stability requirements, the choices for Δx and Δt would have had to be adjusted so that they satisfied both stability requirements.

The solution can now proceed by calculating temperatures at nodes spaced 4 cm apart at times equally spaced on 13.33-s intervals. Temperatures at intermediate locations or times can be determined by interpolation. The temperatures at all internal nodes are evaluated from

$$T_i^{t+\Delta t} = \tfrac{1}{4}\left(T_{i-1}^t + T_{i+1}^t\right) + \tfrac{1}{2}T_i^t$$

and the surface-node temperature is evaluated from the equation

$$T_1^{t+\Delta t} = \frac{1}{2}\left(T_2^t + \frac{T_\infty}{2}\right) + \frac{T_1^t}{4}$$

The resulting temperatures are summarized in the table.

		TEMPERATURES (°C)			
TIME (s)	NODE 1: $x=0$	NODE 2: $x=4$ cm	NODE 3: $x=8$ cm	NODE 4: $x=12$ cm	NODE 5: $x=16$ cm
0.	200.	200.	200.	200.	200.
13.33	175.	200.	200.	200.	200.
26.67	168.8	193.9	200.	200.	200.
40.00	164.1	189.1	198.5	200.	200.
53.33	160.6	185.2	196.5	199.6	200.
66.67	157.8	181.9	194.5	198.9	199.9

Example 3-7 can be solved analytically and the solution is given by Eq. 3-19. As a check on the accuracy of the numerical solution, the table below compares the values for the temperatures given by Eq. 3-19 and the numerical solution at the time 66.67 s.

Temperatures (°C) at Time 66.67 for Example 3-7

NODE	x (cm)	NUMERICAL SOLUTION	EXACT SOLUTION (EQ. 3-19)
1	0	157.8	158.7
2	4	181.9	182.2
3	8	194.5	194.2
4	12	198.9	198.6
5	16	199.9	199.8

Graphical Interpretation of the Numerical Method OMIT

A rather simple graphical solution to transient conduction problems results when the Fourier number is selected equal to one-half. For an interior node, the nodal equation is given by Eq. 3-27. If $Fo = \frac{1}{2}$, the equation simplifies to

$$T_0^{t+\Delta t} = \tfrac{1}{2}(T_1^t + T_2^t) \tag{3-33}$$

This equation states that a future temperature of an interior node is the arithmetic average of the temperatures of the two neighboring nodes at the current time. The process of graphically determining the future temperature at an interior node using Eq. 3-33 is illustrated in Fig. 3-12. The temperature at time $t + \Delta t$ at node 0 is determined by drawing a straight line between points T_1^t and T_2^t.

The graphical construction of the temperatures described here is called the *Binder-Schmidt method*. The method suffers from the disadvantage of giving the same value for the temperature at the nodes for two consecutive time intervals. This unreasonable behavior results from the necessity of selecting the value of the Fourier number exactly equal to the stability limit of one-half. Extensions of the Binder-Schmidt graphical method have been established using values of the Fourier number different from one-half (Ref. 8). These improved methods result in more geometrical construction, but they provide more realistic temperature histories.

The Binder-Schmidt method is limited to one-dimensional geometries. It can provide solutions for composite surfaces and for cases where heat is transferred by convection to the surface of the solid. See Jakob (Ref. 9, chap. 19) for details on these techniques. Even though the graphical

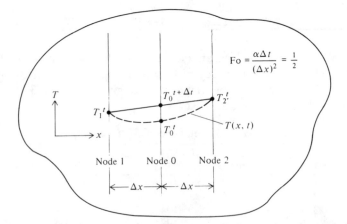

Figure 3-12 Binder-Schmidt method applied to an interior node.

method is limited in its application and is subject to inaccuracies, it is simple to perform and provides a visual history of the temperature distribution. The method is illustrated in the following example.

Example 3-8. A large plate has a thermal diffusivity of 5×10^{-5} m^2/s and a thickness of 0.30 m. The plate is initially isothermal at 0°C. Both surfaces suddenly are placed in direct thermal contact with a 500°C solid. Determine the temperature history of the plate during the first 100 s of contact with the hot solid.

Solution: The plate is drawn to scale and subdivided into six equal subdivisions. The resulting seven nodes are each 5 cm apart. The vertical axis is scaled from the original temperature of 0°C to the maximum temperature of 500°C, as shown in the figure.

The time interval Δt is fixed by the selection of the Fourier number:

$$\text{Fo} = \frac{1}{2} = \frac{\alpha \Delta t}{(\Delta x)^2}$$

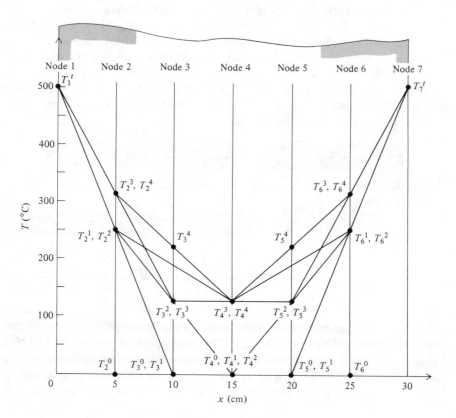

or

$$\Delta t = \frac{(\Delta x)^2}{2(\alpha)} = \frac{(0.05)^2}{2 \times (5 \times 10^{-5})} = 25 \text{ s}$$

The Binder-Schmidt method now proceeds as shown in the figure, where each nodal temperature is identified on the figure. The symbol T_3^2 represents the temperature at node 3 at time $2\Delta t = 50$ s. The temperatures at nodes 1 through 4 at the end of the 100-s time period are:

$$T_1 = 500°C$$
$$T_2 = 312°C$$
$$T_3 = 218°C$$
$$T_4 = 125°C$$

Implicit Method

The major disadvantage of the explicit numerical method is that each nodal equation must satisfy a stability criteria. To satisfy the stability limit it is often necessary to select very small time steps, thereby increasing the number of calculations. This section presents a slightly different numerical method, which is stable for all values of Fourier and Biot numbers.

Consider the case of a two-dimensional geometry and an interior node as shown in Fig. 3-10. The explicit form of the nodal equation for this situation is given by Eq. 3-27. If the energy balance used to derive Eq. 3-27 were modified and written in terms of the temperatures evaluated at $t + \Delta t$, the energy balance for node 0 would become

$$kA \frac{T_1^{t+\Delta t} - T_0^{t+\Delta t}}{\Delta x} + kA \frac{T_2^{t+\Delta t} - T_0^{t+\Delta t}}{\Delta x} = \rho A \Delta x c \frac{T_0^{t+\Delta t} - T_0^{t}}{\Delta t}$$

Rearranging this equation and expressing it in terms of the Fourier number results in

$$[1 + 2(\text{Fo})] T_0^{t+\Delta t} - (\text{Fo})(T_1^{t+\Delta t} + T_2^{t+\Delta t}) - T_0^{t} = 0 \qquad (3\text{-}34)$$

where as usual

$$\alpha = \frac{k}{\rho c_p} \qquad \text{Fo} = \frac{\alpha(\Delta t)}{(\Delta x)^2} = \frac{k(\Delta t)}{\rho c_p (\Delta x)^2}$$

When the nodal equation is written in this form, we see that the future temperature of an interior node depends on the future temperatures of the two adjacent nodes, both of which are unknown. Therefore, all nodal equations must be written and solved simultaneously for the temperature distribution in the solid. The method is called an *implicit method*, as opposed to an explicit method, in which each nodal equation can be solved

$$T_0^{t} = \left[1 + 2(F_0)\right] T_0^{t+\Delta t} - (F_0)\left(T_1^{t+\Delta t} + T_2^{t+\Delta t}\right)$$

explicitly for the local temperature. The method is also referred to as a *backward-difference technique* because the time derivative is approximated on the basis of a time interval which looks backward in time.

The implicit numerical method is stable for all spacing and time intervals. However, the choice of smaller time and spacing intervals will lead to more accurate temperatures, because the truncation errors associated with the difference between derivatives and finite differences will be reduced.

The lack of a stability criteria is a distinct advantage for the implicit method. The disadvantage of the implicit method is the necessity of simultaneously solving a set of algebraic equations. If the number of equations is large, relaxation techniques or matrix-inversion methods that utilize a digital computer are helpful.

Forward- and backward-difference methods have been discussed here. A finite difference based on the arithmetic average of the forward and backward differences is covered in Crank (Ref. 10) and Richtmyer and Morton (Ref. 11). A still more general formulation results from expressing the temperature/time derivative by a weighted average of the forward and backward differences (Ref. 11). "Hopscotch methods," which are a combination of implicit and explicit methods, are presented in References 12, 13, and 14.

The implicit forms of the energy-balance equations for boundary nodes are derived in a manner similar to that used for the explicit method, except that the current value for the nodal temperature is considered to be time $t + \Delta t$. For example, consider a one-dimensional solid with surface node 0 transferring heat by convection to a fluid with ambient temperature T_∞. The convective-heat-transfer coefficient is $\bar{h}_c$ and the geometry is shown in Fig. 3-11. An energy balance for the surface node is

$$kA\frac{T_1^{t+\Delta t} - T_0^{t+\Delta t}}{\Delta x} + \bar{h}_c A\left(T_\infty^{t+\Delta t} - T_0^{t+\Delta t}\right) = \frac{\rho A \, \Delta x c}{2}\frac{\left(T_0^{t+\Delta t} - T_0^t\right)}{\Delta t}$$

$$(3\text{-}35)$$

Comparison of Eqs. 3-28 and 3-35 illustrates the difference between the energy balances for the explicit and implicit approaches. Rearranging Eq. 3-35 produces

$$\left[1 + 2(\text{Fo})(1 + \text{Bi})\right]T_0^{t+\Delta t} - 2(\text{Fo})\left[T_1^{t+\Delta t} + (\text{Bi})T_\infty^{t+\Delta t}\right] - T_0^t = 0$$

$$(3\text{-}36)$$

Nodal equations can be derived for surface nodes with other types of boundary conditions and geometry. A summary of these equations is given in Table 3-5.

Once an energy-balance equation is written for each node and simplified, the set of n equations for the n nodes must be solved for the

Table 3-5   Implicit Nodal Equations for Selected Geometries

SITUATION	GEOMETRY	NODAL EQUATION
One-dimensional geometry, interior node	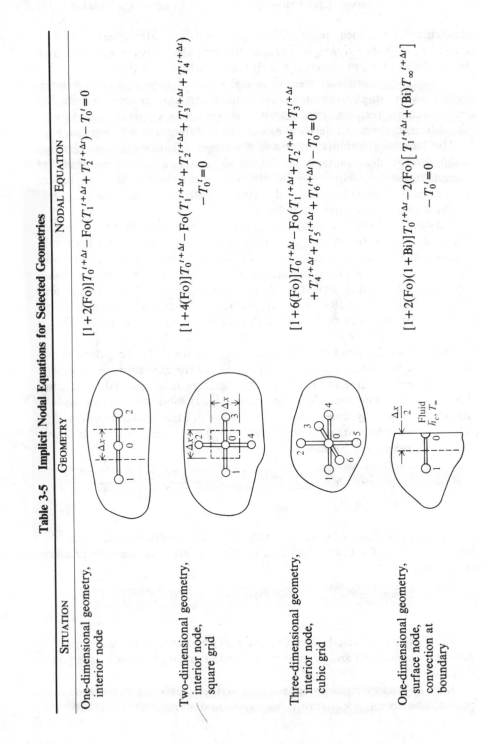	$[1 + 2(\text{Fo})]T_0^{t+\Delta t} - \text{Fo}(T_1^{t+\Delta t} + T_2^{t+\Delta t}) - T_0^t = 0$
Two-dimensional geometry, interior node, square grid		$[1 + 4(\text{Fo})]T_0^{t+\Delta t} - \text{Fo}(T_1^{t+\Delta t} + T_2^{t+\Delta t} + T_3^{t+\Delta t} + T_4^{t+\Delta t})$ $- T_0^t = 0$
Three-dimensional geometry, interior node, cubic grid		$[1 + 6(\text{Fo})]T_0^{t+\Delta t} - \text{Fo}(T_1^{t+\Delta t} + T_2^{t+\Delta t} + T_3^{t+\Delta t}$ $+ T_4^{t+\Delta t} + T_5^{t+\Delta t} + T_6^{t+\Delta t}) - T_0^t = 0$
One-dimensional geometry, surface node, convection at boundary		$[1 + 2(\text{Fo})(1+\text{Bi})]T_0^{t+\Delta t} - 2(\text{Fo})\big[T_1^{t+\Delta t} + (\text{Bi})T_\infty^{t+\Delta t}\big]$ $- T_0^t = 0$

178

Two-dimensional geometry, surface node, convection at boundary

$$[1 + 2(\text{Fo})(2 + \text{Bi})]T_0^{t+\Delta t} - 2(\text{Fo})\left[\frac{T_2^{t+\Delta t}}{2} + \frac{T_3^{t+\Delta t}}{2} + T_1^{t+\Delta t} + (\text{Bi})T_\infty^{t+\Delta t}\right] - T_0^t = 0$$

Two-dimensional geometry, surface node, exterior corner, convection at boundary

$$[1 + 4(\text{Fo})(1 + \text{Bi})]T_0^{t+\Delta t} - 4(\text{Fo})\left[\frac{T_1^{t+\Delta t}}{2} + \frac{T_2^{t+\Delta t}}{2} + (\text{Bi})T_\infty^{t+\Delta t}\right] - T_0^t = 0$$

Two-dimensional geometry, surface node, interior corner, convection at boundary

$$\left[1 + 4(\text{Fo})\left(1 + \frac{\text{Bi}}{3}\right)\right]T_0^{t+\Delta t} - \frac{4(\text{Fo})}{3}\left[\frac{T_1^{t+\Delta t}}{2} + \frac{T_4^{t+\Delta t}}{2} + T_2^{t+\Delta t} + T_3^{t+\Delta t} + (\text{Bi})T_\infty^{t+\Delta t}\right] - T_0^t = 0$$

individual nodal temperatures. The procedure is illustrated in the next two examples. Both problems are solved by a computer program using a matrix-inversion technique. Example 3-9 involves a one-dimensional problem and Example 3-10 concerns a two-dimensional problem. As before, the computer program is written in a general form so that it can be applied to a broad range of problems involving an implicit numerical solution.

The number and complexity of computer programs presented in this chapter has been limited. The emphasis has been on keeping the programs relatively simple at the expense of requiring the user to perform routine energy-balance calculations before the program can be used. Users are therefore introduced to modern computational methods and computers without sacrificing their heat-transfer skills. Readers interested in more advanced heat-transfer programs are referred to Adams and Rogers (Ref. 15) and Schenck (Ref. 16).

A partial list of more advanced computer programs that can be used to solve heat-transfer and related problems is given in Appendix N. The list includes programs that can be applied to very complex problems for which the only practical means for a solution is by a high-speed computer.

A numerical procedure that has not been used in this text but which deserves to be mentioned is the *finite-element method*. This method is a very versatile and powerful numerical technique and it has several advantages over the finite-difference methods illustrated in the examples above. It is possible to write a very general finite-element program that can be applied to a wide variety of heat-transfer problems. A general finite-difference program that is capable of solving the same class of problems would be impractical to write. Irregularly shaped boundaries and mixed boundary conditions pose no particular problems when a finite-element procedure is used. Several excellent references on the finite-element method are Segerlind (Ref. 17), Huebner (Ref. 18), Myers (Ref. 19), and Zeinkiewicz (Ref. 20).

Example 3-9. Work Example 3-7 using an implicit numerical technique. The mass is a semi-infinite solid with

$$k = 40 \text{ W/m·K}$$

$$\alpha = 3 \times 10^{-5} \text{ m}^2/\text{s}$$

$$\bar{h}_c = 500 \text{ W/m}^2\text{·K}$$

$$T_\infty = 100°C$$

$$T_0 = 200°C$$

The geometry is shown in the figure.

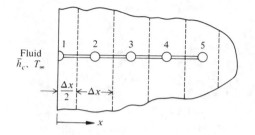

Solution: The plate is subdivided into equal segments, and nodes are placed at the center of each segment as they were in the explicit method. The surface node is denoted as node 1 and the implicit form of the energy balance for node 1 is given by Eq. 3-36:

$$[1+2(\text{Fo})(1+\text{Bi})]T_1^{t+\Delta t} - 2(\text{Fo})[T_2^{t+\Delta t} + (\text{Bi})T_\infty^{t+\Delta t}] - T_1^t = 0$$

The implicit form of the energy-balance equation for an arbitrary interior node i is, by Eq. 3-34:

$$[1+2(\text{Fo})]T_i^{t+\Delta t} - (\text{Fo})(T_{i-1}^{t+\Delta t} + T_{i+1}^{t+\Delta t}) - T_i^t = 0 \qquad (i=2,3,\dots)$$

We will use the matrix-inversion technique to solve the set of algebraic equations, and therefore we must express all the nodal equations in the form

$$\text{AT=B}$$

The boundary node equation written in this form is

$$[1+2(\text{Fo})(1+\text{Bi})]T_1^{t+\Delta t} - 2(\text{Fo})T_2^{t+\Delta t} = 2(\text{Fo})(\text{Bi})T_\infty + T_1^t$$

and for all interior nodes the energy balance in the matrix form is

$$-(\text{Fo})T_{i-1}^{t+\Delta t} + [1+2(\text{Fo})]T_i^{t+\Delta t} - (\text{Fo})T_{i+1}^{t+\Delta t} = T_i^t$$

If the time and space intervals are selected as

$$\Delta x = 0.04 \text{ m}$$
$$\Delta t = 13.333 \text{ s}$$

then

$$\text{Fo} = \frac{\alpha(\Delta t)}{(\Delta x)^2} = \frac{1}{4}$$

$$\text{Bi} = \frac{\bar{h}_c(\Delta x)}{k} = \frac{1}{2}$$

Suppose that we limit the solution to 10 nodes. The 10 nodal equations are then

Node 1: $1.75T_1^{t+\Delta t} - 0.5T_2^{t+\Delta t} = 25 + T_1^t$

Node 2: $-0.25T_1^{t+\Delta t} + 1.5T_2^{t+\Delta t} - 0.25T_3^{t+\Delta t} = T_2^t$

Node 3: $-0.25T_2^{t+\Delta t} + 1.5T_3^{t+\Delta t} - 0.25T_4^{t+\Delta t} = T_3^t$

Node 4: $-0.25T_3^{t+\Delta t} + 1.5T_4^{t+\Delta t} - 0.25T_5^{t+\Delta t} = T_4^t$

Node 5: $-0.25T_4^{t+\Delta t} + 1.5T_5^{t+\Delta t} - 0.25T_6^{t+\Delta t} = T_5^t$

Node 6: $-0.25T_5^{t+\Delta t} + 1.5T_6^{t+\Delta t} - 0.25T_7^{t+\Delta t} = T_6^t$

Node 7: $-0.25T_6^{t+\Delta t} + 1.5T_7^{t+\Delta t} - 0.25T_8^{t+\Delta t} = T_7^t$

Node 8: $-0.25T_7^{t+\Delta t} + 1.5T_8^{t+\Delta t} - 0.25T_9^{t+\Delta t} = T_8^t$

Node 9: $-0.25T_8^{t+\Delta t} + 1.5T_9^{t+\Delta t} - 0.25T_{10}^{t+\Delta t} = T_9^t$

Node 10: $-0.25T_9^{t+\Delta t} + 1.5T_{10}^{t+\Delta t} - 0.25T_{11}^{t+\Delta t} = T_{10}^t$

The 10 equations involve the temperatures of 11 nodes. The temperature of node 11 may be determined by applying a boundary condition which will specify that node 11 will remain at the original temperature of 200°C as long as the time of the solution is not great. We could notice from the solution of Example 3-7 that node 5 did not change temperature appreciably in the first minute of time. Therefore, setting the temperature of node 11 equal to T_0 does not lead to significant errors. The equation from node 10 now becomes

$$-0.25T_9^{t+\Delta t} + 1.5T_{10}^{t+\Delta t} = 50 + T_{10}^t$$

The matrix will be inverted by a computer subprogram called MATINV. The subprogram was used in Example 2-13 and a listing is given in Appendix M. The subprogram MATINV receives values of the matrix A and the number of nodal equations, N. It returns the values of the inverse of the matrix called C. The program listing is given below.

Program Listing for Example 3-9

```
      DIMENSION T(50),A(50,50),B(50),C(50,50)
      READ , N,NTIME,DELX,DELT,TO
      READ , ((A(I,J),J=1,N),I=1,N)
      WRITE (6,10) DELX,DELT,TO
   10 FORMAT(1H ,'*** TRANSIENT TEMPERATURE DISTRIBUTION IN DEGREES' ,/,
      1 'CELSIUS DETERMINED BY AN IMPLICIT NUMERICAL TECHNIQUE ***',/,
      2 'NODE SPACING=' ,F8.4,' METERS' ,/,'TIME INTERVAL=' ,F8.3,
      3 ' SECONDS',/,'ORIGINAL TEMPERATURE=',F8.2,' DEGREES C')
      CALL MATINV (A,N,C)
      DO 15 I=1,N
   15 T(I)=TO
      DO 80 JJ=1,NTIME
   20 B(1)=25.0+T(1)
      B(10)=50.0+T(10)
      KK=N-1
      DO 30 I=2,KK
   30 B(I)=T(I)
      DO 50 I=1,N
```

```
      SUM=0.0
      DO 40 J=1,N
  40  SUM=SUM+C(I,J)*B(J)
  50  T(I)=SUM
      AJ=JJ
      TIME=AJ*DELT
      WRITE (6,70) TIME, (I,T(I),I=1,N)
  70  FORMAT (/,25X,'TIME IS=',F10.3,' SECS',/,4('T(',I2,
     1 ')=',F8.2,2X))
  80  CONTINUE
      STOP
      END
```

The input values to the program are listed in the table below together with their their FORTRAN symbols.

Symbols Used for Input Data for Example 3-9

Symbol	Definition	Unit
N	Integer number of nodes	—
NTIME	Integer number of time steps for which temperatures are calculated	—
DELX	Nodal spacing	m
DELT	Time spacing	s
TØ	Original nodal temperatures	°C
A(I,J)	Elements a_{ij} of the matrix A $(i=1,2,\ldots,N, j=1,2,\ldots,n)$	

The first five input variables are to be placed on a single line. The next N lines contain the elements of the matrix A. Each line contains N values. The values for a_{ij} are given in the nodal equations shown above. The matrix is called a *tridiagonal matrix* because all the terms are zero except three: the diagonal term and two terms on either side of the diagonal.

The input to the program and program output are as follows:

Input to Example 3-9

```
10, 5, 0.04, 13.3333, 200.
1.75, −0.5, 0., 0., 0., 0., 0., 0., 0., 0.,
−0.25, 1.5, −0.25, 0., 0., 0., 0., 0., 0., 0.,
0., −0.25, 1.5, −0.25, 0., 0., 0., 0., 0., 0.,
0., 0., −0.25, 1.5, −0.25, 0., 0., 0., 0., 0.,
0., 0., 0., −0.25, 1.5, −0.25, 0., 0., 0., 0.,
0., 0., 0., 0., −0.25, 1.5, −0.25, 0., 0., 0.,
0., 0., 0., 0., 0., −0.25, 1.5, −0.25, 0., 0.,
0., 0., 0., 0., 0., 0., −0.25, 1.5, −0.25, 0.,
0., 0., 0., 0., 0., 0., 0., −0.25, 1.5, −0.25,
0., 0., 0., 0., 0., 0., 0., 0., −0.25, 1.5
```

Output to Example 3-9

*** TRANSIENT TEMPERATURE DISTRIBUTION IN DEGREES
CELSIUS DETERMINED BY AN IMPLICIT NUMERICAL
TECHNIQUE ***
NODE SPACING = 0.0400 METERS
TIME INTERVAL = 13.333 SECONDS
ORIGINAL TEMPERATURE = 200.00 DEGREES C

TIME IS = 13.333 SECS

T(1) = 184.98	T(2) = 197.42	T(3) = 199.56	T(4) = 199.92
T(5) = 199.99	T(6) = 200.00	T(7) = 200.00	T(8) = 200.00
T(9) = 200.00	T(10) = 200.00		

TIME IS = 26.667 SECS

T(1) = 175.40	T(2) = 193.96	T(3) = 198.65	T(4) = 199.71
T(5) = 199.94	T(6) = 199.99	T(7) = 200.00	T(8) = 200.00
T(9) = 200.00	T(10) = 200.00		

TIME IS = 40.000 SECS

T(1) = 168.90	T(2) = 190.35	T(3) = 197.38	T(4) = 199.35
T(5) = 199.85	T(6) = 199.97	T(7) = 199.99	T(8) = 200.00
T(9) = 200.00	T(10) = 200.00		

TIME IS = 53.333 SECS

T(1) = 164.21	T(2) = 186.92	T(3) = 195.88	T(4) = 198.83
T(5) = 199.69	T(6) = 199.92	T(7) = 199.98	T(8) = 200.00
T(9) = 200.00	T(10) = 200.00		

TIME IS = 66.667 SECS

T(1) = 160.62	T(2) = 183.75	T(3) = 194.24	T(4) = 198.17
T(5) = 199.46	T(6) = 199.85	T(7) = 199.96	T(8) = 199.99
T(9) = 200.00	T(10) = 200.00		

As a check of the accuracy of the implicit numerical method, the temperatures from the computer program are compared in the table below with the exact solution given by Eq. 3-19 for a time of 66.67s.

Comparison of Implicit Solution and Exact Solution to Example 3-9

	Temperatures (°C) at $t = 66.67$ s	
Node	Exact Solution	Implicit Solution
1	158.7	160.6
2	182.2	183.8
3	194.2	194.2
4	198.6	198.2
5	199.9	199.5

Example 3-10. A long, rectangular structural member whose cross section is shown in the figure is to be heated during a manufacturing process. Because of space limitations, the heat can only be applied to one surface of the member. A heater is to be placed in contact with one face until the temperature of the opposite face reaches a minimum value.

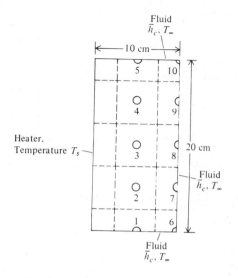

The structural member is originally at a uniform temperature of 50°C. It has a thermal diffusivity of 2.5×10^{-5} m^2/s and a thermal conductivity of 25 W/m·K. The other three surfaces are surrounded by 30°C air and the convective-heat-transfer coefficient on all three surfaces is a constant value of 75 W/m^2·K. The surface temperature of the heater is $T_s = 300$°C and it is placed in good thermal contact with the long side of the rectangle, as shown in the figure. Determine the amount of time that the heater must remain in place in order for the minimum temperature of the opposite face to be 120°C.

Solution: The rectangle is divided into a square grid such that $\Delta x = 0.05$ m and the nodes with unknown temperatures are identified with subscripts between 1 and 10.

The implicit form of the energy equation for all nodes is given in Table 3-5. For all interior nodes,

$$[1+4(\text{Fo})]\,T_i^{t+\Delta t} - \text{Fo}\big(T_{i-1}^{t+\Delta t} + T_{i+1}^{t+\Delta t} + T_{i+5}^{t+\Delta t} + T_s\big) - T_i^t = 0 \qquad (i = 2, 3, 4)$$

The energy equations for boundary nodes that are not on the corners of the solid are:

Node 1:

$$[1+2(\text{Fo})(2+\text{Bi})]\,T_1^{t+\Delta t}$$

$$-2(\text{Fo})\left[\frac{T_6^{t+\Delta t}}{2} + \frac{T_s}{2} + T_2^{t+\Delta t} + (\text{Bi})\,T_\infty\right] - T_1^t = 0$$

Node 5:

$$[1+2(\text{Fo})(2+\text{Bi})]T_5^{t+\Delta t}$$

$$-2(\text{Fo})\left[\frac{T_{10}^{t+\Delta t}}{2}+\frac{T_s}{2}+T_4^{t+\Delta t}+(\text{Bi})T_\infty\right]-T_5^t=0$$

Nodes 7–9:

$$[1+2(\text{Fo})(2+\text{Bi})]T_i^{t+\Delta t}$$

$$-2(\text{Fo})\left[\frac{T_{i-1}^{t+\Delta t}}{2}+\frac{T_{i+1}^{t+\Delta t}}{2}+T_{i-5}^{t+\Delta t}+(\text{Bi})T_\infty\right]-T_i^t=0 \qquad (i=7,8,9)$$

The energy equations for the corner boundary nodes are:
Node 6:

$$[1+4(\text{Fo})(1+\text{Bi})]T_6^{t+\Delta t}$$

$$-4(\text{Fo})\left[\frac{T_1^{t+\Delta t}}{2}+\frac{T_7^{t+\Delta t}}{2}+(\text{Bi})T_\infty\right]-T_6^t=0$$

Node 10:

$$[1+4(\text{Fo})(1+\text{Bi})]T_{10}^{t+\Delta t}$$

$$-4(\text{Fo})\left[\frac{T_5^{t+\Delta t}}{2}+\frac{T_9^{t+\Delta t}}{2}+(\text{Bi})T_\infty\right]-T_{10}^t=0$$

Suppose that we select a time interval of $\Delta t=30$ s; then

$$\text{Fo}=\frac{\alpha(\Delta t)}{(\Delta x)^2}=\frac{(2.5\times10^{-5})\times30}{(0.05)^2}=0.30$$

$$\text{Bi}=\frac{\bar{h}_c(\Delta x)}{k}=\frac{75\times0.05}{25}=0.15$$

Substitution of values for Bi, Fo, T_s, and T_∞ into the energy-balance equations yields 10 nodal equations in the matrix form

$$\mathbf{AT=B}$$

Node 1: $\quad 2.29T_1^{t+\Delta t}-0.6T_2^{t+\Delta t}-0.3T_6^{t+\Delta t}=92.7+T_1^t$

Node 2: $\quad -0.3T_1^{t+\Delta t}+2.2T_2^{t+\Delta t}-0.3T_3^{t+\Delta t}-0.3T_7^{t+\Delta t}=90+T_2^t$

Node 3: $\quad -0.3T_2^{t+\Delta t}+2.2T_3^{t+\Delta t}-0.3T_4^{t+\Delta t}-0.3T_8^{t+\Delta t}=90+T_3^t$

Node 4: $\quad -0.3T_3^{t+\Delta t}+2.2T_4^{t+\Delta t}-0.3T_5^{t+\Delta t}-0.3T_9^{t+\Delta t}=90+T_4^t$

Node 5: $\quad -0.6T_4^{t+\Delta t}+2.29T_5^{t+\Delta t}-0.3T_{10}^{t+\Delta t}=92.7+T_5^t$

Node 6: $\quad -0.6T_1^{t+\Delta t}+2.38T_6^{t+\Delta t}-0.6T_7^{t+\Delta t}=5.4+T_6^t$

Node 7: $\quad -0.6T_2^{t+\Delta t}-0.3T_6^{t+\Delta t}+2.29T_7^{t+\Delta t}-0.3T_8^{t+\Delta t}=2.7+T_7^t$

Node 8: $\quad -0.6T_3^{t+\Delta t}-0.3T_7^{t+\Delta t}+2.29T_8^{t+\Delta t}-0.3T_9^{t+\Delta t}=2.7+T_8^t$

Node 9: $\quad -0.6T_4^{t+\Delta t}-0.3T_8^{t+\Delta t}+2.29T_9^{t+\Delta t}-0.3T_{10}^{t+\Delta t}=2.7+T_9^t$

Node 10: $\quad -0.6T_5^{t+\Delta t}-0.6T_9^{t+\Delta t}+2.38T_{10}^{t+\Delta t}=5.4+T_{10}^t$

Coefficients of these 10 equations determine values for the square matrix A which are input values to the computer program. Values for the column matrix B are evaluated internally.

The program is identical to the one used in Example 3-9 except that the elements of matrix B are different. The values for the N elements of B are calculated by the FORTRAN statements between statements 20 and 30. These are the only statements that differ from the program used in Example 3-9. A listing of the program is given below.

Program Listing for Example 3-10

```
      DIMENSION T(50),A(50,50),B(50),C(50,50)
      READ , N,NTIME,DELX,DELT,TO
      READ , ((A(I,J),J=1,N),I=1,N)
      WRITE (6,10) DELX,DELT,TO
10    FORMAT(1H ,'*** TRANSIENT TEMPERATURE DISTRIBUTION IN DEGREES',/,
     1 'CELSIUS DETERMINED BY AN IMPLICIT NUMERICAL TECHNIQUE ***',//,
     2 'NODE SPACING=',F8.4,' METERS',/,'TIME INTERVAL=',F8.3,
     3 ' SECONDS',/,'ORIGINAL TEMPERATURE=',F8.2,' DEGREES C')
      CALL MATINV (A,N,C)
      DO 15 I=1,N
15    T(I)=TO
      DO 80 JJ=1,NTIME
20    B(1)=92.7+T(1)
      B(5)=92.7+T(5)
      B(6)=5.4+T(6)
      B(10)=5.4+T(10)
      DO 22 I=2,4
22    B(I)=90.0+T(I)
      DO 30 I=7,9
30    B(I)=2.7+T(I)
      DO 50 I=1,N
      SUM=0.0
      DO 40 J=1,N
40    SUM=SUM+C(I,J)*B(J)
50    T(I)=SUM
      AJ=JJ
      TIME=AJ*DELT
      WRITE (6,70) TIME, (I,T(I),I=1,N)
70    FORMAT (/,25X,'TIME IS=',F10.3,' SECS',/,4('T(',I2,
     1 ')=',F8.2,2X))
80    CONTINUE
      STOP
      END
```

The number of the nodes N = 10, the number of time intervals is selected as NTIME = 5, the node spacing DELX = 0.05 m, the time interval DELT = 30.0 s, and the original temperature T0 = 50.0°C. The input to the program and format of the input is identical to that described in Example 3-9. Input to this problem is:

Input to Example 3-10

10, 5, 0.05, 30.0, 50.0
2.29, −0.6, 0., 0., 0., −0.3, 0., 0., 0., 0.,
−0.3, 2.2, −0.3, 0., 0., 0., −0.3, 0., 0., 0.,
0., −0.3, 2.2, −0.3, 0., 0., 0., −0.3, 0., 0.,
0., 0., −0.3, 2.2, −0.3, 0., 0., 0., −0.3, 0.,
0., 0., 0., −0.6, 2.29, 0., 0., 0., 0., −0.3,
−0.6, 0., 0., 0., 0., 2.38, −0.6, 0., 0., 0.,
0., −0.6, 0., 0., 0., −0.3, 2.29, −0.3, 0., 0.,
0., 0., −0.6, 0., 0., 0., −0.3, 2.29, −0.3, 0.,
0., 0., 0., −0.6, 0., 0., 0., −0.3, 2.29, −0.3,
0., 0., 0., 0., −0.6, 0., 0., 0., −0.6, 2.38

The program output is:

Output to Example 3-10

*** TRANSIENT TEMPERATURE DISTRIBUTION IN DEGREES
CELSIUS DETERMINED BY AN IMPLICIT NUMERICAL
TECHNIQUE ***

NODE SPACING = .0500 METERS
TIME INTERVAL = 30.000 SECONDS
ORIGINAL TEMPERATURE = 50.00 DEGREES C

TIME IS = 30.000 SECS

(1) = 96.81	T(2) = 99.48	T(3) = 99.84	T(4) = 99.48
T(5) = 96.81	T(6) = 64.38	T(7) = 66.23	T(8) = 66.52
T(9) = 66.23	T(10) = 64.38		

TIME IS = 60.000 SECS

T(1) = 128.89	T(2) = 134.08	T(3) = 134.94	T(4) = 134.08
T(5) = 128.89	T(6) = 83.96	T(7) = 87.84	T(8) = 88.60
T(9) = 87.84	T(10) = 83.96		

TIME IS = 90.000 SECS

T(1) = 152.09	T(2) = 159.44	T(3) = 160.82	T(4) = 159.44
T(5) = 152.09	T(6) = 103.46	T(7) = 109.36	T(8) = 110.66
T(9) = 109.36	T(10) = 103.46		

TIME IS = 120.000 SECS

T(1) = 169.55	T(2) = 178.67	T(3) = 180.53	T(4) = 178.67
T(5) = 169.55	T(6) = 120.92	T(7) = 128.69	T(8) = 130.52
T(9) = 128.69	T(10) = 120.92		

TIME IS = 150.000 SECS

T(1) = 183.04	T(2) = 193.60	T(3) = 195.89	T(4) = 193.60
T(5) = 183.04	T(6) = 135.83	T(7) = 145.22	T(8) = 147.55
T(9) = 145.22	T(10) = 135.83		

The lowest temperatures in the rectangle occur at the corners (nodes 6 and 10) that are farthest from the heat source and have the largest surface exposed to the cool air. A plot of the temperatures of the most quickly

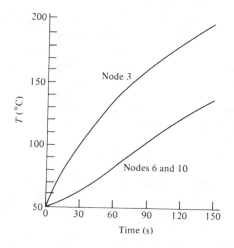

Time (s)

responding node (node 3) and the most slowly responding node (node 6) is shown in the figure. From the figure the time required for the minimum temperature to reach 120°C is approximately 2 min.

REFERENCES

1. H. S. Carslaw and J. C. Jaeger, *Conduction of Heat in Solids*, 2nd ed., Oxford University Press, Inc., New York, 1959.
2. E. R. G. Eckert and R. M. Drake, Jr., *Analysis of Heat and Mass Transfer*, McGraw-Hill Book Company, New York, 1972.
3. V. Arpaci, *Conduction Heat Transfer*, Addison-Wesley Publishing Co., Inc., Reading, Mass., 1966.
4. M. N. Ozisik, *Boundary Value Problems of Heat Conduction*, Intext Publishers Group, New York, 1968.
5. P. J. Schneider, *Conduction Heat Transfer*, Addison-Wesley Publishing Co., Inc., Reading, Mass., 1955.
6. Frank Kreith, *Principles of Heat Transfer*, 3rd ed., Crowell, New York, 1973.
7. R. D. Richtmyer, *Difference Methods for Initial-Value Problems*, John Wiley & Sons, Inc., New York, 1957.
8. B. Paul, "Generalization of the Schmidt Graphical Method for Transient Heat Conduction," *ARS J.*, vol. 32, p. 1098, 1962.
9. M. Jakob, *Heat Transfer*, vol. 2, John Wiley & Sons, Inc., New York, 1949.
10. J. Crank, *The Mathematics of Diffusion*, 2nd ed., Oxford University Press, Inc., New York, 1975.
11. R. D. Richtmyer and K. W. Morton, *Difference Methods for Initial Value Problems*, John Wiley & Sons, Inc., New York, 1967.
12. A. R. Gourlay and G. R. McGuire, "General Hopscotch Algorithm for the Numerical Solution of Partial Differential Equations," *J. Inst. Math. and Its Applications*, vol. 7, pp. 216–227, 1971.

13. A. R. Gourlay, "Hopscotch: A Fast Second-Order Partial Differential Equation Solver," *J. Inst. Math. and Its Applications*, vol. 6, pp. 375–390, 1970.

14. A. R. Gourlay, "Some Recent Methods for the Numerical Solution of Time-dependent Partial Differential Equations," *Proc. Roy. Soc. London*, vol. 323, pp. 219–235, 1971.

15. J. A. Adams and D. F. Rogers, *Computer-aided Heat Transfer Analysis*, McGraw-Hill Book Company, New York, 1973.

16. H. Schenck, Jr., *Fortran Methods in Heat Flow*, The Ronald Press Company, New York, 1963.

17. L. J. Segerlind, *Applied Finite Element Analysis*, John Wiley & Sons, Inc., New York, 1976.

18. K. H. Huebner, *A Finite Element Method for Engineers*, John Wiley & Sons, Inc., New York, 1975.

19. G. E. Myers, *Analytical Methods in Conduction Heat Transfer*, McGraw-Hill Book Company, New York, 1971.

20. O. C. Zeinkiewicz, *The Finite Element Method in Engineering Science*, McGraw-Hill Book Company, New York, 1971.

PROBLEMS

The problems in this chapter are organized in the manner shown in the table. Four problems suggest computer solutions. They are Problems 3-27, 3-29, 3-30, and 3-38. No original programs need to be written. All programs needed for solutions are developed in the examples within the chapter.

PROBLEM NUMBERS	SECTION	SUBJECT
3-1 to 3-7	3-2	Transient conduction with negligible internal resistance
3-8 to 3-11	3-3	Transient conduction in a semi-infinite solid
3-12 to 3-25	3-4	Transient conduction, chart solutions
3-26 to 3-38	3-5	Transient conduction, numerical solutions

3-1 The bulb of a clinical thermometer is 1 cm long and 7 mm in diameter. The thermometer is taken from an alcohol solution and placed in a patient's mouth. Owing to the evaporation of the alcohol solution, the thermometer is originally at a temperature of 18°C. Determine the minimum amount of time that the nurse must wait before removing the thermometer if the thermometer is to be read to within $\frac{1}{2}$°C of its final steady value. The properties of the thermometer bulb may be approximated by the average values between mercury and glass. Determine the time, assuming three values for the average convective-heat-transfer coefficient between the bulb and the patient's mouth of $\bar{h}_c = 10, 50, 100$ W/m²·K.

3-2 A small copper electrical connector is cast and it is removed from a mold at 650°C and cooled in a tank of fluid ($\bar{h}_c = 790$ W/m²·K, $T_\infty = 75$°C). The volume of the connector is 1.75 cm³ and its surface area is 3.5 cm². Determine the amount of time that the connector must remain in the fluid before it cools to 100°C.

3-3 Cylindrical aluminum parts are to be heated as they proceed along a assembly line. The parts are 1 cm long and 1.5 mm in diameter. Before proceeding into the heated section ($\bar{h}_c = 100$ W/m²·K, $T_\infty = 300$°C) of the assembly line, the aluminum parts are at a uniform temperature of 30°C. What should the residence time in the heated section of the line be for each part if the parts are to emerge at a temperature of 200°C? Size the heating element of the heated section assuming that only 30% of the heat is absorbed by the parts and that the production rate of the assembly line is 10,000 parts/h.

3-4 A cubical piece of aluminum 1 cm on a side is to be heated from 50°C to 300°C by a direct flame. How long should the aluminum remain in the flame if the flame temperature is 800°C and the convective-heat-transfer coefficient between the flame and aluminum is 190 W/m²·K?

3-5 Show that the transient temperature of a solid with negligible internal resistance subject to a convective boundary ($\bar{h}_c$, T_∞) and internal generation at a constant rate per unit volume of q_G'' is given by

$$T(t) - T_\infty = \theta(t) = \frac{q_G}{\bar{h}_c A_s}[1 - e^{-(Bi)(Fo)}]$$

The initial condition is $\theta(0) = 0$; that is, the solid is originally at the ambient fluid temperature.

3-6 Suppose that you wish to heat a piece of aluminum wire by passing an electrical current through it. The wire has a diameter of 1 mm and a length of 10 cm, giving it an electrical resistance of 0.2 Ω. If the aluminum is originally at a temperature of 25°C, what is its temperature 1 min. after a steady 1.0-A current is passed through it? Assume that the aluminum is surrounded by air at 25°C and $\bar{h}_c = 20$ W/m²·K. Use the results of Problem 3-5.

3-7 An electrical fuse element is designed in the shape of a cylinder with diameter 0.1 mm and length 0.5 cm. The fuse element is surrounded by air $\bar{h}_c = 10$ W/m²·K, $T_\infty = 30$°C. The fuse properties are $k = 20$ W/m·K, $\alpha = 5 \times 10^{-5}$ m²/s, and electrical resistance is 0.2 Ω. The melting temperature of the fuse material is 900°C. Neglecting radiation and conduction of heat into the supports at the ends of the fuse element, determine the amount of time that it will take to blow the fuse after a constant current of 3 A is passed through it. Use the results of Problem 3-5.

3-8 The surface of a thick flat plate originally at 50°C is suddenly raised to a temperature of 80°C. If the properties of the plate are $\alpha = 2 \times 10^{-6}$ m²/s, $k = 10$ W/m·K, determine (a) the time required for a plane 3 cm from the surface to reach 65°C, and (b) the amount of heat that must be transferred into the plane per m² to reach this temperature.

3-9 Work Problem 3.8 (a) assuming that the heater is replaced by a gas with temperature of 80°C and the convective-heat-transfer coefficient between the gas and the plane is 120 W/m²·K.

3-10 A method of measuring the thermal diffusivity of soil is proposed which involves burying a thermocouple a known distance beneath the surface of the earth. A heater at 90°C is placed in good thermal contact with the earth and the thermocouple records a temperature of 27°C after 15 min of heating. The original soil temperature is 22°C and the thermocouple is buried at a depth of 6 cm. Determine the thermal diffusivity of the soil.

3-11 A large, mild steel plate is 0.25 m thick and it is originally at 45°C. One surface is exposed to 200°C air with $\bar{h}_c = 210$ W/m²·K between the plate and the air. Determine the surface temperature and the temperature at a depth of 2 cm after 3 min of exposure to the hot air.

3-12 Determine the temperature 3 cm below the surface of a 10-cm-diameter hard rubber sphere 10 h after it is dropped into a vat of oil which has a temperature of 100°C. The sphere has a uniform temperature of 200°C before being placed in the oil, and the heat-transfer coefficient between the oil and sphere is 1.5 W/m²·K. Determine the amount of heat that has been removed from the sphere during the 10-h period.

3-13 A 5-cm-diameter PVC sphere is originally at 90°C. It is dropped into a tank of water that is at 20°C. The heat transfer coefficient between the sphere and the water is 20 W/m²·K. Determine the time the sphere must remain in the water until its center temperature reaches 40°C. At this time determine the surface temperature of the sphere. Properties of PVC are: $k = 0.15$ W/m·K and $\alpha = 8 \times 10^{-8}$ m²/s.

3-14 Long metal rods ($k = 45$ W/m·K, $\alpha = 2 \times 10^{-5}$ m²/s) that are 0.5 m in diameter are placed in an oven for heat treatment. The oven temperature is 600°C and the original rod temperature is 60°C. Determine the amount of time the rods must remain in the oven to ensure that the minimum rod temperature reaches 400°C. At this time determine the rod surface temperature. Assume that the convective-heat-transfer coefficient is 120 W/m²·K.

3-15 A 0.2-m-thick wall is to be constructed of fireproof material that has properties $k = 6$ W/m·K and $\alpha = 8 \times 10^{-7}$ m²/s. The fireproof material will not ignite until its temperature reaches 650°C. In tests the wall is exposed to a continuous flame over both surfaces. The flame temperature averages 875°C and the initial wall temperature is 30°C. Determine the minimum and maximum amount of time that the wall can withstand exposure to the flame before igniting, assuming that the limits on the convective-heat-transfer coefficient between the flame and the wall are from 100 W/m²·K to 250 W/m²·K. Determine the wall centerline temperature at the two limits of time.

3-16 Telephone poles must be treated with a tar material to prevent insect and water damage. The tar is cured into the wood under elevated temperatures and pressures. A 0.3-m-diameter pole originally at a temperature of 20°C is placed in the pressurized oven. The pole is to be removed when the tar has had sufficient time to penetrate to a depth of 10 cm. At this time it is determined that the 10-cm depth has reached a temperature of 100°C. If the oven temperature is 350°C and $\bar{h}_c = 145$ W/m²·K, determine the length of time the pole must remain in the oven. The pole properties are $k = 0.20$ W/m·K and $\alpha = 1.1 \times 10^{-7}$ m²/s.

3-17 Oranges can be exposed to freezing temperatures for short periods of time without damage. Suppose that a 0.1-m-diameter orange with properties $\rho = 940$ kg/m³, $c = 3.8 \times 10^3$ J/kg·K, and $k = 0.47$ W/m·K is originally at a temperature of 5°C. The air temperature suddenly drops to -5°C. Determine the time required for the surface temperature of the orange to reach 0°C if $\bar{h}_c = 10$ W/m²·K.

3-18 Determine the centerline temperature of a 3-min egg that is taken from the refrigerator at 5°C and placed in boiling water ($\bar{h}_c = 3000$ W/m²·K). Approximate the egg by a 3.6-cm-diameter sphere with properties $\rho = 1080$ kg/m³, $c = 4.0 \times 10^3$

J/kg·K, and $k=0.75$ W/m·k. Determine the amount of energy required to heat the egg.

3-19 Heaters at 250°C are placed in direct thermal contact on both sides of a large 0.25-m-thick wall made of common building brick. The brick is originally at a temperature of 10°C and it is to be heated until the centerline temperature reaches 140°C. How long should the heaters be left in contact with the wall?

3-20 A raw material must be preheated to a minimum temperature of 200°C before it is used in the production of an assembly line product. The properties of the material are $\rho=4000$ kg/m³, $k=2.0$ W/m·K, and $\alpha=8\times10^{-7}$ m²/s. The material is heated on a conveyor belt which moves through an open furnace in which the temperature is 700°C and the convective-heat-transfer coefficient between the material and the furnace air is 450 W/m²·K. The raw material may be purchased in two different spherical sizes. One size is 8 cm in diameter and the other is 4 cm in diameter. If the material is initially at 30°C before entering the furnace, determine the time that the two different-size spheres must spend in the furnace for the preheat operation. From the standpoint of minimizing the energy consumed in the furnace, is it better to order the small- or the large-diameter spheres? The conveyor belt has a length of 8 m, a width of 1 m and will hold 100 large-diameter and 400 small-diameter spheres in every 1 m² of belt area. Determine the mass production rate of the conveyor belt for the two different-size spheres if the belt speed is adjusted to meet the minimum temperature requirements for both sphere sizes.

3-21 A concrete cylinder, 25 cm long and 10 cm in diameter is initially at 90°C. It is allowed to cool in air that has a temperature of 10°C. Determine the time required for the center temperature to reach 30°C if the heat-transfer coefficient between the air and concrete is 18 W/m²·K.

3-22 You wish to cook a roast that has been rolled into a short cylindrical shape with length 20 cm and diameter 15 cm. The properties of beef are $\rho=960$ kg/m³, $k=0.90$ W/m·K, $c=5\times10^3$ J/kg·K. Determine the time the roast should be cooked to give a rare center, assuming that beef is rare at 75°C. The roast is taken from the refrigerator at a uniform temperature of 10°C. The oven is preheated to 160°C and the convective-heat-transfer coefficient between the roast and air is 30 W/m²·K. Determine the change in cooking time if the roast is allowed to warm to 25°C before placing it in the oven.

3-23 A brick-shaped part is heated in an oven to a uniform temperature of 120°C. It is then removed and placed in 30°C air. The brick is 20×30×50 cm on each side and has the properties $k=2.5$ W/m·K, $\rho=2.8\times10^3$ kg/m³, and $c=800$ J/kg·K and the convective-heat-transfer coefficient between the air and brick is 75 W/m²·K. Determine the temperature of the center of the brick 1 h after it is removed from the oven.

3-24 A short aluminum cylinder 0.6 m long and 0.6 m in diameter is initially at 200°C. It is suddenly subjected to a convection environment of 70°C with $\bar{h}_c=85$

$W/m^2 \cdot K$. Calculate the temperature at a radial position of 10 cm and a distance of 10 cm from one end of the cylinder 1 h after exposure to the environment.

3-25 Ice cream is purchased at the grocery store. The temperature of the ice cream is originally $-10°C$. The air temperature surrounding the ice cream is $25°C$ and $\bar{h}_c = 25 \ W/m^2 \cdot K$. The size of the package is $10 \times 15 \times 20$ cm and the properties of the ice cream are $k = 0.76 \ W/m \cdot K$, $c = 5 \times 10^3 \ J/kg \cdot K$, and $\rho = 845 \ kg/m^3$. Determine the time between removal from the grocery freezer until the ice cream begins to melt. The melting temperature of ice cream is approximately $0°C$. If the ice cream is wrapped immediately in a bag to reduce the heat gain, the convective-heat-transfer coefficient is reduced to $5 \ W/m^2 \cdot K$. Recalculate the melting time under these conditions.

3-26 Using the geometry specified in Problem 2-66, determine the transient temperature of the nine nodes under the following conditions: the solid is initially at a uniform temperature of $20°C$, and at $t = 0$ the solid is subjected to the boundary conditions given in the figure in Problem 2-66. The thermal diffusivity of the solid is $1.2 \times 10^{-6} \ m^2/s$. Use an explicit numerical technique.

3-27 Solve Problem 3-26 by using an implicit numerical technique. Modify the computer program given in Example 3-9 and use it to determine the temperature history of the nine nodes.

3-28 Using the geometry specified in Problem 2-69, determine the transient temperature of the 15 nodes under the following conditions: the solid is initially at a uniform temperature of $250°C$, and at $t = 0$ the solid is subjected to the boundary conditions given in the figure in Problem 2-69. The thermal diffusivity of the solid is $8 \times 10^{-6} \ m^2/s$. Use an explicit numerical technique.

3-29 Solve Problem 3-28 by using an implicit numerical technique. Modify the computer program given in Example 3-9 and use it to determine the temperature history of the 15 nodes.

3-30 Using the geometry specified in Problem 2-72, determine the transient temperature of the 20 nodes under the following conditions: the solid is initially at a uniform temperature of $25°C$, and at $t = 0$ the solid is subjected to the boundary conditions given in the figure in Problem 2-72. The thermal diffusivity of the solid is $2 \times 10^{-6} \ m^2/s$. Use an implicit numerical technique. Modify the computer program given in Example 3-9 and use it to determine the temperature history of the 20 nodes.

3-31 Verify all nodal equations and stability criteria listed in Table 3-4.

3-32 A long rectangular rod is shown in cross section in the figure. The rod is originally at a uniform temperature of $160°C$. Suddenly two of the surfaces are reduced to $100°C$. The other two surfaces are simultaneously insulated. The rod properties are $k = 20 \ W/m \cdot K$ and $\alpha = 3 \times 10^{-5} \ m^2/s$. Determine the temperature history of the six nodes shown in the figure using an explicit numerical technique.

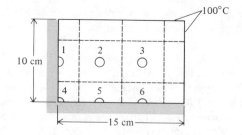

3-33 Verify the implicit forms of the nodal equations listed in Table 3-5.

3-34 Work Problem 3-8(a) using the Binder-Schmidt graphical method. Compare your answer with the one obtained in Problem 3-8.

3-35 Two large pieces of plywood are to be joined together with heat-sensitive glue. Each piece of plywood is 2 cm thick and has thermal diffusivity of $\alpha = 1.1 \times 10^{-7}$ m^2/s. Both pieces of plywood are originally at 20°C, and they are to be held together until the glue reaches 50°C. Two 120°C heaters are brought in direct thermal contact with both faces of the plywood. If the thermal resistance of the glue is neglected, estimate the time that the heat must be maintained for the glue to reach a temperature of 70°C. Use the Binder-Schmidt graphical method.

3-36 A long stainless steel cylinder with a diameter of 0.20 m is originally at a uniform temperature of 20°C and it is insulated around its periphery. The end of the cylinder is brought into direct thermal contact with a 180°C heater. Determine the temperatures at planes 1, 2, and 3 cm from the end of the cylinder at times of 1, 2, and 3 min after the heat is applied. Use the Binder-Schmidt method.

3-37 Work Problem 3-36 by an explicit numerical method.

3-38 Work Problem 3-36 by an implicit numerical method. Modify the computer program given in Example 3-9 and use it to determine the temperature history of the cylinder.

Chapter 4

ANALYSIS
OF CONVECTION
HEAT TRANSFER

4-1 INTRODUCTION

Before attempting to calculate a heat-transfer coefficient, we shall examine the convection process in some detail and relate the convection of heat to the flow of the fluid. Figure 4-1 shows a heated flat plate cooled by a stream of air flowing over it. Also shown are the velocity and the temperature distributions. The first point to note is that the velocity decreases in the direction toward the surface as a result of viscous forces. Since the velocity of the fluid layer adjacent to the wall is zero, the heat transfer between the surface and this fluid layer must be by conduction alone:

$$q_c'' = -k_f \frac{\partial T}{\partial y}\bigg|_{y=0} = \bar{h}_c(T_s - T_\infty) \qquad (4\text{-}1)$$

Although this viewpoint suggests that the process can be viewed as conduction, the temperature gradient at the surface, $(\partial T/\partial y)|_{y=0}$, is determined by the rate at which the fluid farther from the wall can transport the energy into the mainstream. Thus the temperature gradient at the wall depends on the flow field, with higher velocities being able to produce larger temperature gradients and higher rates of heat transfer. At the same time, however, the thermal conductivity of the fluid plays a role. For example, the value of k_f for water is an order of magnitude larger than that of air; thus, as shown in Table 1-2, the convection-heat-transfer coefficient for water is larger than that for air.

197

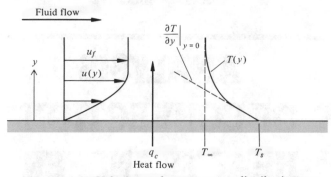

Figure 4-1 Velocity and temperature distributions for forced convection over a heated plate.

The situation is quite similar in free convection, as shown in Fig. 4-2. The principal difference is that in forced convection the velocity approaches the free-stream value imposed by an external force, whereas in free convection the velocity at first increases with increasing distance from the plate because the action of viscosity diminishes rather rapidly while the density differencedecreases more slowly. Eventually, however, the buoyant force decreases as the fluid density approaches the value of the surrounding fluid; this will cause the velocity to reach a maximum and approach zero far away from the heated surface. The temperature fields in free and forced convection have similar shapes, and in both cases the heat-transfer mechanism at the fluid/solid interface is conduction.

The preceding discussion indicates that the convection-heat-transfer coefficient will depend on the density, viscosity, and velocity of the fluid as

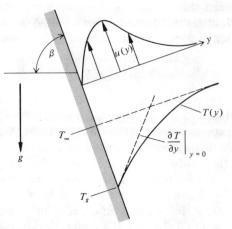

Figure 4-2 Velocity and temperature distributions for free convection over a heated plate inclined on angle β from the horizontal.

well as on its thermal properties (thermal conductivity and specific heat). Whereas in forced convection the velocity is usually imposed on the system by a pump or a fan and can be directly specified, in free convection the velocity will depend on the temperature difference between the surface and the fluid, the coefficient of thermal expansion of the fluid (it determines the density change per unit temperature difference), and the force field, which in systems located on earth is simply the gravitational force.

To gain an understanding of the parameters of significance in forced convection we shall examine the flow field in more detail. Figure 4-3 shows the flow at various distances from the leading edge of a plate. Starting at the leading edge, a region develops in the flow where viscous forces cause the fluid to slow down. These viscous forces depend on the shear stress, τ. In flow over a flat plate the fluid velocity parallel to the plate can be used to define this stress as

$$\tau = \mu \frac{du}{dy} \tag{4-2}$$

where du/dy is the velocity gradient and the constant of proportionality μ is called the *dynamic viscosity*. If the shear stress is expressed in newtons per square meter and the velocity gradient in seconds^{-1}, then μ has the units newton-seconds per square meter ($N \cdot s/m^2$).

The region in the flow near the plate where the velocity of the fluid is slowed down by viscous forces is called the *boundary layer*. The distance from the plate at which the velocity reaches 99 percent of the free-stream velocity is arbitrarily designated as the *boundary-layer thickness*, and the region beyond this point is called the *undisturbed* or *potential flow regime*.

Initially, the flow in the boundary layer is completely laminar. The boundary thickness grows with increasing distance from the leading edge,

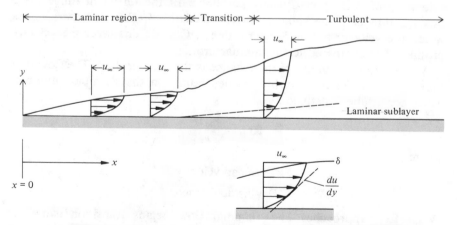

Figure 4-3 Laminar, transition, and turbulent boundary-layer-flow regimes in flow over a flat plate.

and at some critical distance, x_c, the inertial effects become sufficiently large compared to the viscous damping action that small disturbances in the flow begin to grow. As these disturbances become amplified, the regularity of the viscous flow is disturbed and a transition from laminar to turbulent flow takes place. In the turbulent-flow region macroscopic chunks of fluid move across streamlines and transport thermal energy as well as momentum vigorously. As shown in books on fluid mechanics (e.g., Ref. 1) the parameter that quantitatively relates the viscous and inertial forces and whose value determines the transition from laminar to turbulent flow is the *dimensionless Reynolds number*, Re_x, defined as

FLAT PLATE

$$Re_x = \frac{V_\infty x}{\nu_f} \qquad (4\text{-}3)$$

where

$$V_\infty = \text{free-stream velocity}$$

$$x = \text{distance from the leading edge}$$

$$\nu_f = \frac{\mu_f}{\rho_f} = \text{kinematic viscosity of the fluid}$$

The critical value of Re_{x_c} at which transition occurs depends on the surface roughness and the level of turbulent activity, the *turbulence level*, in the mainstream. When large disturbances are present in the main flow, transition begins when $Re_x = 10^5$, but in less disturbed flow fields it will not start until $Re_x = 2 \times 10^5$. The transition regime extends to a Reynolds number about twice the value at which transition began, and beyond this point the boundary layer is turbulent.

Approximate shapes of the velocity profiles in laminar and turbulent flow are sketched in Fig. 4-3. In the laminar range the boundary-layer velocity profile is approximately parabolic. In the turbulent range there exists a thin layer near the surface, called the *laminar sublayer*, across which the velocity profile is nearly linear. Outside this layer the velocity profile is flat compared to the laminar profile.

Another flow geometry of importance is the tube or duct. The flow in a tube can be laminar or turbulent, depending on the Reynolds number, which for a tube is defined as

$$Re_D = \frac{V_m D}{\nu} = \frac{\dot{m} D}{\rho A \gamma}$$

where

$$V_m = \text{mean velocity}$$

$$D = \text{inside diameter}$$

When Re_D approaches 2300, laminar flow begins transition, which is usually complete when Re_D is about 6000. The actual value depends on the roughness of the tube and the turbulence level.

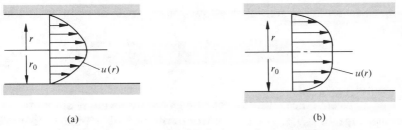

Figure 4-4 Laminar and turbulent velocity profiles in
tube flow: (a) laminar flow; (b) turbulent flow.

Figure 4-4 shows the velocity profiles in laminar and turbulent flow. At
the entrance a boundary layer develops along the inner surface of the tube
and fills more and more of the cross-sectional area with increasing distance
from the inlet. If the flow is laminar, the profile will have a parabolic shape
and the mean velocity will be one-half the value at the center. When the
flow is turbulent, the profile is blunter, as shown, and the mean velocity is
about 83% of the center velocity.

Sometimes it is more convenient to relate the Reynolds number to the
mass velocity G, defined as

$$G = \rho V_m = \frac{\dot{m}}{A_c} \tag{4-4}$$

where

$$\dot{m} = \text{mass rate of flow}$$
$$A_c = \text{cross-sectional area}$$
$$V_m = \text{mean fluid velocity}$$

Equation (4-4) also serves to define the mean velocity. It is easily verified
that the Reynolds number in a duct of diameter D may be written in the
form

$$\text{Re}_D = \frac{GD}{\mu} = \frac{\rho V D}{\mu} \tag{4-5}$$

When the cross-sectional area of a duct is not circular, the appropriate
dimension to use in place of the tube diameter D is the hydraulic diameter
defined by the relation

$$D_H = 4\left(\frac{\text{cross-sectional area}}{\text{wetted perimeter}}\right) \tag{4-6}$$

For example, in flow through an annulus formed between the outside of a
tube of diameter D_o and a larger tube of diameter D_i, the hydraulic
diameter is

$$D_H = 4\frac{\pi(D_i^2 - D_o^2)/4}{\pi(D_i + D_o)} = D_i - D_o$$

If the cross-sectional area of the flow is a rectangle of width w and height h,

$$D_H = 4 \left[\frac{w \times h}{2(w+h)} \right] = 2 \left(\frac{wh}{w+h} \right)$$

When $h \ll w$, $D_H \sim 2h$.

The convective-heat-transfer coefficient varies with position from the leading edge of a flat plate and from entrance of a tube or duct. The parameter that depicts the spacial variation is the local heat-transfer coefficient h_x, where x is the distance from the leading edge of the plate or tube. If we wish to calculate the total heat transfer from a plate, for example, we must know the average heat-transfer coefficient, $\bar{h}_c$. The relationship between the value of h_{cx} and $\bar{h}_c$ is

$$\bar{h}_c = \frac{1}{L} \int_{x=0}^{x=L} \bar{h}_{cx} \, dx \tag{4-7}$$

Other parameters are also a function of location. The *local friction coefficient* is defined as

$$C_{f_x} = \frac{\tau_x}{(\rho u_\infty^2 / 2)} \tag{4-8}$$

The *average friction coefficient* over a plate of length L is

$$\bar{C}_f = \frac{1}{L} \int_{x=0}^{x=L} C_{f_x} \, dx \tag{4-9}$$

The friction coefficient is a dimensionless group that gives a measure of the fluid shear stress or drag on the solid boundary.

4-2 THE CONSERVATION OF MASS, MOMENTUM, AND ENERGY EQUATIONS FOR LAMINAR FLOW OVER A FLAT PLATE

In the classical approach to convection one derives differential equations for the momentum and energy balance in the boundary layer and then solves these equations for the temperature gradient in the fluid at the fluid/wall interface to evaluate the convection-heat-transfer coefficient. A somewhat simpler, but practically more useful approach is to derive integral instead of differential equations and use an approximate analysis to obtain the heat-transfer coefficient. In this section the differential equations governing the flow of a fluid over a flat plate will be derived to illustrate the similarity between heat transfer and momentum transfer and to introduce the Prandtl number, which relates the two processes. Then, the integral equations for flow over a flat surface will be derived and solved to illustrate an analytical approach that will also be used to obtain the heat-transfer coefficients in turbulent flow.

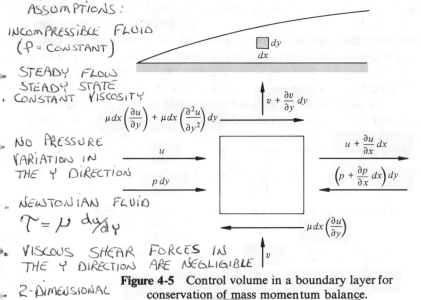

CONSERVATION OF MOMENTUM
  ASSUMPTIONS:

INCOMPRESSIBLE FLUID
  $(\rho = \text{CONSTANT})$

- STEADY FLOW
  STEADY STATE
- CONSTANT VISCOSITY

- NO PRESSURE
  VARIATION IN
  THE Y DIRECTION

- NEWTONIAN FLUID
  $\tau = \mu \, \dfrac{du}{dy}$

- VISCOUS SHEAR FORCES IN
  THE Y DIRECTION ARE NEGLIGIBLE

Figure 4-5 Control volume in a boundary layer for conservation of mass momentum balance.

- 2-DIMENSIONAL

- LAMINAR FLOW

Consider a control volume within the boundary layer as shown in Fig. 4-5 and assume that steady-state conditions prevail and the fluid is incompressible. Then the mass flow rates into and out of the control volume, respectively, in the x direction are

$$\rho u \, dy \quad \text{and} \quad \rho \left(u + \frac{\partial u}{\partial x} dx \right) dy$$

Thus the net mass flow into the element in the x direction is

$$-\rho \frac{\partial u}{\partial x} dx \, dy$$

Similarly, the net mass flow into the control volume in the y direction is

$$-\rho \frac{\partial v}{\partial y} dy \, dx$$

Since the net mass flow rate out of the control volume must be zero, we obtain

$$-\rho \left(\frac{\partial u}{\partial x} + \frac{\partial v}{\partial y} \right) dx \, dy = 0$$

from which it follows that in two-dimensional steady flow, conservation of mass requires that

$$\frac{\partial u}{\partial x} + \frac{\partial v}{\partial y} = 0 \tag{4-10}$$

The conservation of momentum equation is obtained from application of Newton's second law of motion to the element. Assuming that the flow is Newtonian, that there are no pressure gradients in the y direction, and that viscous shear in the y direction is negligible, the rates of momentum flow in the x direction for the fluid flowing across the left- and right-hand vertical faces (see Fig. 4-5) are $\rho u^2\,dy$ and $\rho[u+(\partial u/\partial x)dx]^2\,dy$. It should be noted, however, that flow across the horizontal faces will also contribute to the momentum balance in the x direction. The x momentum flow entering through the bottom face is $\rho uv\,dx$, and the momentum flow per unit width leaving through the upper face is

$$\rho\left(v+\frac{\partial v}{\partial y}\,dy\right)\left(u+\frac{\partial u}{\partial x}\,dx\right)dx$$

The viscous shear force at the bottom face is $-\mu(\partial u/\partial y)dx$ and over the top face is

$$\mu\,dx\left[\frac{\partial u}{\partial y}+\frac{\partial}{\partial y}\left(\frac{\partial u}{\partial y}\right)dy\right]$$

Thus the net viscous shear in the x direction is $\mu\,dx(\partial^2 u/\partial y^2)dy$.

The pressure force over the left face is $p\,dy$ and over the right is $-[p+(dp/dx)dx]dy$. Thus the net pressure force in the direction of motion is $-(\partial p/\partial x)dx\,dy$. Equating the sum of the forces to the momentum flow rate out of the control volume in the x direction gives

$$\rho uv\,dx+\rho\frac{\partial v}{\partial y}u\,dy\,dx+\rho v\frac{\partial u}{\partial x}\,dx+\rho\frac{\partial v}{\partial y}\frac{\partial u}{\partial x}\,dy\,dx=\left(\mu\frac{\partial^2 u}{\partial y^2}-\frac{\partial p}{\partial x}\right)dy\,dx$$

Neglecting second-order differentials and using the conservation of mass equation, the conservation of momentum equation reduces to

$$\rho\left(u\frac{\partial u}{\partial x}+v\frac{\partial u}{\partial y}\right)=\mu\frac{\partial^2 u}{\partial y^2}-\frac{\partial p}{\partial x} \tag{4-11}$$

The conservation of energy equation will be derived on the assumption that all physical properties are temperature-independent and that the flow velocity is sufficiently small that the frictional shear work may be neglected. Figure 4-6 shows the rate at which energy will be conducted and convected into and out of the control volume. There are four convective terms in addition to the conductive terms used previously in Chapter 2, Eq. 2-5. An energy balance requires that the net rate of conduction and the net rate of convection be zero. This yields

$$k\,dx\,dy\left(\frac{\partial^2 T}{\partial x^2}+\frac{\partial^2 T}{\partial y^2}\right)-\left[\rho c_p\left(u\frac{\partial T}{\partial x}+\frac{\partial u}{\partial x}T+\frac{\partial u}{\partial x}\frac{\partial T}{\partial x}\,dx\right)\right]dx\,dy$$

$$-\left[\rho c_p\left(v\frac{\partial T}{\partial y}+\frac{\partial v}{\partial y}T+\frac{\partial v}{\partial y}\frac{\partial T}{\partial y}\,dy\right)\right]dx\,dy=0$$

ENERGY EQUATION
  ASSUMPTIONS
INCOMPRESSIBLE FLUID

. STEADY FLOW

. CONSTANT PROPERTIES
  (Cp, K, VISCOSITY)

. 2-DIMENSIONAL

.. VISCOUS
   DISSIPATION IS NEGLIGIBLE
   (LOW VISCOUS FLOW)

. No HEAT GENERATION
  OR RADIATION

. PROPERTIES ARE TEMPERATURE INDEPENDENT

. LAMINAR
  FLOW

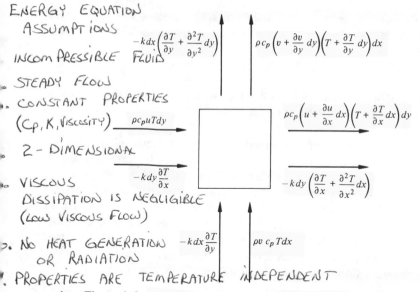

Figure 4-6 Control volume in a boundary layer for energy balance.

Using the conservation of mass equation and neglecting second-order terms, as in the derivation of the conservation of momentum equation, gives the following expression for the energy equation:

$$u\frac{\partial T}{\partial x} + v\frac{\partial T}{\partial y} = \alpha\left(\frac{\partial^2 T}{\partial x^2} + \frac{\partial^2 T}{\partial y^2}\right) \tag{4-12}$$

Under normal conditions, conduction in the x direction is small compared with the other terms and, consequently, the first term in the parentheses of the right-hand side may be neglected. Also, the pressure term in the momentum equation is usually small and can be neglected in comparison with the other terms. Then, the similarity between the momentum and energy equation is apparent:

$$u\frac{\partial u}{\partial x} + v\frac{\partial u}{\partial y} = \nu\left(\frac{\partial^2 u}{\partial y^2}\right) \tag{4-13}$$

$$u\frac{\partial T}{\partial x} + v\frac{\partial T}{\partial y} = \alpha\left(\frac{\partial^2 T}{\partial y^2}\right) \tag{4-14}$$

In the preceding relations, ν is the kinematic viscosity, often called *momentum diffusivity*, μ/ρ, and ν/α is equal to $(\mu/\rho)/(k/\rho c_p)$, which is the *Prandtl number*, Pr. If ν equals α, then Pr is 1 and the momentum and energy equations are identical. For this condition, nondimensional solutions of $u(y)$ and $T(y)$ are identical. Thus it is apparent that the Prandtl number, which is the ratio of fluid properties, controls the relation between the velocity and the temperature distributions. Values of Pr vary from a low of the order of 0.004 for a liquid metal to a high of 4000 for a very viscous oil.

4-3 THE INTEGRAL MOMENTUM AND ENERGY EQUATIONS FOR A LAMINAR BOUNDARY LAYER

To circumvent the problems involved in solving the partial differential equations of the boundary layer, an integral approach is presented below. For that purpose let us consider an elemental control volume that extends from the wall to beyond the limit of the boundary layer in the y direction, is dx thick in the x direction, and has a unit width in the z direction, as shown in Fig. 4-7. To obtain a relationship for the net momentum in-flow and the net energy transport, we proceed in a manner similar to that used to derive the boundary-layer equations in the preceding section.

The momentum flow across the face, AB in Fig. 4-7, will be

$$\int_0^\delta \rho u^2 \, dy.$$

Similarly, the momentum flow across the face CD will be

$$\int_0^\delta \rho u^2 \, dy + (d/dx) \int_0^\delta \rho u^2 \, dy \, dx.$$

However, fluid also enters the control volume across face BD at the rate

$$(d/dx) \int_0^\delta \rho u \, dy \, dx.$$

This quantity is the difference between the rate of flow leaving across face CD and that entering across face AB. Since the fluid entering across BD

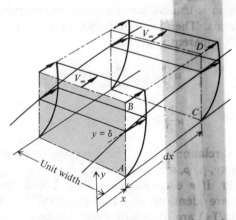

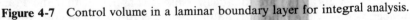

Figure 4-7 Control volume in a laminar boundary layer for integral analysis.

has a velocity component in the x direction equal to the free-stream velocity V_∞, the flow of x momentum into the control volume across the upper face is

$$V_\infty \frac{d}{dx} \int_0^\delta \rho u \, dy \, dx$$

Adding up the x momentum components gives

$$\frac{d}{dx} \int_0^\delta \rho u^2 \, dy \, dx - V_\infty \frac{d}{dx} \int_0^\delta \rho u \, dy \, dx = - \frac{d}{dx} \int_0^\delta \rho u (V_\infty - u) \, dy$$

There will be no shear across the face BD, since this face is outside the boundary layer, where du/dy is equal to zero. There is, however, a shear force τ_w acting at the fluid-solid interface, and there will be pressure forces acting on faces AB and CD. Writing out the net forces acting on the control volume and adding them yields the relation

$$p_x \delta - \left(p_x + \frac{dp_x}{dx} \, dx \right) \delta - \tau_w \, dx = - \delta \frac{dp_x}{dx} \, dx - \tau_w \, dx \qquad (4\text{-}15)$$

For flow over a flat plate the pressure gradient in the x direction may be neglected and the momentum equation can then be written in the form

INTEGRAL
MOMENTUM EQ

ASSUMPTIONS ARE THE SAME AS THE MOMENTUM EQ.
$$\frac{d}{dx} \int_0^\delta \rho u (V_\infty - u) \, dy = \tau_w \qquad (4\text{-}16)$$
ASSUME V_∞ IS NOT A FUNCTION OF X

The integral energy equation may be derived in a similar fashion. In this case, however, a control volume extending beyond the limits of both the temperature and velocity boundary layer must be used in the derivation (see Fig. 4-8). The first law of thermodynamics demands that energy in the form of enthalpy, kinetic energy, and heat, as well as shear work, should be considered. For low velocities, however, the kinetic-energy terms and the

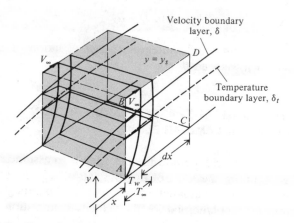

Figure 4-8 Control volume in temperature and velocity boundary layers.

shear work are small compared to the other quantities and may be neglected. Then, the rate at which enthalpy enters across face AB is given by

$$\int_0^{y_s} c_p \rho u T \, dy$$

whereas the rate of enthalpy flow across face CD is

$$\int_0^{y_s} c_p \rho u T \, dy + \frac{d}{dx} \int_0^{y_s} c_p \rho u T \, dy \, dx$$

The enthalpy carried into the control volume across the upper face is given by

$$c_p T_s \frac{d}{dx} \int_0^{y_s} \rho u \, dy \, dx$$

Finally, heat will be conducted across the interface between the fluid and solid surface at the rate

$$-k \, dx \left(\frac{\partial T}{\partial y} \right)_{y=0}$$

Adding up all the energy quantities yields the integral equation for the conservation of energy in the form

$$c_p T_\infty \frac{d}{dx} \int_0^{y_s} \rho u \, dy \, dx - \frac{d}{dx} \int_0^{y_s} \rho c_p T u \, dy \, dx - k \, dx \left(\frac{\partial T}{\partial y} \right)_{y=0} = 0 \quad (4\text{-}17)$$

It should be noted, however, that outside the limit of the temperature boundary layer, the temperature equals the free-stream temperature, T_∞, so that integration need only be taken up to $y = \delta_t$. Equation 4-17 therefore can be simplified to the form

INTEGRAL ENERGY EQ.
ASSUMPTIONS SAME AS

$$\frac{d}{dx} \int_0^{\delta_t} (T_\infty - T) u \, dy - \alpha \left(\frac{\partial T}{\partial y} \right)_{y=0} = 0 \quad (4\text{-}18)$$

ENERGY EQ PLUS NO HEAT CONDUCTION IN X DIRECTION
which is usually known as the integral energy equation of the laminar boundary layer for low-speed flow.

4-4 EVALUATION OF HEAT-TRANSFER AND FRICTION COEFFICIENTS IN LAMINAR FLOW

In the approximate integral method the first step is to assume velocity and temperature contours in the form of polynomials. Then, the coefficient in the polynomial will be evaluated to satisfy the boundary conditions. Assuming a four-term polynomial for the velocity distribution

$$u(y) = a + by + cy^2 + dy^3 \quad (4\text{-}19)$$

the constants are evaluated by applying the boundary conditions

> at $y=0$: $u=0$ and therefore $a=0$
>
> $u=v=0$ and therefore, from Eq. 4-11, $\partial^2 u/\partial y^2=0$
>
> $y=\delta$: $u=V_\infty$ and $\partial u/\partial y=0$

The conditions above provide four equations for the evaluation of the four unknown coefficients in terms of the free-stream velocity and the boundary-layer thickness. It can easily be verified (see Problem 4-14) that the coefficients that satisfy these boundary conditions are

$$a=0, \qquad b=\frac{3}{2}\frac{V_\infty}{\delta}, \qquad c=0, \qquad d=-\frac{V_\infty}{2\delta^3}$$

Substituting these coefficients in Eq. 4-19 and dividing through by the free-stream velocity, V_∞, to dimensionalize the result yields

$$\frac{u}{V_\infty}=\frac{3}{2}\frac{y}{\delta}-\frac{1}{2}\left(\frac{y}{\delta}\right)^3 \tag{4-20}$$

Substituting Eq. 4-20 for the velocity distribution in the integral momentum equation (Eq. 4-16) yields

$$\frac{d}{dx}\int_0^\delta \rho V_\infty^2\left[\frac{3}{2}\frac{y}{\delta}-\frac{1}{2}\left(\frac{y}{\delta}\right)^3\right]\cdot\left[1-\frac{3}{2}\frac{y}{\delta}+\frac{1}{2}\left(\frac{y}{\delta}\right)^3\right]dy=\tau_w=\mu\left(\frac{du}{dy}\right)_{y=0} \tag{4-21}$$

The wall shear stress τ_w can be obtained by evaluating the velocity gradient from Eq. 4-20 at $y=0$ (see Problem 4-14). Substituting for τ_w and performing the integration in Eq. 4-21 yields

$$\frac{d}{dx}\left(\rho V_\infty^2\frac{39\delta}{280}\right)=\frac{3}{2}\mu\frac{V_\infty}{\delta} \tag{4-22}$$

Equation 4-22 may be rearranged and integrated to obtain the boundary-layer thickness in terms of the viscosity, distance from the leading edge, and free-stream velocity distribution:

$$\frac{\delta^2}{2}=\frac{140\nu x}{13V_\infty}+C \tag{4-23}$$

Since $\delta=0$ at the leading edge (i.e., at $x=0$), the coefficient C in the preceding relation must equal 0 and

$$\delta^2=\frac{280\nu x}{13V_\infty}$$

or

$$\frac{\delta}{x}=\frac{4.64}{Re_x^{1/2}} \tag{4-24}$$

δ = BOUNDARY LAYER THICKNESS

To evaluate the friction coefficient substitute Eq. 4-20 into Eq. 4-21.

$$\tau_w = \mu \frac{du}{dy}\bigg|_{y=0} = \mu \frac{3}{2} \frac{V_\infty}{\delta}$$

Substituting for δ from Eq. 4-24 gives

$$\tau_w = \frac{3}{9.28} \frac{\mu V_\infty}{x} \text{Re}_x^{1/2}$$

and the friction coefficient C_{fx} is

$$C_{fx} = \frac{\tau_{w_x}}{\frac{1}{2}\rho V_\infty^2} = \frac{0.647}{\text{Re}_x^{1/2}} \tag{4-25}$$

Example 4-1. Determine the laminar shear stress at a distance 0.2 m from the leading edge for flow of 293K water at a velocity of 1 m/s over a flat plate.

Solution: For water at 293 K, ⟵ from pg 514

$$\mu = 993 \times 10^{-6} \text{N} \cdot \text{s/m}^2$$

$$\text{Re}_x = \frac{\rho V_\infty x}{\mu} = \frac{998 \times 1 \times 0.2}{993 \times 10^{-6}} = 2 \times 10^5$$

and since the flow is laminar,

$$\tau_w = \frac{3}{9.28} \times (993 \times 10^{-6}) \times \left(\frac{1.0}{0.2}\right)(2 \times 10^5)^{1/2} = 0.718 \, \text{N/m}^2$$

We next turn to the energy equation and propose a temperature distribution in the boundary layer of the same form as the velocity distribution:

$$T(y) = e + fy + gy^2 + hy^3 \tag{4-26}$$

The boundary conditions for the temperature distribution are that at $y = 0, T = T_s$; at $y = \delta_t$ (the thickness of the temperature boundary layer), $T = T_\infty$, and $dT/dy = 0$. Also, from Eq. 4-13, d^2T/dy^2 at $y = 0$ must be zero because both u and v are zero at the interface. From these conditions it follows that the constants are (see Problem 4-15)

$$e = T_s, \qquad f = \frac{3}{2}\frac{T_\infty}{\delta_t}, \qquad e = 0, \qquad f = -\frac{T_\infty}{2\delta_t^2}$$

If for convenience the variable in the energy equation is taken as the temperature in the fluid minus the wall temperature, the dimensionless temperature distribution can be written in the form

$$\frac{T - T_s}{T_\infty - T_s} = \frac{3}{2}\frac{y}{\delta_t} - \frac{1}{2}\left(\frac{y}{\delta_t}\right)^3 \tag{4-27}$$

Using Eqs. 4-27 and 4-20, for $(T - T_s)$ and u, respectively, the integral in Eq. 4-18 can be written as

$$\int_0^{\delta_t}(T_\infty - T)u\,dy = \int_0^{\delta_t}[(T_\infty - T_s) - (T - T_s)]u\,dy$$

$$= (T_\infty - T_s)V_\infty \int_0^{\delta_t}\left[1 - \frac{3}{2}\frac{y}{\delta_t} + \frac{1}{2}\left(\frac{y}{\delta_t}\right)^3\right]\left[\frac{3}{2}\frac{y}{\delta} - \frac{1}{2}\left(\frac{y}{\delta}\right)^3\right]dy$$

Performing the multiplication under the integral sign, we obtain

$$(T_\infty - T_s)V_\infty \int_0^{\delta_t}\left(\frac{3}{2\delta}y - \frac{9}{4\delta\delta_t}y^2 + \frac{3}{4\delta\delta_t{}^3}y^4\right.$$

$$\left. - \frac{1}{2\delta^3}y^3 + \frac{3}{4\delta_t\delta^3}y^4 - \frac{1}{4\delta_t{}^3\delta^3}y^6\right)dy$$

which yields, after integrating,

$$(T_\infty - T_s)V_\infty\left(\frac{3}{4}\frac{\delta_t{}^2}{\delta} - \frac{3}{4}\frac{\delta_t{}^2}{\delta} + \frac{3}{20}\frac{\delta_t{}^2}{\delta}\right.$$

$$\left. - \frac{1}{8}\frac{\delta_t{}^4}{\delta^3} + \frac{3}{20}\frac{\delta_t{}^4}{\delta^3} - \frac{1}{28}\frac{\delta_t{}^4}{\delta^3}\right)$$

If we let $\zeta = \delta_t/\delta$, the expression above can be written

$$(T_\infty - T_s)V_\infty\delta\left(\frac{3}{20}\zeta^2 - \frac{3}{280}\zeta^4\right)$$

For fluids having a Prandtl number equal to or larger than unity, ζ is equal to or less than unity and the second term in the parentheses can be neglected compared to the first.* Substituting this approximate form for the integral in Eq. 4-18, we obtain

$$\frac{3}{20}V_\infty(T_s - T_\infty)\zeta^2\frac{\partial\delta}{\partial x} = \alpha\frac{\partial T}{\partial y}\bigg|_{y=0} = \frac{3}{2}\alpha\frac{T_\infty - T_s}{\delta\zeta}$$

or

$$\frac{1}{10}V_\infty\zeta^3\delta\frac{\partial\delta}{\partial x} = \alpha$$

From Eq. 4-24 we obtain

$$\delta\frac{\partial\delta}{\partial x} = 10.75\frac{\nu}{V_\infty}$$

*This assumption is not valid for liquid metals, which have Pr≪1.

δ_T = THICKNESS OF TEMPERATURE BOUNDARY LAYER

and with this expression we get

$$\zeta^3 = \frac{10}{10.75}\frac{\alpha}{\nu}$$

or

$$\delta_t = 0.9\delta \, Pr^{-1/3} \tag{4-28}$$

Except for the numerical constant (0.9 compared with 1.0), the foregoing result is in agreement with the exact calculations of Pohlhausen (Ref. 2).

The rate of heat flow by convection from the plate per unit area is, from Eqs. 4-1 and 4-27,

$$\frac{q}{A} = -k_f \frac{\partial T_f}{\partial y}\bigg|_{y=0} = -\frac{3}{2}\frac{k_f}{\delta_t}(T_\infty - T_s)$$

Substituting Eqs. 4-25 and 4-27 for δ and δ_t yields

$$\frac{q}{A} = -\frac{3}{2}\frac{k}{x}\frac{Pr^{1/3}Re_x^{1/2}}{(0.976)(4.64)}(T_\infty - T_s) = 0.33\frac{k}{x}Re_x^{1/2}Pr^{1/3}(T_s - T_\infty)$$

$$\tag{4-29}$$

and the local Nusselt number, Nu_x, is

LAMINAR
FLOW

$$Nu_x = \frac{h_{cx}x}{k} = \frac{q_x}{A(T_s - T_\infty)}\frac{x}{k} = 0.33 Re_x^{1/2}Pr^{1/3} \tag{4-30}$$

This result is in excellent agreement with the result of the exact analysis of Reference 2.

The foregoing example illustrates the usefulness of the approximate boundary-layer analysis. Guided by a little physical insight and intuition, this technique yields satisfactory results without the mathematical complications inherent in the exact boundary-layer equations. The approximate method has been applied to many other problems, and the results are available in the literature.

$\overline{h_c} = 2 h_{cx}$

$\overline{Nu} = 2\dfrac{h_{cx}L}{k} = 2 Nu_x = .664 Re^{1/2} Pr^{1/3}$ LAMINAR FLOW

4-5 ANALOGY BETWEEN HEAT AND MOMENTUM TRANSFER IN TURBULENT FLOW OVER A FLAT SURFACE

In a majority of practical applications the flow in the boundary layer is turbulent rather than laminar. Qualitatively, the exchange mechanism in turbulent flow can be pictured as a magnification of the molecular exchange in laminar flow. In steady laminar flow, fluid particles follow well-defined streamlines. Heat and momentum are transferred across streamlines only by molecular diffusion and the cross flow is so small that when a colored dye is injected into the fluid at some point, it follows a streamline without appreciable diffusion. In turbulent flow, on the other hand, the color will be distributed over a wide area a short distance

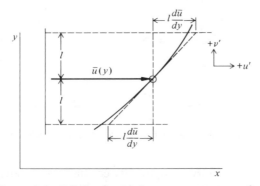

Figure 4-9   Mixing length for momentum transfer.

downstream from the point of injection. The mixing mechanism consists of rapidly fluctuating eddies that transport fluid particles in an irregular manner. Groups of particles collide with each other at random, establish cross flow on a macroscopic scale, and effectively mix the fluid. Since the mixing in turbulent flow is on a macroscopic scale with groups of particles transported in a zigzag path through the fluid, the exchange mechanism is many times more effective than in laminar flow. As a result, the rates of heat and momentum transfer in turbulent flow and the associated friction and heat-transfer coefficients are many times larger than in laminar flow.

If turbulent flow at a point is averaged over a period of time, long as compared with the period of a single fluctuation, the time-mean properties and the velocity of the fluid are constant if the average flow remains steady. It is therefore possible to describe each fluid property and the velocity in turbulent flow in terms of a *mean value* that does not vary with time and a *fluctuating component* which is a function of time. To simplify the problem, consider a two-dimensional flow (Fig. 4-9) in which the mean value of velocity is parallel to the x direction. The instantaneous velocity components u and v can then be expressed in the form

$$u = \bar{u} + u'$$
$$v = v' \tag{4-31}$$

where the bar over a symbol denotes the temporal mean value, and the prime denotes the instantaneous deviation from the mean value. According to the model used to describe the flow,

$$\bar{u} = \frac{1}{\theta^*} \int_0^{\theta^*} u \, d\theta \tag{4-32}$$

where θ^* is a time interval large compared with the period of the fluctuations. Figure 4-10 shows qualitatively the time variation of u and u'. From Eq. 4-32 or from an inspection of the graph it is apparent that the time average of u' is zero (i.e., $\bar{u}' = 0$). A similar argument shows that $\bar{v}'$ and $\overline{(\rho v)}'$ are also zero.

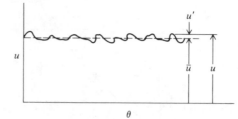

Figure 4-10 Time variation of instantaneous velocity.

The fluctuating velocity components continuously transport mass, and consequently momentum, across a plane normal to the y direction. The instantaneous rate of transfer in the y direction of x momentum per unit area at any point is

$$-(\rho v)'(\bar{u}+u')$$

where the minus sign, as will be shown later, takes account of the statistical correlation between u' and v'.

The time average of the x-momentum transfer gives rise to an *apparent* turbulent shear or Reynolds stress τ_t, defined by

$$\tau_t = -\frac{1}{\theta*}\int_0^{\theta*}(\rho v)'(\bar{u}+u')\,d\theta \tag{4-33}$$

Breaking this term up into two parts, the time average of the first is

$$\frac{1}{\theta*}\int_0^{\theta*}(\rho v)'\bar{u}\,d\theta=0$$

since $\bar{u}$ is a constant and the time average of $(\rho v)'$ is zero. Integrating the second term, Eq. 4-33 becomes

$$\tau_t = -\frac{1}{\theta*}\int_0^{\theta*}(\rho v)'u'\,d\theta = -\overline{(\rho v)'u'} \tag{4-34}$$

or, if ρ is constant,

$$\tau_t = -\rho(\overline{v'u'}) \tag{4-35}$$

where $(\overline{v'u'})$ is the time average of the product of $\bar{u}'$ and $\bar{v}'$.

It is not difficult to visualize that the time averages of the mixed products of velocity fluctuations, such as $\overline{v'u'}$, differ from zero. From Fig. 4-9 we can see that the particles that travel upward ($v'>0$) arrive at a layer in the fluid in which the mean velocity $\bar{u}$ is larger than in the layer from which they come. Assuming that the fluid particles preserve on the average their original velocity $\bar{u}$ during their migration, they will tend to slow down other fluid particles after they have reached their destination and thereby give rise to a negative component u'. Conversely, if v' is negative, the observed value of u' at the new destination will be positive. On the average, therefore, a positive v' is associated with a negative u', and vice versa. The time average of $\overline{u'v'}$ is therefore on the average not zero but a negative quantity. The turbulent shearing stress defined by Eq. 4-35 is thus positive and has the same sign as the corresponding laminar shearing stress,

$$\tau_t = \mu \frac{d\overline{u}}{dy} = \rho\nu \frac{d\overline{u}}{dy}$$

It should be noted, however, that the laminar shearing stress is a true stress, whereas the apparent turbulent shearing stress is simply a concept introduced to account for the effects of the momentum transfer by turbulent fluctuations. This concept allows us to express the total shear stress in turbulent flow as

$$\tau = \frac{\text{viscous force}}{\text{area}} + (\text{turbulent momentum flux}) \qquad (4\text{-}36)$$

To relate the turbulent momentum flux to the time-average velocity gradient, $d\overline{u}/dy$, we postulate that fluctuations of macroscopic fluid particles in turbulent flow are, on the average, similar to the motion of molecules in a gas [i.e., they travel on the average a distance l perpendicular to $\overline{u}$ (Fig. 4-9) before coming to rest in another y plane]. This distance l is known as *Prandtl's mixing length* and corresponds qualitatively to the mean free path of a gas molecule. Assuming that the fluid particles retain their identity and physical properties during the cross motion and that the turbulent fluctuation arises chiefly from the difference in the time-mean properties between y planes spaced a distance l apart, if a fluid particle travels from a layer y to a layer $y + l$,

$$u' \simeq l \frac{d\overline{u}}{dy} \qquad (4\text{-}37)$$

With this model the turbulent shearing stress, τ_t, in a form analogous to the laminar shearing stress is

$$\tau_t = -\rho\overline{v'u'} = \rho\epsilon_M \frac{d\overline{u}}{dy} \qquad (4\text{-}38)$$

where the symbol ϵ_M is called the *eddy viscosity* or the *turbulent exchange coefficient for momentum*. The eddy viscosity ϵ_M is formally analogous to the kinematic viscosity ν, but whereas ν is a physical property, ϵ_M depends on the dynamics of the flow. Combining Eqs. 4-37 and 4-38 shows that $\epsilon_M = -\overline{v'l}$, and Eq. 4-37 gives the total shearing stress in the form

$$\tau = \rho(\nu + \epsilon_M) \frac{d\overline{u}}{dy} \qquad (4\text{-}39)$$

In turbulent flow ϵ_M is much larger than ν and the viscous term may therefore be neglected.

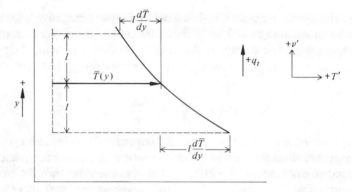

Figure 4-11 Mixing length for energy transfer.

The transfer of energy as heat in a turbulent flow can be pictured in an analogous fashion. Let us consider a two-dimensional time-mean temperature distribution as shown in Fig. 4-11. The fluctuating velocity components continuously transport fluid particles and the energy stored in them across a plane normal to the y direction. The instantaneous rate of energy transfer per unit area at any point in the y direction is

$$(\rho v')(c_p T) \tag{4-40}$$

where $T = \overline{T} + T'$. Following the same line of reasoning that led to Eq. 4-35, the time average of energy transfer due to the fluctuations, called the *turbulent rate of heat transfer*, q_t, is

$$q_t = A\rho c_p \overline{v'T'} \tag{4-41}$$

Using Prandtl's concept of mixing length, we can relate the temperature fluctuation to the time-mean temperature gradient by the equation

$$T' \simeq l\frac{d\overline{T}}{dy} \tag{4-42}$$

This means physically that, when a fluid particle migrates from the layer y to another layer a distance l above or below, the resulting temperature fluctuation is caused chiefly by the difference between the time-mean temperatures in the layers. Assuming that the transport mechanisms of temperature (or energy) and velocity are similar, the mixing lengths in Eqs. 4-37 and 4-42 are equal. The product $\overline{v'T'}$, however, is positive on the average because a positive v' is accompanied by a positive T', and vice versa.

Combining Eqs. 4-41 and 4-42, the turbulent rate of heat transfer per unit area becomes

$$\frac{q_t}{A} = c_p\rho\overline{v'T'} = -c_p\rho\overline{v'}l\frac{d\overline{T}}{dy} \tag{4-43}$$

where the minus sign is a consequence of the second law of thermodynamics (see Chapter 1). To express the turbulent heat flux in a form analogous to the Fourier conduction equation, we define ϵ_H, a quantity called the *turbulent exchange coefficient for temperature, eddy diffusivity of heat*, or *eddy* heat conductivity, by the equation $\epsilon_H = \overline{v'l}$. Substituting ϵ_H for $\overline{v'l}$ in Eq. 4-43 gives

$$\frac{q_t}{A} = -c_p \rho \epsilon_H \frac{d\overline{T}}{dy} \tag{4-44}$$

The total rate of heat transfer per unit area normal to the mean stream velocity can then be written as

$$\frac{q}{A} = \frac{\text{molecular conduction}}{\text{unit area}} + \frac{\text{turbulent transfer}}{\text{unit area}}$$

or in symbolic form as

$$\frac{q}{A} = -c_p \rho (\alpha + \epsilon_H) \frac{d\overline{T}}{dy} \tag{4-45}$$

where $\alpha = k/c_p\rho$, the molecular diffusivity of heat. The contribution to the heat transfer by molecular conduction is proportional to α, and the turbulent contribution is proportional to ϵ_H. For all fluids except liquid metals, ϵ_H is much larger than α in turbulent flow. The ratio of the molecular kinematic viscosity to the molecular diffusivity of heat v/α has previously been named the Prandtl number. Similarly, the ratio of the turbulent eddy viscosity to the eddy diffusivity ϵ_M/ϵ_H could be considered a turbulent Prandtl number Pr_t. According to the Prandtl mixing-length theory, the turbulent Prandtl number is unity, since $\epsilon_M = \epsilon_H = \overline{v'l}$.

Although his treatment of turbulent flow is grossly oversimplified, experimental results indicate it is at least qualitatively correct. Isakoff and Drew (Ref. 3) found that Pr_t for the heating of mercury in turbulent flow inside a tube may vary from 1.0 to 1.6, and Forstall and Shapiro (Ref. 4) found that Pr_t is about 0.7 for gases. The latter investigators also showed that Pr_t is substantially independent of the value of the laminar Prandtl number as well as of the type of experiment. Assuming that Pr_t is unity, the turbulent heat flux can be related to the turbulent shear stress by combining Eqs. 4-38 and 4-44:

$$\frac{q_t}{A} = -\tau_t c_p \frac{d\overline{T}}{d\overline{u}} \tag{4-46}$$

This relation was originally derived in 1874 by the British scientist Osborn Reynolds and is called the *Reynolds analogy*. It is a good approximation whenever the flow is turbulent and can be applied to turbulent boundary layers as well as to turbulent flow in pipes or ducts. However, the Reynolds analogy does not hold in the laminar sublayer. Since this layer offers a large thermal resistance to the flow of heat, Eq. 4-46 does not, in general, suffice for a quantitative solution. Only for fluids having a Prandtl number of unity can it be used directly to calculate the rate of heat transfer. This special case will now be considered.

4-6 REYNOLDS ANALOGY FOR TURBULENT FLOW OVER A FLAT PLATE

To derive a relation between the heat transfer and the skin friction in flow over a plane surface for a Prandtl number of unity, the reader will recall that the laminar shearing stress τ is

$$\tau = \mu \frac{du}{dy}$$

and the rate of heat flow per unit area across any plane perpendicular to the y direction is

$$\frac{q}{A} = -k \frac{dT}{dy}$$

Combining the equations above yields

$$\frac{q}{A} = -\tau \frac{k}{\mu} \frac{dT}{du} \qquad (4\text{-}47)$$

An inspection of Eqs. 4-46 and 4-47 shows that if $c_p = k/\mu$ (i.e., for Pr $= 1$), the same equation of heat flow applies in the laminar and turbulent layers.

To determine the rate of heat transfer from a flat plate to a fluid with Pr $= 1$ flowing over it in turbulent flow, we replace k/μ by c_p and separate the variables in Eq. 4-47. Assuming that q and τ are constant, we get the equation

$$\frac{q_s}{A\tau_s c_p} du = -dT \qquad (4\text{-}48)$$

where the subscript s is used to indicate that both q and τ are taken at the surface of the plate. Integrating Eq. 4-48 between the limits $u = 0$ when $T = T_s$, and $u = V_\infty$ when $T = T_\infty$, yields

$$\frac{q_s}{A\tau_s c_p} V_\infty = (T_s - T_\infty) \qquad (4\text{-}49)$$

But since by definition the local heat-transfer and friction coefficients are

$$h_{cx} = \frac{q_s}{A(T_s - T_\infty)} \quad \text{and} \quad \tau_{sx} = C_{fx} \frac{\rho V_\infty^2}{2}$$

Equation 4-49 can be written $\quad$ Assuming $\Pr \cong 1$

$$\frac{h_{cx}}{c_p \rho V_\infty} = \frac{Nu_x}{Re_x Pr} = \frac{C_{fx}}{2} \tag{4-50}$$

Equation 4-50 is satisfactory for gases in which Pr is approximately unity. Colburn (Ref. 5) has shown that Eq. 4-50 can also be used for fluids having Prandtl numbers ranging from 0.6 to about 50 if it is modified in accordance with experimental results to read $\qquad .6 < \Pr < 50$

CoLBuRN EQuATioN

$$\frac{Nu_x}{Re_x Pr} Pr^{2/3} = St_x Pr^{2/3} = \frac{C_{fx}}{2} \tag{4-51}$$

where the subscript x denotes the distance from the leading edge of the plate.

To apply the analogy between heat transfer and momentum transfer, in practice it is necessary to know the friction coefficient C_{fx}. For turbulent flow over a plane surface the empirical equation for the local friction coefficient

$$C_{fx} = 0.0576 \left(\frac{V_\infty x}{\nu} \right)^{-1/5} \tag{4-52}$$

is in good agreement with experimental results in the Reynolds number range between 5×10^5 and 10^7 as long as no separation occurs. Assuming that the turbulent boundary layer starts at the leading edge, the average friction coefficient over a plane surface of length L can be obtained by integrating Eq. 4-52:

$$\bar{C}_f = \frac{1}{L} \int_0^L C_{fx} \, dx = 0.072 \left(\frac{V_\infty L}{\nu} \right)^{-1/5} \tag{4-53}$$

In reality, however, a laminar boundary layer precedes the turbulent boundary layer between $x=0$ and $x=x_c$. Since the local frictional drag of a laminar boundary layer is less than the local frictional drag of a turbulent boundary layer at the same Reynolds number, the average drag calculated from Eq. 4-53 without correcting for the laminar portion of the boundary layer is too large. The actual drag can be closely estimated, however, by assuming that, behind the point of transition, the turbulent boundary layer behaves as though it had started at the leading edge.

Adding the laminar friction drag between $x=0$ and $x=x_c$ to the turbulent drag between $x=x_c$ and $x=L$ gives, per unit width,

$$\bar{C}_f = \frac{\left[0.072 Re_L^{-1/5} L - 0.072 Re_{x_c}^{-1/5} x_c + 1.33 Re_{x_c}^{-1/2} x_c \right]}{L}$$

For a critical Reynolds number of 5×10^5 this yields

$$\bar{C}_f = 0.072 \left(Re_L^{-1/5} - \frac{0.0464 x_c}{L} \right) \tag{4-54}$$

STANTON # $= S_T = \dfrac{Nu}{Re\, P_R}$

Substituting Eq. 4-52 for C_{fx} in Eq. 4-51 yields the local Nusselt number at any value of x larger than x_c:

FOR $X > X_c$
$$Nu_x = \frac{h_{cx}x}{k} = 0.0288 \ Pr^{1/3}\left(\frac{V_\infty x}{\nu}\right)^{0.8} \tag{4-55}$$

We observe that the local heat-transfer coefficient h_{cx} for heat transfer by convection through a turbulent boundary layer decreases with the distance x as $h_{cx} \propto 1/x^{0.2}$. Equation 4-55 shows that, in comparison with laminar flow, where $h_{cx} \propto 1/x^{1/2}$, the heat-transfer coefficient in turbulent flow decreases less rapidly with x and that the turbulent-heat-transfer coefficient is much larger than the laminar-heat-transfer coefficient at a given value of the Reynolds number.

The average conductance in turbulent flow over a plane surface of length L can be calculated to a first approximation by integrating Eq. 4-55 between $x = 0$ and $x = L$:

$$\bar{h}_c = \frac{1}{L}\int_0^L h_{cx}\,dx$$

In dimensionless form we get

$$\overline{Nu}_L = \frac{\bar{h}_c L}{k} = 0.036 \ Pr^{1/3}Re_L^{0.8} \tag{4-56}$$

Equation 4-56 neglects the existence of the laminar boundary layer and is therefore valid only when $L \gg x_c$. The laminar boundary layer can be included in the analysis if Eq. 4-30 is used between $x = 0$ and $x = x_c$, and Eq. 4-55 between $x = x_c$ and $x = L$ for the integration of h_{cx}. This yields, with $Re_c = 5 \times 10^5$,

$$\overline{Nu}_L = 0.036 \ Pr^{1/3}\left(Re_L^{0.8} - 23{,}200\right) \tag{4-57}$$

Example 4-2. The crankcase of an automobile is approximately 0.6 m long, 0.2 m wide, and 0.1 m deep. Assuming that the surface temperature of the crankcase is 350 K, estimate the rate of heat flow from the crankcase to atmospheric air at 276 K at a road speed of 30 m/s. Assume that the vibration of the engine and the chassis induce the transition from laminar to turbulent flow so near to the leading edge that, for practical purposes, the boundary layer is turbulent over the entire surface. Neglect radiation and use for the front and rear surfaces the same average convective-heat-transfer coefficient as for the bottom and sides.

Solution: Using the properties of air at 313 K from Table G-1, the Reynolds number is

$$Re_L = \frac{\rho V_\infty L}{\mu} = \frac{1.092 \times 30 \times 0.6}{19.123 \times 10^{-6}} = 1.03 \times 10^6$$

From Eq. 4-56 the average Nusselt number is

$$\overline{Nu}_L = 0.036 \, Pr^{1/3} Re_L^{0.8}$$
$$= 0.036(0.71)^{1/3}(1.03 \times 10^6)^{0.8}$$
$$= 2075$$

and the average convective heat-transfer coefficient becomes

$$\bar{h}_c = \frac{\overline{Nu}_L k}{L} = \frac{2075 \times 0.0265}{0.6} = 91.6 \, W/m^2 \cdot K$$

The surface area that dissipates heat is 0.28 m² and the rate of heat loss from the crankcase is therefore

$$q = \bar{h}_c A (T_s - T_\infty) = 91.6 \times 0.28 \times (350 - 276)$$
$$q = 1898 \, W$$

4-7 LAMINAR FORCED CONVECTION IN A TUBE

The laminar-forced-convection coefficient in flow through a tube will be derived for a fully developed flow profile and a constant heat flux at the wall. To derive the conservation of energy equation for this case, consider a small cylindrical element of length dx, inner radius r, and outer radius $r + dr$, as shown in Fig. 4-12. Heat will be conducted into and out of the element in the radial direction, whereas the energy transport in the axial direction is by convection. Then, the rate at which heat is conducted into the element is

$$q_r = - k 2\pi r \, dx \frac{\partial T}{\partial r}$$

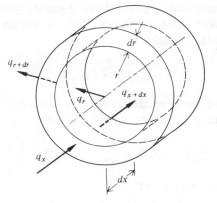

Figure 4-12 Element of laminar flow in a tube.

whereas the net rate at which heat is conducted out of the element is given by

$$\frac{\partial q_r}{\partial r} dr = -k2\pi \, dx \frac{\partial}{\partial r}\left(r\frac{\partial T}{\partial r}\right)dr$$

The axial velocity through the element is constant, but the temperature in the element changes in the axial direction. The rate of convection into the element is

$$q_x = 2\pi r \, dr \, \rho u c_p T$$

and out of the element the rate of convection is

$$2\pi r \, dr \, \rho u c_p \left(T + \frac{\partial T}{\partial x} dx\right)$$

Equating the net rate of conduction to the net rate of convection for steady-state conditions yields the energy equation for laminar flow in a tube:

$$\frac{1}{ur}\frac{\partial}{\partial r}\left(r\frac{\partial T}{\partial r}\right) = \frac{\rho c_p}{k}\frac{\partial T}{\partial x} \tag{4-58}$$

For the constant wall heat flux, q_s'', and constant fluid properties, the temperature of the fluid at any value r must increase linearly in the direction of flow so that (dT/dx) is constant. Other conditions applicable to this system are that at $r=0$ (tube axis), $\partial T/\partial r=0$; and at $r=r_s$, $T=T_s$. Also, at $r=r_s$ the heat flux is related to the temperature gradient according to

$$q_s'' = -k\left(\frac{\partial T}{\partial r}\right)_{r_s}$$

The assumption that (dT/dx) is a constant reduces Eq. 4-58 from a partial differential equation to a total differential equation in which the velocity at any radial distance r is a function of the velocity at the axis of the tube, u_{max}. In fully developed laminar flow the velocity distribution in the tube is parabolic and can be written in dimensionless form as a function of the radial distance,

$$\frac{u}{u_{max}} = 1 - \left(\frac{r}{r_s}\right)^2 \tag{4-59}$$

where r_s is one-half the pipe diameter. Substituting this velocity distribution in Eq. 4-58 yields, after some rearrangement,

$$\frac{\partial}{\partial r}\left(r\frac{\partial T}{\partial r}\right) = \frac{1}{\alpha}\frac{\partial T}{\partial x}u_{max}\left[1 - \left(\frac{r}{r_s}\right)^2\right]r \tag{4-60}$$

which can be integrated to give

$$r\frac{\partial T}{\partial r} = \frac{1}{\alpha}\frac{\partial T}{\partial x}u_{max}\left(\frac{r^2}{2} - \frac{r^4}{4r_s^2}\right) + C_1 \tag{4-61}$$

A second integration gives a temperature distribution as a function of radius in the form

$$T = \frac{1}{\alpha} \frac{\partial T}{\partial x} u_{max} \left(\frac{r^2}{4} - \frac{r^4}{16r_s^2} \right) + C_1 \ln r + C_2 \qquad (4\text{-}62)$$

The constants C_1 and C_2 are obtained from the boundary conditions as follows. Since $\partial T / \partial r = 0$ at $r = 0$, $C_1 = 0$. The other boundary condition is that at $r = r_s$, $T = T_s$, so that

$$T_s = \frac{1}{\alpha} \frac{\partial T}{\partial x} u_{max} \left(\frac{r_s^2}{4} - \frac{r_s^2}{16} \right) + C_2 \qquad (4\text{-}63)$$

and

$$C_2 = T_s - \frac{1}{\alpha} \frac{\partial T}{\partial x} u_{max} \frac{3r_s^2}{16} \qquad (4\text{-}64)$$

Substituting for C_1 and C_2 in Eq. 4-62 gives the temperature distribution in the form

$$T(x,r) = \frac{1}{\alpha} \frac{\partial T}{\partial x} u_{max} r_s^2 \left[\frac{1}{4} \left(\frac{r}{r_s} \right)^2 - \frac{1}{16} \left(\frac{r}{r_s} \right)^4 - \frac{3}{16} \right] + T_s \qquad (4\text{-}65)$$

If we let θ equal to $T(r) - T_s$, the temperature profile can now be cast into a dimensionless form,

$$\frac{\theta}{\theta_a} = 1 - \frac{4}{3} \left(\frac{r}{r_s} \right)^2 + \frac{1}{3} \left(\frac{r}{r_s} \right)^4 \qquad (4\text{-}66)$$

where θ_a is

$$\theta_a = - \frac{3}{16} \left(\frac{1}{\alpha} \frac{\partial T}{\partial x} u_{max} r_s^2 \right) \qquad (4\text{-}67)$$

The heat-transfer coefficient can now be obtained by evaluating the temperature gradient at $r = r_s$. Using the definition of the heat transfer coefficient and substituting the temperature gradient at the wall yields

$$\left(\frac{d\theta}{dr} \right)_{r_s} = \theta_a \left(- \frac{8}{3r_s} + \frac{4}{3r_s} \right) = - \frac{4\theta_a}{3r_s} \qquad (4\text{-}68)$$

and

$$q_s'' = - k \left(\frac{d\theta}{dr} \right)_{r_s} = \frac{4k\theta_a}{3r_s} = \bar{h}_c \theta_a \qquad (4\text{-}69)$$

or

$$\bar{h}_c = \frac{4k}{3r_s} \qquad (4\text{-}70)$$

The relationship above can be expressed in terms of the conventional Nusselt number:

$$\overline{Nu}_D = \frac{\bar{h}_c D}{k} = \frac{8}{3} \qquad (4\text{-}71)$$

Table 4-1 Heat Transfer and Friction for Fully Developed Laminar Flow of a Newtonian Fluid Through Specified Ducts.[a]

GEOMETRY (L/D$_h$ > 100)	Nu$_{H1}$	Nu$_{H2}$	Nu$_T$	fRe	(j$_{H1}$/f)[b]	$\dfrac{Nu_{H1}}{Nu_T}$
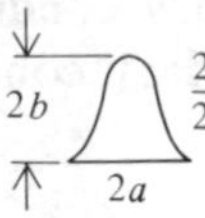 $2b$ $\quad \dfrac{2b}{2a} = \dfrac{\sqrt{3}}{2}$ $\quad 2a$	3.014	1.474	2.39[c]	12.630	0.269	1.26
$60°$ $\quad 2a$ $\dfrac{2b}{2a} = \dfrac{\sqrt{3}}{2}$ $\quad 2b$	3.111	1.892	2.47	13.333	0.263	1.26
$2b$ $\square$ $\dfrac{2b}{2a} = 1$ $\quad 2a$	3.608	3.091	2.976	14.227	0.286	1.21
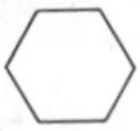 (hexagon)	4.002	3.862	3.34[c]	15.054	0.299	1.20
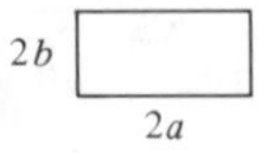 $2b$ $\square$ $\dfrac{2b}{2a} = \dfrac{1}{2} = .5$ $\quad 2a$	4.123	3.017	3.391	15.548	0.299	1.22
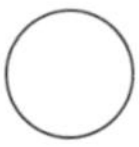 (circle)	4.364	4.364	3.657	16.000	0.307	1.19
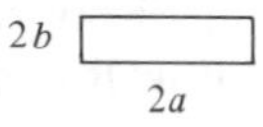 $2b$ $\square$ $\dfrac{2b}{2a} = \dfrac{1}{4} = .25$ $\quad 2a$	5.099	4.35[c]	3.66	18.700	0.307	1.39
$2b$ (ellipse) $\dfrac{2b}{2a} = .9$ $\quad \leftarrow 2a \rightarrow$	5.331	2.930	4.439	18.233	0.329	1.20
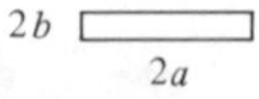 $2b$ $\square$ $\dfrac{2b}{2a} = \dfrac{1}{8} = .125$ $\quad 2a$	6.490	2.904	5.597	20.585	0.353	1.16

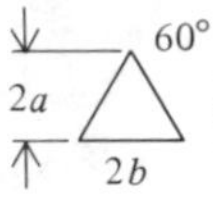

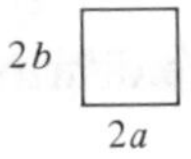

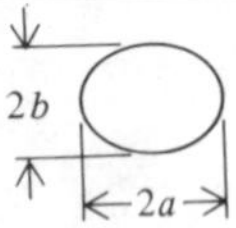

(handwritten annotations)

EQILATERAL TRIANGLE *(label at left of second row)*

FROM EQ 4-6   Pg 201

$$D_H = 4\left(\frac{CROSS\text{-}SECTIONAL\ AREA}{WETTED\ PERIMETER}\right)$$

$\dfrac{2b}{2a}=0$	8.235	8.235	7.541	24.000	0.386	1.09
$\dfrac{b}{a}=0$ insulated	5.385	–	4.861	24.000	0.253	1.11

[a]Subscript $H1$ indicates a uniform heat flux in the flow direction with a uniform wall temperature at a flow cross section; $H2$ indicates a uniform wall heat flux both around the periphery as well as in the flow direction; T denotes a uniform wall temperature at the boundary.
[b]This heading is the same as $\mathrm{Nu}_{H1}\,\mathrm{Pr}^{-1/3}/\mathrm{Re}$ with $\mathrm{Pr}=0.7$.
[c]Interpolated values.
Source: R. K. Shah and A. L. London, *Laminar Flow Forced Convection in Ducts*, Academic Press, New York, N.Y., 1978. See also refs. 6 and 7.

The Nusselt number in the relation above is based on the difference in temperature between the tube axis and the interface at $r_s = D/2$. From a practical point of view, however, it is usually more convenient to base the Nusselt number on the difference between the bulk temperature of the fluid and the temperature of the interface. The bulk temperature of the fluid is usually called the *mixing cup temperature* (see Problem 4-32) and is obtained by collecting the fluid leaving the duct in a cup and thoroughly mixing it. At the inlet, of course, the temperature is uniform, so the mixing cup temperature and the axial temperature are identical. Basing the Nusselt number and the heat-transfer coefficient on the difference in temperature between the bulk of the fluid and the interface yields the following value for the Nusselt number (6):

$$\overline{\mathrm{Nu}_D} = 4.36 \qquad (4\text{-}72)$$

It should be observed that the value of the Nusselt number in fully developed laminar flow is independent of the Reynolds number because in fully developed flow the boundary-layer thickness equals the tube radius. However, near the inlet to the tube the value of the Nusselt number will be larger than that for the fully developed flow condition, and in Chapter 5 equations will be presented for the empirical evaluation of the Nusselt number in the inlet section. Table 4-1 gives values of the Nusselt number of fully developed flow in ducts of various cross section.

Example 4-3. Calculate the heat-transfer coefficient for fully developed laminar flow of glycerine at a bulk temperature of 293 K through a 0.5×0.5 m duct with walls at 400 K, at a velocity of 0.1 m/s.

Solution: From Eq. 4-6, the hydraulic diameter is

$$D_H = 4\left(\frac{0.5 \times 0.5}{4 \times 0.5}\right) = 0.5 \text{ m}$$

From Table 4-1 for a square duct with a constant wall temperature,

$$\text{Nu}_{D_H} = 2.976 = \frac{h_c D_H}{k}$$

This result compares favorably with the value for the Nusselt number given in Eq. 4-71.

From Table F-3 at 293 K, (APPENDIX)

$$k = 0.285 \text{ W/m} \cdot \text{K}$$

Substituting and solving for the heat-transfer coefficient gives

$$h_c = \frac{(2.976)(0.285)}{0.5} = 1.70 \text{ W/m}^2 \cdot \text{K}$$

4-8 REYNOLDS ANALOGY FOR TURBULENT FLOW IN A TUBE

The assumptions necessary for the simple analogy are valid only for fluids having a Prandtl number of unity, but the fundamental relation between heat transfer and fluid friction for flow in ducts can be illustrated for this case without introducing mathematical difficulties. The results of the simple analysis can also be extended to other fluids by means of empirical correction factors, as will be shown in Chapter 5.

The rate of heat flow per unit area in a fluid can be related to the temperature gradient by the equation developed previously:

$$\frac{q}{A\rho c_p} = -\left(\frac{k}{\rho c_p} + \epsilon_H\right)\frac{dT}{dy} \tag{4-45}$$

Similarly, the shearing stress caused by the combined action of the viscous forces and the turbulent momentum transfer is given by

$$\frac{\tau}{\rho} = \left(\frac{\mu}{\rho} + \epsilon_M\right)\frac{du}{dy} \tag{4-39}$$

According to the Reynolds analogy, heat and momentum are transferred by analogous processes in turbulent flow. Consequently, both q and τ vary with y, the distance from the surface, in the same manner. For fully developed turbulent flow in a pipe, the local shearing stress decreases linearly with the radial distance, r. Hence we can write

$$\frac{\tau}{\tau_s} = \frac{r}{r_s} = 1 - \frac{y}{r_s} \tag{4-73}$$

and

$$\frac{q/A}{(q/A)_s} = \frac{r}{r_s} = 1 - \frac{y}{r_s} \tag{4-74}$$

where the subscript s denotes conditions at the inner surface of the pipe. Introducing Eqs. 4-73 and 4-74 into Eqs. 4-39 and 4-45, respectively, yields

$$\frac{\tau_s}{\rho}\left(1 - \frac{y}{r_s}\right) = \left(\frac{\mu}{\rho} + \epsilon_M\right)\frac{du}{dy} \tag{4-75}$$

and

$$\frac{q_s}{A_s \rho c_p}\left(1 - \frac{y}{r_s}\right) = -\left(\frac{k}{\rho c_p} + \epsilon_H\right)\frac{dT}{dy} \tag{4-76}$$

If $\epsilon_H = \epsilon_M$, the brackets on the right-hand side of Eqs. 4-75 and 4-76 are equal, provided that the molecular diffusivity of momentum μ/ρ equals the molecular diffusivity of heat $k/\rho c_p$, that is, when the Prandtl number is unity. Dividing Eq. 4-76 by Eq. 4-75 yields, under these restrictions,

$$\frac{q_s}{A_s c_p \tau_s} du = -dT \tag{4-77}$$

Equation 4-77 can be integrated between the wall, where $u=0$ and $T=T_s$, and the bulk of the fluid, where $u = V_m$ and $T = T_b$. The integration then yields

$$\frac{q_s V_m}{A_s c_p \tau_s} = T_s - T_b \tag{4-78}$$

which can also be written in the form

$$\frac{\tau_s}{\rho V_m^2} = \frac{q_s}{A_s(T_s - T_b)}\frac{1}{c_p \rho V_m} = \frac{\bar{h}_c}{c_p \rho V_m} \tag{4-79}$$

since $\bar{h}_c$ is by definition equal to $q_s/A_s(T_s - T_b)$. Multiplying the numerator and the denominator of the right-hand side of Eq. 4-79 by $D_H \mu k$ and regrouping yields

$$\frac{\bar{h}_c}{c_p \rho V_m}\frac{D_H \mu k}{D_H \mu k} = \left(\frac{\bar{h}_c D_H}{k}\right)\left(\frac{k}{c_p \mu}\right)\left(\frac{\mu}{V_m D_H \rho}\right) = \frac{\overline{Nu}}{Re\, Pr} = \overline{St} \tag{4-80}$$

where St is the *Stanton number*.

To bring the left-hand side of Eq. 4-79 into a more convenient form, we make a force balance on a cylindrical mass of fluid as shown in Fig. 4-13. The pressure difference $p_1 - p_2$ exerts the force $(p_1 - p_2)\pi D^2/4$, which is

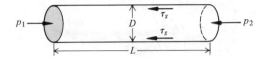

Figure 4-13 Nomenclature for force balance on a fluid element in a tube.

balanced in steady flow by the shear at the wall:

$$(p_1 - p_2)\frac{\pi D^2}{4} = \tau_s \pi DL \qquad (4\text{-}81)$$

Solving for the wall shear per unit area yields

$$\tau_s = \frac{(p_1 - p_2)D}{4L} \qquad (4\text{-}82)$$

Expressing the pressure drop in terms of the *friction factor f* gives

$$p_1 - p_2 = f\frac{L}{D}\frac{\rho V_m^2}{2} \qquad (4\text{-}83)$$

Substituting Eq. 4-83 for $p_1 - p_2$ in Eq. 4-81 gives

$$\tau_s = f\frac{\rho V_m^2}{8} \qquad (4\text{-}84)$$

Substituting Eq. 4-84 for τ_s in Eq. 4-79 finally yields the equation

$$St = \frac{Nu}{Re\,Pr} = \frac{f}{8} \qquad (4\text{-}85)$$

which is known as the *Reynolds analogy* for flow in a tube.* It agrees fairly well with experimental data for heat transfer in gases whose Prandtl number is nearly unity.

According to experimental data for fluids flowing in smooth tubes in the range of Reynolds numbers from 10,000 to 120,000, the friction factor f is given by the empirical relation

$$f = 0.184\,Re_D^{-0.2}$$

Using this relation, Eq. 4-85 can be written as

$$St = \frac{Nu}{Re\,Pr} = 0.023\,Re_D^{-0.2} \qquad (4\text{-}86)$$

or, since Pr was assumed unity, as

$$\frac{h_c\,D}{K} = Nu = 0.023\,Re_D^{0.8} \qquad (4\text{-}87)$$

*The Reynolds analogy can be extended to mass transfer. The analogies among mass, heat, and momentum transfer will be discussed in Chapter 8.

or

$$\bar{h}_c = 0.023 \, V_m^{0.8} D^{-0.2} k \left(\frac{\mu}{\rho}\right)^{-0.8} \tag{4-88}$$

We observe that, in fully established turbulent flow, the convective-unit conductance is directly proportional to the velocity raised to the 0.8 power and inversely proportional to the tube diameter raised to the 0.2 power. For a given flow rate, an increase in the tube diameter reduces the velocity and thereby causes a decrease in $\bar{h}_c$ proportional to $1/D^{1.8}$. The use of small tubes and high velocities is therefore conducive to large heat-transfer coefficients, but at the same time the power required to overcome the frictional resistance is increased. In the design of heat-exchange equipment it is therefore necessary to strike a balance between the gain in heat-transfer rates achieved by the use of ducts having small cross-sectional areas and the accompanying increase in pumping requirements.

REFERENCES

1. H. Schlichting, *Boundary Layer Theory*, 6th ed. (translated by J. Kestin), New York: McGraw-Hill Book Co., Inc., 1968.
2. E. Pohlhausen, "Der Wärmeaustausch zwischen festen Körpern und Flüssigkeiten mit kleiner Reibung und Kleiner Wärmeleitung," ZAMM, Vol. 1, p. 115, 1921.
3. S. E. Isakoff and T. B. Drew, "Heat and Momentum Transfer in Turbulent Flow of Mercury," Inst. Mech. Eng. and ASME, *Proc. General Discussion on Heat Transfer*, pp. 405–409, 1951.
4. W. Forstall, Jr. and A. H. Shapiro, "Momentum and Mass Transfer in Co-axial Jets," *J. Appl. Mech*, Vol. 17, pp. 399, 1950.
5. A. P. Colburn, "A Method of Correlating Forced Convection Heat Transfer Data and a Comparison with Fluid Friction," *Trans. Am. Inst. Chem. Engrs.*, Vol. 29, pp. 174–210, 1933.
6. R. K. Shah, "Laminar Flow Friction and Forced Convection Heat Transfer in Ducts of Arbitrary Geometry," *Int. J. Heat Mass Transfer*, Vol. 18, pp. 849–862, 1975.
7. R. K. Shah and A. L. London, "Thermal Boundary Conditions and Some Solutions for Laminar Duct Flow Forced Convection," *J. Heat Transfer*, Vol. 96, pp. 159–165, 1974.

PROBLEMS

The problems in this chapter are organized in the manner shown in the table.

PROBLEM NUMBERS	SECTIONS	SUBJECT
4-1 to 4-13	4-1 and 4-2	Laminar flow over a flat plate
4-14 to 4-17	4-3	Integral equations for laminar flow
4-18 to 4-25	4-4 and 4-5	Laminar and turbulent flow over a flat plate
4-26 to 4-31	4-6	Reynolds analogy
4-32 to 4-44	4-7 and 4-8	Laminar and turbulent flow in a tube

4-1 A fluid with a temperature of 40°C flows through a circular tube having a diameter of 15 cm. The fluid has an average velocity of 2 m/s. Calculate the Reynolds number if the fluid is:
  (a) Air.
  (b) CO_2.
  (c) Water.
  (d) Engine oil.
Use the properties given in the Appendixes.

4-2 For the conditions given in Problem 4-1, determine if the flow of all the liquids is laminar, transitional, or turbulent.

4-3 Assume that the velocity of a fluid varies parabolically across the cross section of a circular pipe according to the equation

$$\frac{u}{V_{max}} = \left[1 - \left(\frac{r}{R} \right)^2 \right]$$

where $V_{max} = 30$ m/s (the maximum or centerline velocity) and $R = 20$ cm (the inside radius of the pipe). By integrating the velocity profile across the pipe cross-sectional area, determine the average velocity of the fluid.

4-4 Air at 20°C flows over a flat plate with a free-stream velocity 10 m/s. Determine the distance from the leading edge of the plate at which the flow first:
  (a) Enters the transitional regime.
  (b) Enters the turbulent regime.

Assume that the critical Reynolds number for transitional flow is 2×10^5 and for fully turbulent flow is 5×10^5.

4-5 Rework Problem 4-4 assuming that the fluid is (a) water, and (b) mercury.

4-6 A fluid at 40°C flows through a tube with diameter of 7 cm. Calculate the mean fluid velocity necessary to produce both transitional and turbulent flow. Assume that the fluid is:

(a) Air.
(b) Water.
(c) Engine oil.

4-7 Air at 20°C flows through a pipe with 15 cm i.d. The mass flow rate of the air is 1.05×10^{-3} kg/s. Determine the Reynolds number of the air.

4-8 Assume that the temperature profile of air near the surface of an isothermal flat plate is approximated by the equation

$$\frac{T - T_s}{T_\infty - T_s} = \frac{y}{50} \qquad (0 \leqslant y \leqslant 50 \text{ cm})$$

where

$T_s = $ temperature of plate

$T_\infty = $ ambient temperature of fluid

$y = $ distance measured perpendicular to surface of plate, cm

Calculate the convective-heat-transfer coefficient on the plate and the total heat transfer from a plate with a surface area of 2 m² and

$$T_s = 100°C$$
$$T_\infty = 0°C$$

4-9 Water at 60°C flows through a 10×15 cm rectangular duct with a mean velocity of 8 m/s. Calculate the Reynolds number of the water.

4-10 The velocity profile of 60°C air flowing near the surface of a square plate with a surface area exposed to the air of 1.8 m² is approximated by

$$\frac{u}{V_\infty} = 1 - e^{-2y}$$

where y is the distance from the surface of the plate measured in centimeters. Estimate the boundary layer thickness of the air.

4-11 For the conditions stated in Problem 4-10, calculate the shear stress on the plate due to the motion of the air when $V_\infty = 3$ m/s. Also, determine the total drag force on the plate.

4-12 The variation of the local heat-transfer coefficient with distance x measured from the leading edge of a flat plate is given by the dimensionless expression

$$\frac{h_{cx}}{k} = 0.332 \left(\frac{\mu c_p}{k} \right)^{1/3} \mathrm{Re}_x^{1/2}$$

Determine an expression for the average heat-transfer coefficient $\bar{h}$ if the length of the plate in the direction of flow is L. Assume the fluid properties are constant.

4-13 The local heat-transfer coefficient for turbulent flow over a flat plate is given by the expression

$$\frac{h_{cx}}{k} = 0.0288 \left(\frac{\mu c_p}{k} \right)^{1/3} \mathrm{Re}_x^{4/5}$$

where x is measured from the leading edge of the plate. Assuming that turbulent flow begins at the leading edge and flow properties are constant, determine an expression for the average heat-transfer coefficient for a plate of length L in the direction of flow.

4-14 The velocity profile inside the boundary layer for flow over a flat plate assumed to be

$$u = a + by + cy^2 + dy^3$$

where y is the coordinate measured perpendicular to the plate. By applying appropriate boundary conditions solve for the coefficients a, b, c, and d and show that dimensionless velocity profile is

$$\frac{u}{V_\infty} = \frac{3}{2} \frac{y}{\delta} - \frac{1}{2} \left(\frac{y}{\delta} \right)^3$$

Using this velocity profile, determine an expression for the shear stress at the wall in terms of δ.

4-15 The temperature profile inside the boundary layer near a flat plate may be assumed to be the polynomial:

$$T = e + fy + gy^2 + hy^3$$

where y is the coordinate measured perpendicular to the surface of the plate. Apply appropriate boundary conditions and show that the temperature profile equation becomes

$$\frac{T - T_s}{T_\infty - T_s} = \frac{3}{2} \frac{y}{\delta_t} - \frac{1}{2} \left(\frac{y}{\delta_t} \right)^3$$

4-16 Suppose that the velocity profile for flow of a fluid over a flat plate is approximated by the linear profile

$$u = a + by$$

(a) Apply the appropriate boundary conditions and express u in terms of δ and V_∞.

(b) Substitute the velocity profile into the integral momentum equation (Eq. 4-16), and determine an expression for the local boundary layer thickness $\delta(x)$.

(c) Compare your result for $\delta(x)$ with Eq. 4-25, which was obtained with a similar analysis but assuming a more complicated cubic velocity profile.

(d) Using the expression for $\delta(x)$, obtain an expression for C_{fx} and compare your result with the one in Example 4-1.

4-17 Assume a linear temperature profile

$$T = c + dy$$

for flow of a fluid over a flat plate:

(a) Apply the appropriate boundary conditions and express T in terms of δ_t, T_s, and T_∞.

(b) Assuming that the velocity profile is also linear (see Problem 4-16), substitute in the integral energy equation (Eq. 4-18) and obtain an expression for δ/δ_t as a function of the Prandtl number.

(c) Compare your result for δ/δ_t with Eq. 4-28, which was obtained by a more complicated cubic velocity and temperature profile.

(d) Using your expression for δ/δ_t, obtain an expression for Nu_x and compare your result with Eq. 4-30.

4-18 A square flat plate with length of 35 cm is maintained at a temperature of 120°C. Air at 20°C and a free-stream velocity of 22 m/s is forced over the plate. Calculate the heat-transfer rate from the plate.

4-19 A rectangular plate is 120 cm long in the direction of flow and 200 cm wide. The plate is maintained at 80°C when placed in nitrogen that has a velocity of 2.5 m/s and a temperature of 0°C. Determine the following quantities:

(a) The average friction coefficient.
(b) The viscous drag exerted on the plate.
(c) The average convective-heat-transfer coefficient.
(d) The total heat-transfer rate from the plate.

4-20 Air at atmospheric pressure flows over a flat plate that has a length of 3 m and a width of 5 m. The air is at 15°C and the plate is at 65°C. The velocity of the air is 35 m/s. Calculate the following quantities at the end of the plate (i.e., $x = 3$ m):

(a) The local friction coefficient.
(b) The local heat-transfer coefficient.

4-21 Using the information given in Problem 4-20, calculate the required quantities at the local position where the flow becomes turbulent. Assume that $Re_c = 5 \times 10^5$.

4-22 Water flows over a square plate that is 2 m on a side. The ambient temperature of the water is 90°C and it has a velocity of 10 m/s. The plate is isothermal with a temperature of 30°C. Determine the drag on the plate and the rate of heat transfer from its surface.

4-23 A flat electric heater is rated at 400 W. The heater is cooled by 30°C air forced over the heater parallel to the surface with a velocity of 28 m/s. The heater is 75 cm long in the direction of flow and 125 cm long in the direction perpendicular to the flow. Estimate the temperature of the heater. To simplify the calculations, evaluate the properties of air at the ambient temperature.

4-24 Assume turbulent flow begins at the leading edge of a flat plate. Use the expression for local friction coefficient given by Eq. 4-52 and show that the average friction coefficient over a plate of length equal to L is given by Eq. 4-53.

4-25 Verify Eq. 4-57 as the average Nusselt number over a flat plate of length L. Use Eq. 4-30 for the local Nusselt number in the laminar regime and Eq. 4-55 in the turbulent regime. Assume a value for the critical Reynolds number of 5×10^5.

4-26 The friction coefficient is equal to 0.075 when a heated surface is placed in air that has an average temperature of 40°C and a velocity of 20 m/s. Estimate the average convective heat-transfer coefficient for the surface.

4-27 The Stanton number St is defined as

$$St = \frac{Nu}{Re\,Pr}$$

Determine an expression for St in terms of fluid properties, the heat-transfer coefficient, and the fluid velocity.

4-28 A body with a surface area of 0.55 m² is heated to 110°C and placed in water that has a temperature of 10°C. The velocity of the water over the plate is 5 m/s and the drag force on the body is 4.0 N. Calculate the heat-transfer rate from the body.

4-29 A flat electric heater with surface area of 0.10 m² is to be maintained at a constant surface temperature of 80°C. The heater is to operate in 10°C CO_2 that flows across the surface of the heater with a velocity of 20 m/s. If the drag from the CO_2 on the surface of the heater is 0.2 N, calculate the capacity of the heater in watts.

4-30 Measurements made on a body placed in a wind tunnel indicate that the viscous drag on the body is 100 N. The surface area of the body is 0.80 m² and its temperature is 200°C. The air has an ambient temperature of 30°C and velocity of 35 m/s. Estimate the average heat-transfer coefficient and the total heat-transfer rate from the body.

4-31 An electronic instrument package used to measure properties of the water is towed behind a ship. When the velocity of the ship is 24 m/s the tension in the towing cable is 370 N. The instrument package has a surface area of 3 m² and a power consumption 8500 W. Estimate the surface temperature of the package if the ambient water temperature is 15°C.

4-32 The mixing cup temperature or mean temperature of a fluid flowing through a pipe is obtained by hypothetically removing the fluid from the pipe and adiabatically mixing the fluid until it reaches a uniform temperature:

(a) Show that the mean temperature for flow of a fluid with velocity $u(r)$ through a pipe with internal radius r_w is

$$T_m = \frac{\int_{r=0}^{r_w} \rho(r)c_p(r)u(r)T(r)r\,dr}{\int_{r=0}^{r_w} \rho(r)c_p(r)u(r)r\,dr}$$

(b) Show that this expression can be reduced to

$$T_m = \frac{\int_{r=0}^{r_w} u(r)T(r)r\,dr}{\int_{r=0}^{r_w} u(r)r\,dr}$$

for a fluid with constant properties.

(c) Show that for slug flow (u=constant) the expression for T_m reduces to

$$T_m = \frac{2\pi}{A} \int_{r=0}^{r_w} T(r)r\,dr$$

where A is the cross-sectional area of the pipe.

4-33 Calculate the mean temperature of a constant property fluid flowing through a 6-cm-diameter pipe if the temperature variation with radius is given by

$$\frac{T-50}{100} = 1-\left(\frac{r}{3}\right)^2$$

where T is in °C and r is in centimeters. Assume slug flow and use the results given in Problem 4-32.

4-34 Fully developed laminar flow of saturated liquid Freon 12 occurs through a tube with a diameter of 1 cm. The tube wall is isothermal with a temperature of 0°C, and the mean temperature of the Freon is 20°C. Determine the heat-transfer rate from the Freon if the length of tube is 1.7 m.

4-35 Engine oil with a mean temperature of 100°C is drained from an engine through a 5-mm-tube that has a wall temperature of 40°C. Assuming fully developed laminar flow, calculate the convective-heat-transfer coefficient for the flow and the heat-transfer rate from the oil per meter length of the tube.

4-36 Carbon dioxide flows through a duct with a triangular cross section. The duct wall is isothermal with a temperature of 20°C and the mean temperature of the CO_2 is 120°C. The length of each side of the duct is 1.20 cm. Calculate the heat-transfer rate per unit axial length of the duct assuming fully developed laminar flow.

4-37 Water flows through a 10×4 mm rectangular duct. The average mean temperature of the water is 10°C and the flow is laminar and fully developed. Estimate the convective heat-transfer coefficient for the water assuming:

(a) Constant duct temperature.

(b) Uniform heat flux in the flow direction and uniform duct temperature at a given flow cross section.

(c) Uniform heat flux over the entire duct area.

4-38 Nitrogen at atmospheric pressure and a mean temperature of 40°C flows through a smooth tube with a 15-cm i.d. and a length of 20 m. The mass flow rate of the N_2 is 120 g/s. The tube surface is maintained at a constant temperature of 100°C. Determine:

(a) The friction coefficient for the flow.

(b) The pressure drop of the nitrogen over the length of the tube.

(c) The convective-heat-transfer coefficient for the nitrogen.

(d) The heat-transfer rate into the N_2 from the tube surface.

4-39 The friction coefficient for flow of steam entering a circular tube at atmospheric pressure is 0.02. The diameter of the tube is 18 cm and its length is 40 m. The average mean temperature of the steam is 200°C and the tube wall is isothermal at 154°C. Determine the heat-transfer rate between the tube and water vapor. Calculate the pressure at the outlet of the tube.

4-40 A pipe in a chemical processing plant is used to transport turpentine at a mass flow rate of 23 kg/s. The pipe is 10 m long and it has an internal diameter of 13 cm. The average mean temperature of the turpentine in the tube is 80°C and it enters the pipe at 100°C. If the pipe surface is maintained at a uniform temperature of 30°C, calculate the heat-transfer rate from the turpentine and estimate the temperature of the turpentine as it leaves the tube.

4-41 Hydrogen at atmospheric pressure and a Reynolds number of 1.7×10^4 flows through a smooth 1-m-long tube with a diameter of 1.5 cm. The hydrogen enters at 20°C and the tube wall is maintained at 40°C. Calculate the heat-transfer rate from the tube and determine the temperature of the H_2 as it leaves the tube.

4-42 A steam line in a factory has a 6-cm i.d. and a length of 100 m. The pressure drop over the 100-m length of pipe is 6000 N/m^2. The steam velocity is 24 m/s. The average mean temperature of the steam is 140°C and the temperature of the pipe is 70°C. Calculate the heat-transfer rate from the steam.

4-43 Liquid Freon 12 flows through a smooth 1.3-cm-diameter tube. The Freon velocity and average bulk temperature are 2.9 m/s and -10°C, respectively. Calculate the heat-transfer rate from the Freon per unit length of tube assuming that the temperature of the tube is $+10$°C.

4-44 Ethylene glycol leaves a chemical processing unit at 45°C and enters a 4-cm-diameter smooth pipe with a velocity of 15 m/s. The pipe temperature is uniform at 20°C. Estimate the length of pipe necessary for the fluid temperature to drop to 35°C.

Chapter 5

ENGINEERING RELATIONS FOR CONVECTION HEAT TRANSFER

5-1 INTRODUCTION

As shown in Chapter 4, convection heat transfer can be treated analytically to obtain the value of the heat-transfer coefficient. However, analytic solutions of convection heat transfer are restricted to relatively simple geometries, and in engineering practice, the heat-transfer coefficients for real systems are evaluated from empirical relations determined by a combination of dimensional analysis and experiments.

5-2 DIMENSIONLESS PARAMETERS FOR CORRELATING CONVECTION DATA

In general, the selection of the physical parameter of significance for a given convection problem requires some prior physical insight into the process. But once the significant physical quantities have been identified, dimensional analysis makes it possible to obtain a relation between the important physical variables of the process in terms of a few dimensionless groups whose precise functional relationship can then be determined by experiments. To illustrate the procedure we will derive the conventional dimensionless groups which determine the value of the Nusselt number in forced convection in a long, smooth tube.

The dependent variable for the process is the convection-heat-transfer coefficient, $\bar{h}_c$. For incompressible low-speed flow the independent variables that determine the heat-transfer coefficient are the fluid velocity, V, a linear dimension (e.g., the pipe diameter, D), and the fluid properties of thermal conductivity, k, viscosity, μ, specific heat, c_p, and density, ρ.

The independent dimensional quantities to be used in the analysis are mass, M; length, L; time, θ; and temperature, T. The variables, their symbols, and their dimensional equations are listed in Table 5-1.

Table 5-1   Variables for Convection in a Tube

Variable	Symbol	Dimensional Equation
Tube diameter	D	$[L]$
Thermal conductivity of the fluid	k	$[ML/\theta^3 T]$
Velocity of the fluid	V	$[L/\theta]$
Density of the fluid	ρ	$[M/L^3]$
Viscosity of the fluid	μ	$[M/L\theta]$
Specific heat at constant pressure	c_p	$[L^2/\theta^2 T]$
Heat-transfer coefficient	$\bar{h}_c$	$[M/\theta^3 T]$

There are seven physical quantities and four primary dimensions. We therefore expect that three dimensionless groups will be required to correlate experimental data. To determine these dimensionless groups, write π as a product of each of the variables raised to an unknown power:

$$\pi = D^a k^b V^c \rho^d \mu^e c_p{}^f \bar{h}_c{}^g \tag{5-1}$$

and then substitute the dimensional formulas from the tabulation above:

$$\pi = [L]^a \left[\frac{ML}{\theta^3 T}\right]^b \left[\frac{L}{\theta}\right]^c \left[\frac{M}{L^3}\right]^d \left[\frac{M}{L\theta}\right]^e \left[\frac{L^2}{\theta^2 T}\right]^f \left[\frac{M}{\theta^3 T}\right]^g \tag{5-2}$$

For π to be dimensionless, the exponents of each primary dimension must separately add to zero. Equating the sum of the exponents of each primary dimension to zero yields the following set of equations:

$$\begin{aligned}
b + d + e + g &= 0 \quad &\text{for } M \\
a + b + c - 3d - e + 2f &= 0 \quad &\text{for } L \\
-3b - c - e - 2f - 3g &= 0 \quad &\text{for } \theta \\
-b - f - g &= 0 \quad &\text{for } T
\end{aligned} \tag{5-3}$$

Since there are seven unknowns but only four equations, three of the exponents can be arbitrarily selected for each dimensionless group. The only restriction is that each of the arbitrarily selected exponents be independent of the other. This requirement is met if the determinant formed by the coefficients of the remaining terms does not equal zero.

Since the convection heat-transfer coefficient is the dependent variable that we eventually want to evaluate, set its exponent g equal to unity in the first dimensionless group, π_1. To simplify the algebraic manipulations, arbitrarily set $c=d=0$. This means that in the first dimensionless group velocity and density will not appear.

Solving the equations above simultaneously, we obtain $a=1$, $b=-1$, and $e=f=0$. The first dimensionless group is therefore

$$\pi_1 = \frac{\bar{h}_c D}{k} \tag{5-4}$$

which we recognize from the analysis in Chapter 4 as the Nusselt number, Nu_D.

For the second dimensionless group, we set $g=0$ to avoid having the dependent variable $\bar{h}_c$ appear again. Arbitrarily, we let $a=1$ and $f=0$. Simultaneous solution of the equations above with these choices yields the second parameter,

$$\pi_2 = \frac{VD\rho}{\mu} \tag{5-5}$$

which we recognize as the Reynolds number, Re_D, with the tube diameter as the pertinent length dimension.

For the third dimensionless group, we wish to have the specific heat, which has not appeared in any previous group, to be included and therefore set $f=1$. Similarly, we do not wish to have the heat-transfer coefficient or the diameter appear again and therefore let a and g equal to 0. This yields the third dimensionless group,

$$\pi_3 = \frac{c_p \mu}{k} \tag{5-6}$$

which we recognize from Chapter 4 as the Prandtl number, Pr. Thus the heat-transfer coefficient in π_1 can be functionally related to the dimensionless groups π_2 and π_3 in the form

$$\mathrm{Nu}_D = f(\mathrm{Re}_D, \mathrm{Pr}) \tag{5-7}$$

Although the dimensional analysis cannot obtain the functional relationship, the analysis has reduced the number of variables from seven to three and provides the basis for experiments to determine the relation between the three parameters in Eq. 5-7.

Correlation of Data

Equation 5-7 shows that the dimensionless heat-transfer coefficient, that is, the Nusselt number, depends on the Reynolds number and the Prandtl number. A convenient and relatively simple relation for the correlation of experimental data is to assume an equation of the form

$$\mathrm{Nu} = C\,\mathrm{Re}^m \mathrm{Pr}^n \tag{5-8}$$

where C, m, and n are constants that must be determined experimentally.

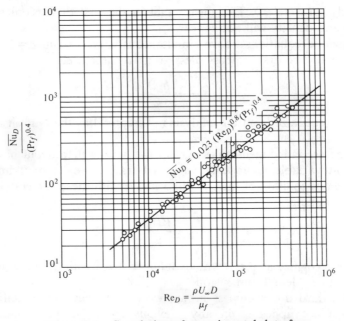

Figure 5-1   Correlation of experimental data for convection heat transfer in flow through a tube.

To obtain values for C, m, and n it is convenient to plot experimental data for a given fluid to obtain, first, a relation between the Nusselt number and Reynolds number, that is, to find the appropriate value for the exponent m. This is easily accomplished on a log plot with data for a single fluid at a relatively constant temperature to eliminate the influence of the Prandtl number. Then, using the estimate for the exponent m, the data for several fluids are plotted as the logarithm of $(\mathrm{Nu}_D/\mathrm{Re}^m)$ versus the logarithm of the Prandtl number and a value for the exponent n is determined. Finally, using this value of n, data for a variety of fluids are plotted once again as the logarithm of $\mathrm{Nu}_D/\mathrm{Pr}^n$ versus the logarithm of the Reynolds number and a revised value of the exponent m as well as an appropriate value for the constant C is determined. An example of such a correlation is shown in Fig. 5-1 for turbulent forced-convection flow in smooth tubes. Final correlation equations for heat transfer can usually represent experimental data to within plus or minus 25%.

5-3 CONVECTION HEAT TRANSFER IN FLOW
THROUGH TUBES AND DUCTS

In practical problems it is usually necessary to determine the temperature change when a fluid at a given velocity passes through a duct, with a specified inlet temperature and given duct temperature. For flow in a pipe

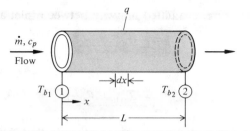

Figure 5-2 Schematic diagram of heat balance
for convection in flow through a tube.

of length L and at temperature T_s, the rate of heat transfer to the fluid can
be written in the form

$$q_c = c_p \rho V_{av} \frac{\pi D^2}{4} (T_{b_2} - T_{b_1}) \qquad (5\text{-}9)$$

where the temperatures T_{b_2} and T_{b_1} are the average, or *mixing cup*,
temperatures at the outlet and inlet, respectively. In terms of the heat-
transfer coefficient, the rate of heat transfer over a differential length dx
(Fig. 5-2) is related to the change in bulk temperature as well as to the
difference in temperature between wall, $T_s(x)$, and fluid bulk, $T_b(x)$, at that
location as shown below:

$$dq_c = \dot{m} c_p \, dT = \bar{h}_c (\pi D) \, dx (T_s - T_b) \qquad (5\text{-}10)$$

The total heat-transfer coefficient for flow through a duct can then be
expressed in the form

$$q_c = \bar{h}_c A (T_s - T_b)_{av} \qquad (5\text{-}11)$$

where A is the total surface contact area between the heat-transfer surface
and the fluid. Obviously, both T_s and T_b can vary along the length of the
tube and a suitable averaging process must be developed to use Eq. 5-11 in
practice. In this chapter emphasis will be placed on determining the value
of the heat-transfer coefficient, and the temperature averaging for different
industrial equipment will be taken up in Chapter 7.

Turbulent Flow in Tubes and Ducts   TURBULENT IF $Re_D > 6000$

The experimental data for turbulent flow through a long tube for fluids
having a Prandtl number between 0.5 and 100 can be correlated by the
relation (Ref. 1)

$$Nu_D = 0.023 \, Re_D^{0.8} \, Pr^{0.33} \; = \; \frac{h_c D}{K} \qquad (5\text{-}12)$$

In this equation all physical properties should be evaluated at a mean film
temperature, $\bar{T}_f$, equal to the average between the wall temperature and the

ASSUME : 3. FULLY DEVELOPED FLOW
          4. SMOOTH PIPE

bulk fluid temperature, evaluated halfway between inlet and outlet:

$$\overline{T}_f = \frac{T_s + T_{b,av}}{2} \tag{5-13}$$

where

$$T_{b,av} = \frac{T_{b,in} + T_{b,out}}{2}$$

The reason for taking an average temperature is that the fluid properties vary as a result of heat transfer. Another method of taking the variation of fluid properties has been proposed by Sieder and Tate (Ref. 2), who recommend the following relation for the evaluation of the Nusselt number in flow through a long duct:

$$\text{Nu}_D = 0.027\,\text{Re}_D{}^{0.8}\,\text{Pr}^{0.33}\left(\frac{\mu_b}{\mu_s}\right)^{0.14} \tag{5-14}$$

where

μ_b = viscosity at average bulk temperature, $T_{b,av}$

$T_{b,av} = (T_{b,in} + T_{b,out})/2$

μ_s = viscosity at wall temperature, T_s

All other fluid properties are evaluated at $T_{b,av}$.

Equations 5-12 and 5-14 apply to fully developed turbulent flow in tubes. They can also be applied to fully developed turbulent flow in ducts with other than circular cross sections if the diameter is replaced by the hydraulic diameter discussed in Chapter 4. $D_H = A\,\dfrac{CROSS\text{-}SECT.\ AREA}{WETTED\ PERIMETER}$

Example 5-1. Air at a pressure of 2 atm and a temperature of 490 K is flowing through a 2-cm-i.d. tube at a velocity of 10 m/s. Calculate the heat-transfer coefficient if the tube temperature is 510 K and estimate the rate of heat transfer per unit length if a uniform heat flux is maintained.

Solution: From Table G-1 the properties of the air at the average temperature between the wall and the bulk (500 K) are

$$\mu = 29.37 \times 10^{-6}\ \text{N}\cdot\text{s/m}^2$$
$$k = 0.0386\ \text{W/m}\cdot\text{K}$$
$$c_p = 1038\ \text{J/kg}\cdot\text{K}$$
$$\rho = 0.689 \times 2 = 1.378\ \text{kg/m}^3$$
$$\text{Pr} = 0.71$$

The Reynolds number is

$$\text{Re}_D = \frac{(1.378)(10)(0.02)}{29.37 \times 10^{-6}} = 9384$$

$$Re_D = \frac{\rho V(iD)}{\mu}$$

Thus the flow is turbulent and the heat-transfer coefficient is given by Eq. 5.12:

$$\bar{h}_c = \frac{k}{D} 0.023 \text{Re}_D{}^{0.8} \text{Pr}^{0.33}$$

$$= (0.0386/0.02)0.023(9384)^{0.8}(0.71)^{0.33}$$

$$= 59.7 \text{ W/m}^2 \cdot \text{K}$$

If a uniform heat flux were maintained, $(T_s - T_b)$ would remain constant but the bulk temperature would rise. Ignoring changes in the heat-transfer coefficient caused by changes in the fluid properties, the rate of heat transfer per meter of the tube length is

$$q' = \bar{h}_c \pi D (T_s - T_b) = (59.7)(\pi \times 0.02)(20) = 75.0 \text{ W/m}$$

In numerous practical problems the tubes and ducts are not sufficiently long, however, to have fully developed flow. A relation that contains a correction factor to take care of the entrance region is (Ref. 3)

$$\text{Nu}_{D_H} = 0.036 \text{Re}_{D_H}{}^{0.8} \text{Pr}^{0.33} \left(\frac{D_H}{L}\right)^{0.055} \quad \left(\text{valid for } 10 < \frac{L}{D_H} < 400\right) \quad \text{(5-15)}$$

TURBULENT

where L is the length of the tube and D_H is the hydraulic diameter of the duct. Properties in the equations above are to be evaluated at the mean film temperature given by Eq. 5-13.

Laminar Flow in Tubes and Ducts

LAMINAR IF $\text{Re}_D < 2000$

smooth pipe

The heat-transfer coefficient for laminar flow in tubes and ducts for $(\text{Re}_D \text{Pr} D/L) > 10$ can be evaluated from an empirical correlation suggested by Sieder and Tate (Ref. 2):

$$\text{Nu}_D = 1.86 (\text{Re}_D \text{Pr})^{0.33} \left(\frac{D}{L}\right)^{0.33} \left(\frac{\mu_b}{\mu_s}\right)^{0.14} \quad \text{(5-16)}$$

μ_s EVALUATED AT T_s

where all properties are evaluated at the bulk temperature T_b and the empirical correction factor $(\mu_b/\mu_s)^{0.14}$ is introduced to account for the effect that temperature variations have on the viscosity. In liquids the viscosity decreases with increasing temperature, while in gases the reverse trend exists. When a liquid is heated, the fluid near the wall is less viscous than the fluid in the center. Consequently, the velocity of the heated liquid is larger than for an unheated liquid at the same average velocity and temperature. The distortion of the parabolic profile would be in the opposite direction, with liquid cooling as shown in Fig. 5-3. For gases the conditions are reversed because the viscosity increases with increasing temperature, but in addition to the temperature, variations in density can introduce distortion in the profile.

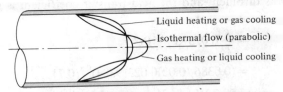

Figure 5-3 Velocity distribution in laminar flow of a liquid with heating and cooling.

Example 5-2. Water at an inlet temperature of 333 K flows at a velocity of 0.2 m/s through a 0.3-m-long capillary tube, 2.54×10^{-3} m i.d. Assuming that the tube temperature is maintained at 353 K, calculate the outlet temperature of the water.

Solution: The properties of water at 333 K are from Table F-1:

$$\rho = 983 \text{ kg/m}^3$$
$$c_p = 4181 \text{ J/kg} \cdot \text{K}$$
$$\mu = 4.72 \times 10^{-4} \text{ N} \cdot \text{s/m}^2$$
$$k = 0.658 \text{ W/m} \cdot \text{K}$$
$$\text{Pr} = 3.00$$

To ascertain if the flow is laminar, evaluate the Reynolds at the inlet bulk temperature,

$$\text{Re}_D = \frac{\rho V D}{\mu} = \frac{(983)(0.2)(0.00254)}{4.72 \times 10^{-4}} = 1058$$

The flow is laminar and because

$$\text{Re}_D \text{Pr} \frac{D}{L} = \frac{(1058)(3.00)(0.00254)}{0.3} = 26.9 > 10$$

equation 5-16 can be used to evaluate the heat-transfer coefficient. But since the mean bulk temperature is not known, we shall evaluate all the properties first at the inlet bulk temperature, T_{b1}, then determine an exit bulk temperature, and make a second iteration to obtain a more precise value. Designating inlet and outlet conditions with the subscripts 1 and 2, respectively, the energy balance becomes

$$q = \bar{h}_c \pi D L \left(T_s - \frac{T_{b1} + T_{b2}}{2} \right) = \dot{m} c_p (T_{b2} - T_{b1}) \tag{a}$$

At the wall temperature of 353 K, $\mu_s = 3.52 \times 10^{-4}$ N·s/m² from Table F-1. From Eq. 5-16 we can calculate the average Nusselt number:

$$\overline{\text{Nu}}_D = 1.86 \left[\frac{(1058)(3.00)(0.00254)}{0.3} \right]^{0.33} \left(\frac{4.72}{3.52} \right)^{0.14} = 5.74$$

and thus

$$\bar{h}_c = \frac{k\overline{Nu}_D}{D} = \frac{(0.658)(5.74)}{0.00254} = 1487 \text{ W/m}^2\cdot\text{K}$$

The mass flow rate is

$$\dot{m} = \rho \frac{\pi D^2}{4} V = \frac{(983)\pi(0.00254)^2(0.2)}{4} = 0.996 \times 10^{-3} \text{ kg/s}$$

Inserting the calculated values for $\bar{h}_c$ and $\dot{m}$ into Eq. (a), along with $T_{b1} = 333$ K and $T_s = 353$ K gives

$$(1487)\pi(0.00254)(0.3)\left(353 - \frac{333 + T_{b2}}{2}\right) = (0.996 \times 10^{-3})(4181)(T_{b2} - 333)$$

(b)

Solving for T_{b2} gives

$$T_{b2} = 345 \text{ K}$$

For the second iteration we shall evaluate all properties at the new average bulk temperature

$$\bar{T}_b = \frac{345 + 333}{2} = 339 \text{ K}$$

At this temperature we get, from Table F-1,

$$\rho = 980 \text{ kg/m}^3$$
$$c_p = 4185 \text{ J/kg}\cdot\text{K}$$
$$\mu = 4.36 \times 10^{-4} \text{ N}\cdot\text{s/m}^2$$
$$k = 0.662 \text{ W/m}\cdot\text{K}$$
$$\text{Pr} = 2.78$$

Recalculating the Reynolds number with properties based on the new mean bulk temperature gives

$$\text{Re}_D = \frac{\rho V D}{\mu} = \frac{(980)(0.2)(2.54 \times 10^{-3})}{4.36 \times 10^{-4}} = 1142$$

With this value of Re_D, the heat-transfer coefficient can now be calculated. One obtains on the second iteration $\text{Re}_D \text{Pr}(D/L) = 25.9, \overline{Nu}_D = 5.61$, and $\bar{h}_c = 1490 \text{ W/m}^2\cdot\text{K}$. Substituting the new value of $\bar{h}_c$ in Eq. (b) gives $T_{b2} = 345$ K. Further iterations will not affect the results appreciably in this example because of the small difference between bulk and wall temperature. In cases where the temperature difference is large, a second iteration may be necessary.

Equation 5-16 can obviously not be used for long tubes since it would yield a zero heat-transfer coefficient. The following relation (Ref. 5) has been found to be useful for ordinary liquids and gases when (L/D) is less than 0.0048Re_D in tubes and when (L/D_H) is less than 0.0021Re_{D_H} in

ASSUMPTIONS

ducts of rectangular cross section for both constant wall-temperature conditions and constant heat-flux conditions in the tube:

$$\overline{Nu}_{D_H} = \frac{Re_{D_H} Pr D_H}{4L} \ln \left\{ \frac{1}{1 - \left[2.654/Pr^{0.167} (Re_{D_H} Pr D_H/L)^{0.5} \right]} \right\} \quad (5\text{-}17)$$

For very long ducts the Nusselt number should be obtained from Table 4-1.

The Nusselt number obtained from Eqs. 5-16 and 5-17 is averaged with respect to the length of the tube according to

$$\overline{Nu}_D = \frac{D}{kL} \int_0^L h_{cx} \, dx = \frac{\bar{h}_c D}{k} \quad (5\text{-}18)$$

where x refers to the distance from the entrance. This Nusselt number is often called the *log-mean Nusselt number* because it can be used directly in the design equations for heat exchangers considered in Chapter 7.

Liquid Metals

The preceding relations for heat-transfer coefficients in laminar and turbulent flow apply to fluids having Prandtl numbers larger than 0.5. The only fluids that have Prandtl numbers considerably below 0.5 are liquid metals. Liquid metals have small Prandtl numbers because they possess large thermal conductivities. This makes it possible to remove larger heat fluxes than with other liquids or gases, and liquid metals are therefore extensively used in nuclear reactors, which can generate extremely high heat fluxes in their core. Liquid metals are difficult to handle because many of them are corrosive and they can also produce violent chemical reactions when they come in contact with water or air. But, despite their disadvantages, liquid metals have been used in high-heat-flux systems. Empirical relations to determine the Nusselt number for liquid metals have been summarized by Lubarsky and Kaufman (Ref. 6) and Lyon (Ref. 7). In liquid-metal heat transfer the Nusselt number has been found to depend on the product of the Reynolds number and the Prandtl number, called the *Peclet number*, Pe. For fully developed turbulent flow in tubes with uniform heat flux the experimental data are, according to Reference 6, correlated by the relation

$$\overline{Nu}_D = 0.625 Pe^{0.4} \quad (5\text{-}19)$$

if all properties are evaluated at the average bulk temperature. The preceding relation is valid for Peclet numbers between 100 and 10,000 and length/diameter ratios larger than 60. For constant wall temperatures the relationship

IF $T_{WALL} = $ CONSTANT

$$\overline{Nu}_D = 5.0 + 0.025 Pe^{0.8} \quad (5\text{-}20)$$

$$Pe = Re_D \, Pr \qquad \overline{Nu}_D = \frac{\bar{h}_c D}{K}$$

is recommended in Reference 8 for Peclet numbers larger than 100 and length/diameter ratios larger than 60 if all properties are evaluated at the average bulk temperature. There are still many unresolved questions in liquid-metal heat transfer, and for more detailed information the reader is referred to References 6 and 7.

Example 5-3. A liquid metal flows at a mass rate of 3 kg/s through a constant heat flux 5-cm-i.d. tube in a nuclear reactor. The fluid at 473 K is to be heated with the tube wall 30 K above the fluid temperature. Determine the length of the tube required for a 1-K rise in bulk fluid temperature, using the following properties:

$$\rho = 7.7 \times 10^3 \text{ kg/m}^3$$
$$\nu = 8.0 \times 10^{-8} \text{ m}^2/\text{s}$$
$$c_p = 130 \text{ J/kg·K}$$
$$k = 12 \text{ W/m·K}$$
$$\text{Pr} = 0.011$$

Solution: The rate of heat transfer per unit temperature rise is

$$q = \dot{m} c_p \, \Delta T = (3.0)(130)(1) = 390 \text{ W}$$

The Reynolds number is

$$\text{Re}_D = \frac{\dot{m}D}{\rho A \nu} = \frac{3 \times 0.05}{7.7 \times 10^3 \left[\pi \dfrac{(0.05)^2}{4} \right] 8.0 \times 10^{-8}} = 1.24 \times 10^5$$

The heat-transfer coefficient is obtained from Eq. 5-19:

$$\bar{h}_c = \frac{k}{D} \times 0.625 (\text{Re}_D \text{Pr})^{0.4}$$

$$= \frac{12}{0.05} \times 0.625 (1.24 \times 10^5 \times 0.011)^{0.4}$$

$$= 2692 \text{ W/m}^2 \cdot \text{K}$$

The surface area required is

$$A = \pi D L = \frac{q}{\bar{h}_c (T_s - T_b)}$$

$$= \frac{390}{(2692)(30)} = 0.00483 \text{ m}^2$$

Finally, the required length is

$$L = \frac{A}{\pi D} = \frac{0.00483}{\pi \times 0.05} = 0.0307 \text{ m}$$

Forced Convection in Transition Flow

The mechanism of heat transfer in fluid flow in the transition regime (Re_{D_H} between 2000 and 6000) varies considerably from system to system. In this region the flow may be unstable and fluctuations in pressure drop and heat transfer have been observed. There exists a large uncertainty in the basic heat-transfer and friction flow performance, and designers are advised to design their equipment if at all possible to operate outside the transition regime. For the purpose of estimating the Nusselt number in the transition regime, the curves of Fig. 5-4 may be used, but actual performance may deviate considerably from that predicted on the basis of these curves (Ref. 5).

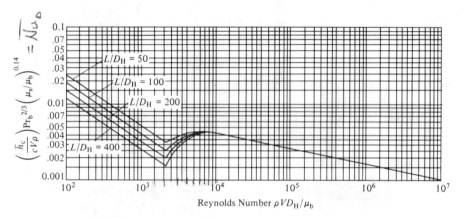

Figure 5-4 Nusselt number versus Reynolds number in transition regime.

5-4 FORCED-CONVECTION HEAT TRANSFER IN EXTERNAL FLOW

In addition to flow through ducts, in many engineering systems it is important to evaluate the rate of heat transfer by convection in flow over flat plates, cylinders, spheres, and tube banks.

Flat Plates

An analysis of heat transfer in flow over flat plates was presented in Chapter 4. There, it was shown that for a given fluid, the average Nusselt number depends primarily on the Reynolds number, which for flow over a plate is based on the free-stream velocity and on the length of the plate in the direction of flow. In some cases the local heat-transfer coefficient is sought and then the appropriate length dimension in the Nusselt and Reynolds numbers is the distance from the leading edge. For engineering calculations the local Nusselt number in laminar flow over a flat plate

$(\text{Re}_x < 5 \times 10^5)$ is given by   LAMINAR

$$\text{Nu}_x = \frac{h_{cx} x}{k} = 0.332 \text{Re}_x^{1/2} \text{Pr}^{1/3} \tag{5-21}$$

whereas the average Nusselt number is

$$\overline{\text{Nu}_L} = \frac{\bar{h}_c L}{k} = 0.664 \text{Re}_L^{1/2} \text{Pr}^{1/3} \tag{5-22}$$

The average heat-transfer coefficient in Eq. 5-22 is obtained by integrating, or

$$\bar{h}_c = \frac{1}{L} \int_0^L h_{cx} \, dx \tag{5-23}$$

In turbulent flow $(\text{Re}_L > 5 \times 10^5)$ there will be a portion of the plate near the leading edge over which the flow is laminar and another portion over which the flow is turbulent. The local Nusselt number at any value of x larger than the transition point, that is, $x > x_c$, is given by

$$\text{Nu}_x = \frac{h_{cx}}{k} = 0.0288 \left(\frac{\rho V_\infty x}{\mu} \right)^{0.8} \text{Pr}^{1/3} \tag{5-24}$$

whereas the average value with transition at $\text{Re}_x = 5 \times 10^5$ is

$$\overline{\text{Nu}_L} = \frac{\bar{h}_c L}{k} = 0.036 \text{Pr}^{1/3} \left(\text{Re}_L^{0.8} - 23{,}200 \right) \tag{5-25}$$

Single Cylinders and Spheres

The principal difference in flow over a cylinder or a sphere, compared to the flow over a flat plate, is that the boundary layer in flow over cylinders and spheres may not only undergo a transition from laminar to turbulent flow, but also usually separates somewhere in the rear from the interface between the object and the fluid. The reason for this separation is the increasing pressure in the direction of flow, which causes a separated flow region to develop on the back of a cylinder or a sphere if the free-stream velocity is sufficiently large. The development of a separated flow region in flow over a cylinder is shown schematically in Fig. 5-5, and Fig. 5-6 is a

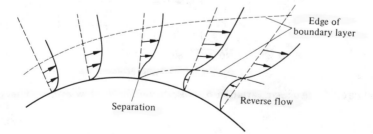

Figure 5-5 Schematic diagram showing the development of flow separation.

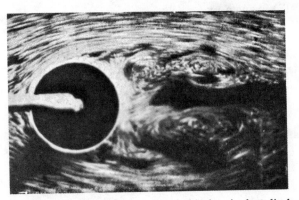

Figure 5-6 Separated flow region behind a single cylinder.

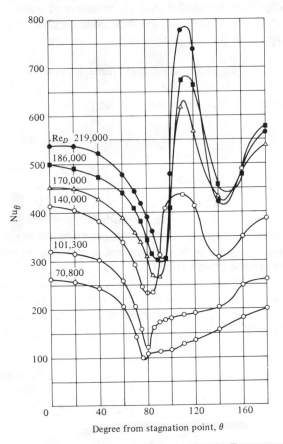

Figure 5-7 Angular variation of Nusselt number in flow over a cylinder.

photograph of the flow with separation. Obviously, regions in which the boundary layer has separated from the surface will exhibit considerably different Nusselt number characteristics from those regions where the boundary layer is attached. This is illustrated for free stream Reynolds numbers between 70,000 and 220,000 in Fig. 5-7, where the local Nusselt number, $h_{c,\theta}D/k$, for flow over a cylinder is plotted versus the angular distance θ from the stagnation point. It should be observed that at first, as in laminar flow over a plate, the local Nusselt number decreases with distance from the leading edge, but increases abruptly as the fluid undergoes transition from laminar to turbulent flow, and then decreases again in the turbulent boundary layer. However, in the back of the cylinder, where the flow is separated, the Nusselt number increases again. At the two lowest Reynolds numbers (70,000 and 100,000), separation occurs before the flow undergoes transition and a minimum point of heat-transfer coefficient occurs approximately at the separation point (Ref. 9).

In normal engineering practice, it is not necessary to evaluate a local value of the Nusselt number, but average values of the heat-transfer coefficient will suffice. The average Nusselt number, $\bar{h}_c D/k$, can be related to the free-stream Reynolds number, $\rho V_\infty D/\mu$, and the Prandtl number, $c_p \mu/k$, by an empirical correlation equation similar to that derived previously for flow through a duct, but the length dimension in the Reynolds and Nusselt numbers must be referred to the appropriate outside dimension of the object, which for flow over cylinders and spheres is the diameter, D. For gases and ordinary liquids, the average heat-transfer coefficient for flow across a single cylinder may be calculated from the relation

FLOW NORMAL TO CYLINDER AXIS

$$\overline{Nu}_D = \frac{\bar{h}_c D}{k} = C\left(\frac{\rho V_\infty D}{\mu}\right)^n Pr^{1/3} \tag{5-26}$$

where V_∞ is the free-stream velocity and the values of the coefficient C and the exponent n are presented in Table 5-2 for various ranges of Re_D.

Table 5-2   Constants for Use with Eq. 5-26

$Re_{D,f}$	C	n
0.4–4	0.989	0.330
4–40	0.911	0.385
40–4,000	0.683	0.466
4,000–40,000	0.193	0.618
40,000–400,000	0.0266	0.805

All properties in Eq. 5-26 should be evaluated at the arithmetic mean between the surface temperature and the fluid temperature. For flow over a noncircular cylinder, the recommended values of C and n are presented alongside the appropriate shapes in Table 5-3.

$T_f = \dfrac{T_s + T_\infty}{2}$

Table 5-3 Constants for Use in Eq. 5-26 When Applied to Heat Transfer in Flow over Noncircular Cylinders

GEOMETRY	$Re_{D,f}$	C	n
$\xrightarrow{V_\infty}$ ◇ d	$5 \times 10^3 - 10^5$	0.246	0.588
$\xrightarrow{V_\infty}$ ▢ d	$5 \times 10^3 - 10^5$	0.102	0.675
$\xrightarrow{V_\infty}$ ⬡ d	$5 \times 10^3 - 1.95 \times 10^4$ $1.95 \times 10^4 - 10^5$	0.160 0.0385	0.638 0.782
$\xrightarrow{V_\infty}$ ⬠ d	$5 \times 10^3 - 10^5$	0.153	0.638
$\xrightarrow{V_\infty}$ | d	$4 \times 10^3 - 1.5 \times 10^4$	0.228	0.731

Source: M. Jakob, *Heat Transfer*, vol. 1, John Wiley & Sons, Inc., New York, 1949.

For flow over spheres Whitaker (Ref. 13) has developed a single correlation equation:

$$\overline{Nu}_D = 2 + \left(0.4\,Re_D^{1/2} + 0.06 Re_D^{2/3}\right) Pr^{0.4} \left(\mu_\infty / \mu_s\right)^{0.25} \qquad (5\text{-}27)$$

which is valid for the range $3.5 < Re_D < 8 \times 10^4$ and for Prandtl numbers between 0.7 and 380. All properties, except μ_s in the equation above should be evaluated at the free-stream temperature. (T_∞)

For flow of a liquid metal over a sphere, the heat-transfer coefficient can be obtained from the relation (Ref. 14)

$$\overline{Nu}_D = 2.0 + 0.386 (Re_D Pr)^{0.5} \qquad (5\text{-}28)$$

in the Reynolds number range between 3×10^4 and 1.5×10^5.

Example 5-4. Determine the rate of heat transfer for atmospheric air at 358 K flowing at a velocity of 5 m/s across a 0.5-m-diameter 10-m-long duct whose surface temperature is 373 K.

Solution: First we evaluate, from Table G-1, the physical properties of the air at the average film temperature $T_f = (T_s + T_\infty)/2 = 365.5$ K, by interpolation:

$$\rho = 0.936 \text{ kg/m}^3$$
$$\mu = 21.3 \times 10^{-6} \text{ N·s/m}^2$$
$$k = 0.0302 \text{ W/m·K}$$
$$\text{Pr} = 0.71$$

The Reynolds number is calculated next:

$$\text{Re}_{Df} = \frac{\rho V_\infty D}{\mu} = \frac{0.936 \times 5 \times 0.5}{21.3 \times 10^{-6}} = 1.1 \times 10^5$$

From Table 5-2 we obtain

$$C = 0.0266 \quad \text{and} \quad n = 0.805$$

The average heat-transfer coefficient is then evaluated from Eq. 5-26:

$$\bar{h}_c = \frac{0.0302}{0.5} \times 0.0266(1.1 \times 10^5)^{0.805}(0.71)^{0.33} = 16.4 \text{ W/m}^2\text{·K}$$

and the rate of heat transfer is

$$q = \bar{h}_c(\pi DL)(T_s - T_\infty) = 16.4\pi(0.5 \times 10)15$$
$$= 3864 \text{ W}$$

$$q = \bar{h}_c \cdot A \, (\Delta T)$$

Tube Banks

Heat transfer in flow over a bank of tubes is important in heat-exchanger design, to be discussed more fully in Chapter 7. An extensive correlation for heat transfer in flow over tube banks has been developed by Grimison (Ref. 15), utilizing the same form as Eq. 5-26, which was used previously for flow over a single tube. However, the values of the constant C and exponent n depend on the distance between adjacent tubes and the distance between rows of tubes in the flow direction, and also on whether the tubes are in line or staggered, as shown in Fig. 5-8. Table 5-4 presents the values of C and n to be used in Eq. 5-26 for various combinations of tube banks with 10 or more rows in the direction of flow.

For fewer rows the ratio of $\bar{h}_c$ for N rows deep to that for 10 rows is given in Table 5-5. The Reynolds number, Re_{max}, in flow through tube banks is based on the tube diameter and the maximum velocity in the tube bank (i.e., the velocity through the minimum flow area).

For liquid metals the heat-transfer coefficients in flow over tube banks are correlated by the relationship (Ref. 17)

$$\overline{\text{Nu}}_D = 4.03 + 0.228(\text{Re}_{max}\text{Pr})^{0.67} \tag{5-29}$$

in the Reynolds number range between 20,000 and 80,000.

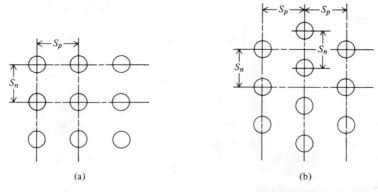

Figure 5-8 Nomenclature for use with Table 5-4.
(a) in-line; (b) staggered tube arrangement.

Table 5-4   Heat-Transfer Correlation for Flow over Tube Banks of 10 or More Rows

S_p/D	S_n/D							
	1.25		1.5		2.0		3.0	
S_p/D	C	n	C	n	C	n	C	n
	In line							
1.25	0.386	0.592	0.305	0.608	0.111	0.704	0.0703	0.752
1.5	0.407	0.586	0.278	0.620	0.112	0.702	0.0753	0.744
2.0	0.464	0.570	0.332	0.602	0.254	0.632	0.220	0.648
3.0	0.322	0.601	0.396	0.584	0.415	0.581	0.317	0.608
	Staggered							
0.6	—	—	—	—	—	—	0.236	0.636
0.9	—	—	—	—	0.495	0.571	0.445	0.581
1.0	—	—	0.552	0.558				
1.125	—	—			0.531	0.565	0.575	0.560
1.25	0.575	0.556	0.561	0.554	0.576	0.556	0.579	0.562
1.5	0.501	0.568	0.511	0.562	0.502	0.568	0.542	0.568
2.0	0.448	0.572	0.462	0.568	0.535	0.556	0.498	0.570
3.0	0.344	0.592	0.395	0.580	0.488	0.562	0.467	0.574

Source: E. D. Grimson, "Correlation and Utilization of New Data on Flow Resistance and Heat Transfer for Cross Flow of Gases over Tube Banks," *Trans. ASME*, vol. 59, pp. 583–594, 1937.

FOR GASES & ORDINARY LIQUIDS   EQ (5-26)

$$\overline{Nu}_D = \frac{\bar{h}_c D}{K} = C \left(\frac{\rho\, V_{max}\, D}{\mu} \right)^N P_R^{1/3}$$

$$Re_{max} = \frac{\rho\, V_{max}\, D}{\mu}$$

Table 5-5   Ratio of $\bar{h}_c$ for N Rows to That for 10 Rows

N	1	2	3	4	5	6	7	8	9	10
Ratio for staggered tubes	0.68	0.75	0.83	0.89	0.92	0.95	0.97	0.98	0.99	1.0
Ratio for in-line tubes	0.64	0.80	0.87	0.90	0.92	0.94	0.96	0.98	0.99	1.0

Source: W. M. Kays and R. K. Lo, Basic Heat Transfer and Friction Data for Gas Flow Normal to Banks of Staggered Tubes: Use of a Transient Technique, *Stanford Univ. Tech. Rept. 15*, Navy Contract N6-ONR251, T.O. 6, 1952.

The pressure drop in N/m^2 for flow of gases over a bank of tubes may be calculated from the relation

$$\Delta p = \frac{f' G_{max}^2 N}{\rho} \left(\frac{\mu_s}{\mu_b} \right)^{0.14} \tag{5-30}$$

where

1 ATM = 101 325 N/m²

G_{max} = mass velocity at minimum flow area, $kg/s \cdot m^2$ = ρV_{max}

ρ = density evaluated at free-stream conditions, kg/m^3

N = number of transverse rows

The empirical friction factor f' is (Ref. 18)

$$f' = 2 \left\{ 0.25 + \frac{0.118}{[(S_n - D)/D]^{1.08}} \right\} Re_{max}^{-0.16} \qquad \text{STAGGERED} \tag{5-31}$$

for staggered tube arrangements, and

$$f' = 2 \left\{ 0.044 + \frac{0.08 S_p/D}{[(S_n - D)/D]^{0.43 + 1.13 D/S_p}} \right\} Re_{max}^{-0.15} \qquad \text{IN-LINE} \tag{5-32}$$

for in-line tube arrangements.

For the purpose of preliminary <u>estimates in turbulent flow over tube</u> banks with 10 or more rows ($Re_{max} > 6000$), irrespective of whether they are in-line or staggered, the relationship

$$\frac{\bar{h}_c D}{k_f} = 0.33 \left(\frac{G_{max} D}{\mu_f} \right)^{0.6} Pr_f^{0.3} \tag{5-33}$$

has been found to correlate experimental data with good accuracy (Ref. 5).

5-5 NATURAL CONVECTION

In natural convection, temperature and density gradients are produced by heat transfer to or from a body in a stagnant fluid at a different temperature than the body. In contrast to forced convection, where the velocity of the fluid is determined by external forces, the fluid motion in natural convection is the result of buoyancy forces arising from temperature and density variations within the fluid and cannot be specified explicitly. As in forced convection, the fluid motion generated by the buoyancy forces may be laminar or turbulent. However, the boundary layer in free convection flow has a zero fluid velocity both at the solid/fluid interface and at the outer limit far away from the body. The fluid motion may be upward, as when a flat plate is heated in air, or it may be a sinking downward motion, as when a cold cylinder is placed into a stagnant pool of warm water.

Before making a dimensional analysis of natural convection it is necessary to consider the nature of the driving force. If ρ_∞ is the density of the cold undisturbed fluid and ρ is the density of the warmer fluid, the buoyancy force per unit volume in the earth's gravitational field is $(\rho_\infty - \rho)g$, where g is the acceleration of gravity.

The density difference can be expressed as a function of the coefficient of volume expansion of the fluid β according to the definition

$$\beta = \frac{1}{v}\left(\frac{\partial v}{\partial T}\right)_p = \frac{1}{v_\infty}\frac{v - v_\infty}{T - T_\infty} = \frac{\rho_\infty - \rho}{\rho(T - T_\infty)} \tag{5-34}$$

Substituting in the relation above for $\rho_\infty - \rho$, the buoyancy force becomes $\rho g \beta(T_s - T_\infty)$ for an object at a temperature T_s. Recognizing that the buoyancy force can be expressed in terms of the variables β, g, and $(T_s - T_\infty)$, it is apparent that these three variables, as well as a linear dimension characteristic of the system, L, and the pertinent fluid properties ρ, μ, c_p, and k must be considered in a dimensional analysis. However, under normal circumstances the gravitational field is constant and β and g can therefore be combined into a single variable (βg) for the dimensional analysis.

Writing out the variables, raising each of them to an unknown power, and performing a dimensional analysis, as shown in Chapter 1 (see also Problem 5-2), it can be shown that the following three-dimensional parameters are sufficient to correlate data for heat transfer in natural convection:

$$\pi_1 = \frac{\bar{h}_c L}{k} = \overline{\mathrm{Nu}}_L$$

$$\pi_2 = \frac{c_p \mu}{k} = \mathrm{Pr}$$

$$\pi_3 = \frac{\beta g \rho^2 (T_s - T_\infty) L^3}{\mu^2} = \mathrm{Gr}_L$$

The first parameter is the Nusselt number, the second parameter is the Prandtl number, and the third parameter is known as the *Grashof number*. Gr_L is the ratio of the buoyancy force to the shear force. The buoyancy force in natural convection replaces the momentum force in forced convection. It should also be noted that since $\beta g \rho(T_s - T_\infty)$ is the buoyancy force per unit volume, $[\beta g \rho(T_s - T_\infty) L]$ would be a buoyancy force per unit area. Thus the ratio of buoyancy to shear force per unit area is $\beta g \rho(T_s - T_\infty)L/(\mu V/L)$. However, the velocity V is a dependent variable proportional to $(\mu/\rho L)$, so the ratio of buoyancy to shear force becomes

$$Gr_L = \frac{\beta g \rho^2 (T_s - T_\infty)L^3}{\mu^2} \qquad (5\text{-}35)$$

On the basis of numerous experiments it has been found that the average Nusselt number can be related to the Grashof and Prandtl numbers by an equation of the form

$$\overline{Nu}_f = C(Gr_f Pr_f)^n \qquad (5\text{-}36)$$

where the subscript f indicates that all physical properties should be evaluated at $T_f = (T_s + T_\infty)/2$. The product, GrPr, is known as the *Rayleigh number*, Ra. In engineering practice, however, the results are generally given in terms of the product GrPr, because it is often necessary to iterate on the Grashof number at a given Prandtl number, since the temperature difference, $T_s - T_\infty$, in the Grashof number may not be explicitly known.

Laminar and turbulent flow regimes have been observed in natural convection and transition generally occurs in the range $10^7 < GrPr < 10^9$, depending on the geometry of the system. Figure 5-9 shows the correlation of the experimental results for free convection to or from a vertical flat plate of height L as a plot of $\overline{Nu}_{L,f}$ versus $(Gr_L Pr)_f$.

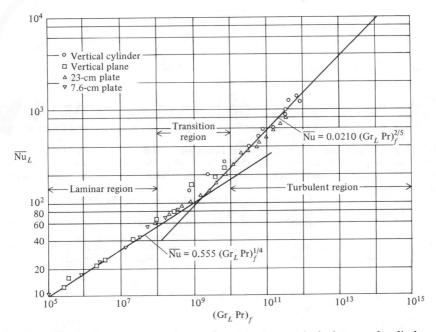

Figure 5-9 Correlation of free-convection data for vertical plates and cylinders. (From B. Gebhart, *Heat Transfer*, 2nd ed., McGraw-Hill, New York, 1970.)

Table 5-6 Constants for Use with Eq. 5-36

Geometry	$Gr_f Pr_f$	C	n	Reference
Vertical planes and vertical cylinders	$10^{-1}-10^4$	Use Fig. 5-9	Use Fig. 5-9	[a]
	10^4-10^9	0.59	$\frac{1}{4}$	[a]
	10^9-10^{13}	0.021	$\frac{2}{5}$	[b]
	10^9-10^{13}	0.10	$\frac{1}{3}$	[c,d]
Horizontal cylinders	$0-10^{-5}$	0.4	0	[a]
	$10^{-5}-10^4$	Use Fig. 5-10	Use Fig. 5-10	[a]
	10^4-10^9	0.53	$\frac{1}{4}$	[a]
	10^9-10^{12}	0.13	$\frac{1}{3}$	[a]
Upper surface of warm plates or lower surface of cool plates	$2\times10^4-8\times10^6$	0.54	$\frac{1}{4}$	[e,f]
Upper surface of warm plates or lower surface of cool plates	$8\times10^6-10^{11}$	0.15	$\frac{1}{3}$	[e,f]
Lower surface of warm plates or upper surface of cool plates	10^5-10^{11}	0.58	$\frac{1}{5}$	[e,g]

[a] W. H. McAdams, *Heat Transmission*, 3rd ed., McGraw-Hill Book Company, New York, 1954.

[b] E. R. G. Eckert and T. W. Jackson, Analysis of Turbulent Free Convection Boundary Layer on a Flat Plate, *NACA Rept. 1015*, 1951.

[c] C. Y. Warner and V. S. Arpaci, "An Experimental Investigation of Turbulent Natural Convection in Air at Low Pressure Along a Vertical Heated Flat Plate," *Int. J. Heat Mass Transfer*, vol. 11, p. 397, 1968.

[d] F. J. Bayley, "An Analysis of Turbulent Free Convection Heat Transfer," *Proc. Inst. Mech. Eng.*, vol. 169, no. 20, p. 361, 1955.

[e] T. Fujii and H. Imura, "Natural Convection Heat Transfer from a Plate with Arbitrary Inclination," *Int. J. Heat Mass Transfer*, vol. 15, p. 755, 1972.

[f] J. R. Lloyd and W. R. Moran, Natural Convection Adjacent to Horizontal Surface of Various Planforms, *ASME Paper 74-WA/HT-66*.

[g] S. N. Singh, R. C. Birkebak, and R. M. Drake, "Laminar Free Convection Heat Transfer from Downward-facing Horizontal Surfaces of Finite Dimensions," *Prog. Heat Mass Transfer*, vol. 2, p. 87, 1969.

Vertical Planes and Cylinders EQ 5-36

For vertical surfaces of height L and vertical cylinders having a diameter-to-height ratio, D/L, larger than $(35/\mathrm{Gr}_L^{0.25})$ the constants to be used in the evaluation of the average Nusselt number in Eq. 5-36 are presented in Table 5-6 for isothermal surfaces. Since the same type of correlation equation can be used for horizontal plates, also the appropriate constants for this geometry are presented in the same table. The equations for laminar free convection from isothermal vertical plates may also be applied to surfaces with uniform heat flux if the surface temperature is taken at the midpoint of the plate (Ref. 5).

When a heated plate is inclined a small angle θ from the vertical the body force along the x axis becomes $g\beta(T_w - T_\infty)\cos\theta$, and the average Nusselt number for the upper surface can be obtained by means of Eq. 5-36 using an effective Grashof number, $\mathrm{Gr}_{\mathrm{eff}}$, given by

$$\mathrm{Gr}_{\mathrm{eff}} = \frac{\beta g \rho^2 \cos\theta (T_s - T_\infty) L^3}{\mu^2} \tag{5-37}$$

Horizontal Cylinders, Spheres, and Blocks

The average Nusselt number for free convection to and from horizontal cylinders can be calculated from Eq. 5-36 with the appropriate constants from Table 5-6 or from the equation (Ref. 19)

$$\overline{\mathrm{Nu}}_{D,f} = \left[0.60 + 0.387 \left\{ \frac{\mathrm{Gr}_D \mathrm{Pr}_f}{\left[1 + (0.56/\mathrm{Pr}_f)^{9/16} \right]^{16/9}} \right\}^{1/6} \right]^2 \tag{5-38}$$

in the Rayleigh number range between 10^{-5} and 10^{12}. Figure 5-10 shows the correlation of experimental data for fluids with $\mathrm{Pr} > 0.5$ graphically.

Heat transfer from horizontal cylinders to liquid metals can be calculated from the relation (Ref. 20)

$$\overline{\mathrm{Nu}}_{D,f} = 0.53 \left(\mathrm{Gr}_{D,f} \mathrm{Pr}^2 \right)^{0.25} \tag{5-39}$$

For Grashof number larger than unity, free convection to and from spheres of diameter D is correlated by the relation (Ref. 21)

NOT FOR
LIQUID METALS

$$\overline{\mathrm{Nu}}_{D,f} = 2.0 + 0.45 \left(\mathrm{Gr}_{D,f} \mathrm{Pr}_f \right)^{0.25} \tag{5-40}$$

As the Grashof number approaches zero, the Nusselt number approaches the value 2, which corresponds to pure conduction through a stagnant layer of fluid surrounding a sphere.

The heat-transfer coefficient from vertical rectangular blocks of horizontal dimension L_h and vertical dimension L_v, respectively, can be estimated (Ref. 22) from Eq. 5-36 for the range $10^4 < \mathrm{Gr}_L \mathrm{Pr} < 10^9$ with $C = 0.60$, $n = 0.25$, and $1/L = 1/L_h + 1/L_v$.

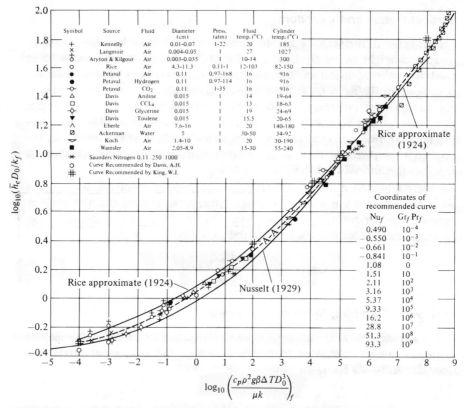

Figure 5-10 Correlation of experimental data for free convection to and from horizontal cylinders. (From W. H. McAdams, *Heat Transmission*, 3rd ed., McGraw-Hill, 1954.)

Example 5-5. Determine the rate of heat loss from a 0.3-m-o.d., 10-m-long horizontal steam pipe at an outer surface temperature of 510 K in a room with ambient air at 300 K.

Solution: From Table G-1 we use physical properties to calculate the product $Gr_D Pr_f$, with

$$T_f = \frac{T_s + T_\infty}{2} = \frac{510 + 300}{2} = 405 \text{ K}$$

The Rayleigh number is obtained by interpolation from Table G-1:

$$Gr_D Pr_f = \frac{g\beta(T_s - T_\infty)D^3 Pr}{\nu^2}$$

$$= (0.373 \times 10^8) \times (510 - 300) \times 0.3^3 \times 0.71$$

$$= 1.5 \times 10^8$$

From Table 5-6, $C = 0.53$ and $n = 0.25$. We then obtain

$$\overline{\mathrm{Nu}}_D = 0.53(\mathrm{Gr}_D \mathrm{Pr}_f)^{0.25} = 0.53(1.5 \times 10^8)^{0.25}$$

$$= 58.7$$

$$\overline{h}_c = \frac{k\overline{\mathrm{Nu}}_D}{D} = \frac{0.0327 \times 58.7}{0.3} = 6.40 \ \mathrm{W/m^2 \cdot K}$$

The rate of heat loss is then

$$q_c = \overline{h}_c A (T_s - T_\infty)$$

$$= 6.40 \times \pi \times 0.3 \times 10 \times 210$$

$$= 12.7 \ \mathrm{kW}$$

Free Convection in Enclosed Spaces

Free convection in enclosed spaces is important in energy-conservation systems such as double-glazed windows. Free-convection-flow phenomena inside enclosed spaces are complex because a number of different flow configurations can exist. The simplest case is a fluid confined between two large horizontal plates spaced a distance b apart. If the lower plate has a lower surface temperature than the upper plate, the process is one of pure conduction across the fluid layer, neglecting end effects, and, if the driving temperature difference is taken to be the difference in the plate temperatures, the Nusselt number is unity:

$$\overline{\mathrm{Nu}}_b = \frac{\overline{h}_c b}{k} = 1.0 \tag{5-41}$$

For air confined between horizontal plates with the lower plate hotter than the upper plate, Jakob (Ref. 23) recommends the following correlations for the average Nusselt number, where the characteristic dimension is the plate spacing b. The properties are evaluated at the average of the two plate temperatures, and the driving-temperature difference is the difference between the surface temperatures of the plates:

$$\overline{\mathrm{Nu}}_b = 0.195 \mathrm{Gr}_{b,f}^{1/4} \tag{5-42}$$

for $10^4 < \mathrm{Gr}_{b,f} < 3.7 \times 10^5$, and

$$\overline{\mathrm{Nu}}_b = 0.068 \mathrm{Gr}_{b,f}^{1/3} \tag{5-43}$$

for $3.7 \times 10^5 < \mathrm{Gr}_{b,f} < 10^7$.

For liquids confined between two parallel horizontal plates with the hotter plate below, Globe and Dropkin (Ref. 24) correlated data from experiments with water, silicone oils, and mercury with the relation

$$\overline{\mathrm{Nu}}_b = 0.069(\mathrm{Gr}_b \mathrm{Pr})_f^{1/3} \mathrm{Pr}_f^{0.074} \tag{5-44}$$

for $1.5 \times 10^5 < \mathrm{Ra} < 10^9$.

For vertical air spaces between two plates spaced a distance b apart and of height L in the vertical direction, Jakob (Ref. 23) correlated the results of several investigators. For Grashof numbers based on the plate spacing b, the recommended Nusselt number is

$$\overline{Nu}_b = \frac{\bar{h}_c b}{k} = 1.0 \tag{5-45}$$

for $Gr_{b,f} < 2000$;

$$\overline{Nu}_b = \frac{0.18 Gr_{b,f}^{1/4}}{(L/b)^{1/9}} \tag{5-46}$$

for $2 \times 10^4 < Gr_{b,f} < 2 \times 10^5$; and

$$\overline{Nu}_b = \frac{0.065 Gr_{b,f}^{1/3}}{(L/b)^{1/9}} \tag{5-47}$$

for $2 \times 10^5 < Gr_{b,f} < 10^7$. In using Eqs. 5-45 through 5-47, the Nusselt number and Grashof number are formed using the plate spacing b as the characteristic dimension, the fluid properties are evaluated at the average of the two surface temperatures, the appropriate driving temperature difference is the surface temperature difference, and it is required that

$$\frac{L}{b} > 3 \tag{5-48}$$

For values of $L/b < 3$, it is expected that the surface coefficients for a vertical surface freely convecting in a large expanse of fluid apply, at least approximately, to each of the two surfaces.

Emery and Chu (Ref. 25) have investigated enclosed vertical layers of fluids with Prandtl numbers between 3 and 30,000 and recommend

$$\overline{Nu}_b = 1$$

for $Ra_b < 10^3$, and

$$\overline{Nu}_b = \frac{0.280 Ra_b^{1/4}}{(L/b)^{1/4}} \tag{5-49}$$

for $10^3 < Ra_b < 10^7$.

5-6 COMBINED FREE AND FORCED CONVECTION

When the Reynolds number is small, especially if large temperature differences exist or if the passage geometry has a large hydraulic diameter, free convection effects may be important. The effect of superimposed free convection is primarily important in laminar flow through ducts. The subject of combined free and forced convection is complex and here only limited information on free convection superimposed on forced convection is presented for horizontal and vertical circular tubes.

Combined Free and Forced Convection in Horizontal Circular Tubes

In *horizontal tubes*, free convection sets up secondary flows at any cross section that aids the convection process. Hence the heat-transfer coefficient and Nusselt number for the combined convection are higher than those for pure forced convection. The maximum heat transfer occurs at the bottom of the tube. When the free-convection effects are significant in laminar flow, large temperature gradients exist near the wall, and the temperature variation in the vertical direction is significant. The measured velocity distributions in horizontal and vertical directions are also markedly different from the velocity distribution for Poiseuille flow. Metais and Eckert (Ref. 26) have classified free-, mixed-, and forced-convection regimes, as shown in Fig. 5-11, for horizontal pipes with the axially constant wall-temperature boundary condition. The limits of the forced- and mixed-convection regimes are defined in such a manner that free-convection effects contribute only about 10 percent to the heat flux. Figure 5-11 may therefore be used as a guide to determine whether or not free convection is important.

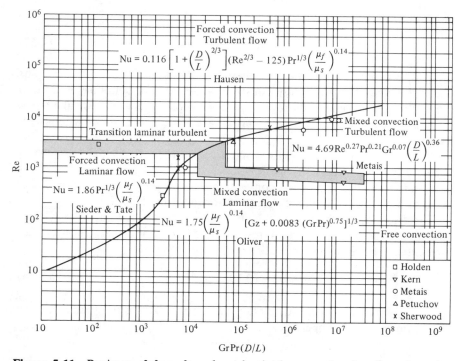

Figure 5-11 Regimes of free, forced, and mixed convection for flow through horizontal tubes $[10^{-2} < \Pr(d/L) < 1]$. (From B. Metais and E. R. G. Eckert, "Forced, Mixed and Free Convection Regimes," *J Heat Transfer*, pp. 295–296, 1964.)

Brown and Thomas (Ref. 27) measured experimentally laminar combined free and forced convection to water flowing through a horizontal tube with *constant wall temperature*. Their experimental results are correlated by the relationship

$$\overline{\mathrm{Nu}}_D = 1.75\left[\mathrm{Gz} + 0.012\left(\mathrm{GzGr}_D^{1/3}\right)^{4/3}\right]^{1/3}\left(\frac{\mu_s}{\mu_m}\right)^{-0.14} \tag{5-50}$$

where

$$\mathrm{Gz} = \pi D\,\mathrm{RePr}/4L. \tag{5-51}$$

All the fluid properties in Nu_D, Gz, Gr_D, and μ_m are evaluated at the fluid bulk mean temperature. This equation correlates the experimental data for water within $\pm 8\%$ and agrees within $\pm 50\%$ with most published experimental data for water, ethyl alcohol, and viscous oils.

Morcos and Bergles (Ref. 28) propose the following correlation for the combined convection in a horizontal circular tube for an *axially constant wall heat flux, q''*

$$\overline{\mathrm{Nu}}_h = \left\{(4.36)^2 + \left[0.145\left(\mathrm{Gr}^*\mathrm{Pr}^{1.35}K_p^{0.25}\right)^{0.265}\right]^2\right\}^{1/2} \tag{5-52}$$

where

$$\mathrm{Gr}^* = \mathrm{GrNu}_h = \frac{g\beta D_H^4 q''}{\nu^2 k} \tag{5-53}$$

and $K_p = k_w a'/kD_H$ is the peripheral heat-conduction parameter. The suffix h stands for the axially constant wall-heat-flux boundary condition with finite heat conduction in the peripheral direction. All the fluid properties used in Nu_h, Gr^*, Pr, and K_p are evaluated at the *film* temperature.

The thermal entrance length is significantly reduced by the presence of free-convection effects. Thus for most cases when free convection is superimposed, the combined convection flow is fully developed and Eq. 5-50 or 5-52 gives the Nusselt number for any tube length.

The friction factors are also higher for the combined-convection case. Based on the ethylene glycol test data, Morcos and Bergles (Ref. 28) presented the following correlation:

$$\frac{f}{f_{\mathrm{iso}}} = \left[1 + (0.195\,\mathrm{Ra}^{0.15})^{15}\right]^{1/15} \tag{5-54}$$

where f_{iso} is the isothermal friction factor $= 16/\mathrm{Re}$. Both f and Ra are based on the fluid properties evaluated at the *film* temperature.

Effects of Superimposed Free Convection in Vertical Circular Tubes

Unlike horizontal tubes, the effect of superimposed free convection in *vertical tubes* depends on the flow direction and whether or not the fluid is heated or cooled. For fluid heating with upward flow or fluid cooling with

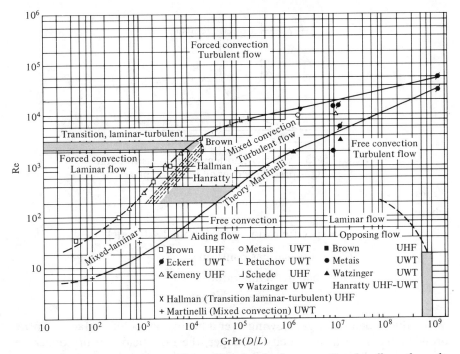

Figure 5-12 Regimes of free, forced, and mixed convection for flow through vertical tubes $[10^{-2} < \Pr(d/L) < 1]$. (From B. Metais and E. R. G. Eckert, "Forced, Mixed and Free Convection Regimes," *J. Heat Transfer*, pp. 295–296, 1964.)

downward flow, free convection aids forced convection, and the resultant heat-transfer coefficient is higher than the pure forced-convection coefficient. However, for fluid cooling with upward flow or fluid heating with downward flow, free-convection counters forced convection and a lower heat-transfer coefficient results. The flow-regime chart of Metais and Eckert for vertical tubes, shown in Fig. 5-12, provides guidelines to determine the significance of the superimposed free convection. The results of Fig. 5-12 are applicable for both constant heat-flux and constant wall-temperature boundary conditions.

5-7 HEAT TRANSFER IN HIGH-SPEED FLOW

Convection heat transfer in high-speed flow is important for systems such as aircraft and missiles when the velocity approaches or exceeds the velocity of sound. For a perfect gas the acoustical velocity, a, can be obtained from the relation

$$a = \sqrt{\gamma R_u T / \mathfrak{M}} \tag{5-55}$$

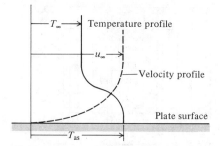

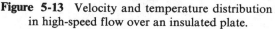

Figure 5-13 Velocity and temperature distribution in high-speed flow over an insulated plate.

where

$$\gamma = \text{specific ratio, } c_p/c_v \text{ (about 1.4 for air)}$$

$$R_u = \text{universal gas constant}$$

$$T = \text{absolute temperature}$$

$$\mathcal{M} = \text{molecular weight of gas}$$

When the velocity of a gas flowing over a heated or cooled surface is of the order of the acoustical velocity or larger, the flow field can no longer be described solely in terms of the Reynolds number, but the ratio of the gas flow velocity to the acoustical velocity (i.e., the Mach number $M = V_\infty/a_\infty$) must also be considered. When the gas velocity in a flow system reaches a value of about one-half of the speed of sound, the effects of viscous dissipation in the boundary layer become important. Under such conditions the temperature of a surface over which a gas is flowing can actually exceed the free-stream temperature. For flow over an adiabatic surface (e.g., a perfectly insulated wall), Fig. 5-13 shows the velocity and temperature distributions schematically. The high temperature at the surface is the combined result of the heating due to viscous dissipation and the temperature rise of the fluid as the kinetic energy of the flow is converted to internal energy while the flow decelerates through the boundary layer. The actual shape of the temperature profile depends on the relation between the rate at which viscous shear work increases the internal energy of the fluid and the rate at which heat is conducted toward the free stream.

Although the processes in a high-speed boundary layer are not adiabatic, it is general practice to relate them to adiabatic processes. The conversion of kinetic energy in a gas being slowed down adiabatically to zero velocity is described by the relation

$$i_0 = i_\infty + \frac{V_\infty^2}{2} \qquad (5\text{-}56)$$

where i_0 is the stagnation enthalpy and i_∞ the enthalpy of the gas in the free stream. For an ideal gas Eq. 5-56 becomes

$$T_0 = T_\infty + \frac{V_\infty^2}{2c_p}$$

or, in terms of the Mach number,

$$\frac{T_0}{T_\infty} = 1 + \frac{\gamma - 1}{2} M_\infty^2 \tag{5-57}$$

where T_0 is the stagnation temperature and T_∞ the free-stream temperature.

In a real boundary layer the fluid is not brought to rest reversibly because the viscous shearing process is thermodynamically irreversible. To account for the irreversibility in a boundary-layer flow, we define a *recovery factor r* as

$$r = \frac{T_{as} - T_\infty}{T_0 - T_\infty} \tag{5-58}$$

where T_{as} is the adiabatic surface temperature.

Experiments have shown that in laminar flow

$$r = Pr^{1/2} \tag{5-59}$$

whereas in turbulent flow

$$r = Pr^{1/3} \tag{5-60}$$

When a surface is not insulated, the rate of heat transfer by convection between a high-speed gas and that surface is governed by the relation

$$\frac{q_c}{A} = -k \frac{\partial T}{\partial y}\bigg|_{y=0}$$

The influence of heat transfer to and from the surface on the temperature distribution is illustrated in Fig. 5-14. We observe that in high-speed flow

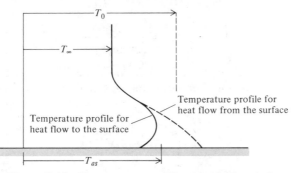

Temperature profile for heat flow from the surface

Temperature profile for heat flow to the surface

Figure 5-14 Temperature profiles in a high-speed boundary layer for heating and cooling.

heat can be transferred to the surface even when the surface temperature is above the free-stream temperature. This phenomenon is the result of viscous shear, often called *aerodynamic heating*. The rate of heat transfer can be calculated with the same relations as those used for low-speed flow if the average convection heat-transfer coefficient is redefined by the relation

$$\frac{q_c}{A} = \bar{h}_c(T_s - T_{as}) \tag{5-61}$$

which will yield a zero heat flow when the surface temperature T_s equals the adiabatic surface temperature.

Since in high-speed flow the temperature gradients in a boundary layer are large, variations in the physical properties of the fluid will also be substantial. However, the constant property heat-transfer equations can still be used if all the properties are evaluated at a reference temperature T^* given by the relation (Ref. 29)

$$T^* = T_\infty + 0.5(T_s - T_\infty) + 0.22(T_{as} - T_\infty) \tag{5-62}$$

The local values of the heat-transfer coefficient, defined by the relation

$$h_{c_x} = \frac{q/A}{T_s - T_{as}}$$

can be obtained from the following equations:

Laminar boundary layer ($\mathrm{Re}_x{}^* < 10^5$):

$$\mathrm{St}_x{}^* = \left(\frac{h_{c_x}}{c_p \rho V_\infty}\right)^* = 0.332(\mathrm{Re}_x{}^*)^{-1/2}(\mathrm{Pr}^*)^{-2/3} \tag{5-63}$$

Turbulent boundary layer ($10^5 < \mathrm{Re}_x{}^* < 10^7$):

$$\mathrm{St}_x{}^* = \left(\frac{h_{c_x}}{c_p \rho V_\infty}\right)^* = 0.0288(\mathrm{Re}_x{}^*)^{-1/5}(\mathrm{Pr}^*)^{-2/3} \tag{5-64}$$

Turbulent boundary layer ($10^7 < \mathrm{Re}_x{}^* < 10^9$):

$$\mathrm{St}_x{}^* = \left(\frac{h_{c_x}}{c_p \rho V_\infty}\right)^* = \frac{2.46}{(\ln \mathrm{Re}_x{}^*)^{2.584}}(\mathrm{Pr}^*)^{-2/3} \tag{5-65}$$

because experimental data for local friction coefficients in high-speed gas flow in the Reynolds number range between 10^7 and 10^9 are correlated by the relation

$$C_{fx} = \frac{4.92}{(\ln \mathrm{Re}_x{}^*)^{2.584}} \tag{5-66}$$

If the average value of the heat-transfer coefficients is to be determined, the expressions above must be integrated between $x=0$ and $x=L$ for

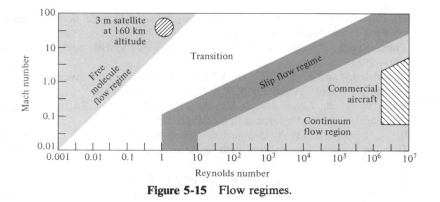

Figure 5-15 Flow regimes.

low-speed flow. However, the integration has to be done numerically in most practical cases because the reference temperature T^* is not the same for the laminar and turbulent portions of the boundary layer, as shown by Eqs. 5-59 and 5-60.

In some situations (e.g., extremely high altitudes) the fluid density may be so small that the distance between gas molecules becomes of the same order of magnitude as the boundary layer. In such cases the fluid cannot be treated as a continuum and it is necessary to subdivide the flow processes into regimes. These flow regimes are characterized by the ratio of the molecular free path to a significant physical scale of the system, the *Knudsen number*, Kn. Continuum flow corresponds to small values of Kn, while at larger values of Kn, molecular collisions occur primarily at the surface and in the main stream. Since energy transport is by free motion of molecules between the surface and the main stream, this regime is called *free-molecule*. Between the free-molecule and the continuum regime is a transition range, called the *slip-flow* regime because it is treated by assuming temperature and velocity "slip" at fluid/solid interfaces. Figure 5-15 shows a map of these flow regimes. For a treatment of heat transfer and friction in these specialized flow systems the reader is referred to References 30, 31, and 32.

REFERENCES

1. A. P. Colburn, "A Method of Correlating Forced Convection Heat Transfer Data and a Comparison with Fluid Friction," *Trans. AIChE*, vol. 29, pp. 174–210, 1933.
2. E. N. Sieder and C. E. Tate, "Heat Transfer and Pressure Drop of Liquids in Tubes," *Ind. Eng. Chem.*, vol. 28, p. 1429, 1936.
3. W. Nusselt, "Der Wärmeaustausch zwischen Wand und Wasser im Rohr," *Forsch. Geb. Ingenieurwes.*, vol. 2, p. 309, 1931.

4. H. Hausen, "Darstellung des Wärmeuberganges in Rohren durch verallgemeinerte Potenzbeziehungen," *VDI Z.*, no. 4, p. 91, 1943.

5. Frank Kreith, *Principles of Heat Transfer*, 3rd ed., Harper and Row, New York, 1973.

6. B. Lubarsky and S. J. Kaufman, "Review of Experimental Investigations of Liquid-Metal Heat Transfer," *NACA Tech. Note 3336*, 1955.

7. R. D. Lyon, ed., *Liquid Metals Handbook*, 3rd ed., Atomic Energy Commission and U.S. Navy Department, Washington, D.C., 1952.

8. R. A. Seban and T. T. Shimazaki, "Heat Transfer to a Fluid Flowing Turbulently in a Smooth Pipe with Walls at Constant Temperature," *Trans. ASME*, vol. 73, p. 803, 1951.

9. W. H. Giedt, "Investigation of Variation of Point Unit-Heat-Transfer Coefficient Around a Cylinder Normal to an Air Stream," *Trans. ASME*, vol. 71, pp. 375–381, 1949.

10. R. Hilpert, "Wärmeabgabe von geheizen Drahten und Rohren," *Forsch. Geb. Ingenieurwes.*, vol. 4, p. 220, 1933.

11. J. D. Knudsen and D. L. Katz, *Fluid Dynamics and Heat Transfer*, McGraw-Hill Book Company, New York, 1958.

12. M. Jakob, *Heat Transfer*, vol. 1, John Wiley & Sons, Inc., New York, 1949.

13. S. Whitaker, "Forced Convection Heat-Transfer Correlations for Flow in Pipes, Past Flat Plates, Single Cylinders, Single Spheres, and Flow in Packed Beds and Tube Bundles," *AIChE J.*, vol. 18, p. 361, 1972.

14. L. C. Witte, "An Experimental Study of Forced-Convection Heat Transfer from a Sphere to Liquid Sodium," *J. Heat Transfer*, vol. 90, p. 9, 1968.

15. E. D. Grimison, "Correlation and Utilization of New Data on Flow Resistance and Heat Transfer for Cross Flow of Gases over Tube Banks," *Trans. ASME*, vol. 59, pp. 583–594, 1937.

16. W. M. Kays and R. K. Lo, "Basic Heat Transfer and Flow Friction Data for Gas Flow Normal to Banks of Staggered Tubes: Use of a Transient Technique," *Stanford Univ. Tech. Rept. 15*, Navy Contract N6-ONR251 T.O. 6, 1952.

17. S. Kalish and O. E. Dwyer, "Heat Transfer to NaK Flowing Through Unbaffled Rod Bundles," *Int. J. Heat Mass Transfer*, vol. 10, pp. 1533–1558, 1967.

18. M. Jacob, "Heat Transfer and Flow Resistance in Cross Flow of Gases over Tube Banks," *Trans. ASME*, vol. 60, p. 384, 1938.

19. S. W. Churchill and H. H. S. Chu, "Correlating Equations for Laminar and Turbulent Free Convection from a Horizontal Cylinder," *Int. J. Heat Mass Transfer*, vol. 18, p. 1049, 1975.

20. S. C. Hyman, C. F. Bonilla, and S. W. Ehrlich, "Heat Transfer to Liquid Metals from Horizontal Cylinders," *AIChE Symp. Heat Transfer*, Atlantic City, N.J., p. 21, 1953.

21. T. Yuge, "Experiments on Heat Transfer from Spheres Including Combined Natural and Forced Convection," *J. Heat Transfer*, ser. C, vol. 82, p. 214, 1960.

22. W. J. King, "The Basic Laws and Data of Heat Transmission," *Mech. Eng.*, vol. 54, p. 347, 1932.

23. M. Jakob, "Free Heat Convection Through Enclosed Plane Gas Layers," *Trans. ASME*, vol. 68, pp. 189–94, April 1946.

24. S. Globe and D. Dropkin, "Natural Convection Heat Transfer in Liquids Confined by Two Horizontal Plates and Heated from Below," *J. Heat Transfer*, vol. 81, pp. 24–29, 1959.

25. A. Emery and N. C. Chu, "Heat Transfer Across Vertical Layers," *J. Heat Transfer*, vol. 87, no. 1, pp. 110–16, 1965.

26. B. Metais and E. R. G. Eckert, "Forced, Mixed, and Free Convection Regions," *J. Heat Transfer*, vol. 86, ser. C, pp. 295–296, 1964.

27. A. R. Brown and M. A. Thomas, "Combined Free and Forced Convection Heat Transfer for Laminar Flow in Horizontal Tubes," *J. Mech. Eng. Sci.*, vol. 7, pp. 440–448, 1965.

28. S. M. Morcos and A. E. Bergles, "Experimental Investigation of Combined Forced and Free Laminar Convection in Horizontal Tubes," *J. Heat Transfer*, vol. 97, ser. C, pp. 212–219, 1975.

29. E. R. A. Eckert, "Engineering Relations for Heat Transfer and Friction in High-Velocity Laminar and Turbulent Boundary Layer Flow over Surface with Constant Pressure and Temperature," *Trans. ASME*, vol. 78, pp. 1273–1284, 1956.

30. E. R. Van Driest, "Turbulent Boundary Layer in Compressible Fluids," *J. Aeronautical Sci.*, vol. 18, no. 3, pp. 145–161, 1951.

31. A. K. Oppenheim, "Generalized Theory of Convective Heat Transfer in a Free-Molecule Flow," *J. Aero. Sci.*, vol. 20, pp. 49–57, 1953.

32. W. D. Hayes and R. F. Probstein, *Hypersonic Flow Theory*, Academic Press, New York, 1959.

PROBLEMS

The problems in this chapter are organized in the manner shown in the table.

PROBLEM NUMBERS	SECTION	SUBJECT
5-1 to 5-4	5-1	Dimensionless parameters
5-5 to 5-15	5-2	Convection through tubes and ducts
5-16 to 5-38	5-3	Forced convection in external flow
5-39 to 5-58	5-4	Natural convection
5-59 to 5-61	5-5	Combined free and forced convection
5-62 to 5-65	5-6	Heat transfer in high-speed flow

5-1 The convection-heat-transfer rate for short ducts is known to be a function of the following quantities:

V–fluid velocity
D–duct diameter
L–length of duct
ρ–fluid density
c_p–fluid specific heat
k–fluid thermal conductivity

Show by using dimensional analysis that a dimensionless quantity involving these quantities is the Graetz number, defined as

$$Gz = \frac{D}{L}\frac{VD\rho c_p}{k} = \frac{D}{L}\,\text{Re}_D\text{Pr}$$

5-2 In free convection the quantities known to affect the heat transfer are:

ρ–fluid density
μ–fluid viscosity
β–fluid coefficient of thermal expansion
g–acceleration of gravity
D–characteristic dimension of the body

$T_s - T_\infty$–temperature difference between
the body and ambient fluid

Use dimensional analysis to show that a dimensionless group involving these quantities is the Grashof number, defined as

$$Gr = \frac{\rho^2 g \beta (T_s - T_\infty) D^3}{\mu^2}$$

5-3   The drag on a body inserted in a stream of fluid is known to be a function of the following quantities:

ρ–fluid density
μ–fluid viscosity
V_∞–free-stream velocity
D–characteristic dimension of body
τ_s–shear stress on the surface of the
body

Use the Buckingham pi theorem to show that the dimensionless drag

$$\frac{\tau_s}{\rho V_\infty^2}$$

can be expressed as a function of the Reynolds number:

$$\frac{\rho V_\infty D}{\mu}$$

5-4   Suppose that you have measured the value for $\bar{h}_c$ for the case of air in forced convection over a sphere of diameter D. When the data are plotted on a graph of $\overline{Nu}_d$ as a function of $Re_d Pr$, they appear as shown in the sketch.

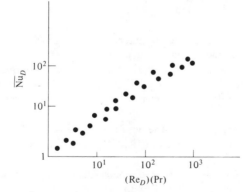

Write an appropriate dimensionless correlation for the average Nusselt number for these data and state any limitations to your equation.

5-5   Water flowing at the rate of 4 kg/s through a 5-cm-i.d. tube is to be heated from 30 to 50°C. If the tube wall temperature is maintained at 80°C, determine the length of the tube required.

5-6   Dry air at a pressure of 1400 kN/m^2 flows through a duct 7.5 cm i.d. and 5 m long at a rate of 0.5 kg/s. If the duct wall is maintained at an average temperature of 200°C and the average air temperature in the duct is 260°C, calculate the heat-transfer coefficient and the decrease in air temperature in the duct.

$$R_{E_D} = \frac{4\,\dot{m}}{\pi\,D\,\mu}$$

5-7   Water at an average temperature of 10°C flows through a 2.5-cm-i.d. tube at a rate of 0.5 kg/s. If a constant heat flux is imposed by an electric heating element and the average wall temperature is 50°C, determine the heat-transfer coefficient. Then estimate the change in temperature of the water flowing through a 5.6-m-long tube.

5-8   Water enters a 2.5-cm-i.d. 1.5-m-long tube at a bulk temperature of 20°C. If the flow rate of the water is 1.0 kg/s, estimate the exit-water temperature if the tube-wall temperature is maintained constant at 50°C.

5-9   Atmospheric air at 10°C enters a 6m long rectangular duct having a cross section 7.5 × 15 cm. All four surfaces of this duct are maintained by solar insolation at 70°C. Calculate the air flow rate to achieve an average exit-air temperature 20°C above the inlet temperature.

5-10   Water at an average temperature of 20°C flows at a mean velocity of 0.1 m/s through a tube having an inside diameter of 3 mm. Assuming that the tube is electrically heated and generates a constant wall heat flux of 1.0 kW/m^2, determine the temperature of the water as a function of radial distance.

5-11   Atmospheric air at a temperature of 120°C enters a pipe having an inside diameter of 7 cm at a flow rate of $0.22 \text{ m}^3/\text{s}$. Assuming that the wall of that pipe is maintained at a constant temperature of 20°C, estimate the length of the pipe necessary for the air to be cooled to 80°C.

5-12   Experimental measurements made on a pipe with a ellipsoidal cross section have shown that the convective-heat-transfer coefficient in air is $35 \text{ W/m}^2\cdot\text{K}$ when the ambient air velocity is 15 m/s and the hydraulic diameter of the pipe is 0.4 m. What would the heat-transfer coefficient be if a geometrically similar pipe with a hydraulic diameter of 0.2 m were placed in air with a velocity of 30 m/s?

5-13   Mercury enters a 3.4-cm tube with a velocity of 3 m/s and a temperature of 80°C. Estimate the heat-transfer coefficient if the tube is 6 m long.

5-14   Liquid sodium enters a 2-cm-diameter tube with a mass flow rate of 0.2 kg/s. The tube adds a constant heat flux to the sodium, and the temperature of the tube is 10°C above the sodium temperature. The sodium temperature increases by 2.8°C while in the tube and it enters at 378 K. Determine the convection-heat-transfer rate to the sodium.

5-15   A duct with an equilateral triangular cross section is 4 m long and 3 cm on a side. Liquid sodium enters the duct at 478 K with a mass flow rate of 3.6 kg/s.

The duct is isothermal at a temperature of 525 K. Evaluate all properties at the inlet temperature and determine:

(a) The heat-transfer rate to the sodium.

(b) The outlet sodium temperature.

5-16 Assume that the velocity distribution over a flat plate is given by the dimensionless expression

$$\frac{u}{V_\infty} = \frac{y}{4x} \text{Re}_x^{1/2}$$

where x is the distance measured in the plane and y the distance perpendicular to the plane. Water with an average temperature of 60°C ($T_s = 80$°C, $T_\infty = 40$°C) and $V_\infty = 5$ m/s flows over one side of a 3×3 m plate. Determine the following information:

(a) Total drag on the plate.

(b) The expression for the local boundary layer thickness $\delta(x)$.

(c) The boundary-layer thickness at the end of the plate ($x = 3$ m).

5-17 Assume that the local Nusselt number for flow over a surface is given by the equation

$$\text{Nu}_x = 0.5 \text{Re}_x^{1/3} \text{Pr}^{1/3}$$

Determine an expression for the average Nusselt number $\overline{\text{Nu}}_L$ in terms of Re_L.

5-18 Air at a temperature of 22°C blows across an asphalt street with a velocity of 4 m/s. The temperature of the asphalt is 58°C. The width of the street is 12 m. Estimate the convective-heat-transfer rate from the surface of the asphalt.

5-19 A flat horizontal roof has a temperature of 10°C in the winter. The roof is square and measures 40×40 m. Determine the heat-transfer rate from the roof to air at -10°C when the wind velocity is (a) 2 m/s, (b) 6 m/s, and (c) 12 m/s.

5-20 A steel plate measuring 2×20 m leaves a rolling process and travels along a conveyor at a velocity of 4 m/s. The temperature of the steel is 835 K and the ambient air is at 305 K. Determine the heat-transfer rate by the convection mode from the upper surface of the steel.

5-21 A flat plate is 3 m long and 6 m wide. The temperature of the plate is 160°C. Atmospheric air at 40°C is forced over the plate parallel to the 3-m side. Calculate the convective-heat-transfer rate from the plate when the air has a velocity of (a) 0.1 m/s, (b) 10 m/s, and (c) 20 m/s.

5-22 The average Nusselt number for forced convection of a gas over a sphere is given by the relationship

$$\overline{\text{Nu}}_d = 0.37 \text{Re}_d^{0.6} \quad \text{for} \quad 17 < \text{Re}_d < 70,000$$

Determine the rate at which heat must be supplied to a 4-cm-diameter sphere to maintain the surface temperature at 90°C when placed in 30°C (a) air, and (b) water. The fluid velocity for both is 1.0 m/s.

5-23 A spherical water droplet 4 mm in diameter falls freely in still air. Calculate the average convection-heat-transfer coefficient when the droplet has reached a velocity of 10 m/s. Assume that the water droplet is at 10°C and the air has a temperature of 30°C. Neglect evaporation from the droplet. If you had not neglected mass transfer, would the heat transfer to the drop increase or decrease? Explain your reasoning.

5-24 The shape of a hot-air balloon can be approximated by a sphere with a diameter of 18 m. The total weight of the gondola, burner, occupants, and balloon material is 1050 N. Estimate the capacity of a gas burner used to provide hot air for the interior of the balloon, in kW. Assume that the balloon is operating at an altitude where the air temperature is 18°C and the wind velocity is 0.1 m/s. Neglect the thermal resistance of the balloon material.

5-25 A viscous liquid is to be pumped between two buildings of a chemical plant in an aboveground pipe. To reduce the viscosity of the fluid and thereby reduce the energy required to pump the liquid, it is heated to 40°C. The pipe has an o.d. of 22 cm and a length of 110 m. The flow rate of the liquid in the pipe is 20 kg/s. The wind direction is perpendicular to the axis of the pipe with a velocity of 14 m/s. Neglect the thermal resistance of the pipe and estimate the rate at which the liquid cools on a day when the air temperature is 0°C. Determine the temperature drop of the liquid in the pipe if the liquid has a specific heat of 1.3 kJ/kg·K.

5-26 A storage tank for liquefied natural gas (LNG) is spherical in shape with a diameter of 5.0 m. The temperature of the outer surface of the tank is measured and found to be 10°C. The ambient air temperature surrounding the tank is 30°C. Neglecting radiation effects, calculate the heat-transfer rate to the LNG when the average wind velocity is 7 m/s.

5-27 Air flows over the exterior surface of a single 5-cm-diameter cylinder in a direction perpendicular to the cylinder axis. The air has a velocity of 35 m/s and an ambient temperature of 100°C. If the temperature of the cylinder is 200°C, determine the ratio of the average heat-transfer coefficient to the local heat-transfer coefficient at the stagnation point, assuming that the Nusselt number at the stagnation point is given by

$$\mathrm{Nu}_d = \sqrt{\mathrm{Re}_d}$$

5-28 Toluene at 20°C flows over a single 4-cm-diameter pipe at a velocity of 2 m/s. The flow direction is perpendicular to the axis of the pipe. Estimate the heat-transfer coefficient for this situation.

5-29 Water at 20°C approaches a series of tubes with a velocity of 8 m/s. The tubes are spaced on 8-cm centers and their outside diameter is 4 cm. Determine a value for the maximum Reynolds number, Re_{max}.

5-30 Freon 12 at 5°C approaches an array of staggered tubes with an outside diameter of 1.8 cm. The velocity of the Freon prior to reaching the array of tubes is

10 m/s. The distance between the centerline of adjacent tubes in the same row is 3.5 cm and the distance between the centerlines of adjacent rows is 3.0 cm. Calculate the maximum Reynolds number, Re_{max}.

5-31 Oxygen at 1 atm pressure and 100°C enters a heat exchanger with an average velocity of 4 m/s. The heat exchanger consists of 12 transverse in-line rows of 3-cm-o.d. tubes. The centerline spacing of the tubes is 6 cm, the distance between the rows of tubes in the flow direction is 9 cm, and the tubes are 6 m long. The surface temperature of the tubes is 50°C. Estimate the following quantities:
  (a) The convective-heat-transfer coefficient for oxygen flow over the tubes.
  (b) The heat-transfer rate from the oxygen per tube.
  (c) The pressure of the oxygen as it leaves the tube bank.

5-32 Rework Problem 5-31 for tubes that are staggered instead of in-line.

5-33 Mercury enters a tube bank consisting of 3-cm-o.d. tubes on 6-cm centers. The centerline spacing between transverse rows is 4.5 cm and the tubes are arranged in-line. The entering mercury has a temperature of 423 K and a uniform velocity of 7 cm/s. Estimate the convective-heat-transfer coefficient.

5-34 Sodium enters a tube bank with a temperature of 644 K and a velocity of 60 cm/s. The tube diameter is 0.8 cm and the centerline spacing transverse to the flow direction is 3.1 cm. The tubes are staggered and the centerline distance between rows in the flow direction is 3.8 cm. Estimate the value for $\bar{h}$.

5-35 Exhaust gases from a furnace are used to heat water that flows internally through a series of staggered tubes. The total water flow rate is 25 kg/s. The exhaust gases enter the tube bank with a velocity of 35 m/s and a temperature of 320°C. The centerline spacing of the tubes perpendicular to the gas flow is 2.4 cm and parallel to the flow direction is 1.6 cm. The tube bank consists of 36 tubes that are 5 m long with a 1.6-cm OD arranged in six rows of six tubes each. The surface temperature of the tubes averages 80°C. The properties of the exhaust gases can be estimated to be the same as air at one atmosphere. Estimate the following quantities:
  (a) The convective-heat-transfer coefficient for the gases.
  (b) The convective-heat-transfer rate to the water.
  (c) The temperature rise of the water.
  (d) The pressure drop of the exhaust gases across the tube bank.

5-36 It has been proposed to tow an antarctic iceburg to populated arid areas for a freshwater supply. The iceburg can be approximated by a cylinder that is 320 m deep and 730 m in diameter. The average towing speed is 1.0 m/s through water that has an average temperature of 10°C. Estimate the amount of ice that is lost during a 3-month trip due to convection from the side and bottom of the iceburg. The latent heat of fusion for water is 3.35×10^5 J/kg.

5-37 An in-line tube bank consists of 50 rows of 50 tubes that are 4 m long. The outside diameter of the tubes is 1.1 cm. The centerline spacing between adjacent tubes and adjacent rows is both 1.65 cm. Air enters the tube bank with a velocity of 6 m/s and a temperature of 180°C. If the tubes have a uniform temperature of 20°C, determine the heat-transfer rate to the tubes and the pressure drop across the tube bank.

5-38 Air at 40°C with a maximum Reynolds number of 4×10^5 flows over a tube bank consisting of 5 rows of staggered 1 cm o.d. tubes. Estimate the convective-heat-transfer coefficient for the air.

5-39 A vertical flat plate 25 cm high is heated to a temperature of 90°C. A fluid with an ambient temperature of 30°C is placed in contact with the vertical plate. Calculate the Grashof number if the fluid is (a) air at atmospheric pressure, (b) water, (c) engine oil, and (d) mercury.

5-40 For the conditions given in Problem 5-39, calculate the free convection-heat-transfer rate for the four fluids if the plate has a surface area of 0.5 m².

5-41 A steam pipe passes vertically through a room. The exterior surface of the pipe is 60°C and the still air temperature in the room is 20°C. The outside diameter of the pipe is 6.8 cm and the exposed length of pipe is 3.4 m. Determine the heat-transfer rate by convection from the pipe. Assuming that the exterior surface of the room is completely insulated and the volume of the room is 900 m³, calculate how long it would take for the steam to raise the temperature of the room air 1°C.

5-42 A 0.3×0.3 m electric hot pad is rated at 12.5 W. Determine the temperature the hot pad will reach if it is suspended vertically in still 25°C air. Assume that heat is transferred equally from both sides of the hot pad.

5-43 For the conditions given in Problem 5-42, calculate the temperature of the hot pad if it is placed horizontally on a well-insulated surface.

5-44 A sheet-metal heating duct carries warm air vertically from one floor to another. The temperature of the duct is 55°C and the ambient air surrounding the duct is 25°C. The duct has a cross section of 0.8×0.8 m and it is 4 m long. Calculate the heat loss from the warm air in the duct.

5-45 A single-room cabin has a 12×14 m floor plan and the floor-to-ceiling height is 2.6 m. The temperature of the air in the room is 25°C, and the temperature of the interior surface of the walls is 15°C. Estimate the heat-transfer rate through the wall surfaces.

5-46 Calculate the heat-transfer coefficient for a 2.8-cm-o.d. horizontal cylinder at 300°C surrounded by sodium at 110°C.

5-47 The outside diameter of a horizontal steam pipe is 4.6 cm. The pipe is located in a room where the ambient temperature is 20°C. The exterior surface temperature of the pipe is 40°C. Determine the heat-transfer rate from the pipe per unit length of pipe.

5-48 The electric element of a charcoal lighter is 1 cm in diameter, 70 cm in length, and is rated at 425 W. The element is placed outdoors in a horizontal position where the ambient air temperature is 28°C. The emissivity of the element is approximately 0.30, and the wind velocity can be neglected. Estimate the surface temperature of the element if it is plugged in and left until it reaches steady-state conditions. Does your answer confirm why the manufacturer warns that the lighter should not be left on more than 8 min?

5-49 A 100×120 m indoor ice rink has the temperature of the ice surface at −4°C. The ambient air temperature in the rink is 20°C. Assuming that the only significant mode of heat transfer from the ice surface is convection to the air, estimate the cooling capacity of the refrigeration unit used to maintain the ice at −4°C. Assume that the air over the ice is calm, with no ambient velocity.

5-50 A 40-W fluorescent light bulb has outside diameter 3.2 cm and length 136 cm. The bulb has a surface temperature of 36°C when the ambient air temperature is 24°C. What percent of the total heat dissipation from the bulb leaves by the convection mode of heat transfer? Assume that the bulb is suspended horizontally in still air.

5-51 A horizontal 14-cm-o.d. steam pipe traverses the still air in a manufacturing plant. The outer surface of the pipe has a temperature of 55°C and the temperature of the ambient air in the plant is 25°C. Calculate the heat-transfer rate from the steam if the pipe is 20 m long.

5-52 A thin-walled stainless steel tank used to fry potato chips is placed in still air at 25°C. The tank has a square cross section and it is filled with vegetable oil that is heated to 205°C by gas burners on the bottom of the tank. The tank is 3×3 m and the depth of the oil is 1 m. Estimate the gas consumption rate in m^3/s at standard conditions when the tank reaches steady conditions. The heating value for natural gas is 3.7×10^7 J/m^3 of gas at standard pressure and temperature. If natural gas costs $0.07/$m^3$ of gas, estimate the fuel costs per hour due to convection heat loss through the walls of the tank.

5-53 Bricks emerge from a kiln at a temperature of 420°C, and they are cooled in still air that has a temperature of 30°C. The vertical dimension of the brick is 10 cm and the horizontal dimensions are 14×7 cm. Calculate the convection cooling rate of the bricks when they first leave the kiln. Assume the bottom surface of the brick rests on an insulated surface.

5-54 An electronics package consumes 75 W of power, and for protection it is covered by a cubic tin box that is 28 cm on a side. Estimate the surface temperature of the box if the air temperature surrounding the box is 20°C. The bottom of the box sits on a well-insulated surface. Neglect radiation from the box.

5-55 A double-pane insulated window unit is installed in a store. The unit consists of two 1-cm-thick pieces of glass separated by a 1.8-cm layer of air. The unit is 3 m high and 4 m wide. Determine the rate of heat transfer through the unit when the extreme outside surface of the glass is at −10°C and the inside surface of the other pane of glass is at 20°C.

5-56 The top of a furnace consists of two horizontal metal plates separated by still air. The temperature of the lower and upper plates are 380°C and 120°C, respectively. The spacing between the plates is 2 cm. Determine the convective heat-transfer coefficient between the two plates.

5-57 A large tank containing a heated liquid has a dual metallic wall separated by a trapped layer of air. One side of the tank has a vertical dimension of 12 m and a horizontal dimension of 15 m. The separation distance between the two walls is 3.5 cm. Calculate the heat-transfer rate between the walls if the inner wall is at 90°C and the outer wall is 30°C.

5-58 Two large vertical planes are separated by a 3-cm-thick layer of air. The temperature of one plane is 10°C and the other is at 70°C. Determine the heat-transfer rate between the planes if the surface area of the planes is 20 m² and the vertical dimension is 5.6 m.

5-59 Water at 30°C enters a horizontal tube with a velocity of 0.15 m/s. The tube wall is isothermal at 90°C. The tube has an i.d. of 3 mm and a length of 250 cm. Considering a combined free-forced model, estimate the heat-transfer coefficient. Compare your result with the value for a purely forced convection model.

5-60 A horizontal tube 1 m long has a surface temperature of 100°C and an inside diameter of 5 cm. Oxygen enters at atmospheric pressure, a temperature of 60°C, and a velocity of 0.4 m/s. Assuming that the flow is a combination of free and forced convection, estimate the heat-transfer coefficient.

5-61 Saturated liquid ammonia enters a horizontal tube with a temperature of 40°C, and the tube surface is isothermal at 0°C. The tube is 85 cm long and has an inside diameter of 0.25 cm. The velocity of the ammonia is 13.5 cm/s. Calculate the heat-transfer coefficient assuming:
  (a) Pure forced convection.
  (b) Combined free and forced convection.

5-62 A flat plate is placed in air that has a Mach number of 2.0, a pressure of 25,000 N/m², and an ambient temperature of −15°C. The temperature of the plate cannot exceed 120°C. If the plate is 30 cm long in the direction of flow, calculate the cooling rate per unit area that is required to maintain the plate temperature below this limit.

5-63 A satellite reenters the earth's atmosphere at a velocity of 2700 m/s. Estimate the maximum temperature the heat shield would reach if the shield material is not allowed to ablate and radiation effects are neglected. The temperature of the upper surface of the atmosphere is −50°C.

5-64 A scale model of an airplane wing section is placed in a wind tunnel and tested at a Mach number of 1.5. The air pressure and temperature at the test section is 20,000 N/m² and −30°C, respectively. Assume that the wing section is to be cooled to a temperature of 60°C. Calculate the heat flux that must be removed from the wing section to maintain this temperature. Assume the wing approximates a flat plate that is 30 cm long in the flow direction.

5-65 Air at 0°C and 23,000 N/m² flows with a velocity of 500 m/s over a flat plate heated to a temperature of 80°C. The plate is 20 cm long in the direction of flow and 40 cm wide in the direction perpendicular to the flow. Estimate the following quantities:

    (a) The adiabatic surface temperature.

    (b) The distance from the leading edge of the plate where the flow becomes turbulent, x_{cr}.

    (c) The local heat-transfer coefficient at x_{cr}.

    (d) The local friction coefficient at the end of the plate.

    (e) The local heat-transfer rate at the end of the plate.

Chapter 6

RADIATION

6-1 INTRODUCTION

The *radiative mode* of heat transfer is characterized by energy transported in the form of electromagnetic waves. The waves travel at the speed of light. The transport of energy by radiation can occur between surfaces that are separated by a vacuum. The sun, for example, transfers energy to the earth entirely by the radiation mode across millions of kilometers of evacuated space. Also, the only way to transfer energy to or from an orbiting spacecraft is by the radiative mode.

Both the wave theory and the particle theory are useful in helping to explain the behavior of thermal radiation. The wave theory states that radiation can be imagined to be a wave oscillating with a frequency v and a wavelength λ. The product of the frequency and wavelength is the velocity of propagation, which is the velocity of light, c:

$$c = \lambda v \tag{6-1}$$

The particle theory assumes that radiant energy is transported as packets of energy called *photons*. Each photon travels with the speed of light with a distinct energy level given by

$$e = hv$$

where h is Planck's constant. The photons with higher frequencies possess more energy than those with lower frequencies. When a body is heated, free electrons can jump to higher or excited levels. As the electron returns to its lower energy level, it emits a photon whose energy is equal to the difference in energy between its excited state and equilibrium state. For any surface, numerous electrons are experiencing a change in energy level at any given instant, and therefore the energy leaving the surface is distributed over a spectrum of frequencies. The energy is emitted solely by virtue of the temperature of the body. The energy leaving the surface in this manner is called *thermal radiation*.

There are ways other than heating a surface that can cause photons to be emitted from the body. At the short-wavelength end of the spectrum, for example, are the x rays, which can be produced by subjecting a piece of metal to a stream of electrons. On the other end of the spectrum are the radio waves, with long wavelengths, that can be produced by electronic equipment and crystals. Between these two extremes is thermal radiation, which is emitted from a body as a result solely of its temperature. The entire range of all wavelengths is called the *electromagnetic spectrum*.

The electromagnetic spectrum is subdivided into a number of wavelength ranges, as shown in Fig. 6-1(a). Thermal radiation that is emitted from a surface due solely to its temperature exists between wavelengths of 10^{-7} m and 10^{-4} m. Because of its importance in a text of heat transfer, the thermal-radiation portion of the spectrum is detailed in Fig. 6-1(b).

The human eye is able to detect electromagnetic waves between wavelengths of approximately 3.8×10^{-7} m and 7.6×10^{-7} m, and radiation between these wavelengths is called *visible radiation*. The visible part on the spectrum is a very small portion of the entire spectrum and lies completely within the thermal radiation range. Various units of length are used to measure the wavelengths of waves. They are:

$$1\,\text{Å} = 1 \text{ angstrom} = 10^{-10}\,\text{m} = 10^{-8}\,\text{cm} = 10^{-4}\,\mu\text{m}$$

$$1\,\mu\text{m} = 1 \text{ micrometer} = 1 \text{ micron} = 10^{-6}\,\text{m} = 10^{-4}\,\text{cm} = 10^{4}\,\text{Å}$$

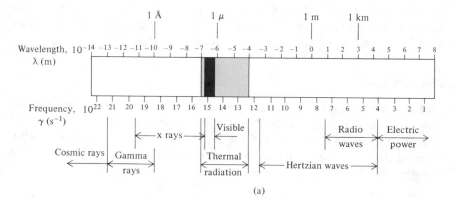

(a)

Figure 6-1 (a) Electromagnetic spectrum.

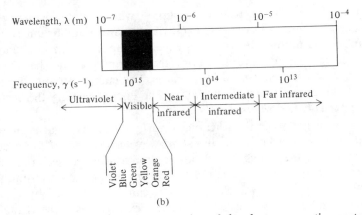

(b)

Figure 6-1 (b) Thermal-radiation portion of the electromagnetic spectrum.

The next section deals with the physics of thermal radiation. It introduces the concept of a thermal blackbody, and it presents the energy distribution as a function of wavelength and temperature for a blackbody radiator.

6-2 PHYSICS OF RADIATION

Concept of a Blackbody

Not all surfaces emit or absorb the same amount of radiant energy when they are heated to the same temperature. A body that emits and absorbs the maximum amount of energy at a given temperature is a black surface or simply a *blackbody*. A blackbody is a standard that can be approached in practice by coating the surface of the body or by modifying the shape of the surface. A blackbody is a standard with which all other radiators can be compared. In this respect it is much like an isentropic pump or turbine used as a standard of comparison in thermodynamics.

Planck's Law

When a blackbody is heated to a temperature T, photons are emitted from the surface of the body. The photons have a definite distribution of energy, depending upon the surface temperature T. Max Planck in 1900 showed that the energy emitted at a wavelength λ from a blackbody at a temperature T is

$$E_{b_\lambda}(T) = \frac{C_1}{\lambda^5(e^{C_2/\lambda T} - 1)} \qquad (6-2)$$

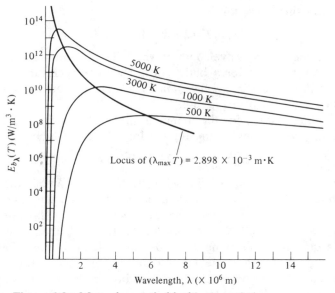

Figure 6-2 Monochromatic blackbody emissive power.

where

$E_{b\lambda}$ = monochromatic or spectral emissive power of a blackbody at temperature T, W/m^3

C_1 = first radiation constant

= 3.7418×10^{-16} W·m²

C_2 = second radiation constant

= 1.4388×10^{-2} m·K

The variation of monochromatic blackbody emissive power with temperature and wavelength given by Eq. 6-2 is known as *Planck's law*.

A plot of the monochromatic blackbody emissive power as given by Eq. 6-2 is shown in Fig. 6-2. The figure indicates that the radiative energy given off by a black surface increases as its temperature increases. Also, the emissive power reaches a maximum value at a wavelength which decreases as the temperature of the surface increases.

Wien's Displacement Law

The wavelength at which the blackbody emissive power reaches a maximum value for a given temperature can be determined from Planck's law by satisfying the condition for a maximum value:

$$\frac{dE_{b_\lambda}}{d\lambda} = \frac{d}{d\lambda}\left[\frac{C_1}{\lambda^5(e^{C_2/\lambda T}-1)}\right]_{T=\text{const}} = 0$$

The result of this operation is

$$\lambda_{max}T = 2.898 \times 10^{-3} \text{ m} \cdot \text{K} \qquad (6\text{-}3)$$

where λ_{max} denotes the wavelength at which the maximum monochromatic emissive power occurs for a black surface with temperature T. Equation 6-3 is called *Wien's displacement law*. The locus of points described by Wien's law is included in Fig. 6-2.

The maximum value for the monochromatic blackbody emissive power can be obtained by substituting Eq. 6-3 into Eq. 6-2, which results in

$$\left(E_{b_\lambda}\right)_{max} = 1.287 \times 10^{-5} T^5 \text{ W/m}^3 \qquad (6\text{-}4)$$

We should already be familiar with the results of Wien's displacement law. Suppose that an electric current is passed through a thin filament, causing its temperature to increase. At relatively low filament temperatures, below about 900 K, the wavelength at which the emissive power reaches a maximum is around 3.2×10^{-6} m, which is in the infrared-wavelength range. We can sense the radiant energy being emitted by the filament by noticing the temperature rise in our skin when we place our hands beside the filament. However, our eyes are unable to detect visible radiation emitted by the filament, because an insignificant amount of the energy falls in the visible-wavelength range.

As the temperature of the filament is increased, the amount of radiant energy increases and more of the energy is emitted at shorter wavelengths. Above about 1000 K a small portion of the energy falls in the long-wavelength or red end of the visible spectrum. Our eyes are able to detect this radiation and the filament appears to be a dull red color. As the temperature increases further, more of the energy falls in the visible range and above about 1600 K all visible wavelengths are included, so the filament appears "white" hot at this temperature.

An example of an energy source that is at a high temperature is the sun. The outer surface of the sun has a temperature of approximately 5800 K. According to Wien's law, the value for λ_{max} at this temperature is 5.2×10^{-7} m, which is near the center of the visible range. The human eye is perfectly adapted to sensing the maximum monochromatic energy that is emitted from our sun.

Our eyes are reliable detectors of radiant energy that falls in the visible wavelength range. An object that appears white when placed in the sun reflects nearly all and absorbs practically none of the radiation in the visible range. On the other hand, a blackbody absorbs all visible radiation while reflecting no radiation and therefore appears black to our eyes.

Since the human eye does not respond to radiant energy outside the visible range, it can only predict surface behavior over a very small wavelength range. We will see later in the chapter when we discuss monochromatic properties that some surfaces will behave as good absorbers in the visible range and therefore appear dark in color to our eyes. In the infrared range their behavior can change and they can become poor absorbers. On the other hand, some surfaces can be poor absorbers in the visible range and appear white to the eye. These same surfaces can become excellent absorbers for wavelengths outside the visible range.

Stefan-Boltzmann Law

The total amount of radiative energy per unit area leaving a surface with absolute temperature T over all wavelengths is called the *total emissive power*. If the surface is a blackbody, the total emissive power is given by the integral of Planck's distribution over all wavelengths:

$$E_b(T) = \int_0^\infty E_{b_\lambda}(T)d\lambda = \int_0^\infty \frac{C_1}{\lambda^5(e^{C_2/\lambda T}-1)}\,d\lambda \qquad (6\text{-}5)$$

When the integration is carried out, the result is

$$E_b(T) = \sigma T^4 \qquad (6\text{-}6)$$

which is known as the *Stefan-Boltzmann law*. The symbol σ is the *Stefan-Boltzmann constant*, which has a value of

$$\sigma = \left(\frac{\pi}{C_2}\right)^4 \frac{C_1}{15} = 5.67 \times 10^{-8} \text{ W/m}^2 \cdot \text{K}^4 \qquad (6\text{-}7)$$

The constants C_1 and C_2 are those used in Planck's law, and E_b has units of heat flux, W/m^2.

We see from the Stefan-Boltzmann law that the effects of radiation in most circumstances are insignificant at low temperatures, owing to the small value for σ. At room temperature, or about 300 K, the total emissive power of a black surface is only approximately 460 W/m^2. This value is only about one-tenth of the heat flux transferred from a surface to a fluid by convection when the convective-heat-transfer coefficient and temperature difference are reasonably low values of 100 $\text{W/m}^2 \cdot \text{K}$ and 50 K, respectively. At low temperatures, therefore, we are often justified in neglecting radiative effects. However, we must certainly include radiation effects at high temperatures because the emissive power increases with the fourth power of the absolute temperature.

Radiation Functions

If the monochromatic blackbody emissive power given by Planck's law is integrated over the wavelength range from $\lambda=0$ to $\lambda=\lambda_1$, the result would be the total amount of radiative energy between the wavelengths 0 and λ_1 emitted from a black surface with a temperature T. It can be shown by carrying out the integration that the result is only a function of the

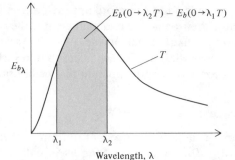

$E_b(0 \rightarrow \lambda_2 T) - E_b(0 \rightarrow \lambda_1 T)$

T

$E_{b\lambda}$

λ_1 λ_2

Wavelength, λ

Figure 6-3   Significance of radiation functions.

product of $\lambda_1 T$. The integral is denoted by $E_b(0 \rightarrow \lambda_1 T)$:

$$\int_0^{\lambda_1} E_{b_\lambda}(T) d\lambda = E_b(0 \rightarrow \lambda_1 T) \qquad (6\text{-}8)$$

If we wish to determine the total amount of radiative energy emitted between two wavelengths λ_1 and λ_2 for a black surface at a temperature T, we could take the difference between the two integrals

$$\int_0^{\lambda_2} E_{b_\lambda}(T) d\lambda - \int_0^{\lambda_1} E_{b_\lambda}(T) d\lambda = E_b(0 \rightarrow \lambda_2 T) - E_b(0 \rightarrow \lambda_1 T) \qquad (6\text{-}9)$$

The physical significance of the integrals in Eq. 6-9 is shown in Fig. 6-3 as the area of under the $E_{b_\lambda}(T)$ curve between wavelengths λ_1 and λ_2.

If we wish to know the percentage of the total black-body energy emitted over the entire spectrum that lies in a wavelength interval, say $\lambda_1 < \lambda < \lambda_2$, we need to divide Eq. 6-9 by

$$\int_0^\infty E_{b_\lambda}(T) d\lambda = \sigma T^4$$

which results in

$$\begin{bmatrix} \text{percentage of total} \\ \text{blackbody radia-} \\ \text{ive energy that lies} \\ \text{in wavelength in-} \\ \text{terval } \lambda_1 < \lambda < \lambda_2 \end{bmatrix} = \frac{E_b(0 \rightarrow \lambda_2 T) - E_b(0 \rightarrow \lambda_1 T)}{\sigma T^4} \qquad (6\text{-}10)$$

Dunkle (Ref. 1) has tabulated values for

$$\frac{E_b(0 \rightarrow \lambda T)}{\sigma T^4} \qquad (6\text{-}11)$$

as a function of the product of λT. These values in SI units taken from Reference 2 appear in Table 6-1. Values for the term in Eq. 6-11 are

Table 6-1 Blackbody Radiation Functions

λT $(\mathrm{mK} \times 10^3)^a$	$\dfrac{E_b(0 \to \lambda T)}{\sigma T^4}$	λT $(\mathrm{mK} \times 10^3)^a$	$\dfrac{E_b(0 \to \lambda T)}{\sigma T^4}$
0.2	0.341796×10^{-26}	6.2	0.754187
0.4	0.186468×10^{-11}	6.4	0.769282
0.6	0.929299×10^{-7}	6.6	0.783248
0.8	0.164351×10^{-4}	6.8	0.796180
1.0	0.320780×10^{-3}	7.0	0.808160
1.2	0.213431×10^{-2}	7.2	0.819270
1.4	0.779084×10^{-2}	7.4	0.829580
1.6	0.197204×10^{-1}	7.6	0.839157
1.8	0.393449×10^{-1}	7.8	0.848060
2.0	0.667347×10^{-1}	8.0	0.856344
2.2	0.100897	8.5	0.874666
2.4	0.140268	9.0	0.890090
2.6	0.183135	9.5	0.903147
2.8	0.227908	10.0	0.914263
3.0	0.273252	10.5	0.923775
3.2	0.318124	11.0	0.931956
3.4	0.361760	11.5	0.939027
3.6	0.403633	12	0.945167
3.8	0.443411	13	0.955210
4.0	0.480907	14	0.962970
4.2	0.516046	15	0.969056
4.4	0.548830	16	0.973890
4.6	0.579316	18	0.980939
4.8	0.607597	20	0.985683
5.0	0.633786	25	0.992299
5.2	0.658011	30	0.995427
5.4	0.680402	40	0.998057
5.6	0.701090	50	0.999045
5.8	0.720203	75	0.999807
6.0	0.737864	100	1.000000

[a]See page 510 for the convention used when tabular values are listed with column headings multiplied by powers of 10.

usually referred to as *radiation functions*. The use of these values will be illustrated in the example that follows.

Example 6-1. Assume that the sun ($T = 5800\,\mathrm{K}$) and an incandescent bulb ($T = 2800\,\mathrm{K}$) are both blackbodies. Calculate for both of these sources of radiant energy the following information:

a. The total emissive power.
b. The maximum monochromatic blackbody emissive power.
c. The wavelength at which the maximum emissive power occurs.
d. The percent of total emitted energy that lies in the visible wavelength range.

Solution

a. The total emissive power is given by the Stefan-Boltzmann law (Eq. 6-6):

$$E_b(T) = \sigma T^4$$

For the sun

$$E_b(T) = (5.67 \times 10^{-8})(5800)^4 = 6.42 \times 10^7 \, \text{W/m}^2$$

For the bulb

$$E_b(T) = (5.67 \times 10^{-8})(2800)^4 = 3.49 \times 10^6 \, \text{W/m}^2$$

b. The maximum monochromatic blackbody emissive power is given by Eq. 6-4. For the sun

$$(E_{b_\lambda})_{max} = (1.287 \times 10^{-5})(5800)^5 = 8.45 \times 10^{13} \, \text{W/m}^3$$

For the bulb

$$(E_{b_\lambda})_{max} = (1.287 \times 10^{-5})(2800)^5 = 2.21 \times 10^{12} \, \text{W/m}^3$$

c. The value for λ_{max} is determined by Wien's law (Eq. 6-3). For the sun

$$\lambda_{max} = \frac{2.898 \times 10^{-3}}{5800} = 5.00 \times 10^{-7} \, \text{m}$$

which is in the visible wavelength range. For the bulb

$$\lambda_{max} = \frac{2.898 \times 10^{-3}}{2800} = 1.04 \times 10^{-6} \, \text{m}$$

which is in the infrared-wavelength range.

d. The percent of total emitted energy that falls in the visible ($3.8 \times 10^{-7} \, \text{m} < \lambda < 7.6 \times 10^{-7} \, \text{m}$) range can be determined by using the radiation functions in Table 6-1. For the sun

$$\lambda_1 T = (3.8 \times 10^{-7}) \times 5800 = 2.204 \times 10^{-3} \, \text{m} \cdot \text{K}$$

$$\lambda_2 T = (7.6 \times 10^{-7}) \times 5800 = 4.408 \times 10^{-3} \, \text{m} \cdot \text{K}$$

$$\begin{bmatrix} \text{percent of total} \\ \text{emitted energy} \\ \text{in visible range} \end{bmatrix} = \frac{E_b(0 \to \lambda_2 T) - E_b(0 \to \lambda_1 T)}{\sigma T^4}$$

$$= 0.5500 - 0.1017 = 0.4483 = 44.83\%$$

For the bulb

$$\lambda_1 T = (3.8 \times 10^{-7}) \times 2800 = 1.064 \times 10^{-3} \, \text{m} \cdot \text{K}$$

$$\lambda_2 T = (7.6 \times 10^{-7}) \times 2800 = 2.128 \times 10^{-3} \, \text{m} \cdot \text{K}$$

$$\begin{bmatrix} \text{percent of total} \\ \text{emitted energy} \\ \text{in visible range} \end{bmatrix} = \frac{E_b(0 \rightarrow \lambda_2 T) - E_b(0 \rightarrow \lambda_1 T)}{\sigma T^4}$$

$$= 0.0886 - 0.0009 = 0.0877 = 8.77\%$$

Notice that the peak of the sun's energy is in the visible portion of the spectrum and greater than 40 percent of its total emitted energy can be detected by the human eye. The incandescent bulb, on the other hand, emits less than 10 percent of its total energy in the visible segment of the spectrum. The remaining 90 percent of the energy leaves the bulb as infrared energy, and that portion cannot be detected by our eyes but can be sensed by our skin as heat.

6-3 RADIATION PROPERTIES

Radiative properties are those properties which quantitatively describe how radiant energy interacts with the surface of the material. Specifically the radiative properties describe how the surface emits, reflects, absorbs, and transmits radiant energy.

In general, the radiative properties are functions of wavelength. For example, a surface may be a good reflector in the visible-wavelength range and a poor reflector in the infrared range. The properties that describe how a surface behaves as a function of wavelength are called *monochromatic* or *spectral properties*. The radiative properties are also a function of the direction at which the radiation is incident upon the surface. Properties that describe how the distribution of energy varies with angle are called *directional properties*.

If we wish to perform an energy balance on a surface to determine its temperature, for example, we must know the radiative properties of the surface and all other surfaces that exchange energy with that surface. Even when the spectral and directional properties of all these surfaces are known, the analysis is extremely involved. The complexity of the problem and more often the complete lack of the detailed properties suggest that we should search for a simplified approach. The simplified approach involves using a single radiative property value that is an average value over all wavelengths and all directions. The properties that are averaged over all wavelengths and angles are called *total properties*. The use of total properties in a radiative-heat-transfer analysis often results in answers that are accurate enough for most engineering purposes, and they certainly reduce a very complex problem to a more simplified one.

Even though we will almost exclusively use the total radiative properties in this text, it is important to be aware of the spectral and directional characteristics of surfaces so that their variations can be accounted for in problems in which these effects are significant.

Our discussion of radiant properties will be in order of increasing complexity. We will discuss the total properties first, followed by spectral and finally by directional properties.

Total Radiation Properties

Consider a beam of radiant energy incident on a surface as shown in Fig. 6-4. The total incident energy is referred to as the *total irradiation* and is given the symbol G. When the irradiation strikes a surface, a portion of the energy is absorbed within the material, a portion is reflected from the surface, and the remainder is transmitted through the body. Three of the radiative properties—the absorptivity, reflectivity, and the transmissivity—describe how the incident energy is distributed into these three categories.

The *absorptivity*, α, of the surface is the fraction of incident energy absorbed by the body. The *reflectivity*, ρ, of the surface is defined as the fraction of incident energy reflected from the surface. The *transmissivity*, τ, of the body is the fraction of incident energy that is transmitted through the body. If an energy balance is made on the body in Fig. 6-4, we know that the irradiation must either be absorbed by, reflected from, or transmitted through the body. The energy balance may be expressed mathematically as

$$\alpha G + \rho G + \tau G = G$$

or simply

$$\alpha + \rho + \tau = 1 \qquad (6\text{-}12)$$

Often a surface is opaque. That is, it will not transmit any of the incident radiant energy. For an opaque surface

$$\tau = 0$$

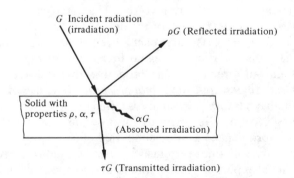

G Incident radiation (irradiation)

ρG (Reflected irradiation)

Solid with properties ρ, α, τ

αG (Absorbed irradiation)

τG (Transmitted irradiation)

Figure 6-4 Definition of total radiation properties.

and Eq. 6-12 for an opaque surface reduces to

$$\alpha + \rho = 1.0 \tag{6-13}$$

If a surface is said to be a perfect reflector, all irradiation is reflected, or

$$\rho = 1.0 \tag{6-14}$$

and the energy balance for a perfectly reflecting surface implies that

$$\tau = \alpha = 0 \tag{6-15}$$

A blackbody absorbs the maximum amount of incident energy, or

$$\alpha = 1.0 \tag{6-16}$$

and therefore

$$\tau = \rho = 0 \tag{6-17}$$

for a blackbody.

Another very important total radiation property is the *emissivity* of a body. The emissivity of a surface is defined as the total emitted energy divided by the total energy emitted by a blackbody at the same temperature. The mathematical definition of the total emissivity ϵ is then

$$\epsilon = \frac{E(T)}{E_b(T)} = \frac{E(T)}{\sigma T^4} \tag{6-18}$$

Since a blackbody emits the maximum amount of radiation at a given temperature, the emissivity of a surface is always between zero and one. When a surface is a blackbody, $E(T) = E_b$ and $\epsilon = \alpha = 1.0$ for a blackbody.

Kirchhoff's Law

An important relationship exists between the absorptivity and emissivity of a material. We can easily derive this relationship by placing a test body with absorptivity of α and an emissivity of ϵ in an isothermal enclosure as shown in Fig. 6-5. Assume that the body and enclosure are in thermal equilibrium; that is, the temperature of the test body and enclosure are the same. At equilibrium the absorbed energy must equal the emitted energy, or if G is the irradiation on the test body, then

$$\alpha_1 G = E_1 \tag{6-19}$$

The test body is now imagined to be removed and it is replaced by a second test body of equal size while the conditions of the enclosure remained unchanged. When equilibrium conditions are once again reached, the second test body reaches the same temperature as the first test body. Suppose that the second test body is a blackbody. Then equilibrium requires that

$$\alpha_2 G = E_2 = E_b \tag{6-20}$$

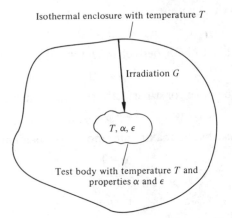

Isothermal enclosure with temperature T

Irradiation G

T, α, ϵ

Test body with temperature T and
properties α and ϵ

Figure 6-5 Situation used to derive Kirchhoff's law.

Taking the ratio of Eqs. 6-19 and 6-20, we see that

$$\frac{\alpha_1}{\alpha_2} = \frac{E_1}{E_b}$$

Since material 2 is a blackbody, $\alpha_2 = 1.0$, so

$$\alpha_1 = \frac{E_1}{E_b} \tag{6-21}$$

Equation 6-21 is identical to the definition of the emissivity of surface 1, so

$$\alpha_1 = \epsilon_1 \tag{6-22}$$

This result is *Kirchhoff's law*. It states that *at thermal equilibrium* the absorptivity of a body is equal to its emissivity. The result is not valid when a body is not at equilibrium with its surroundings.

Kirchhoff's law implies that a good absorber, that is, a body which has a high value of absorptivity, will also be a good emitter of thermal radiation. This result suggests that a simple way of approaching a blackbody radiator would be to shape an isothermal enclosure so that it has a small hole in its surface, as in Fig. 6-6. Any energy that passes through the opening will be reflected numerous times from the internal surface of the enclosure, and regardless of the surface conditions of the enclosure, the incident energy will be largely absorbed. Therefore, practically all the irradiation on the opening will be absorbed, and the hole in the isothermal enclosure will approach blackbody behavior. The ability of an isothermal enclosure to absorb practically all the incident energy on a small opening in its surface is called the *cavity effect*. The effective absorptivity of a number of common shapes such as spherical, rectangular, and cylindrical cavities has been reported in the literature (Refs. 3 to 5).

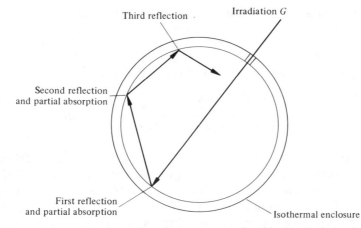

Third reflection

Irradiation G

Second reflection
and partial absorption

First reflection
and partial absorption

Isothermal enclosure

Figure 6-6   An isothermal enclosure approaches blackbody behavior.

Isothermal enclosures or cavities can be used in laboratory work to provide a source of black-body radiation. A cavity can be heated to a uniform temperature T. The interior of the cavity is then uniformly filled with blackbody radiation with emissive power equal to σT^4. Also, any energy escaping a small hole in the cavity will have an emissive power of σT^4.

Monochromatic Radiation Properties

The definitions of the radiation properties made thus far are *total* properties, indicating that they are values averaged over the entire electromagnetic spectrum. We could define monochromatic properties which apply at only a single wavelength. The monochromatic absorptivity, α_λ, would then be the absorbed irradiation at wavelength λ divided by the irradiation on the surface at wavelength λ. The relationship between the monochromatic absorptivity of a body and its total absorptivity α would be

$$\alpha = \frac{\int_0^\infty \alpha_\lambda G_\lambda \, d\lambda}{\int_0^\infty G_\lambda \, d\lambda} \qquad (6\text{-}23)$$

where the subscript λ refers to a monochromatic value. Similar expressions can be written for the relationships between the monochromatic and total reflectivity and transmissivity.

If an energy balance on a surface is taken on a monochromatic basis, Eq. 6-12 becomes

$$\alpha_\lambda + \rho_\lambda + \tau_\lambda = 1.0 \qquad (6\text{-}24)$$

The *monochromatic* or *spectral emissivity* is defined as

$$\epsilon_\lambda = \frac{E_\lambda(T)}{E_{b_\lambda}(T)} \tag{6-25}$$

The relationship between the monochromatic and total emissivity is

$$\epsilon = \frac{\int_0^\infty \epsilon_\lambda E_{b_\lambda}(T)\,d\lambda}{\int_0^\infty E_{b_\lambda}(T)\,d\lambda} = \frac{1}{\sigma T^4}\int_0^\infty \epsilon_\lambda E_{b_\lambda}(T)\,d\lambda \tag{6-26}$$

When Eqs. 6-23 and 6-26 are compared, we notice an important difference between the total absorptivity and total emissivity. Both the absorptivity and emissivity are properties of the surface and they are functions of the type of material, its surface condition, and its temperature. In addition, the absorptivity is a function of all surrounding surfaces that contribute to the irradiation G_λ, as can be seen in Eq. 6-23. The emissivity of a surface as defined in Eq. 6-26 is not a function of other surfaces. It is a function only of its own surface material, the condition of the surface, and the temperature. This fact is illustrated in the next example.

Example 6-2. A surface has a constant monchromatic absorptivity of 0.6 for wavelengths $4\times10^{-7} \leqslant \lambda \leqslant 4\times10^{-6}$ m. The absorptivity is zero for all other wavelengths. Assume that the monochromatic emissivity and monochromatic absorptivity are equal. Calculate the total absorptivity and total emissivity if the surface is at 3000 K for the following two cases:

a. Irradiation is from a black surface at 3000 K.
b. Irradiation is from a black surface at 1000 K.

Solution: The total emissivity of the surface is

$$\epsilon = \frac{1}{\sigma T^4}\int_0^\infty \epsilon_\lambda E_{b_\lambda}(T)\,d\lambda = \frac{0.6}{\sigma T^4}\int_{4\times10^{-7}}^{4\times10^{-6}} E_{b_\lambda}(T)\,d\lambda$$

$$= 0.6\left[\frac{E_b(0\rightarrow\lambda_2 T)}{\sigma T^4} - \frac{E_b(0\rightarrow\lambda_1 T)}{\sigma T^4}\right]$$

$$= 0.6(0.9452 - 0.0021) = 0.566$$

where the two definite integrals are evaluated at $\lambda T = 1.2\times10^{-3}$ m·K and 1.2×10^{-2} m·K by using the radiation functions given in Table 6-1. The emissivity of the surface is not a function of the irradiation, so the emissivity for parts (a) and (b) is the same. The absorptivity of the surface for black-body irradiation at 3000 K is

$$\alpha = \frac{\int_0^\infty \alpha_\lambda G_\lambda\,d\lambda}{\int_0^\infty G_\lambda\,d\lambda} = \frac{0.6\int_{4\times10^{-7}}^{4\times10^{-6}} E_{b_\lambda}(T)\,d\lambda}{\int_0^\infty E_{b_\lambda}(T)\,d\lambda}$$

or

$$\alpha = 0.6\left[\frac{E_b(0\to\lambda_2 T)}{\sigma T^4} - \frac{E_b(0\to\lambda_1 T)}{\sigma T^4}\right]$$

The integrals are the same as those used to evaluate the emissivity, so

$$\alpha = 0.6(0.9452 - 0.0021) = 0.566$$

The absorptivity for blackbody irradiation at 1000 K is

$$\alpha = 0.6\left[\frac{E_b(0\to\lambda_2 T)}{\sigma T^4} - \frac{E_b(0\to\lambda_1 T)}{\sigma T^4}\right]$$

$$= 0.6(0.4809 - 0) = 0.289$$

where the λT products are $\lambda_2 T = (4\times 10^{-6})\times 1000 = 4\times 10^{-3}$ m·K and $\lambda_1 T = (4\times 10^{-7})\times 1000 = 4\times 10^{-4}$ m·K.

Notice that the emissivity is not a function of the spectral distribution of irradiation, while the absorptivity is a function of G_λ. We can see also that even when the monochromatic emissivity and absorptivity are identical, the total values for α and ϵ can be different. In the case of the irradiation originating from a black surface with a temperature equal to that of the receiving surface, however, the emissivity of the receiving surface is equal to its absorptivity. This result was predicted by Kirchhoff's law (Eq. 6-22).

Example 6-3. An automobile is parked in the sun. The windshield has a transmissivity of 0.92 for wavelengths between 3×10^{-7} m and 3×10^{-6} m and it is opaque at all other wavelengths. The interior of the car can be considered to be a blackbody at 300 K. The sun is a black source at 5800 K producing an irradiation of 1100 W/m^2 on the windshield. Calculate:

a. The total transmissivity of the windshield for the sun's irradiation.
b. The total transmissivity of the windshield for the irradiation from the interior of the car.
c. The rate at which the sun's radiant energy is transmitted through the windshield.
d. The rate at which the radiant energy from the interior of the car is transmitted through the windshield.

Solution

a. The total transmissivity is

$$\tau = \frac{\int_0^\infty \tau_\lambda G_\lambda \, d\lambda}{\int_0^\infty G_\lambda \, d\lambda}$$

When the irradiation is from a blackbody, $G_\lambda = E_{b\lambda}$, or

$$\tau = \frac{\int_0^\infty \tau_\lambda E_{b_\lambda} \, d\lambda}{\sigma T^4} = \frac{0.92 \int_{3\times10^{-7}}^{3\times10^{-6}} E_{b_\lambda} \, d\lambda}{\sigma T^4}$$

Using the radiation functions for a source at 5800 K,

$$\lambda_2 T = (3\times10^{-6})\times5800 = 1.74\times10^{-2} \text{ m} \cdot \text{K}$$

$$\lambda_1 T = (3\times10^{-7})\times5800 = 1.74\times10^{-3} \text{ m} \cdot \text{K}$$

and

$$\tau = 0.92(0.977 - 0.033) = 0.868$$

b. When the source of irradiation is a blackbody at 300 K, we find

$$\lambda_2 T = (3\times10^{-6})\times300 = 9\times10^{-4} \text{m} \cdot \text{K}$$

$$\lambda_1 T = (3\times10^{-7})\times300 = 9\times10^{-5} \text{m} \cdot \text{K}$$

and

$$\tau = 0.92(1.686\times10^{-4} - 0) = 1.55\times10^{-4}$$

c. The transmitted sun energy is

$$(\tau G)_{\text{sun}} = 0.868\times1100 = 955 \text{ W}/\text{m}^2$$

d. The transmitted energy from the interior of the car is

$$(\tau G)_{\text{interior}} = (\tau E_b)_{\text{interior}} = (\tau\sigma T^4)_{\text{interior}}$$

$$= (1.55\times10^{-4})(5.67\times10^{-8})(300)^4$$

$$= 0.0712 \text{ W}/\text{m}^2$$

Anyone who has entered a car after it has been parked in the sun can understand these results. The windshield transmits 87 percent of the sun's energy into the passenger compartment while it transmits practically none of the energy traveling on the opposite direction. The characteristics of glass for transmitting a large percentage of the short-wavelength solar energy and blocking the transmission of the low-temperature infrared radiation is called the *greenhouse effect*. It explains how the interior of an automobile or greenhouse can remain warm by trapping solar energy even during a cold winter day.

Kirchhoff's law, which was previously derived on a total property basis, can also be derived on a monochromatic basis. The result is

$$\alpha_\lambda = \epsilon_\lambda \tag{6-27}$$

The monochromatic form of Kirchhoff's law is not restricted to conditions of thermal equilibrium. Therefore, the monochromatic emissivity and absorptivity of a body are equal, even though all surfaces that contribute irradiation on the body are not all at the same temperature.

Concept of a Gray Body

A useful concept that greatly simplifies radiation calculations is that of a *gray body*. A gray body is a surface for which the monochromatic properties are constant over all wavelengths. When a surface is gray,

$$\epsilon_\lambda = \text{constant}$$

$$\alpha_\lambda = \text{constant}$$

The other monocromatic properties, such as the transmissivity and reflectivity, are also constant for a gray body.

The total absorptivity for a gray body is

$$\alpha = \frac{\int_0^\infty \alpha_\lambda G_\lambda \, d\lambda}{\int_0^\infty G_\lambda \, d\lambda} = \frac{\alpha_\lambda \int_0^\infty G_\lambda \, d\lambda}{\int_0^\infty G_\lambda \, d\lambda} = \alpha_\lambda \qquad (6\text{-}28)$$

That is, the total absorptivity and monochromatic absorptivity of a gray surface are equal. Similar expressions for the other properties can be derived for a gray body. They are

$$\epsilon = \epsilon_\lambda$$

$$\rho = \rho_\lambda$$

$$\tau = \tau_\lambda \qquad (6\text{-}29)$$

If Kirchhoff's law in the form of Eq. 6-27 is applied to a gray body, then

$$\alpha_\lambda = \epsilon_\lambda$$

By substituting the results of Eq. 6-29, we find that a gray surface is one for which

$$\alpha = \epsilon \qquad (6\text{-}30)$$

The total emissivity and absorptivity are equal for a graybody even if the body is not in thermal equilibrium with its surroundings. This is an important characteristic of a gray body. If we know the emissivity of a gray body at one wavelength, we also know the total absorptivity and the total emissivity of the surface.

The concept of a gray body is an idealized one because gray bodies do not exist in practice. Typical monochromatic absorptivities of several real surfaces are shown in Fig. 6-7. Even though real surfaces can only approach gray behavior, radiation calculations can often be based on the gray assumption with satisfactory accuracy. If the real surface has large variations in the monochromatic properties, the total properties can be approximated by dividing the electromagnetic spectrum into segments over which the properties can be assumed constant. The process is illustrated in Fig. 6-8, where the monochromatic emissivity of the real surface is shown with the approximated distribution used to evaluate the total emissivity. Using the nomenclature in Fig. 6-8, the total emissivity for the surface

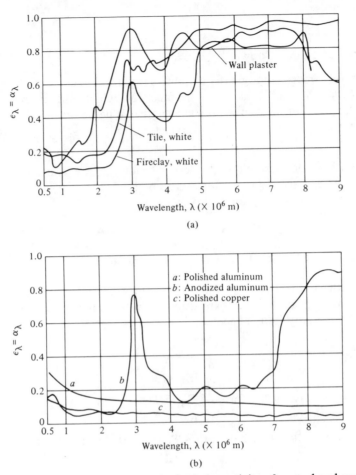

Figure 6-7 Monochromatic emissivity and absorptivity of several real surfaces: (a) electrical nonconductors; (b) electrical conductors.

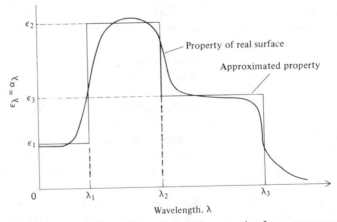

Figure 6-8 Method of determining the total properties for a nongray surface.

would be

$$\epsilon = \frac{\int_0^\infty \epsilon_\lambda E_{b_\lambda}\, d\lambda}{\int_0^\infty E_{b_\lambda}\, d\lambda} = \epsilon_1 \left[\frac{E_b(0 \to \lambda_1 T)}{\sigma T^4} \right]$$

$$+ \epsilon_2 \left[\frac{E_b(0 \to \lambda_2 T)}{\sigma T^4} - \frac{E_b(0 \to \lambda_1 T)}{\sigma T^4} \right] + \epsilon_3 \left[\frac{E_b(0 \to \lambda_3 T)}{\sigma T^4} - \frac{E_b(0 \to \lambda_2 T)}{\sigma T^4} \right]$$

where the radiation functions given in Table 6-1 can be used to evaluate the terms in brackets.

Examination of Fig. 6-7(b) illustrates another important point concerning the radiative properties of materials. The monochromatic emissivity and absorptivity of polished and anodized aluminum are greatly different, even though the base metal is identical for both. Since the radiant energy is emitted and absorbed within a very thin layer at the surface, the radiative properties are determined largely by the surface condition of the material. As a result, oxide layers, paints, surface contamination such as oil or grease, and the general surface condition such as roughness of finish can greatly affect the radiative properties.

The total emissivity of many engineering materials are listed in Tables I1 to I3. Many of the materials are classified according to their surface finish. The reader should notice the large variations in properties depending upon surface condition of the materials.

Directional Radiation Properties

Thus far we have assumed that the radiation properties are functions only of wavelength and conditions that describe the surface of the sending and receiving areas. To complicate matters further, the properties are also functions of the direction in which the energy is incident or leaves a surface. The properties that describe the angular variation are called *directional properties*. Several examples of directional emissivities are shown in the polar plots of Fig. 6-9. The angle θ in the figure is the angle between the normal to the surface and the directional energy emitted from the surface.

The directional emissivity for electrical conductors is characteristically higher for large θ angles than for small θ values. A conductor would therefore emit more energy at grazing angles than at angles more normal to its surface. Electrical nonconductors behave quite differently. They emit more strongly in directions close to the normal while their emissivity drops to zero as θ approaches $90°$.

Before we can relate directional properties to total properties, we need to discuss two quantities. The first quantity is a solid angle which is a measure of an angle in a solid geometry. Consider a differential area dA shown in Fig. 6-10 which subtends a differential solid angle $d\omega$ at point O.

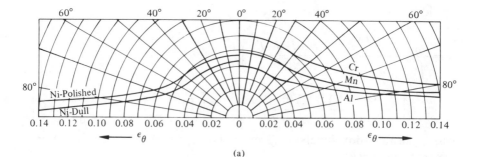

(a)

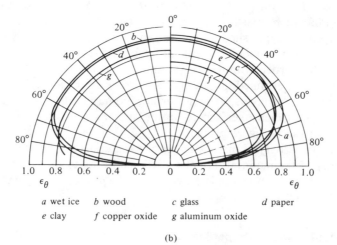

| a wet ice | b wood | c glass | d paper |
| e clay | f copper oxide | g aluminum oxide | |

(b)

Figure 6-9 Directional emissivity of several real surfaces: (a) electrical conductors; (b) electrical nonconductors. (From E. Schmidt and E. Eckert, Uber die Richtungsverteilung der Warmestrahlung, *Forsch. Gebeite Ingenieurwesen*, vol. 6, 1935.)

The solid angle is a dimensionless quantity defined as the normal projection of dA divided by the square of the radius between point O and the projected area. The solid angle $d\omega$ is then

$$d\omega = \frac{dA_N}{r^2} = \frac{dA \cos\theta}{r^2} \qquad (6\text{-}31)$$

The unit of the solid angle is a steradian, abbreviated sr.

We should note the similarity between a plane angle and a solid angle. The plane angle is an angle subtended by the normal projection of a line divided by the radius to the line. The plane angle is dimensionless and is measured in radians. There are 2π radians subtended by a closed line like a circle. The number of steradians subtended by a closed surface like a sphere can be determined by integrating Eq. 6-31 over a sphere. Referring

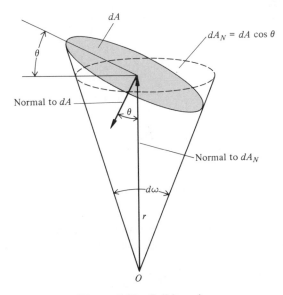

Figure 6-10 Solid angle.

to Fig. 6-11, we see that the number of steradians in a sphere is

$$\omega_{sphere} = \int_{sphere} d\omega = \int_{sphere} \frac{dA_N}{r^2} = \int_{\phi=0}^{2\pi} \int_{\theta=0}^{\pi} \frac{r\,d\theta \cdot r\sin\theta\,d\phi}{r^2} = 4\pi \qquad (6\text{-}32)$$

The second quantity that must be introduced when discussing the directional properties is the *intensity* of radiation. The intensity of radiation is given the symbol I. The intensity is defined as the radiant energy per

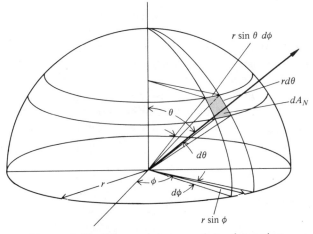

Figure 6-11 Nomenclature used to determine the solid angle subtended by a sphere.

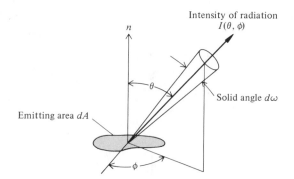

Figure 6-12   Intensity of radiation.

unit time, per unit solid angle, per unit area projected in a direction normal to the surface. Referring to Fig. 6-12, the intensity of radiation $I(\theta,\phi)$ is

$$I(\theta,\phi) = \frac{dq''}{\cos\theta\, d\omega} \tag{6-33}$$

where the symbol q'' is used as in other chapters to denote energy per unit time and area. Units of intensity are $W/m^2 \cdot sr$.

Once the distribution of intensity is known, the emissive power which leaves a plane surface can be determined by integrating Eq. 6-33 over all solid angles subtended by a hemisphere placed over the surface. The emissive power of a surface is therefore

$$E = q'' = \int_{\text{hemisphere}} I(\theta,\phi)\cos\theta\, d\omega \tag{6-34}$$

Substituting the results of Eq. 6-32 for $d\omega$ gives

$$E = \int_{\phi=0}^{2\pi} \int_{\theta=0}^{\pi/2} I(\theta,\phi)\sin\theta\cos\theta\, d\theta\, d\phi \tag{6-35}$$

The integration of Eq. 6-35 cannot be completed until we know the distribution of intensity over the angles θ and ϕ. The simplest distribution of intensity with angle would be to assume that the intensity is constant. A surface that emits with equal intensity over all angles is called a *diffuse surface* or sometimes a surface that obeys Lambert's cosine law because the energy leaving a diffuse surface in a given direction varies as the cosine of the angle to the surface normal. Therefore, a diffuse surface is one for which

$$I(\theta,\phi) = \text{constant}$$

and the emissive power of a diffuse surface is

$$E = I\int_{\phi=0}^{2\pi} \int_{\theta=0}^{\pi/2} \sin\theta\cos\theta\, d\theta\, d\phi = \pi I \tag{6-36}$$

A black surface is also a diffuse surface because if it did not emit with equal intensity in all directions it would not be emitting with maximum energy for its given temperature. Therefore, the blackbody emissive power and blackbody intensity are related by Eq. 6-36:

$$E_b = \pi I_b \tag{6-37}$$

Directional radiative properties cannot be defined in terms of emissive power because the emissive power is not dependent upon the direction from the surface. Directional radiative properties must be defined in terms of the intensity. For example, the directional emissivity $\epsilon(\theta, \phi)$ is defined as the intensity of radiation emitted from the surface in the direction specified by the angle θ, ϕ divided by the blackbody intensity in the same direction:

$$\epsilon(\theta, \phi) = \frac{I(\theta, \phi)}{I_b} \tag{6-38}$$

The total emissivity of a surface was defined in Eq. 6-18 as

$$\epsilon = \frac{E}{E_b} \tag{6-39}$$

If the surface is not diffuse, the emitted intensity is a function of direction. Substituting Eq. 6-35 for the numerator and Eq. 6-37 for the denominator, we arrive at

$$\epsilon = \frac{\int_{\phi=0}^{2\pi} \int_{\theta=0}^{\pi/2} I(\theta, \phi) \sin\theta \cos\theta \, d\theta \, d\phi}{\pi I_b} \tag{6-40}$$

Now substituting Eq. 6-38 for the intensity, the relationship between total emissivity and directional emissivity becomes

$$\epsilon = \frac{1}{\pi} \int_{\phi=0}^{2\pi} \int_{\theta=0}^{\pi/2} \epsilon(\theta, \phi) \sin\theta \cos\theta \, d\theta \, d\phi \tag{6-41}$$

If the directional emissivity is not a function of the angle ϕ, the directional properties become like those shown in Fig. 6-9. When ϵ depends only on θ, the integration of Eq. 6-41 over the variable ϕ can be carried out, resulting in

$$\epsilon = 2 \int_{\theta=0}^{\pi/2} \epsilon(\theta) \sin\theta \cos\theta \, d\theta \tag{6-42}$$

Example 6-4. The directional emissivity of oxidized copper can be approximated by the expression

$$\epsilon(\theta) = 0.70 \cos\theta$$

Determine the amount of radiant energy per unit time emitted by 0.5 m^2 of oxidized copper when heated to 800 K.

Solution: The total rate of energy emitted by the surface is

$$q = EA = \epsilon E_b A = \epsilon \sigma T^4 A$$

where ϵ is the total emissivity of the surface. The directional emissivity is related to the total emissivity by Eq. 6-42:

$$\epsilon = 2 \int_0^{\pi/2} 0.70 \cos^2 \theta \sin \theta \, d\theta = \left. \frac{-1.4}{3} \cos^3 \theta \right|_0^{\pi/2}$$

Substituting the limits of integration results in the total emissivity of

$$\epsilon = \frac{1.4}{3} = 0.467$$

The rate of emitted radiant energy is therefore

$$q = \epsilon A \sigma T^4 = (0.467)(0.5)(5.67 \times 10^{-8})(800)^4$$
$$= 5422 \text{ W}$$

The directional absorptivity, reflectivity, and transmissivity depend upon the distribution of incoming as well as outgoing intensity. These directional properties are referred to as bidirectional properties, indicating that both angles must be specified to determine these directional properties. Bidirectional properties are discussed in References 7 and 8.

Two types of limiting reflective surfaces are defined for the purpose of simplifying radiation exchange calculations. The first is a *diffusely reflecting* surface. A diffusely reflecting surface reflects a single incident ray such that the energy has an equal intensity over all reflected angles. When surfaces reflect diffusely, it is not necessary that we trace each ray of energy as it reflects between participating surfaces, because once a ray reflects, it loses its identity. The assumption of diffusely reflecting surfaces therefore offers a great simplification to radiation exchange calculations.

The other limiting case of a reflector is that of a *specular reflector*. Like a mirror, a specular reflector redirects an incident ray so that the angle of incidence is equal to the angle of reflection. A surface will behave specularly when the surface roughness is very small compared to the wavelength

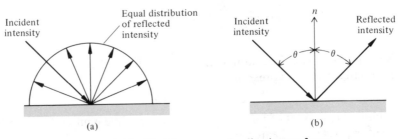

Figure 6-13 Limiting cases of reflecting surfaces:
(a) diffuse reflector; (b) specular reflector.

of incident radiation. A more detailed discussion of specular properties and calculation of radiant exchange rates between diffuse and specular reflectors can be found in References 9 and 10. The behavior of diffuse and specular reflectors is illustrated in Fig. 6-13.

6-4 RADIATION SHAPE FACTOR

To calculate radiative-heat-transfer rates between two surfaces, we must determine the percentage of the total radiant energy that leaves one surface and arrives directly on a second surface. Let us define the radiation *shape factor* $F_{1\rightarrow2}$ as the fraction of total radiant energy that leaves surface 1 which arrives directly on surface 2. The shape factor is a dimensionless quantity. In some texts it is also called a *view factor* or *configuration factor*.

An expression for the shape factor can be derived by considering the geometry in Fig. 6-14, where dA_1 is shown as the emitting area and dA_2 is the receiving area. The rate of radiant energy per unit area of dA_1 that leaves dA_1 and arrives on dA_2 is given by Eq. 6-33:

$$dq''_{1\rightarrow2} = I_1 \cos\theta_1 \, d\omega_{1\rightarrow2} \tag{6-43}$$

The subscript $1\rightarrow2$ refers to energy leaving surface 1 arriving at surface 2. The symbol $d\omega_{1\rightarrow2}$ represents the solid angle subtended by the area dA_2 at dA_1. Substituting Eq. 6-31 for the solid angle gives

$$dq''_{1\rightarrow2} = \frac{I_1 \cos\theta_1 \cos\theta_2 \, dA_2}{r^2} \tag{6-44}$$

If we assume that the emitting area is diffuse, the intensity of radiation leaving dA_1 is independent of direction. The total radiant flux that leaves dA_1 must arrive on a hemisphere placed over dA_1, and the total flux

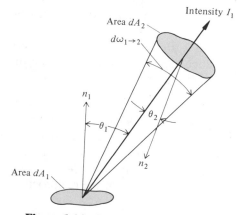

Figure 6-14 Radiation shape factor.

leaving the diffuse area dA_1 is given by Eq. 6-36:

$$q''_{1 \to \text{hemisphere}} = \pi I_1 \tag{6-45}$$

The radiation shape factor between the two differential areas dA_1 and dA_2 is then the ratio of Eq. 6-44 and Eq. 6-45:

$$F_{dA_1 \to dA_2} = \frac{\cos\theta_1 \cos\theta_2 \, dA_2}{\pi r^2} \tag{6-46}$$

We very seldom wish to determine radiation exchange rates between two infinitesimal areas, because we are more concerned about radiation exchange between finite surfaces. Expressions for the shape factors for finite areas can be obtained by integrating Eq. 6-46 over both the emitting and receiving areas.

The shape factor between a diffuse differential emitting area and finite receiving area is

$$F_{dA_1 \to A_2} = \int_{A_2} F_{dA_1 \to dA_2} = \int_{A_2} \frac{\cos\theta_1 \cos\theta_2 \, dA_2}{\pi r^2} \tag{6-47}$$

The shape factor for a finite diffusely emitting area A_1 and a finite receiving area A_2 is

$$F_{A_1 \to A_2} = \frac{1}{A_1} \int_{A_1} \int_{A_2} \left(F_{dA_1 \to dA_2} \right) dA_1 \tag{6-48}$$

or

$$F_{A_1 \to A_2} = \frac{1}{A_1} \int_{A_1} \int_{A_2} \frac{\cos\theta_1 \cos\theta_2 \, dA_2 \, dA_1}{\pi r^2} \tag{6-49}$$

The shape-factor expressions given in Eqs. 6-46, 6-47, and 6-49 are restricted to diffusely emitting surfaces. The assumption of a diffuse surface is a great simplification, because for this condition the shape factors are only functions of the geometry and not of the distribution of radiation intensity.

If the subscripts 1 and 2 are interchanged on the terms in Eq. 6-49 by assuming that the surface denoted by 2 is the emitting surface and 1 denotes the receiving area, the result is

$$A_1 F_{1 \to 2} = A_2 F_{2 \to 1} \tag{6-50}$$

The subscript $A_1 \to A_2$ in Eq. 6-49 has been simplified to $1 \to 2$. The simplified subscript will be used in future shape-factor expressions, and it implies a shape factor between two diffuse finite areas 1 and 2. Equation 6-50 is called the *reciprocity relationship*. Reciprocity can be applied to any two surfaces i and j. The general form of the reciprocity relationship would be

$$A_i F_{i \to j} = A_j F_{j \to i} \tag{6-51}$$

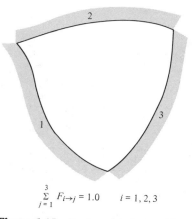

$$\sum_{j=1}^{3} F_{i \to j} = 1.0 \qquad i = 1, 2, 3$$

Figure 6-15 Enclosure relationship.

An additional relationship between shape factors can be derived when the participating surfaces form an enclosure. Consider the three surfaces which form an enclosure in Fig. 6-15. All the energy that leaves surface 1, for example, must be directly incident upon one of the three surfaces which form the enclosure:

$$\begin{pmatrix} \text{energy emitted} \\ \text{by surface 1} \end{pmatrix} = \begin{pmatrix} \text{energy arriving} \\ \text{at surface 1} \end{pmatrix} + \begin{pmatrix} \text{energy arriving} \\ \text{at surface 2} \end{pmatrix}$$

$$+ \begin{pmatrix} \text{energy arriving} \\ \text{at surface 3} \end{pmatrix} \tag{6-52}$$

If Eq. 6-52 is divided by the quantity on the left-hand side of the equation, each term becomes a shape factor:

$$F_{1 \to 1} + F_{1 \to 2} + F_{1 \to 3} = 1.0 \tag{6-53}$$

This equation is known as the *enclosure relationship*, and in the form of Eq. 6-53 it is limited to three surfaces. The enclosure relationship written for the general case of n surfaces forming an enclosure becomes

$$\sum_{j=1}^{n} F_{i \to j} = 1.0 \qquad i = 1, 2, \ldots, n \tag{6-54}$$

The term $F_{i \to i}$ must be included in the enclosure relationship whenever surface i is concave. A concave surface can "see" itself and some of the energy leaving a portion of the surface will be incident upon another part of the surface. If the ith surface is convex or flat so that it cannot see itself, then

$$F_{i \to i} = 0$$

The reciprocity and enclosure relationships are very valuable equations. When they are applied to the problem of evaluating shape factors, they can

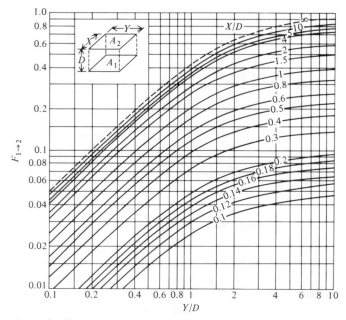

Figure 6-16 Radiation shape factor for directly opposed diffuse rectangles.

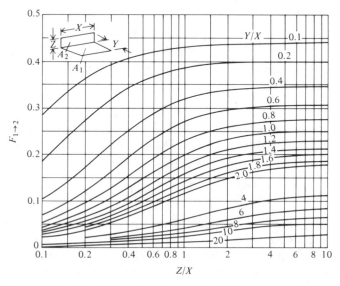

Figure 6-17 Radiation shape factor for diffuse perpendicular rectangles sharing a common edge.

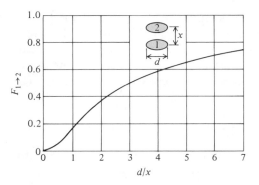

Figure 6-18   Radiation shape factor for directly opposed diffuse circles.

often save us time by offering an alternative to carrying out detailed integration.

The radiation shape factors for many surfaces encountered in engineering work have been evaluated and are provided in graphical form. Several examples of these shape factors are given in Figs. 6-16 through 6-18. A more extensive list of shape factors is given in Hamilton and Morgan (Ref. 11), and Siegel and Howell (Ref. 12).

Shape-Factor Algebra

The charts for shape factors can be used to determine values for a much larger class of geometries by a process known as *shape-factor algebra*. Suppose that we wish to evaluate the shape factor $F_{1\rightarrow2}$ where the areas 1 and 2 are shown in Fig. 6-19. The shape factor $F_{1\rightarrow3}$ is plotted in Fig. 6-17

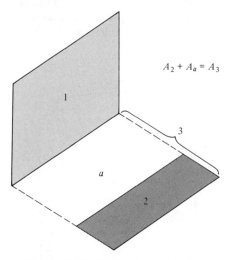

Figure 6-19   Shape-factor algebra.

where area 3 is the sum of areas 2 and a. Shape-factor algebra is simply a statement of conservation of energy. That is, the energy leaving surface 1 in Fig. 6-19 and arriving on surface 3 must be equal to the sum of the energies leaving surface 1 and arriving on the subareas that comprise surface 3. Since $A_3 = A_2 + A_a$, conservation of energy requires that

$$F_{1 \to 3} = F_{1 \to a} + F_{1 \to 2} \tag{6-55}$$

Both $F_{1 \to 3}$ and $F_{1 \to a}$ are given in Fig. 6-17, so $F_{1 \to 2}$ can be determined from Eq. 6-55.

If reciprocity is applied to each term in Eq. 6-55, the results are

$$A_3 F_{3 \to 1} = A_a F_{a \to 1} + A_2 F_{2 \to 1} \tag{6-56}$$

which means that the total energy arriving on surface 1 from surface 3 is equal to the sum of the energies from subareas a and 2 arriving on surface 1. Notice that the areas are included in Eq. 6-56 and not in Eq. 6-55. The areas remain in Eq. 6-56 because the emitting areas are not equal. The areas cancel from Eq. 6-55 because the emitting areas are all equal. The techniques of shape-factor algebra are illustrated in the examples at the end of this section.

Crossed-String Method

A very useful and simple method for evaluating shape factors exists when the surfaces are two-dimensional, infinite in extent in one direction and characterized by all cross sections normal to the infinite direction being identical. The method is called the *crossed-string method* and a proof is given in Hottel and Sarofim (Ref. 13). Two surfaces that satisfy these geometric restrictions are shown in Fig. 6-20. The method states that the shape factor $F_{1 \to 2}$ is equal to the sum of the lengths of the cross strings stretched between the ends of the two surfaces minus the sum of the uncrossed strings divided by twice the length of A_1. The shape factor $F_{1 \to 2}$

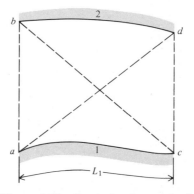

Figure 6-20 Crossed-string method.

is therefore

$$F_{1\to2} = \frac{1}{2L_1}\left[\left(\overline{ad}+\overline{cb}\right)-\left(\overline{ab}+\overline{cd}\right)\right] \qquad (6\text{-}57)$$

Example 6-5. Evaluate the shape factor $F_{1\to2}$ for the geometry in the accompanying figure. The surfaces are diffuse.

$$A_3 = A_a + A_1$$
$$A_4 = A_b + A_2$$

Solution: Applying the principles developed in Eqs. 6-55 and 6-56 results in

$$A_3 F_{3\to4} = A_a F_{a\to b} + A_a F_{a\to2} + A_1 F_{1\to b} + A_1 F_{1\to2}$$
$$A_3 F_{3\to b} = A_a F_{a\to b} + A_1 F_{1\to b}$$
$$F_{a\to4} = F_{a\to b} + F_{a\to2}$$

Combining these three equations and solving for $F_{1\to2}$ gives

$$F_{1\to2} = \frac{1}{A_1}\left(A_3 F_{3\to4} - A_a F_{a\to4} - A_3 F_{3\to b} + A_a F_{a\to b}\right)$$

The shape factors on the right-hand side of this equation are plotted in Fig. 6-17. The values are

$$F_{3\to4} = 0.19$$
$$F_{a\to4} = 0.32$$
$$F_{3\to b} = 0.08$$
$$F_{a\to b} = 0.19$$

Therefore, $F_{1\to2}$ is

$$F_{1\to2} = \frac{1}{30}\left[50(0.19) - 20(0.32) - 50(0.08) + 20(0.19)\right]$$

$$F_{1\to2} = 0.097$$

That is, 9.7 percent of the diffuse energy that leaves surface 1 will be directly incident on surface 2.

Example 6-6. An enclosure consists of three diffuse infinitely long planes whose cross section forms a right triangle like shown in the figure. Determine the shape factors $F_{1\rightarrow2}$, $F_{2\rightarrow1}$, $F_{1\rightarrow3}$, and $F_{3\rightarrow1}$.

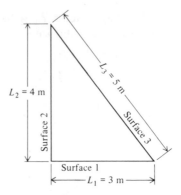

Solution: The two-dimensional surfaces are infinite in extent in and out of the plane of the paper, and each cross-sectional area is identical, so the crossed-string method can be used to determine $F_{1\rightarrow2}$.

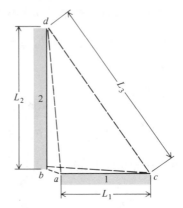

The two areas A_1 and A_2 are drawn slightly separated so that the locations of the strings can be shown more clearly. According to Eq. 6-57, the shape factor $F_{1\rightarrow2}$ is

$$F_{1\rightarrow2} = \frac{1}{2L_1}\left[(\overline{ad}+\overline{cb})-(\overline{ab}+\overline{cd})\right]$$

When the two planes are placed in contact, points a and b are identical and

$$ab = 0$$
$$cb = L_1$$
$$ad = L_2$$
$$cd = L_3$$

and the shape factor becomes

$$F_{1\to2}=\frac{1}{2L_1}(L_2+L_1-L_3)$$

Substituting values for the lengths gives

$$F_{1\to2}=\tfrac{1}{3}$$

The remaining shape factors will be evaluated using the enclosure and reciprocity relationships. Applying the enclosure relationship,

$$F_{1\to1}+F_{1\to2}+F_{1\to3}=1.0$$

But surface 1 is plane, so $F_{1\to1}=0$. Therefore,

$$F_{1\to3}=1-F_{1\to2}=1-\tfrac{1}{3}=\tfrac{2}{3}$$

Since $F_{1\to2}$ is known, reciprocity can be used to provide a value for $F_{2\to1}$.

$$F_{2\to1}=\frac{A_1}{A_2}F_{1\to2}=\tfrac{3}{4}\left(\tfrac{1}{3}\right)=\tfrac{1}{4}$$

Applying reciprocity to the shape factor $F_{1\to3}$ gives

$$F_{3\to1}=\frac{A_1}{A_3}F_{1\to3}=\tfrac{3}{5}\left(\tfrac{2}{3}\right)=\tfrac{2}{5}$$

6-5 RADIATIVE EXCHANGE BETWEEN BLACK SURFACES

In this section we will develop techniques to determine radiative-heat-transfer rates between black surfaces. It is particularly convenient to start our analysis of heat-transfer rates with black surfaces because a black surface does not have the complicating feature of reflecting a portion of incoming energy. In the next section we will consider gray surfaces and the analysis will be more complex.

The analysis that follows is based on several simplifying assumptions. We assume steady-state conditions exist and all participating surfaces are black and isothermal. Any surface that is not isothermal is subdivided until the smaller surfaces are approximately at constant temperature. The participating surfaces are separated by a medium that does not emit any radiant energy and it does not absorb or scatter the radiant intensity that passes between the black surfaces. Finally, we assume that the intensity of radiation is uniform over each surfaces and all surfaces are diffuse emitters and reflectors. The assumptions assure us that the expressions for the radiation shape factors developed in the previous section are valid. The assumption of diffuse surfaces poses no restriction when we consider only black surfaces because a black body is always a diffuse surface.

The definitions of several heat-exchange rates are important in radiation work. Three of the more important ones are defined as:

$q_{i \rightarrow j}$ = rate of radiant energy that is emitted by surface i and absorbed by surface j (6-58)

$(q_i)_{net}$ = net rate of energy that must be added to surface i to maintain its temperature at a constant value (6-59)

$q_{i \rightleftarrows j}$ = net rate of radiant energy transfer between surfaces i and j (6-60)

We will develop some general techniques for evaluating these heat-transfer rates by considering a simple geometry shown in Fig. 6-21. The geometry consists of two black planes and a third surface, which represents the surroundings. The three surfaces form an enclosure. The two plane surfaces are at known temperatures of T_1 and T_2, while the surroundings are assumed to be at 0 K. The surroundings therefore do not contribute any radiant energy on the two planes, and any energy that leaves either plane that is incident upon the surroundings can never return to either plane because it is totally absorbed by the black surroundings.

Considering a system of surfaces that forms an enclosure has several advantages. When we analyze radiation in an enclosure, we are assured of accounting for energy coming from all directions. Even when a problem involves surfaces that do not form an enclosure, we will find it advantageous to construct imaginary surfaces to complete the enclosure.

The enclosure can be formed by drawing imaginary surfaces that stretch between solid participating surfaces. Each imaginary surface is assigned an equivalent temperature and properties such that the radiative characteristics of the imaginary surfaces simulate the conditions stated in the actual problem. For example, suppose that two relatively small solid surfaces are

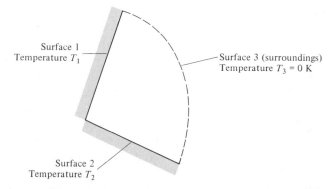

Figure 6-21 Radiant exchange among three black surfaces.

placed in a large room that has a known temperature T_s. To simulate the conditions in this problem, an enclosure could be constructed of the two solid surfaces plus an imaginary surface between the two surfaces which completes the enclosure. The imaginary surface will replace the room and it will be assigned a temperature and emissivity such that it behaves radiatively like the room.

The problem geometry is stated so that any radiation leaving the two solid surfaces which escapes to the surroundings has a very small chance of reflecting back to the original emitting surface. Therefore, the large room can be replaced by an imaginary black surface with an equivalent temperature of T_s. The black surface would not reflect any energy back to the emitter, and therefore it would behave like the large room. Assigning a temperature of T_s to the imaginary black surface would also ensure that the emitted energy from the room that is incident on the two planes would be accurately simulated.

The rate of radiant energy emitted by surface 1 in Fig. 6-21 that arrives on and is absorbed by surface 2 is

$$q_{1\rightarrow2} = A_1 F_{1\rightarrow2} E_{b_1} \tag{6-61}$$

and the energy emitted by surface 2 that is absorbed by surface 1 is

$$q_{2\rightarrow1} = A_2 F_{2\rightarrow1} E_{b_2} \tag{6-62}$$

The net radiative-heat-transfer rate between surfaces 1 and 2 is the difference between Eqs. 6-61 and 6-62:

$$\begin{aligned} q_{1\rightleftarrows2} &= q_{1\rightarrow2} - q_{2\rightarrow1} \\ &= A_1 F_{1\rightarrow2} E_{b_1} - A_2 F_{2\rightarrow1} E_{b_2} \end{aligned} \tag{6-63}$$

When reciprocity is applied, Eq. 6-63 becomes

$$q_{1\rightleftarrows2} = A_1 F_{1\rightarrow2} \left(E_{b_1} - E_{b_2} \right) \tag{6-64}$$

Equation 6-64 can be written in the general form of Ohm's law as a potential difference divided by a thermal resistance. The general expression for the net radiative-heat-transfer between surfaces i and j is

$$q_{i\rightleftarrows j} = \frac{E_{b_i} - E_{b_j}}{1/A_i F_{i\rightarrow j}} \tag{6-65}$$

The thermal resistance is caused by the fact that all energy emitted by surface i does not reach surface j. The resistance $1/A_i F_{i\rightarrow j}$ is thus called a *geometric resistance*. The potentials in the thermal circuit are the blackbody emissive powers of both surfaces, and they are related to the surface temperature by the Stefan-Boltzmann law. If Eq. 6-65 is applied to all three surfaces forming the enclosure, a partial thermal circuit can be drawn as shown in Fig. 6-22.

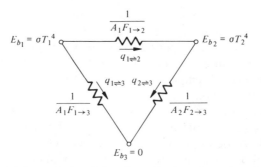

Figure 6-22 Partial thermal circuit for three black surfaces forming an enclosure.

The net rate of energy that must be added to surface 1 to maintain its temperature at a steady value is the difference between the emitted energy and the absorbed irradiation on surface 1. The net heat-transfer rate added to surface 1 is defined as $(q_1)_{net}$ in Eq. 6-59.

The energy emitted by surface 1 is

$$q_{emitted} = A_1 E_{b_1} \qquad (6\text{-}66)$$

and the energy absorbed by surface 1 is that energy incident from surface 2:

$$q_{absorbed} = A_2 F_{2 \to 1} E_{b_2} \qquad (6\text{-}67)$$

No energy is incident from surface 3 because it has a temperature of absolute zero and no energy can be incident from surface 1 because it is plane and cannot "see" its own surface. The net radiant energy rate for surface 1, $(q_1)_{net}$, is therefore the difference between Eqs. 6-66 and 6-67:

$$(q_1)_{net} = A_1 E_{b_1} - A_1 F_{1 \to 2} E_{b_2} \qquad (6\text{-}68)$$

Using a similar procedure, we can show that the expression for $(q_2)_{net}$ is

$$(q_2)_{net} = A_2 E_{b_2} - A_2 F_{2 \to 1} E_{b_1} \qquad (6\text{-}69)$$

The terms $(q_1)_{net}$ and $(q_2)_{net}$ can be included in the circuit diagram started in Fig. 6-22. The heat-transfer rate $(q_1)_{net}$ is the amount of energy that must be added per unit time to surface 1 in order to maintain its temperature at a constant value. The net energy added can be simulated electrically by a voltage source which is connected to the potential junction represented by E_{b_1}. Adding the terms $(q_1)_{net}$, $(q_2)_{net}$, and $(q_3)_{net}$, the completed electrical circuit for the geometry in Fig. 6-21 is illustrated in Fig. 6-23.

The thermal circuit is a convenient and compact way to express the heat-transfer rates. At the intersection represented by E_{b_1} in Fig. 6-23, the

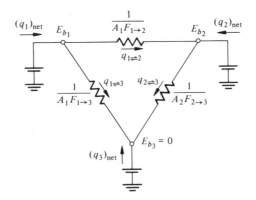

Figure 6-23 Complete thermal circuit for three
black surfaces forming an enclosure.

conservation of energy for steady-state conditions specifies that all heat
flows must sum to zero:

$$(q_1)_{net} = q_{1\rightleftarrows2} + q_{1\rightleftarrows3} \tag{6-70}$$

This equation is equivalent to Kirchhoff's current law applied to the steady
flow of current in electrical circuits. Substituting Eq. 6-65 for both terms
on the right-hand side of Eq. 6-70, applying reciprocity and the enclosure
relationship and setting $E_{b_3} = 0$, will produce the result shown in Eq. 6-68.
A similar conservation of energy at the branch point E_{b_2} in the thermal
circuit can be reduced to Eq. 6-69. Also, by examining the circuit in Fig.
6-23, we can determine that

$$(q_1)_{net} + (q_2)_{net} + (q_3)_{net} = 0 \tag{6-71}$$

$$q_{2\rightleftarrows3} = A_2 F_{2\rightarrow3} E_{b_2} \tag{6-72}$$

$$q_{1\rightleftarrows3} = A_1 F_{1\rightarrow3} E_{b_1} \tag{6-73}$$

The technique of analyzing radiation problems by drawing a thermal
circuit can be extended to any number of black surfaces that form an
enclosure. For example, the thermal circuit for four black surfaces forming
an enclosure is shown in Fig. 6-24. Application of Kirchhoff's current law
shows that

$$(q_i)_{net} = \sum_{j=1}^{4} (q_{i\rightleftarrows j}) \qquad (i = 1, 2, 3, 4) \tag{6-74}$$

where the term $j = i$ is not included in the summation term. Application of
Ohm's law to the circuit in Fig. 6-24 requires that

$$q_{i\rightleftarrows j} = \frac{E_{b_i} - E_{b_j}}{1/A_i F_{i\rightarrow j}} \tag{6-75}$$

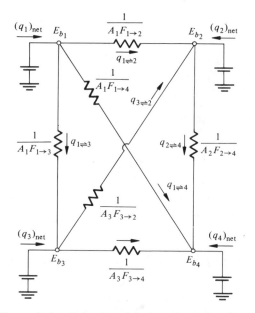

Figure 6-24 Thermal circuit for four black surfaces forming an enclosure.

Refractory Surfaces

Occasionally conditions are such that the net heat flow to a particular surface in a radiating system is zero:

$$(q_i)_{net} = 0$$

When this condition occurs, the surface is called a *refractory* or *reradiating surface*. If a refractory surface only exchanges heat by radiation and heat transfer by the other modes is negligible, the incident energy or irradiation must equal the energy leaving the surface.

We can easily modify the thermal circuit to account for any number of refractory surfaces by simply setting $(q_i)_{net}$ equal to zero at each circuit branch point representing a refractory surface. The refractory branch point then becomes a floating potential whose black-body emissive power, and therefore its temperature, is determined by the temperatures of all the other participating surfaces. The thermal circuit for two black surfaces (surfaces 1 and 2) and a refractory surface (surface 3) that form an enclosure is shown in Fig. 6-25.

The temperature of the refractory surface can be determined from the thermal circuit by noticing that

$$q_{1\rightleftarrows3} = q_{3\rightleftarrows2} \qquad (6\text{-}76)$$

or

$$\frac{E_{b_1} - E_{b_3}}{1/A_1 F_{1\rightarrow3}} = \frac{E_{b_3} - E_{b_2}}{1/A_2 F_{2\rightarrow3}} \qquad (6\text{-}77)$$

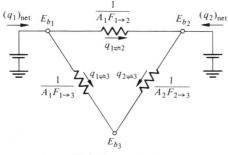

Figure 6-25 Thermal circuit for two black surfaces and a refractory surface forming an enclosure.

Equation 6-77 can be used to determine T_3 once the temperatures of T_1 and T_2 are known and the geometry is specified.

Example 6-7. A black plane has a surface area of 2 m^2 and is buried so that it is level with the surface of the earth. The lower surface of the plane is insulated. The upper surface is exposed to 300 K air and the convective-heat-transfer coefficient between the plane and air is 10 W/m$^2 \cdot$K. Calculate the equilibrium temperature of the plane for two conditions:

a. The plane radiates to a clear night sky and the effective radiation temperature of the sky is 100 K.

b. The plane radiates into a cloudy night sky and the effective sky temperature is 250 K.

Solution: The system is shown in the figure. The plane is called surface 1 and the sky is called surface 2. The sky can be assumed to be a black body because it will reflect a negligible amount of energy back to the plane and it will absorb practically all the energy emitted by the plane. The radiation system therefore consists of an enclosure formed by two black surfaces that are exchanging heat. The thermal circuit for the two black surfaces shown below is a simplified version of Fig. 6-23.

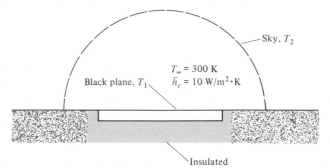

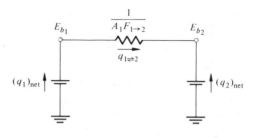

All energy that leaves the plane must arrive at the sky, so

$$F_{1\rightarrow2}=1.0$$

This result assumes that the plane is placed in an open area, and no radiation is incident from surrounding buildings, trees, and so on.

The net heat flow into the plane is due to heat being convected from the air, so

$$(q_1)_{net}=\bar{h}_cA_1(T_\infty-T_1) \tag{6-78}$$

and according to the thermal circuit, the heat flow to the plane by convection must equal the net heat flow from the plane to the sky by radiation or

$$(q_1)_{net}=q_{1\rightleftarrows2}=A_1F_{1\rightarrow2}(E_{b_1}-E_{b_2})$$

Substituting Eq. 6-78 and the Stefan-Boltzmann law for the emissive power gives

$$\bar{h}_c(T_\infty-T_1)=F_{1\rightarrow2}\sigma(T_1^4-T_2^4) \tag{6-79}$$

a. When $T_2=100$ K, Eq. 6-79 becomes

$$10(300-T_1)=5.67\times10^{-8}(T_1^4-100^4)$$

Solving for T_1 gives

$$T_1=270.2 \text{ K}=-3°C$$

b. When $T_2=250$ K, solving Eq. 6-79 results in an equilibrium plane temperature of

$$T_1=284.8 \text{ K}=11.6°C$$

These results show that a surface which has a good view of the sky on a cool, clear night can easily reach a temperature below the freezing point of water. Water vapor in the air can then condense on the surface and freeze, forming a layer of frost, even though the air temperature is above freezing.

A layer of clouds increases the effective temperature of the sky and will help prevent the frost from forming. Surrounding buildings or trees that obstruct the view of the sky on a clear night will have a similar effect of increasing the effective sky temperature. This explains why the windshield of a car parked in the open might have a layer of frost on it in the morning while a car parked near a building or under trees may not have any frost, even though the air temperature is the same for both cases.

Example 6-8. Two parts on a spacecraft can be approximated by directly opposed parallel rectangles as shown in the figure. The rectangles are black and they are maintained at temperatures of 500 K and 400 K by external means. Incident energy from other sources is negligible and the backs of both surfaces are insulated. Calculate:

a. The net rate of energy that must be added to the warmer surface to maintain its temperature.
b. The net rate of energy that must be added to the cooler surface to maintain its temperature.
c. The net rate of energy that must be added to the surroundings to maintain its temperature.
d. The net heat-transfer rate between the two surfaces and the surroundings.

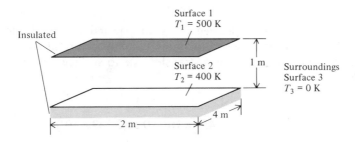

Solution: The thermal circuit for the problem is identical to the one shown in Fig. 6-23. Since the incident radiation from the surroundings is negligible, the surroundings can be assumed to be a black surface at 0 K, which form an enclosure with the two rectangles.

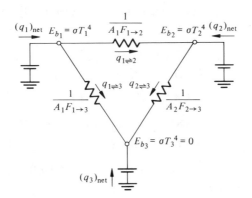

The three potentials in the circuit are:

$$E_{b_1} = \sigma T_1{}^4 = 3.543 \times 10^3 \text{ W/m}^2$$

$$E_{b_2} = \sigma T_2{}^4 = 1.451 \times 10^3 \text{ W/m}^2$$

$$E_{b_3} = \sigma T_3{}^4 = 0$$

The three shape factors in the thermal circuit can be determined by using data from Fig. 6-16 along with using the enclosure relationship and the symmetry of the geometry.

$$F_{1 \to 2} = 0.52$$

$$F_{1 \to 3} = 1 - F_{1 \to 2} = 0.48$$

$$F_{2 \to 3} = F_{1 \to 3} = 0.48$$

a. The net radiant heat-transfer rate to surface 1 is given by Eq. 6-68:

$$(q_1)_{\text{net}} = A_1 \left(E_{b_1} - F_{1 \to 2} E_{b_2} \right)$$

$$= 8 \left[3.543 \times 10^3 - 0.52(1.451 \times 10^3) \right]$$

$$= 2.231 \times 10^4 \text{ W}$$

That is, surface 1 must be *heated* by an external source of energy at a rate of 22.31 kW in order to maintain its temperature at 500 K.

b. An expression for the net heat-transfer rate to surface 2 is given by Eq. 6-69:

$$(q_2)_{\text{net}} = A_2 \left(E_{b_2} - F_{2 \to 1} E_{b_1} \right)$$

$$= 8 \left[1.451 \times 10^3 - 0.52(3.543 \times 10^3) \right]$$

$$= -3.13 \times 10^3 \text{ W}$$

Surface 2 must be *cooled* at a rate of 3.13 kW in order to maintain its temperature at 400 K. The negative sign on $(q_2)_{\text{net}}$ indicates that the net heat flow to surface 2 is away from the surface or it must be cooled by external means.

c. By examining the thermal circuit an expression for the net heat-transfer rate to the surroundings is

$$(q_3)_{\text{net}} = -q_{1 \rightleftarrows 3} - q_{2 \rightleftarrows 3}$$

$$= -A_1 F_{1 \to 3} E_{b_1} - A_2 F_{2 \to 3} E_{b_2}$$

$$= -A_1 F_{1 \to 3} \left(E_{b_1} + E_{b_2} \right)$$

$$= -8 \times 0.48(3.543 \times 10^3 + 1.451 \times 10^3)$$

$$= -1.918 \times 10^4 \text{ W}$$

The surroundings must be cooled at a rate of 19.18 kW or its temperature will tend to increase.

As a check on these calculations we note that the net transfer of heat to all three surfaces must be zero as predicted by Eq. 6-71.

$$(q_1)_{net} + (q_2)_{net} + (q_3)_{net} = 2.231 \times 10^4 - 3.13 \times 10^3 - 1.918 \times 10^4 = 0$$

d. The net heat-transfer rates between the surfaces are given by Eq. 6-64:

$$q_{1 \rightleftarrows 2} = A_1 F_{1 \rightarrow 2} (E_{b_1} - E_{b_2}) = 8.703 \times 10^3 \text{ W}$$

$$q_{1 \rightleftarrows 3} = A_1 F_{1 \rightarrow 3} E_{b_1} = 1.361 \times 10^4 \text{ W}$$

$$q_{2 \rightleftarrows 3} = A_2 F_{2 \rightarrow 3} E_{b_2} = 5.572 \times 10^3 \text{ W}$$

As a final check on the solution we should be able to verify that

$$(q_1)_{net} = q_{1 \rightleftarrows 2} + q_{1 \rightleftarrows 3}$$

$$(q_2)_{net} = -q_{1 \rightleftarrows 2} + q_{2 \rightleftarrows 3}$$

$$(q_3)_{net} = -q_{1 \rightleftarrows 3} - q_{2 \rightleftarrows 3}$$

6-6 RADIATIVE EXCHANGE BETWEEN GRAY SURFACES

Heat-transfer calculations involving gray surfaces are more complicated than those involving only black surfaces. When radiation is incident on a gray surface, a portion of the radiation is reflected and it must be accounted for in any energy balance. The previous section did not account for the reflected energy because we considered only black surfaces there.

We will first develop general equations for steady heat transfer between gray surfaces by considering three gray, isothermal surfaces forming an enclosure. The enclosure is filled with a radiatively nonparticipating medium. All surfaces are isothermal and the irradiation on each surface is distributed uniformly. The gray surfaces are assumed diffuse, and since they are gray, Kirchhoff's law states that the surface emissivity is equal to its absorptivity. Furthermore, we assume that all surfaces are opaque, so Eq. 6-13 states that

$$\alpha = \epsilon = 1 - \rho \tag{6-80}$$

If an imaginary plane is placed above the gray surface as in Fig. 6-26, an energy balance on the plane at steady state requires that the net energy

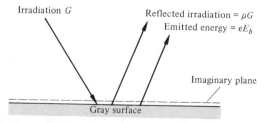

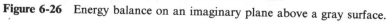

Figure 6-26 Energy balance on an imaginary plane above a gray surface.

that must be supplied to the surface to maintain its temperature is the difference between the energy leaving and energy arriving on the surface.

The energy leaving the gray surface is called the *radiosity*. The radiosity is given the symbol J and it is equal to the sum of the emitted energy per unit area and reflected irradiation:

$$J = \epsilon E_b + \rho G \tag{6-81}$$

The net heat flow to a gray surface designated by the subscript i will be the difference between the radiosity and irradiation on the surface:

$$(q_i)_{net} = A_i(J_i - G_i) \tag{6-82}$$

Substituting Eqs. 6-80 and 6-81 into Eq. 6-82 gives

$$(q_i)_{net} = \frac{E_{b_i} - J_i}{\rho_i / \epsilon_i A_i} \tag{6-83}$$

When Eq. 6-83 is considered in the form of Ohm's law, the denominator is a thermal resistance which separates the two potentials E_{b_i} and J_i as shown in Fig. 6-27. The resistance, $\rho_i / \epsilon_i A_i$, is due to the fact that the gray surface reflects a portion of the irradiation, and it is referred to as a *surface resistance*. Note that the surface resistance will be zero for a black surface for which $\rho = 0$.

If the surface resistance is added to a thermal circuit like shown in Fig. 6-23 formulated for black surfaces, the circuit can be modified to account for exchange between gray surfaces. The basic structure of the circuit developed for black surfaces should remain unchanged when we consider gray surfaces, because the resistances are geometric resistances that are functions of geometry only and not of surface properties. The completed thermal circuit for radiation between three gray surfaces forming an enclosure is shown in Fig. 6-28.

The thermal circuit is a particularly convenient way to analyze radiation problems. By applying Ohm's law across each resistance and Kirchhoff's current law at every branch point in the circuit, a large number of equations relating heat flow and properties can be condensed into a small

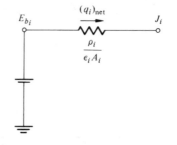

Figure 6-27 Surface resistance for a gray surface.

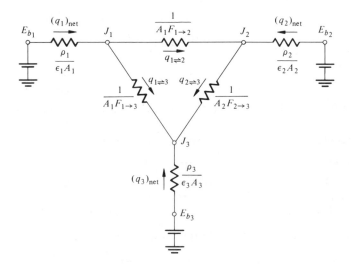

Figure 6-28 Thermal circuit for three gray surfaces forming an enclosure.

diagram. The application of the thermal circuit is illustrated in the examples at the end of this section.

The concept of a refractory surface was introduced in the previous section as a surface for which $(q_i)_{net} = 0$. When the net heat flow in the thermal circuit in Fig. 6-27 is set equal to zero, we can see that a refractory surface is one for which

$$J_i = E_{b_i} \qquad (6\text{-}84)$$

Also, when $(q_i)_{net}$ is set equal to zero in Eq. 6-82, then

$$J_i = G_i \qquad (6\text{-}85)$$

for a refractory surface.

Equations 6-84 and 6-85 show that the irradiation, radiosity, and blackbody emissive power are all equal for a refractory surface. A surface for which the arriving and leaving energy are equal can be visualized as a perfectly reflecting surface. The refractory is therefore simply a surface that redirects the incident energy and does not absorb any of the incident energy. The refractory surface will attain an equilibrium temperature determined by the temperature of the other participating surfaces such that the blackbody emissive power of the refractory surface equals the surface radiosity and irradiation.

Two Gray Surfaces Forming an Enclosure

Often we are faced with problems that involve only two gray, opaque, diffuse surfaces that form an enclosure. For this situation the thermal circuit is shown in Fig. 6-29. Several examples of geometries that satisfy

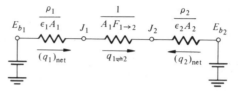

Figure 6-29 Thermal circuit for two gray surfaces forming an enclosure.

the assumptions of two surfaces forming an enclosure would be:

1. Two infinitely large parallel plates.
2. Two long concentric cylinders.
3. Two spheres, one enclosing the second.
4. A small body enclosed by a second closed surface.

The thermal circuit in Fig. 6-29 for these geometries requires that

$$(q_1)_{net} = q_{1 \rightleftarrows 2} = -(q_2)_{net} \tag{6-86}$$

and

$$(q_1)_{net} = \frac{E_{b_1} - E_{b_2}}{\rho_1/\epsilon_1 A_1 + 1/A_1 F_{1 \to 2} + \rho_2/\epsilon_2 A_2} \tag{6-87}$$

If the geometry consists of two infinite parallel plates of equal area, then $A_1 = A_2$ and $F_{1 \to 2} = 1.0$. Substituting these results into Eq. 6-87 gives

$$(q_1)''_{net} = \frac{E_{b_1} - E_{b_2}}{1 + \rho_1/\epsilon_1 + \rho_2/\epsilon_2} \tag{6-88}$$

If the two surfaces that form an enclosure are infinitely long cylinders with cylinder 1 enclosed by cylinder 2, then $F_{1 \to 2} = 1.0$ and the expression for $(q_1)''_{net}$ is

$$(q_1)''_{net} = \frac{E_{b_1} - E_{b_2}}{1 + \rho_1/\epsilon_1 + A_1 \rho_2/A_2 \epsilon_2} \tag{6-89}$$

Equation 6-89 also applies to the case of two spheres with sphere 1 completely enclosed by sphere 2.

If surface 1 is a small gray body that is completely enclosed by a large closed surface called surface 2, then $A_1/A_2 \to 0$ and Eq. 6-89 can be further simplified to

$$(q_1)''_{net} = q''_{1 \to 2} = \frac{E_{b_1} - E_{b_2}}{(\epsilon_1 + \rho_1)/\epsilon_1} = \epsilon_1 (E_{b_1} - E_{b_2}) \tag{6-90}$$

Three Gray Surfaces Forming an Enclosure

When an enclosure consists of three gray surfaces, the expressions for the radiative-heat-transfer rates are more complicated than for two surfaces, but they are obtained by a similar procedure. Suppose that we

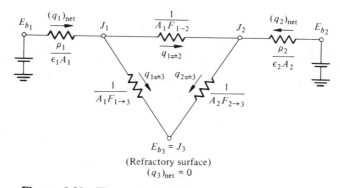

Figure 6-30 Thermal circuit for two gray surfaces and a refractory surface forming an enclosure.

consider two gray opaque surfaces and one refractory surface that form an enclosure. The thermal circuit for this case is shown in Fig. 6-30, where surface 3 has been selected as the refractory surface. Conservation of energy requires that

$$(q_1)_{net} = -(q_2)_{net} \qquad (6\text{-}91)$$

Also, the net heat flux to surface 1 is equal to the potential difference $(E_{b_1} - E_{b_2})$ divided by the total resistance between these potentials:

$$(q_1)_{net} = \frac{E_{b_1} - E_{b_2}}{\rho_1/\epsilon_1 A_1 + R_{equiv} + \rho_2/\epsilon_2 A_2} \qquad (6\text{-}92)$$

where R_{equiv} is the equivalent resistance of the geometric resistance between J_1 and J_2 placed in parallel with the two geometric resistances between J_1 and J_3 and between J_2 and J_3. The equivalent resistance is

$$R_{equiv} = \frac{\dfrac{1}{A_1 F_{1\to2}}\left(\dfrac{1}{A_1 F_{1\to3}} + \dfrac{1}{A_2 F_{2\to3}}\right)}{\dfrac{1}{A_1 F_{1\to2}} + \dfrac{1}{A_1 F_{1\to3}} + \dfrac{1}{A_2 F_{2\to3}}} \qquad (6\text{-}93)$$

Substituting Eq. 6-93 into Eq. 6-92 and simplifying yields

$$(q_1)_{net} = \frac{A_1(E_{b_1} - E_{b_2})}{\dfrac{\rho_1}{\epsilon_1} + \dfrac{A_1}{A_2}\dfrac{\rho_2}{\epsilon_2} + \dfrac{1}{F_{1\to2} + \dfrac{A_2 F_{1\to3} F_{2\to3}}{A_1 F_{1\to3} + A_2 F_{2\to3}}}} \qquad (6\text{-}94)$$

Example 6-9. A long, gray metallic pipe carries a heated liquid from one processing machine to another in a manufacturing plant. The pipe has an external diameter of 0.8 m and a surface emissivity of 0.5. The pipe is surrounded by a second gray pipe that has an i.d. of 1.0 m and an emissivity of 0.3. The space between the two pipes is evacuated to minimize the heat losses from the liquid. The inside pipe has a temperature of 550 K, and the outside pipe has a temperauture of 300 K.

Estimate the heat loss from the liquid per meter of pipe length. Determine the percent reduction in heat loss from the liquid if the internal pipe is coated with a gray paint that has an emissivity of 0.15.

Solution: The space between the two pipes is evacuated so that the only way in which the liquid can lose energy is by radiation between the inner and outer pipe surfaces. The two concentric pipes are gray, opaque, and form an enclosure, so Eq. 6-89 can be used to determine the heat loss from the liquid.

Calling the interior pipe surface 1 and the exterior pipe surface 2, the net heat loss by the liquid is

$$(q_1)''_{net} = \frac{(q_1)_{net}}{A_1} = \frac{E_{b_1} - E_{b_2}}{1 + \dfrac{\rho_1}{\epsilon_1} + \dfrac{A_1}{A_2}\dfrac{\rho_2}{\epsilon_2}}$$

where

$$E_{b_1} = \sigma T_1^4 = (5.67 \times 10^{-8})(550)^4 = 5187 \text{ W/m}^2$$

$$E_{b_2} = \sigma T_2^4 = (5.67 \times 10^{-8})(300)^4 = 459 \text{ W/m}^2$$

$$\rho_1 = 1 - \epsilon_1 = 0.5$$

$$\rho_2 = 1 - \epsilon_2 = 0.7$$

$$\frac{A_1}{A_2} = \frac{\pi d_1 L}{\pi d_2 L} = \frac{0.8}{1.0} = 0.8$$

and the area of the inner pipe per unit length is

$$\pi d_1 = 0.8\pi$$

Substituting these values into the expression for $(q_1)_{net}$ gives the net heat flux per unit length of

$$(q_1)'_{net} = \frac{0.8\pi(5187 - 459)}{1 + 0.5/0.5 + 0.8(0.7/0.3)}$$

$$= 3073 \text{ W/m}$$

If ϵ_1 is changed to 0.15, the heat loss from the liquid becomes

$$(q_1)'_{net} = 1393 \text{ W/m}$$

or a reduction of 54.7%.

Example 6-10. A gray circular heater 10 cm in diameter is placed parallel to and 5 cm from a second gray circular receiver as shown in the figure. The power input to the heater is 300 W. The backs of both the heater and

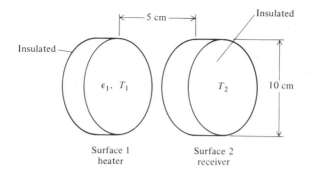

receiver are insulated. Convection from both surfaces is negligible. The heater and receiver are placed in a large room that is maintained at 350 K. The heater has an emissivity of 0.8. Calculate:

a. The heater temperature.
b. The receiver temperature.
c. The net radiative-heat-transfer rate to the surroundings.
d. The net radiative-heat-transfer rate between the heater and the receiver.

Solution: The given information in the problem includes the net heat-transfer rate to the heater and the temperature of the surroundings. If we denote the heater as surface 1, the receiver as surface 2 and the surroundings as surface 3, then

$$(q_1)_{net} = 300 \text{ W}$$
$$T_3 = 350 \text{ K}$$

Both the temperature of the receiver and the net heat-transfer rate of the receiver are unknown quantities. However, the problem states that the receiver is insulated and the convection to it is negligible. There is no external energy supplied to the receiver, and therefore the net radiative heat-transfer rate to it must be zero. The receiver is simply a refractory surface for which

$$(q_2)_{net} = 0$$

and, from Eq. 6-84,

$$J_2 = E_{b_2}$$

Since the surroundings are large compared to the size of the heater and receiver, they act like a blackbody, so

$$\epsilon_3 = 1.0$$

and

$$J_3 = E_{b_3}$$

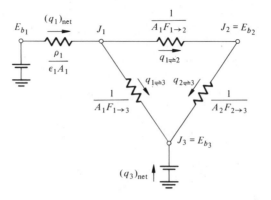

The thermal circuit for this problem is similar to the one in Fig. 6-28 except that the surface resistance for the surroundings is zero because it is a blackbody and the net heat flow to surface 2 is zero. The shape factors are evaluated from Fig. 6-18:

$$F_{1\to2}=0.37$$
$$F_{1\to3}=F_{2\to3}=1-F_{1\to2}=0.63$$

The area of the heater and receiver is

$$A_1=A_2=\pi r^2=\pi(0.05)^2=7.85\times10^{-3}\text{ m}^2$$

The emissive power of the surroundings is

$$E_{b_3}=\sigma T_3^4=5.67\times10^{-8}(350)^4=851\text{ W/m}^2$$

Referring to the thermal circuit, we see that conservation of energy on surface 2 requires

$$\frac{J_1-J_2}{1/A_1F_{1\to2}}=\frac{J_2-E_{b_3}}{1/A_2F_{2\to3}}$$

and at surface 1

$$(q_1)_{\text{net}}=\frac{J_1-E_{b_3}}{1/A_1F_{1\to3}}+\frac{J_1-J_2}{1/A_1F_{1\to2}}$$

The only unknowns in these two equations are J_1 and J_2. Solving them simultaneously gives

$$J_1=45,134\text{ W/m}^2$$
$$J_2=17,234\text{ W/m}^2$$

a. The temperature of the heater can be determined from the expression

$$(q_1)_{\text{net}}=\frac{E_{b_1}-J_1}{\rho_1/\epsilon_1A_1}$$

or

$$T_1 = \left(\frac{E_{b_1}}{\sigma}\right)^{1/4} = \left[\frac{(\rho_1/\epsilon_1 A_1)(q_1)_{net} + J_1}{\sigma}\right]^{1/4}$$

$$= \left[\frac{\left[\dfrac{1-0.8}{0.8 \times (7.85 \times 10^{-3})}\right]300 + 45,134}{5.67 \times 10^{-8}}\right]^{1/4}$$

$$= 991 \text{ K} = 718°C$$

b. The temperature of the receiver can be determined from the blackbody emissive power of the receiver:

$$J_2 = E_{b_2} = \sigma T_2^4$$

$$T_2 = \left(\frac{J_2}{\sigma}\right)^{1/4} = \left(\frac{17,234}{5.67 \times 10^{-8}}\right)^{1/4}$$

$$= 743 \text{ K} = 470°C$$

c. The net radiative-heat-transfer rate to the surroundings as determined from the thermal circuit is

$$(q_3)_{net} = \frac{E_{b_3} - J_1}{1/A_1 F_{1\to3}} + \frac{E_{b_3} - J_2}{1/A_2 F_{2\to3}}$$

$$= A_1 F_{1\to3}(2E_{b_3} - J_1 - J_2)$$

$$= (7.85 \times 10^{-3}) \times 0.63(2 \times 851 - 45,134 - 17,234)$$

$$= -300 \text{ W}$$

As a check on this value, we know that the sum of the net heat-transfer rates to the three surfaces must be zero if the total energy of the system is to be conserved.

$$(q_1)_{net} + (q_2)_{net} + (q_3)_{net} = 300 + 0 - 300 = 0 \text{ W}$$

d. The net radiative-heat-transfer rate between the heater and receiver is

$$q_{1\rightleftharpoons2} = A_1 F_{1\to2}(J_1 - J_2) = (7.85 \times 10^{-3}) \times 0.37(45,134 - 17,234)$$

$$= 81.0 \text{ W}$$

6-7 MATRIX METHODS

When a radiation problem involves more than about four or five participating surfaces, the amount of effort and time required to determine either the heat fluxes or temperatures of the surfaces by methods described in the previous sections becomes large enough to warrant a computer solution. As the number of surfaces increases, it becomes necessary to organize the governing equations in a systematic fashion and apply standard mathematical techniques for the solution of the equations. One of the simplest and yet most powerful techniques that can be used to analyze radiation problems of this type is the matrix method.

We have seen that the equations relating the heat flow to surface temperature, Eqs. 6-65, 6-74, and 6-83, for example, are all algebraic equations. If the heat flux equations can be organized into the form of a matrix, then standard matrix techniques such as matrix inversion can be applied to determine the unknown parameters. When the number of surfaces involved in a problem becomes large, a computer program can be used to invert the matrix and solutions that would normally require a large amount of time to solve by hand can be determined very rapidly.

The matrix equations used when the surfaces have known temperatures differ slightly from the equations used when the surfaces have known values for heat flux. We will consider these two cases separately, looking at the problem of surfaces with known temperatures first.

Surfaces with Known Temperatures

The first task in the matrix method is to organize the governing equations into a matrix form. We will assume that the temperature of all surfaces are known and that the values for the net heat flux to all surfaces are desired. All surfaces are opaque, gray, and the distribution of radiant energy over the surfaces is uniform. The surfaces form an enclosure and each surface is isothermal. Any surface that is not isothermal must be divided into smaller areas until the subdivided areas are reasonably isothermal. One distinct advantage of the matrix technique coupled with a computer solution is that the addition of more surfaces into the problem does not significantly increase the amount of work required for a solution.

For simplicity, we will develop the form of the matrix equations by considering only three surfaces that form an enclosure. The proper thermal circuit for this case is drawn in Fig. 6-28. The net heat flux for each of the three surfaces can be obtained by combining Eqs. 6-82 and 6-83. The result for surface 1 is

$$(q_1)''_{net} = \frac{\epsilon_1}{\rho_1}(E_{b_1} - J_1) = J_1 - G_1 \qquad (6\text{-}95)$$

The irradiation G_1 on surface 1 is due to energy leaving surface 1 (if it is concave and can "see" itself) and the other two surfaces which form the enclosure:

$$A_1 G_1 = A_1 F_{1\to 1} J_1 + A_2 F_{2\to 1} J_2 + A_3 F_{3\to 1} J_3 \qquad (6\text{-}96)$$

Applying reciprocity to Eq. 6-96 gives

$$G_1 = F_{1\to 1} J_1 + F_{1\to 2} J_2 + F_{1\to 3} J_3 \qquad (6\text{-}97)$$

Substituting Eq. 6-97 into Eq. 6-95 gives

$$(q_1)''_{net} = \frac{\epsilon_1}{\rho_1}(E_{b_1} - J_1) = J_1 - (F_{1\to 1} J_1 + F_{1\to 2} J_2 + F_{1\to 3} J_3) \qquad (6\text{-}98)$$

which can be rewritten as

$$\left(1-F_{1\rightarrow1}+\frac{\epsilon_1}{\rho_1}\right)J_1+(-F_{1\rightarrow2})J_2+(-F_{1\rightarrow3})J_3=\frac{\epsilon_1}{\rho_1}E_{b_1} \qquad (6\text{-}99)$$

Similar expressions can be written for surfaces 2 and 3. They are:

$$(-F_{2\rightarrow1})J_1+\left(1-F_{2\rightarrow2}+\frac{\epsilon_2}{\rho_2}\right)J_2+(-F_{2\rightarrow3})J_3=\frac{\epsilon_2}{\rho_2}E_{b_2} \qquad (6\text{-}100)$$

$$(-F_{3\rightarrow1})J_1+(-F_{3\rightarrow2})J_2+\left(1-F_{3\rightarrow3}+\frac{\epsilon_3}{\rho_3}\right)J_3=\frac{\epsilon_3}{\rho_3}E_{b_3} \qquad (6\text{-}101)$$

Equations 6-99, 6-100, and 6-101 can be condensed into a matrix form of

$$AJ=B \qquad (6\text{-}102)$$

For the general case of n surfaces forming an enclosure, A is a $n \times n$ matrix of the form

$$A=\left\{\begin{array}{l} a_{11}a_{12}\cdots a_{1n} \\ a_{21}a_{22}\cdots a_{2n} \\ \vdots \\ a_{n1}a_{n2}\cdots a_{nn} \end{array}\right\} \qquad (6\text{-}103)$$

while J and B are both column matrices consisting of n elements

$$J=\left\{\begin{array}{l} J_1 \\ J_2 \\ \vdots \\ J_n \end{array}\right\} \qquad (6\text{-}104)$$

$$B=\left\{\begin{array}{l} b_1 \\ b_2 \\ \vdots \\ b_n \end{array}\right\} \qquad (6\text{-}105)$$

By examining Eqs. 6-99 through 6-101 we can see that the off-diagonal elements of the matrix A are

$$a_{ij}=-F_{i\rightarrow j} \qquad (i\neq j) \qquad (6\text{-}106a)$$

and the diagonal terms are

$$a_{ii}=1-F_{ii}+\frac{\epsilon_i}{\rho_i} \qquad (6\text{-}106b)$$

The elements of the matrix B are

$$b_i=\frac{\epsilon_i E_{b_i}}{\rho_i} \qquad (6\text{-}107)$$

When a surface in the enclosure is black, a special form of the matrix equation, Eq. 6-99, must be considered. For a black surface

$$J_i = E_{b_i} \qquad (6\text{-}108)$$

Therefore, the elements of the matrices A and B for a blackbody are

$$a_{ij} = 0 \qquad (i \neq j) \qquad (6\text{-}109\text{a})$$

$$a_{ii} = 1.0 \qquad (6\text{-}109\text{b})$$

$$b_i = E_{b_i} \qquad (6\text{-}110)$$

When the temperatures of all surfaces are known, the problem of determining the net heat flux of all surfaces is approached in the following manner. Elements of the matrices A and B are evaluated first. For all nonblack surfaces, the elements of the matrix A are given by Et_i. 6-106. If the ith surface is black, then the ith row of elements in the matrix A is given by Eq. 6-109. Notice that all elements of the matrix are known once the geometry and surface properties are specified.

The elements of the column matrix B are given by Eq. 6-107 for all nonblack surfaces and by Eq. 6-110 when the surface is a blackbody. The elements of the matrix B are functions of the surface properties and the known temperatures.

Once the matrices A and B are determined, the matrix A is inverted. The inverse of A is C:

$$A^{-1} = C = \begin{Bmatrix} c_{11} c_{12} \cdots c_{1n} \\ c_{21} c_{22} \cdots c_{2n} \\ \vdots \\ c_{n1} c_{n2} \cdots c_{nn} \end{Bmatrix} \qquad (6\text{-}111)$$

where all the elements of C are known values. The elements of J may now be determined from the solution to Eq. 6-102:

$$J = CB \qquad (6\text{-}112)$$

which can be written in long-hand form as

$$J_1 = c_{11} b_1 + c_{12} b_2 + \cdots + c_{1n} b_n$$
$$\vdots \qquad \vdots \qquad \qquad \vdots \qquad\qquad (6\text{-}113)$$
$$J_n = c_{n1} b_1 + c_{n2} b_2 + \cdots + c_{nn} b_n$$

Once values for the irradiation J_i on all surfaces are known from Eq. 6-113, then Eq. 6-95 can be used to determine values for the net heat flux to all surfaces. The net heat flux for a nonblack surface i is

$$(q_i)''_{net} = \frac{\epsilon_i}{\rho_i} (E_{b_i} - J_i) \qquad (6\text{-}114)$$

When surface i is black, the expression for the net heat flux is a combination of Eqs. 6-95 and 6-97.

$$(q_i)''_{net} = J_i - G_i = J_i - \sum_{j=1}^{n} F_{i \to j} J_j \tag{6-115}$$

Surfaces with Known Net Heat Flux

Assume that the temperatures of all surfaces that form the enclosure are unknown, but the net heat flux of all surfaces are given. Equation 6-98 can be modified into the matrix form in terms of known quantities by eliminating the unknown E_b and retaining the known $(q)''_{net}$. The result for surface 1 of a three-surface enclosure is

$$(1 - F_{1 \to 1})J_1 + (-F_{1 \to 2})J_2 + (-F_{1 \to 3})J_3 = (q_1)''_{net} \tag{6-116}$$

Similar expressions can be written for the other two surfaces.

For a problem in which the temperatures are unknown and the net heat flux of each surface is given, Eq. 6-116 indicates that the off-diagonal elements of the matrix A in the matrix Eq. 6-102 are

$$a_{ij} = -F_{i \to j} \qquad (i \neq j) \tag{6-117a}$$

and the diagonal elements are

$$a_{ii} = 1 - F_{i \to i} \tag{6-117b}$$

The elements of the B matrix are

$$b_i = (q_i)''_{net} \tag{6-118}$$

Written in these forms the elements of both matrices A and B are known.

Once the elements of A and B are determined, the solution proceeds as follows. The matrix A is inverted and the resulting matrix is called C. The elements of the radiosity matrix are then determined by the equation

$$J = CB$$

The temperatures of the surfaces can be determined from a form of Eq. 6-95

$$E_{b_i} = \sigma T_i^4 = \frac{\rho_i}{\epsilon_i} (q_i)''_{net} + J_i$$

or

$$T_i = \left[\frac{\dfrac{\rho_i}{\epsilon_i} (q_i)''_{net} + J_i}{\sigma} \right]^{1/4} \tag{6-119}$$

If the ith surface is a black body ($\rho_i = 0$), the expression for its temperature reduces to

$$T_i = \left(\frac{J_i}{\sigma} \right)^{1/4}$$

The equations and techniques described above can also be used when a problem involves a mixture of boundary conditions. When some of the gray surfaces have known temperatures, but unknown heat fluxes, the individual rows in the A and B matricies are written in the form of Eqs. 6-106 and 6-107, respectively. The net heat flux for each of these surfaces is given by Eq. 6-114. When the surface has a known net heat flux but unknown temperature, an individual row of the A or B matrices is given by Eqs. 6-117 and 6-118, respectively. The temperature of the surface is given by Eq. 6-119.

The matrix method is a convenient and powerful technique when it is used to solve radiative heat transfer problems. When the technique is programmed for use on a digital computer using standard library sub-routines for matrix inversion, problems involving a large number of participating surfaces can be solved with a relatively simple program.

Since the matrix method is so helpful in solving radiation problems, we will devote some effort to developing a computer program based on the matrix method. The language used in the program is FORTRAN IV. The program is quite general and it can be used to calculate either surface temperature or the net heat flux of each participating surface in an enclosure. The program is written so that it can analyze a large number of isothermal, gray, opaque surfaces that form an enclosure. If the temperature of the surface is given, the program determines the net heat flux supplied to the surface. If the net heat flux is specified, the program calculates the temperature of the surface.

The program uses the subroutine called MATINV for matrix inversion. A copy of MATINV is given in Appendix E. The subroutine MATINV receives the matrix A and the number of surfaces forming the enclosure, N, as input and returns the inverse of A as the matrix C.

A copy of the radiation program based on the matrix method is shown below.

Program Listing for a Matrix-Method Solution to Radiation Problems

```
      DIMENSION A(50,50),B(50),C(50,50),EMIS(50),F(50,50),G(50),J(50),
     1 QNET(50),T(50)
      REAL J
      SIGMA=5.67E-08
      READ , N,L
      M=N-L
      LL=L+1
      READ , ((F(I,K),K=1,N),I=1,N),(EMIS(I),I=1,N),
     1 (T(I),I=1,L),(QNET(I),I=LL,N)
      DO 20 I=1,N
      DO 20 K=1,N
```

```
20    A(I,K) = -F(I,K)
      DO 60 I = 1,L
      IF (EMIS(I).EQ. 1.0) GO TO 40
      A(I,I) = 1.0 - F(I,I) + EMIS(I)/(1.0 - EMIS(I))
      GO TO 60
40    DO 50 K = 1,N
50    A(I,K) = 0.0
      A(I,I) = 1.0
60    CONTINUE
      DO 80 I = 1,L
      B(I) = SIGMA*T(I)**4
      IF (EMIS(I) .EQ. 1.0) GO TO 80
      B(I) = EMIS(I)*B(I)/(1.0 - EMIS(I))
      GO TO 80
80    CONTINUE
      DO 85 I = LL,N
      A(I,I) = 1.0 - F(I,I)
      B(I) = QNET(I)
85    CONTINUE
      CALL MATINV (A,N,C)
      DO 100 I = 1,N
      SUM = 0.0
      DO 90 K = 1,N
90    SUM = SUM + C(I,K)*B(K)
100   J(I) = SUM
      DO 120 I = 1,N
      SUM = 0.0
      DO 110 K = 1,N
110   SUM = SUM + F(I,K)*J(K)
120   G(I) = SUM
      DO 125 I = 1,L
125   QNET(I) = J(I) - G(I)
      DO 140 I = LL,N
      IF (EMIS(I) .EQ. 0.0) GO TO 130
      T(I) = (((1.0 - EMIS(I))*QNET(I)/EMIS(I) + J(I))/SIGMA)**0.25
      GO TO 140
130   T(I) = (J(I)/SIGMA)**0.25
140   CONTINUE
      WRITE (6,150)
150   FORMAT (1H ,20X,'RADIATION ANALYSIS BY MATRIX METHOD',//,
     1 1X,'SURFACE',1X,'EMISSIVITY',2X,'TEMPERATURE',5X,'QNET',6X,
     2 'RADIOSITY',2X,'IRRADIATION',/,22X,'(KELVIN)',5X,'(W/SQ.M)'
     3 ,5X,'(W/SQ.M)',4X,'(W/SQ.M)',/)
      WRITE (6,160) (I,EMIS(I),T(I),QNET(I),J(I),G(I),I = 1,N)
160   FORMAT (4X,I2,6X,F5.3,5X,F7.2,4X,E11.5,2X,E11.5,2X,E11.5)
      STOP
      END
```

The program is quite general and it can be applied to a large variety of radiation problems. The user can copy the program exactly and add the subroutine MATINV by copying it from Appendix E; or a local library subroutine could be used in place of MATINV to perform the inversion of A.

Input to the program and the order in which it must be entered into the data deck is as follows:

$N =$ an integer value equal to the total number of surfaces that form the enclosure

$L =$ an integer value equal to the number of surfaces that have known temperatures but unknown net heat fluxes

Either the temperature or the net heat flux must be given for all N surfaces. That is, $0 \leqslant L \leqslant N$, and when L is less than N, the program assumes that $(N - L)$ surfaces have given values for net heat flux and unknown temperatures.

$F(I, J) =$ a two-dimensional array of N^2 shape-factor values arranged in the order $F_{1 \to 1}, F_{1 \to 2}, \ldots, F_{1 \to n}, F_{2 \to 1}, \ldots, F_{2 \to n}, \ldots,$ $F_{n \to 1}, \ldots, F_{n \to n}.$

$EMIS(I) =$ a one-dimensional array of N emissivity values for all N surfaces

$T(I) =$ a one-dimensional array of L temperature values for the L surfaces that have known temperatures

$QNET(I) =$ a one-dimensional array of $(N - L)$ net heat flux values for the $(N - L)$ surfaces that have known net heat fluxes

The program assumes that the surfaces are numbered such that the first L surfaces have known temperatures and the remaining $(N - L)$ surfaces have known net heat flux values. For example, assume that an enclosure consists of 20 surfaces and the temperatures of only 5 of the surfaces are known. The remaining 15 surfaces have known net heat flux values. For this case $N = 20$ and $L = 5$. The first five surfaces are designated as those with known temperatures. The surface with known heat flux are designated as surfaces 6 through 20.

Symbols used in the program are compared with nomenclature used in the text in the table below. The table also includes the units that are used in the program.

Symbols Used In Radiation Program

SYMBOL		DEFINITION	UNITS
A(I, J)	a_{ij},	elements of matrix A, defined in Eq. 6-103	—
B(I)	b_i,	elements of matrix B, defined in Eq. 6-105	W/m^2
C(I, J)	c_{ij},	elements of matrix C, defined in Eq. 6-111	—
EMIS(I)	ϵ_i,	emissivity of all N surfaces	—
F(I, J)	$F_{i \to j}$	shape factors (total of N^2 values)	—
G(I)	G_i,	irradiation per unit area on all surfaces	W/m^2
J(I)	J_i,	radiosity per unit area for all surfaces	W/m^2
L		number of surfaces with known temperatures $(0 \leqslant L \leqslant N)$	—
N		total number of surfaces that form the enclosure	—
QNET(I)	$(q_i)''_{net}$,	the net heat flux to $(N - L)$ surfaces	W/m^2
SIGMA	σ	5.67×10^{-8}	$W/m^2 \cdot K^4$
T(I)	T_i,	the temperature of L surfaces	K

The program is used in the following two examples. The problems illustrate the matrix method and they also show how the program can be used to analyze problems that are more involved than those previously discussed.

Example 6-11. A long structural member on a spacecraft has a rectangular cross section as shown in the figure. Telemetry data indicate that the temperatures of the four surfaces are $T_1 = 200$ K, $T_2 = 300$ K, $T_3 = 400$ K, and $T_4 = 500$ K. The emissivities of the four surfaces are $\epsilon_1 = 0.2$, $\epsilon_2 = 0.3$, $\epsilon_3 = 0.4$, and $\epsilon_4 = 0.5$. Calculate the net radiative heat transfer per unit area of each of the four surfaces.

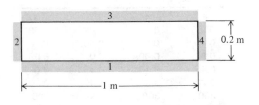

Solution: Convection is unimportant in outer space and we will assume that the conduction between the members is negligible. Therefore, the only mode of heat transfer among the four surfaces is radiation.

Determination of the heat-transfer rates for the four surfaces requires the simultaneous solution of four algebraic equations. We will use the general FORTRAN IV program just described to solve for the heat fluxes.

A total of 16 shape factors are needed as input values to the program. The shape factor for a rectangular geometry is given in Fig. 6-16.

$$F_{1 \to 3} = F_{3 \to 1} = 0.82$$

By symmetry

$$F_{1 \to 2} = F_{1 \to 4} = F_{3 \to 2} = F_{3 \to 4}$$

and by the enclosure relationship

$$F_{1 \to 2} + F_{1 \to 3} + F_{1 \to 4} = F_{1 \to 3} + 2F_{1 \to 4} = 1.0$$

Therefore,

$$F_{1 \to 4} = F_{1 \to 2} = \tfrac{1}{2}(1 - F_{1 \to 3}) = 0.09$$

Also,

$$F_{1 \to 1} = F_{2 \to 2} = F_{3 \to 3} = F_{4 \to 4} = 0$$

By reciprocity and symmetry

$$F_{2 \to 1} = F_{4 \to 1} = F_{2 \to 3} = F_{4 \to 3} = \frac{A_1}{A_2} F_{1 \to 2} = \frac{100}{20}(0.09) = 0.45$$

$$F_{2 \to 4} = F_{4 \to 2} = 1 - F_{2 \to 1} - F_{2 \to 3} = 1.0 - 0.45 - 0.45 = 0.10$$

When the shape factors are organized in an array $F_{i \to j}$, they are

	$j=1$	$j=2$	$j=3$	$j=4$
$i=1$	0	0.09	0.82	0.09
$i=2$	0.45	0	0.45	0.10
$i=3$	0.82	0.09	0	0.09
$i=4$	0.45	0.10	0.45	0

Notice that the enclosure relationship requires that each row of the array sum to unity.

The input data to the program is

Program Input for Example 6-11

4,	4		
0.0,	0.09,	0.82,	0.09
0.45,	0.0,	0.45,	0.10
0.82,	0.09,	0.0,	0.09
0.45,	0.10,	0.45,	0.0
0.20,	0.30,	0.40,	0.50
200.,	300.,	400.,	500.

The first row of data contains two integer values. The first value is N, the total number of surfaces that form the enclosure. The second value is L, the number of surfaces that have specified temperature. In this problem both values are 4.

The next four lines of data contain the 16 shape factor values given above. A free field FORMAT statement was used for input so any number of shape factors may be entered on a single line, but the user must be careful to enter the values in consecutive order $F_{1 \to 1}, F_{1 \to 2}, \ldots, F_{1 \to n}, F_{2 \to 1}, F_{2 \to 2}, \ldots$.

The next line contains the emissivity of the four surfaces in the order ϵ_1, ϵ_2, ϵ_3, and ϵ_4.

The final line of data contains the temperatures in kelvin for L of the surfaces, where the value of L is specified in the first line of input. In this problem $L=4$, and therefore the temperatures of the four surfaces are entered in the order T_1, T_2, T_3, and T_4. The number of temperature values must always equal L.

When $N = L$, as it does in this problem, we simply omit the $(q)''_{net}$ values because there are no surfaces with known net heat flux values. When $L=0$, the temperature data values are omitted and the final line of data consists of N values of $(q)''_{net}$.

The output of the program is shown below.

Program Output for Example 6-11

RADIATION ANALYSIS BY MATRIX METHOD

	EMISSIVITY	TEMPERATURE (KELVIN)	QNET (W/SQ.M)	RADIOSITY (W/SQ.M)	IRRADIATION (W/SQ.M)
1	.200	200.00	$-.26190E+03$	$.11383E+04$	$.14002E+04$
2	.300	300.00	$-.26654E+03$	$.10812E+04$	$.13477E+04$
3	.400	400.00	$.82610E+02$	$.13276E+04$	$.12450E+04$
4	.500	500.00	$.11630E+04$	$.23808E+04$	$.12177E+04$

The program output identifies each surface and lists its emissivity, temperature, net heat flux, radiosity, and irradiation. The units of each quantity are listed in the output.

We can easily check the output by comparing the values of $(q_i)''_{net}$ with Eq. 6-82.

Example 6-12. Two long, gray parallel planes are placed opposite each other as shown in the figure. The planes are diffuse surfaces and they are opaque. They are made of a poorly conducting material, and therefore there is a significant temperature gradient across both planes. Perfectly reflecting (refractory) surfaces are placed between the ends of both planes. The emissivity of the upper surface is 0.6 and the emissivity of the lower surface is 0.2.

Our analysis of radiation problems has always assumed that each participating surface is isothermal. In an attempt to satisfy this assumption, we will subdivide both planes into three strips of equal width. The strips are numbered from 1 to 6 as shown and the refractory surfaces are called surfaces 7 and 8. Local temperature measurements on the six subdivided areas of the planes result in the values

$$T_1 = T_3 = 400 \text{ K}$$

$$T_2 = 500 \text{ K}$$

$$T_4 = T_6 = 700 \text{ K}$$

$$T_5 = 800 \text{ K}$$

Determine the total net heat flux to the upper and lower planes and the temperatures of the two refractory end planes.

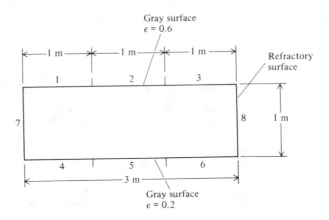

Solution: For this problem the number of surfaces that form the enclosure is eight and the amount of effort required to solve the problem by hand justifies the use of a computer program. We must first put the given data in the form of input information for the program described in Example 6-11. The surfaces are labeled so that the surfaces with known temperature are numbered first. The last surfaces are those which have known values of net heat flux. The refractory surfaces, numbered 7 and 8, have zero net heat flux:

$$(q_7)''_{net} = (q_8)''_{net} = 0$$

By using shape-factor algebra, the reciprocity relationship, the enclosure relationship, and the values given in Fig. 6-16, the shape-factor array $F_{i \rightarrow j}$

is

	$j=1$	$j=2$	$j=3$	$j=4$	$j=5$	$j=6$	$j=7$	$j=8$
$i=1$	0	0	0	0.42	0.20	0.05	0.29	0.04
$i=2$	0	0	0	0.20	0.42	0.20	0.09	0.09
$i=3$	0	0	0	0.05	0.20	0.42	0.04	0.29
$i=4$	0.42	0.20	0.05	0	0	0	0.29	0.04
$i=5$	0.20	0.42	0.20	0	0	0	0.09	0.09
$i=6$	0.05	0.20	0.42	0	0	0	0.04	0.29
$i=7$	0.29	0.09	0.04	0.29	0.09	0.04	0	0.16
$i=8$	0.04	0.09	0.29	0.04	0.09	0.29	0.16	0

The input to the program is:

Program Input for Example 6-12

8,	6						
0.0,	0.0,	0.0,	0.42,	0.20,	0.05,	0.29,	0.04
0.0,	0.0,	0.0,	0.20,	0.42,	0.20,	0.09,	0.09
0.0,	0.0,	0.0,	0.05,	0.20,	0.42,	0.04,	0.29
0.42,	0.20,	0.05,	0.0,	0.0,	0.0,	0.29,	0.04
0.20,	0.42,	0.20,	0.0,	0.0,	0.0,	0.09,	0.09
0.05,	0.20,	0.42,	0.0,	0.0,	0.0,	0.04,	0.29
0.29,	0.09,	0.04,	0.29,	0.09,	0.04,	0.0,	0.16
0.04,	0.09,	0.29,	0.04,	0.09,	0.29,	0.16,	0.0
0.6,	0.6,	0.6,	0.2,	0.2,	0.2,	0.0,	0.0
400.,	500.,	400.,	700.,	800.,	700.		
0.0,	0.0						

The first line contains two integers. The first value is the total number of surfaces and the second value is the number of surfaces with known temperatures. The next eight lines contain the $F_{i \to j}$ values. The next line of data contains the emissivities of the eight surfaces listed in order. The next-to-last line lists the temperatures of the six surfaces with known temperature listed in order. The final line of data gives the two values of known net heat flux for the two refractory surfaces.

The output of the program is given below.

Program Output for Example 6-12

		RADIATION ANALYSIS BY MATRIX METHOD			
SURFACE	EMISSIVITY	TEMPERATURE	QNET	RADIOSITY	IRRADIATION
		(KELVIN)	(W/SQ.M)	(W/SQ.M)	(W/SQ.M)
1	.600	400.00	$-.28190E+04$	$.33307E+04$	$.61498E+04$
2	.600	500.00	$-.19153E+04$	$.48205E+04$	$.67353E+04$
3	.600	400.00	$-.28190E+04$	$.33307E+04$	$.61498E+04$
4	.200	700.00	$.18814E+04$	$.60872E+04$	$.42060E+04$
5	.200	800.00	$.37905E+04$	$.80619E+04$	$.42713E+04$
6	.200	700.00	$.18814E+04$	$.60872E+04$	$.42060E+04$
7	0.000	547.11	0.	$.50802E+04$	$.50802E+04$
8	0.000	547.11	0.	$.50802E+04$	$.50802E+04$

The total net heat flux to the upper surface is

$$(q)''_{net} = (-0.28190 - 0.19153 - 0.28190) \times 10^4$$
$$= -7553 \text{ W/m}^2$$

The total net heat flux to the lower surface is

$$(q)''_{net} = (0.18814 + 0.37905 + 0.18814) \times 10^4$$
$$= 7553 \text{ W/m}^2$$

Notice that the total net heat flux to all surfaces is zero, as it should be to satisfy conservation of energy for the entire enclosure. The computer output shows the expected symmetry for surfaces 1 and 3 and for surfaces 4 and 6. The temperatures of both refractory surfaces are

$$T_7 = T_8 = 547.11 \text{ K}$$

which also indicates the expected symmetry for the two end surfaces.

6-8 RADIATION THROUGH ABSORBING, TRANSMITTING MEDIA

In the previous sections we assumed that all radiating surfaces were opaque so that none of the energy incident on the surface was transmitted through the surface. This assumption greatly simplifies radiation calculations. Many substances of engineering importance such as glasses, some plastics, and gases are not opaque, and we must therefore extend our analysis of radiation problems if we wish to account for transmission of radiant energy through such materials.

To develop a general technique, let us consider a simple case of radiant exchange between two very large gray opaque surfaces separated by a transmitting medium. Furthermore, let us assume that the partially absorbing medium is isothermal and gray and that it does not reflect any of the incident radiation. This assumption is particularly valid when dealing with gases, although glasses and other partly transparent solids can reflect a portion of the incident radiation. The geometry is shown in Fig. 6-31. The two solid surfaces are called surfaces 1 and 2, while the properties of the transmitting gas are denoted with a subscript g. Assume that the properties and temperatures of the two solid surfaces are known and that we wish to determine the radiative-heat-transfer rates between the three materials and the temperature of the gas.

Kirchhoff's law applied to the transmitting gray gas requires that

$$\alpha_g = \epsilon_g$$

and since the reflectivity of the medium is zero,

$$\tau_g = 1 - \alpha_g = 1 - \epsilon_g$$

We will derive the equations for the heat-transfer rates between the surfaces by developing the electric analog for the problem. The expressions

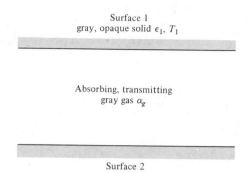

Surface 1
gray, opaque solid ϵ_1, T_1

Absorbing, transmitting
gray gas α_g

Surface 2
gray, opaque solid ϵ_2, T_2

Figure 6-31 Two infinite planes separated by an absorbing, transmitting gas.

for radiosity from surface 1 (Eq. 6-81) and net heat-transfer rate (Eq. 6-82) are not affected by the presence of the gas; therefore, the expression for the net heat-transfer rate to surface 1 is given by Eq. 6-83:

$$(q_1)_{\text{net}} = \frac{E_{b_1} - J_1}{\rho_1/\epsilon_1 A_1} \tag{6-120}$$

The irradiation on surface 1 is the sum of radiosity leaving surface 2 that is transmitted through the gas and the contribution due to emission from the gas:

$$G_1 = \tau_g J_2 + \epsilon_g E_{b_g} \tag{6-121}$$

The expression for $(q_1)_{\text{net}}$ is

$$(q_1)_{\text{net}} = A_1(J_1 - G_1) \tag{6-122}$$

and substitution of Eq. 6-121 into Eq. 6-122 gives

$$(q_1)_{\text{net}} = A_1\left(J_1 - \tau_g J_2 - \epsilon_g E_{b_g}\right)$$

which can be rearranged into the form

$$(q_1)_{\text{net}} = A_1\left[\tau_g(J_1 - J_2) + \epsilon_g(J_1 - E_{b_g})\right]$$

$$= \frac{J_1 - J_2}{1/A_1\tau_g} + \frac{J_1 - E_{b_g}}{1/A_1\epsilon_g} \tag{6-123}$$

A similar analysis for surface 2 will result in an expression for $(q_2)_{\text{net}}$ which is

$$(q_2)_{\text{net}} = \frac{E_{b_2} - J_2}{\rho_2/\epsilon_2 A_2} = \frac{J_2 - J_1}{1/A_2\tau_g} + \frac{J_2 - E_{b_g}}{1/A_2\epsilon_g} \tag{6-124}$$

Equations 6-120, 6-123, and 6-124 form the basis for the electric analog for the problem being considered. The electric analog is shown in Fig. 6-32.

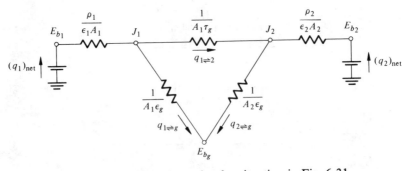

Figure 6-32 Electric analog for situation in Fig. 6-31.

The thermal circuit in Fig. 6-32 shows that the blackbody emissive power E_{b_g} of the gas is a floating potential, and therefore the temperature of the gas becomes a function of the gas properties and the temperature and surface properties of the two opaque surfaces. The gas is effectively a refractory substance because there is no external energy supply to the gas, or $(q_{net})_g = 0$.

Example 6-13. Two large gray plates ($\epsilon_1 = 0.2$, $T_1 = 600\,\mathrm{K}$, $\epsilon_2 = 0.6$, $T_2 = 800\,\mathrm{K}$) are separated by a gray gas ($\epsilon_g = 0.1$). Calculate:

a. The net heat-transfer rate to surface 1 and the net radiative-heat-transfer rate between the two opaque surfaces when the gas is present.
b. The temperature of the gas.
c. The heat-transfer rates in part (a) when the gas is replaced by a vacuum.

Solution

a. From the thermal circuit in Fig. 6-32, the expression for $(q_1)''_{net}$ in terms of known quantities is

$$(q_1)''_{net} = \frac{E_{b_1} - E_{b_2}}{\dfrac{\rho_1}{\epsilon_1} + \dfrac{\rho_2}{\epsilon_2} + \dfrac{(1/\tau_g)(2/\epsilon_g)}{1/\tau_g + 2/\epsilon_g}}$$

$$= \frac{(5.67 \times 10^{-8})(600^4 - 800^4)}{\dfrac{0.8}{0.2} + \dfrac{0.4}{0.6} + \dfrac{\left(\dfrac{1}{0.9}\right)\left(\dfrac{2}{0.1}\right)}{1/0.9 + 2/0.1}} = -2776\,\mathrm{W/m^2}$$

Surface 1 must be cooled at a rate of 2.776 kW/m² in order to maintain its temperature at 600 K. The value for $q_{1 \rightleftarrows 2}$ can be determined once values for J_1 and J_2 are known.

$$(q_1)''_{net} = -(q_2)''_{net} = \frac{E_{b_1} - J_1}{\rho_1/\epsilon_1} = \frac{J_2 - E_{b_2}}{\rho_2/\epsilon_2}$$

Substituting values for properties, $(q_1)''_{net}$, E_{b_1}, and E_{b_2} results in

$$J_1 = 18{,}450 \, \text{W/m}^2$$

$$J_2 = 21{,}370 \, \text{W/m}^2$$

The net radiative exchange between surfaces 1 and 2 is

$$q''_{1 \rightleftarrows 2} = \frac{J_1 - J_2}{1/\tau_g} = \frac{18{,}450 - 21{,}370}{1/0.9}$$

$$= -2628 \, \frac{\text{W}}{\text{m}^2}$$

b. The temperature of the gas is given by the equation

$$q''_{1 \rightleftarrows g} = (q_1)''_{net} - q''_{1 \rightleftarrows 2} = \frac{J_1 - E_{b_g}}{1/\epsilon_g}$$

$$-2776 + 2628 = \frac{18{,}450 - 5.67 \times 10^{-8} T_g^4}{1/0.1}$$

$$T_g = 770 \, \text{K}$$

c. When the gas is not present, the values for $(q_1)''_{net}$ and $q''_{1 \rightleftarrows 2}$ are equal and they are given by Eq. 6-88.

$$(q_1)''_{net} = q''_{1 \rightleftarrows 2} = \frac{E_{b_1} - E_{b_2}}{1 + \rho_1/\epsilon_1 + \rho_2/\epsilon_2} = \frac{(5.67 \times 10^{-8})(600^4 - 800^4)}{1 + 0.8/0.2 + 0.4/0.6}$$

$$= -2802 \, \text{W/m}^2$$

Notice that the net heat-transfer rate to surface 1 is reduced when the gas is present, owing to emission from the gas. The net radiative-heat-transfer rate between the surfaces is also decreased when the gas is present, because the gas absorbs some of the radiant energy traveling between the plates.

In Example 6-13 we considered two infinite plates for which $F_{1 \rightarrow 2}$ is unity and therefore the view factors did not enter the discussion. When the surfaces are finite in size, the thermal circuit must be modified to account for the shape factors not being equal to 1. For example, suppose that we wish to calculate the heat-transfer rates for two finite opaque parallel planes separated by a gray isothermal, nonreflecting gas. The expression

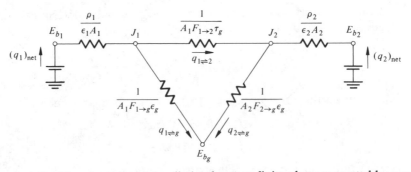

Figure 6-33 Electric analog for radiation between finite plates separated by a gas.

for the irradiation on surface 1 is

$$A_1 G_1 = A_2 F_{2 \to 1} \tau_g J_2 + A_g F_{g \to 1} \epsilon_g E_{b_g} \qquad (6\text{-}125)$$

Substitution of Eq. 6-125 into Eq. 6-122 results in

$$(q_1)_{\text{net}} = A_1 J_1 - A_2 F_{2 \to 1} \tau_g J_2 - A_g F_{g \to 1} \epsilon_g E_{b_g} \qquad (6\text{-}126)$$

After applying reciprocity and rearranging Eq. 6-126 can be transformed into

$$(q_1)_{\text{net}} = \frac{J_1 - J_2}{1 / A_1 F_{1 \to 2} \tau_g} + \frac{J_1 - E_{b_g}}{1 / A_1 F_{1 \to g} \epsilon_g} \qquad (6\text{-}127)$$

Equations 6-120 and 6-127 are in the form of Ohm's law and can be displayed as a thermal circuit as shown in Fig. 6-33. Note the similarity with the thermal circuit for infinite planes in Fig. 6-32.

6-9 RADIATIVE PROPERTIES OF GASES

Emitted radiation from gases is quite different from the emitted radiation from a solid substance. While the monochromatic emissive power for a solid substance is relatively continuous for all wavelengths, the emission and absorption of gases occurs in narrow-wavelength bands. The monochromatic absorptivity of water vapor, for example, is shown in Fig. 6-34.

The absorptance behavior of water vapor is typical of other gases. Emission and absorption is significant over small wavelength bands, while emission and absorption may drop to zero over adjacent bands. Gases with symmetrical molecular structures such as O_2, N_2, and H_2 are not strongly absorbing or emitting substances. For most situations that involve temperatures below values required to ionize these gases, emission from the symmetrical gases can be neglected. On the other hand, emission and absorption from gases with nonsymmetrical structures can be significant.

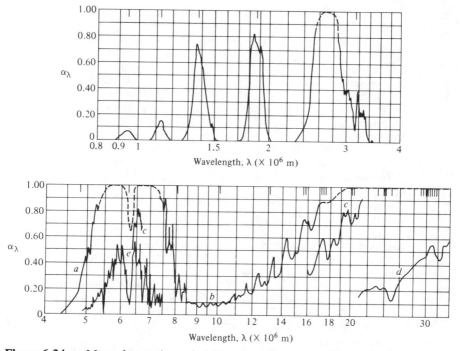

Figure 6-34   Monochromatic emissivity of water vapor. From Frank Kreith, *Principles of Heat Transfer*, 3rd ed., T. Y. Crowell (Harper & Row), 1973.

For engineering purposes, the most important nonsymmetrical gases are H_2O, CO_2, CO, SO_2, NH_3, and hydrocarbons. We will limit our discussion to just two of these gases; H_2O and CO_2. Properties for other gases can be found in References 14 through 19.

Another major difference between the radiative properties of opaque solids and gases is the fact that the shape of the gas affects the properties while the properties of an opaque solid are not a function of the shape of the solid. Thick layers of a gas will absorb more energy than a thin layer, and a thick layer of gas will transmit less energy than a thin layer of the gas. Therefore, in addition to specifying properties that fix the state of the gas such as the pressure and temperature, we must also specify a characteristic length of the gas mass before we can determine the radiating properties of the gas. The characteristic length of the gas is called the *mean beam length*. The mean beam length of various common gas shapes are given in Table 6-2. For gas shapes other than listed in Table 6-2, the mean beam length L may be approximated by

$$L = 3.6 \left(\frac{V}{A} \right) \qquad (6\text{-}128)$$

where V is the volume of the gas and A the surface area of the gas.

Table 6-2 Mean Beam Length of Various Gas Shapes

GEOMETRY	L
Sphere	$\frac{2}{3}$ (diameter)
Infinite cylinder	Diameter
Infinite parallel planes	2 (distance between planes)
Semiinfinite cylinder, radiating to center of base	Diameter
Right circular cylinder, height equal to diameter:	
Radiating to center of base	Diameter
Radiating to whole surface	$\frac{2}{3}$ (diameter)
Infinite cylinder of half-circular cross section radiating to spot in middle of flat side	Radius
Rectangular parallelepipeds:	
Cube	$\frac{2}{3}$ (edge)
1:1:4 radiating to 1×4 face	0.9 (shortest edge)
radiating to 1×1 face	0.86 (shortest edge)
radiating to all faces	0.891 (shortest edge)
Space outside infinite bank of tubes with centers on equilateral triangles:	
Tube diameter = clearance	3.4 (clearance)
Tube diameter = 1/2 (clearance)	4.44 (clearance)

Source: W. M. Rohsenow and J. P. Hartnett, eds., *Handbook of Heat Transfer*, McGraw-Hill Book Company, New York, 1973.

Hottel (Refs. 20, 21, and 22) has measured the emissivity of a number of gases as a function of temperature, total pressure, and mean beam length. Curves for the emissivity of water vapor and carbon dioxide are shown in Figs. 6-35 and 6-36, respectively. In these two figures the symbols P_{H_2O} and P_{CO_2} represent the partial pressures of the gases. The total pressure for both figures is 1 atm. When the total gas pressure is other than 1 atm the values from Figs. 6-35 and 6-36 must be multiplied by a correction factor. Correction factors C_{H_2O} and C_{CO_2} are plotted in Figs. 6-37 and 6-38, respectively. The emissivity of H_2O and CO_2 at a total pressure P_T other than 1 atm are given by the expressions

$$\left(\epsilon_{H_2O}\right)_{P_T} = C_{H_2O}\left(\epsilon_{H_2O}\right)_{P_T=1}$$

$$\left(\epsilon_{CO_2}\right)_{P_T} = C_{CO_2}\left(\epsilon_{CO_2}\right)_{P_T=1}$$

When both H_2O and CO_2 exist in a mixture, the emissivity of the mixture may be calculated by adding the emissivity of the gases determined by assuming that each gas exists alone and then subtracting a

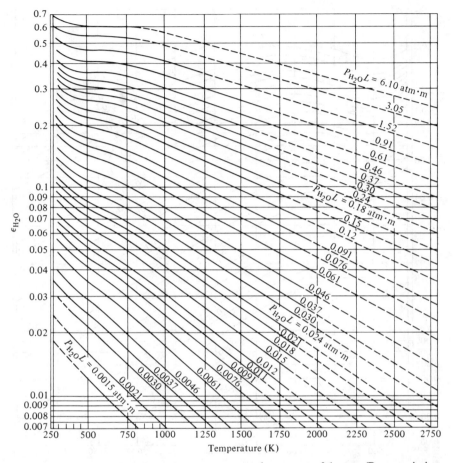

Figure 6-35 Emissivity of water vapor at a total pressure of 1 atm. (By permission from H. C. Hottel, in W. H. McAdams, *Heat Transmission*, 3rd ed., chap. 4, McGraw-Hill Book Company, New York, 1954.)

factor $\Delta\epsilon$, which accounts for emission in overlapping wavelength bands. The factor $\Delta\epsilon$ for H_2O and CO_2 is plotted in Fig. 6-39. The emissivity of a mixture of H_2O and CO_2 is therefore given by the expression

$$\epsilon_{mix} = C_{H_2O}(\epsilon_{H_2O})_{P_T=1} + C_{CO_2}(\epsilon_{CO_2})_{P_T=1} - \Delta\epsilon \qquad (6\text{-}129)$$

Example 6-14. Determine the emissivity of a gas mixture consisting of N_2, H_2O, and CO_2 at a temperature of 800 K existing in a sphere with diameter of 0.4 m. The partial pressures of the gases are $P_{N_2}=1$ atm, $P_{H_2O}=0.4$ atm, and $P_{CO_2}=0.6$ atm.

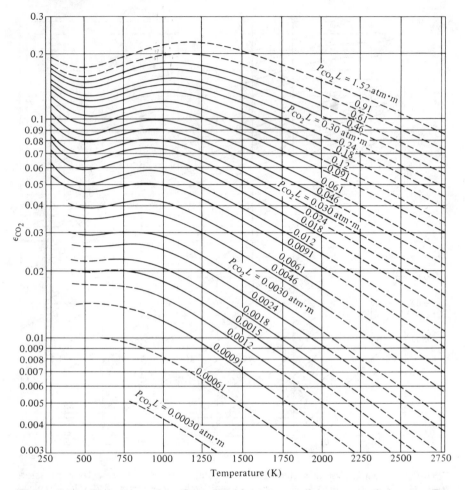

Figure 6-36 Emissivity of carbon dioxide at a total pressure of 1 atm. (By permission from H. C. Hottel, in W. H. McAdams, *Heat Transmission*, 3rd ed., chap. 4, McGraw-Hill Book Company, New York, 1954.)

Solution: From Table 6-2 the mean beam length for a spherical mass of gas is

$$L = (2/3)D = 0.27\,\text{m}$$

(If Eq. 6-128 is used, L is equal to 0.24 m.) Values for the parameters used in Figs. 6-35 and 6-36 are

$$T = 800\,\text{K}$$
$$P_{H_2O}L = 0.104\,\text{atm} \cdot \text{m}$$
$$P_{CO_2}L = 0.156\,\text{atm} \cdot \text{m}$$

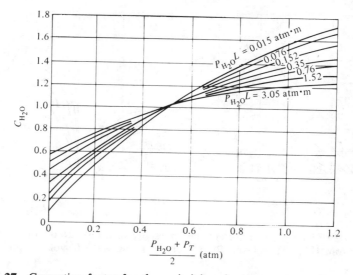

Figure 6-37 Correction factor for the emissivity of water vapor at pressures other than 1 atm. (From H. C. Hottel and R. B. Egbert, *Trans. AIChE*, vol. 38, pp. 531–565, 1942; R. B. Egbert, Sc.D. thesis in chemical engineering, MIT, Cambridge, Mass., 1941.)

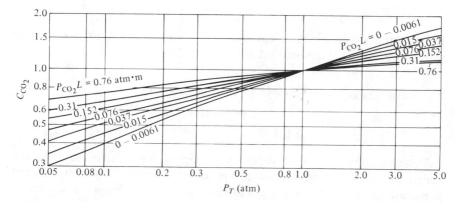

Figure 6-38 Correction factor for the emissivity of CO_2 at pressures other than 1 atm. (From H. C. Hottel and R. B. Egbert, *Trans. AIChE*, vol. 38, pp. 531–565, 1942.)

The emissivity values for 1 atm total pressure are

$$(\epsilon_{H_2O})_{P_T=1} = 0.15$$

$$(\epsilon_{CO_2})_{P_T=1} = 0.125$$

We will assume that the N_2 does not emit an appreciable amount of energy at 800 K.

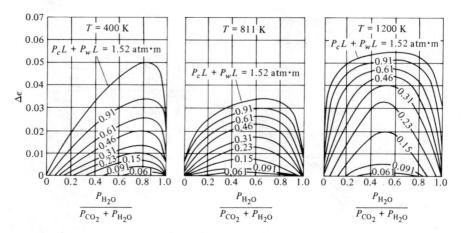

Figure 6-39 Factor $\Delta\epsilon$ to correct the emissivity of a mixture of water vapor and CO_2. (From H. C. Hottel and R. B. Egbert, *Trans. AIChE*, vol. 38, pp. 531–565, 1942.)

Since the total gas pressure is 2 atm, we must correct the 1-atm values for ϵ. Values for C_{H_2O} and C_{CO_2} from Figs. 6-37 and 6-38 are

$$C_{H_2O} = 1.62$$

$$C_{CO_2} = 1.12$$

Finally, the value for $\Delta\epsilon$ used to correct for emission in overlapping wavelength bands is determined from Fig. 6-39:

$$\Delta\epsilon = 0.005$$

The emissivity of the mixture is given by Eq. 6-129:

$$\epsilon_{mix} = 1.62 \times 0.15 + 1.12 \times 0.125 - 0.005$$

$$= 0.378$$

Determination of gas absorptivity is slightly more involved than the emissivity. The procedure involves the use of the emissivity charts shown previously, but modifying the parameters used in the charts. As an example, consider the case of water vapor at a temperature of T_{H_2O} when incident radiation is from a surface at a temperature of T_s. The absorptivity of the H_2O can be approximated by the equation

$$\alpha_{H_2O} = C_{H_2O}\epsilon'_{H_2O}\left(\frac{T_{H_2O}}{T_s}\right)^{0.45} \tag{6-130}$$

where the value for C_{H_2O} comes from Fig. 6-37 and ϵ'_{H_2O} is the value for the emissivity of water vapor from Fig. 6-35 evaluated at a temperature of T_s and a pressure/mean beam length product of $P_{H_2O}L(T_s/T_{H_2O})$.

The value for absorptivity of CO_2 is determined in a similar procedure according to the equation

$$\alpha_{CO_2} = C_{CO_2} \epsilon'_{CO_2} \left(\frac{T_{CO_2}}{T_s} \right)^{0.65} \tag{6-131}$$

where the value for C_{CO_2} is given in Fig. 6-38 and the value for ϵ'_{CO_2} is evaluated from Fig. 6-36 at $P_{CO_2} L (T_s / T_{CO_2})$.

When water vapor and carbon dioxide are present in a mixture, the absorptivity is

$$\alpha_{mix} = \alpha_{H_2O} + \alpha_{CO_2} - \Delta\alpha$$

where α_{H_2O} and α_{CO_2} are determined in Eqs. 6-130 and 6-131, respectively, and $\Delta\alpha = \Delta\epsilon$, which is evaluated from Fig. 6-39 at a temperature of T_s.

Example 6-15. Determine the absorptivity of a mixture of O_2 and water vapor with a total pressure of 2.0 atm and a temperature of 400 K. The mean beam length of the gases is 1.5 m and the incident radiation is emitted from a surface at 800 K. The partial pressure of the H_2O is 0.2 atm.

Solution: We will assume that the oxygen does not absorb an appreciable amount of the incident energy and the absorptivity of the mixture is equal to that of the water vapor alone. The absorptivity of the H_2O is given by Eq. 6-130.

$$\alpha_{H_2O} = C_{H_2O} \epsilon'_{H_2O} \left(\frac{T_{H_2O}}{T_s} \right)^{0.45}$$

The parameters used to determine C_{H_2O} and ϵ'_{H_2O} are

$$P_{H_2O} L = 0.3 \text{ atm} \cdot \text{m}$$

$$\tfrac{1}{2}(P_T + P_{H_2O}) = 1.1 \text{ atm}$$

$$P_{H_2O} L \left(\frac{T_s}{T_{H_2O}} \right) = 0.6 \text{ atm} \cdot \text{m}$$

From Fig. 6-37,

$$C_{H_2O} = 1.45$$

and from Fig. 6-35,

$$\epsilon'_{H_2O} = 0.33$$

The absorptivity of the water vapor is, therefore,

$$\alpha_{H_2O} = 1.45 \times 0.33 \left(\frac{400}{800} \right)^{0.45} = 0.350$$

6-10 SOLAR RADIATION

The sun is a small star in our galaxy which produces energy by a fusion reaction. It is estimated that the power radiated by the sun is 3.8×10^{26} W, of which about 1.7×10^{17} W is intercepted by the earth. The average flux of solar energy impinging on the outer fringes of the earth's atmosphere at the mean distance between the earth and the sun of 1.495×10^{11} m (1 astronomical unit) is called the solar constant. The solar constant is approximately 1353 W/m². Owing to the fact that the motion of the earth around the sun is elliptical, the solar radiation incident on the atmosphere varies with time of year, as shown in Fig. 6-40. Early in January, when the sun is closest to the earth, extraterrestrial solar radiation increases to 1.43 kW/m², and during June, when it is farthest away, it decreases to 1.33 kW/m².

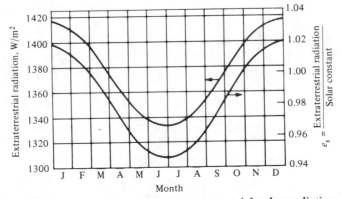

Figure 6-40 Variation of the ratio of extraterrestrial solar radiation to the solar constant, and of the extraterrestrial radiation, with time of year.

The sun radiates approximately as a black body at 5762 K. For more precise calculations, Table 6-3 gives the NASA standard spectral irradiance at the mean sun-to-earth distance on the outer fringes of the atmosphere. Approximately 99 percent of the total direct radiation emitted by the sun falls at wavelengths between 0.1 and 4 μm. However, when solar radiation passes through the atmosphere, some energy is scattered and some is absorbed in the atmosphere. The solar energy arriving at the earth's surface is therefore composed of a direct (or beam) and a diffuse component. The sum of the direct and diffuse component is called the *total* (or *global*) solar radiation and the rate at which solar energy is received on a horizontal surface is called *insolation*.

Table 6-3 Extraterrestrial Solar Irradiance[a]

λ (μm)	E_λ^b (W/m²·μm)	D_λ^c (%)	λ (μm)	E_λ (W/m²·μm)	D_λ (%)	λ (μm)	E_λ (W/m²·μm)	D_λ (%)
0.115	0.007	1×10^{-4}	0.43	1639	12.47	0.90	891	63.37
0.14	0.03	5×10^{-4}	0.44	1810	13.73	1.00	748	69.49
0.16	0.23	6×10^{-4}	0.45	2006	15.14	1.2	485	78.40
0.18	1.25	1.6×10^{-3}	0.46	2066	16.65	1.4	337	84.33
0.20	10.7	8.1×10^{-3}	0.47	2033	18.17	1.6	245	88.61
0.22	57.5	0.05	0.48	2074	19.68	1.8	154	91.59
0.23	66.7	0.10	0.49	1950	21.15	2.0	103	93.49
0.24	63.0	0.14	0.50	1942	22.60	2.2	79	94.83
0.25	70.9	0.19	0.51	1822	24.01	2.4	62	95.86
0.26	130	0.27	0.52	1833	25.38	2.6	48	96.67
0.27	232	0.41	0.53	1842	26.74	2.8	39	97.31
0.28	222	0.56	0.54	1783	28.08	3.0	31	97.83
0.29	482	0.81	0.55	1725	29.38	3.2	22.6	98.22
0.30	514	1.21	0.56	1695	30.65	3.4	16.6	98.50
0.31	689	1.66	0.57	1712	31.91	3.6	13.5	98.72
0.32	830	2.22	0.58	1715	33.18	3.8	11.1	98.91
0.33	1059	2.93	0.59	1700	34.44	4.0	3.5	99.06
0.34	1074	3.72	0.60	1666	35.68	4.5	5.9	99.34
0.35	1093	4.52	0.62	1602	38.10	5.0	3.8	99.51
0.36	1068	5.32	0.64	1544	40.42	6.0	1.8	99.72
0.37	1181	6.15	0.66	1486	42.66	7.0	1.0	99.82
0.38	1120	7.00	0.68	1427	44.81	8.0	0.59	99.88
0.39	1098	7.82	0.70	1369	46.88	10.0	0.24	99.94
0.40	1429	8.73	0.72	1314	48.86	15.0	0.0048	99.98
0.41	1751	9.92	0.75	1235	51.69	20.0	0.0015	99.99
0.42	1747	11.22	0.80	1109	56.02	50.0	0.0004	100.00

[a]Solar constant = 1353 W/m².

[b]E_λ is the solar spectral irradiance averaged over a small bandwidth centered at λ.

[c]D_λ is the percentage of the solar constant associated with wavelengths shorter than λ.

Source: Adapted from M. P. Thekaokaw, "Solar Energy Outside the Earth's Atmosphere," Solar Energy, vol. 14, pp. 109–127, 1973; by permission of the publishers.

As solar radiation passes through the earth's atmosphere, its intensity is reduced by:

1. Dry air molecular absorption and scattering, called *Rayleigh scattering*.
2. Absorption and scattering from dust.
3. Selective absorption by water vapor, carbon monoxide, and carbon dioxide absorption.
4. Reflection and absorption in cloud layers.

Figure 6-41 illustrates these processes for clear sky conditions; Fig. 6-42 shows the spectrum of a blackbody at 5762 K and the spectral distribution of insolation on the top of the atmosphere and on the earth's surface for an air mass of 1. The air mass is the path length of radiation through the atmosphere and equals unity at sea level when the sun is directly overhead. The exact amount of radiation received at the earth depends on the height above sea level and the sun's position in the sky relative to the point of observation, which determines the distance through the atmosphere that the radiation must penetrate before reaching the earth. At midlatitudes when the sky is clear, the radiation received on a surface placed perpendicular to the sun within ± 4 h of noon will average about 70% of the solar constant (i.e., approximately 1 kW/m^2).

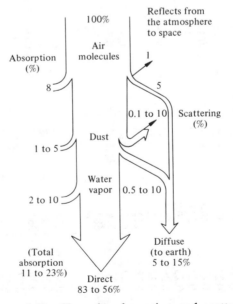

Figure 6-41 Clear sky absorption and scattering. (All values are in percent of incoming flux.)

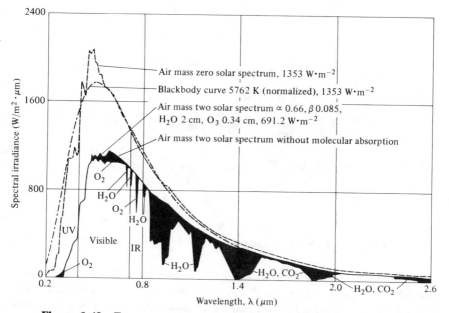

Figure 6-42 Extraterrestrial solar radiation spectral distribution; also shown are equivalent blackbody and atmosphere-attenuated spectra.

Example 6-16. Calculate the fraction of solar radiation within the visible part of the spectrum (i.e., between 0.40 and 0.75 μm).

Solution: The first column in Table 6-3 gives the wavelength. The second column gives the averaged solar spectral irradiance in a band centered at the wavelength in the first column. The third column, D_λ, gives the percentage of solar total radiation at wavelengths shorter than the value of λ in the first column. At a value of 0.40 μm, 8.7% of the total radiation occurs at shorter wavelength. At a wavelength of 0.75 μm, 51.7% of the radiation occurs at shorter wavelengths. Consequently, 43% of the total radiation lies within the band between 0.40 and 0.75 μm, and the total energy received outside the earth's atmosphere within that spectral range is 582 W/m^2.

In order to calculate the solar radiation incident on a surface of given inclination and orientation, it is necessary to define some basic terms. The significance of these terms will become clear as one is working a given specific problem.

The *altitude*, α, of the sun is the angle between a direct ray from the sun and a horizontal surface at a particular place on earth (see Fig. 6-43). The altitude at a given day of the year and time of day varies with location on earth.

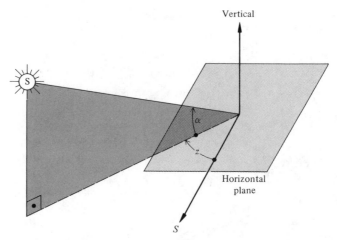

Figure 6-43 Solar altitude angle α and solar azimuth angle α_s.

The *azimuth* of the sun, a_s, is the angle between the horizontal component of a ray from the sun and a line toward the south in the northern hemisphere. It is measured in angular displacement east and west of south in the horizontal plane (see Fig. 6-43).

The *declination* of the sun, δ_s, is the angular displacement from the plane of the earth's equator. As shown in Figs. 6-44 and 6-45, because the axis of rotation of the earth is tilted at an angle of $23\frac{1}{2}$ degrees relative to the axis of the plane in which it orbits around the sun, the value of the declination will vary between $+23\frac{1}{2}$ degrees and $-23\frac{1}{2}$ degrees throughout the year.

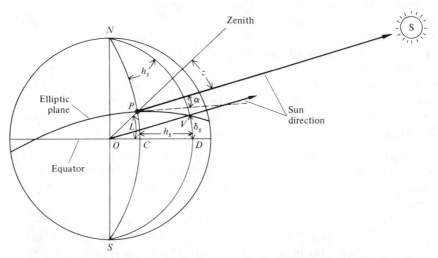

Figure 6-44 Definition of solar-hour angle h_s (CND), solar declination δ_s (VOD), and latitude L (POC); P, location of observer on earth; S, sun.

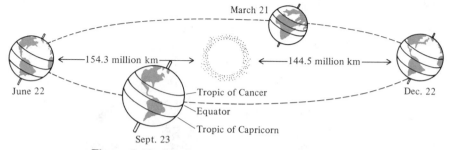

Figure 6-45 Motion of the earth around the sun.

Declination is usually expressed in radians (or degrees) north or south of the equator. Declinations north of the equator (i.e., during the summer in the northern hemisphere) are positive.

The *ecliptic plane* is the plane in which the sun moves during a day as seen by a stationary observer at P in Fig. 6-44.

Sun time is the time in hours before or after noon, with noon being defined as the time when the sun is highest in the sky.

The *solar hour angle*, h_s, is the angular displacement of the sun from noon (1 h corresponds to $\pi/12$ rad or 15 minutes of angular displacement). Values east of due south (i.e., morning values) are positive.

The *latitude* of a place on earth, L, is its angular displacement above or below the plane of the equator measured from the center of the earth, as shown in Fig. 6-44. It is positive north of the equator.

Longitude is the angle which the semiplane through the poles at a particular place on the surface of the earth makes with a plane through Greenwich. The semiplane through Greenwich is an arbitrary zero and the line it makes in cutting the earth's surface is called the *Greenwich meridian.* Longitude is measured east and west of Greenwich. Latitude and longitude are coordinates that locate a point on the surface of the earth.

The *zenith angle*, z, is the angle subtended by the zenith and the line of sight of the sun. It is the compliment of the altitude.

The availability of solar energy on earth depends on latitude, season, and weather. The reason for the dependency of the first two variables becomes apparent by observing the trajectory of the earth around the sun, shown schematically in Fig. 6-45. The path of the earth around the sun is an ellipse, with the sun at one focus. The distance between the sun and the earth varies throughout the year. The earth is closest to the sun in December (1.445×10^{11} m) and farthest in June (1.543×10^{11} m). As shown in Fig. 6-40, this relatively small variation in distance makes an appreciable difference in the insolation, which is a function of the square of the distance.

Figure 6-46 shows the relationship between the earth and the sun as seen from the point of view of an observer fixed on earth with the sun as the

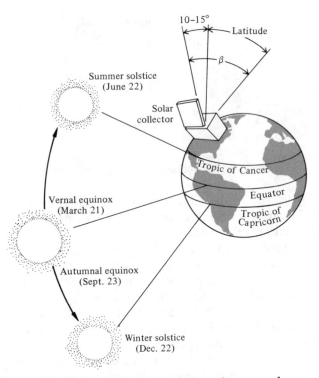

Figure 6-46 Seasonal relations between the sun and earth and solar collector tilt angle β.

moving body. From this viewpoint, often called the *Ptolomaic view*, the sun moves farthest to the north on June 22, when it is directly overhead at the Tropic of Cancer. The autumnal and vernal equinoxes occur when the sun crosses the equator on September 23 and March 21, respectively. The sun is never directly overhead anywhere in the continental United States, but reaches a maximum solar altitude at the summer solstice.

The solar altitude, which is the angle the sun rises above the horizon, is important for two reasons. First, at higher solar altitudes the solar radiation travels a shorter distance in traversing the atmosphere, whereas at lower solar altitudes the radiation from the sun is forced to travel through a much larger air mass. The attenuating effects of the intervening air mass reduce the solar insolation, and higher solar altitudes also provide more daylight hours during which solar energy can be collected. The relation between season and solar altitude is illustrated schematically in Fig. 6-47, where the solar altitude and insolation are shown as a function of time at 40 degrees latitude for January, September, and June. Figure 6-48 shows the annual variation in daily solar energy received by a horizontal surface located at selected latitudes in the northern hemisphere outside the atmosphere.

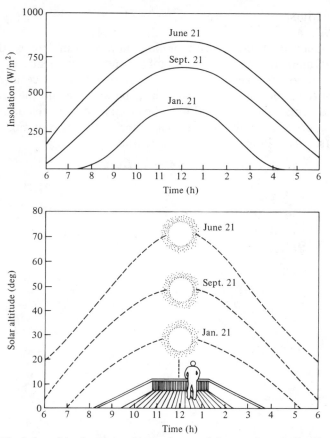

Figure 6-47 Solar altitude (June, September, and January) and cloudless insolation for a horizontal surface located at 40°N latitude versus time during a day.

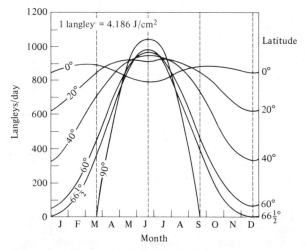

Figure 6-48 Annual variation in daily insolation at selected latitudes in the northern hemisphere. 1 langley = 4.186 J/cm². (From A. N. Strahler, *The Earth Sciences,* Harper & Row, Publishers, New York, 1963.)

The influence of the relative motion between sun and earth can be expressed mathematically in terms of α and z by means of spherical geometry to obtain the insolation at a horizontal surface at the outer fringes of the atmosphere. However, the solar altitude and the azimuth are not fundamental quantities and must be calculated from relations containing the hour angle, latitude, and solar declination. From Fig. 6-44 one obtains (see Ref. 25 for a complete derivation)

$$\sin \alpha = \sin L \sin \delta_s + \cos L \cos \delta_s \cos h_s \qquad (6\text{-}132)$$

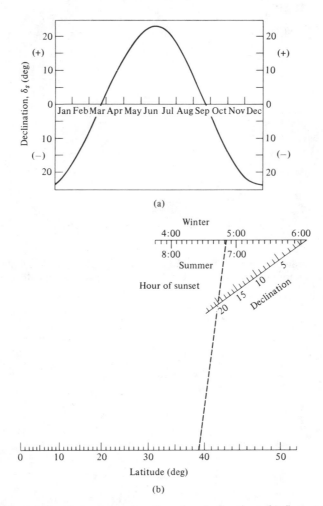

(a)

(b)

Figure 6-49 (a) Graph to determine the solar declination. (b) Sunset nomograph example showing determination of sunset time for summer (7:08 P.M.) and winter (4:52 P.M.) when the latitude is 39°N and the solar declination angle is 20°. (From A. Whillier, "Solar Radiation Graphs," *Solar Energy*, vol. 9, p. 164. Copyright 1965 by Pergamon Press. By permission of the publishers.)

Table 6-4   Summary Solar Emphemeris[a]

DATE		DECLI-NATION °	'	EQUATION OF TIME min	s	DATE		DECLI-NATION °	'	EQUATION OF TIME min	s
Jan.	1	−23	4	−3	14	Feb.	1	−17	19	−13	34
	5	22	42	5	6		5	16	10	14	2
	9	22	13	6	50		9	14	55	14	17
	13	21	37	8	27		13	13	37	14	20
	17	20	54	9	54		17	12	15	14	10
	21	20	5	11	10		21	10	50	13	50
	25	19	9	12	14		25	9	23	13	19
	29	18	8	13	5						
Mar.	1	−7	53	−12	38	Apr.	1	+4	14	−4	12
	5	6	21	11	48		5	5	46	3	1
	9	4	48	10	51		9	7	17	1	52
	13	3	14	9	49		13	8	46	−0	47
	17	1	39	8	42		17	10	12	+0	13
	21	−0	5	7	32		21	11	35	1	6
	25	+1	30	6	20		25	12	56	1	53
	29	3	4	5	7		29	14	13	2	33
May	1	+14	50	+2	50	June	1	+21	57	+2	27
	5	16	2	3	17		5	22	28	1	49
	9	17	9	3	35		9	22	52	1	6
	13	18	11	3	44		13	23	10	+0	18
	17	19	9	3	44		17	23	22	−0	33
	21	20	2	3	34		21	23	27	1	25
	25	20	49	3	16		25	23	25	2	17
	29	21	30	2	51		29	23	17	3	7
July	1	+23	10	−3	31	Aug.	1	+18	14	−6	17
	5	22	52	4	16		5	17	12	5	59
	9	22	28	4	56		9	16	6	5	33
	13	21	57	5	30		13	14	55	4	57
	17	21	21	5	57		17	13	41	4	12
	21	20	38	6	15		21	12	23	3	19
	25	19	50	6	24		25	11	2	2	18
	29	18	57	6	23		29	9	39	1	10
Sept.	1	+8	35	−0	15	Oct.	1	−2	53	+10	1
	5	7	7	+1	2		5	4	26	11	17
	9	5	37	2	22		9	5	58	12	27
	13	4	6	3	45		13	7	29	13	30
	17	2	34	5	10		17	8	58	14	25
	21	+1	1	6	35		21	10	25	15	10
	25	−0	32	8	0		25	11	50	15	46
	29	2	6	9	22		29	13	12	16	10
Nov.	1	−14	11	+16	21	Dec.	1	−21	41	+11	16
	5	15	27	16	23		5	22	16	9	43
	9	16	38	16	12		9	22	45	8	1
	13	17	45	15	47		13	23	6	6	12
	17	18	48	15	10		17	23	20	4	17
	21	19	45	14	18		21	23	26	2	19
	25	20	36	13	15		25	23	25	+0	20
	29	21	21	11	59		29	23	17	−1	39

[a]Since each year is 365.25 days long, the precise value of declination varies from year to year. *The American Ephemeris and Nautical Almanac*, published each year by the U.S. Government Printing Office contains precise values for each day of each year.

and

$$\sin \alpha_s = \frac{\cos \delta_s \sin h_s}{\cos \alpha} \tag{6-133}$$

The declination of the sun is shown in the nomograph of Fig. 6-49 as a function of time of year, but for more accurate values of δ_s a solar ephemeris should be used. Table 6-4 is an abbreviated version of an ephemeris. The nomograph also gives the number of hours of sun during a day. For engineering calculations the declination can be taken as constant during any one day.

Example 6-17. Calculate the solar altitude and azimuth angles at 10 A.M. solar time on December 25, Denver, Colorado.

Solution: Denver is at 40°N latitude. From Fig. 6-49(a), $\delta_s = -23.5°$ on December 25. The hour angle in Fig. 6-49(b), h_s, is $+30°$ at 2 h before noon. Substituting in Eqs. 6-132 and 6-133 gives

$$\alpha = 20.6° \quad \text{and} \quad \alpha_s = +29.3°$$

Tilted Surfaces

The amount of beam radiation intercepted by a surface, $I_{b,c}$, depends on the *incidence angle*, i, defined as the angle between the surface-normal and the rays of the sun: consider first a south-facing surface in the northern hemisphere, tilted at an angle β from the horizontal. The insolation can be

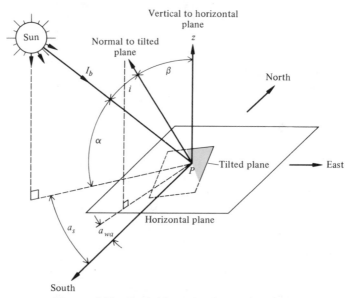

Figure 6-50 Definition of solar and surface angles for a non-south-facing tilted surface.

divided into two components, one perpendicular and one parallel to the tilted surface. Only the perpendicular component impinges on the surface, and if the incidence angle between the surface normal to the collector and the direction of the sun is i as shown in Fig. 6-50, the effective component of insolation $I_{b,c}$ is given by

$$I_{b,c} = I_b \cos i \qquad (6\text{-}134)$$

where I_b is the beam, or direct, component of solar radiation and

$$\cos i = \sin \delta_s \sin (L - \beta) + \cos \delta_s \cos (L - \beta) \cos h_s \qquad (6\text{-}135)$$

where L is the latitude.

If a non-south-facing tilted surface in the northern hemisphere faces a direction other than due south, Eqs. 6-136 and 6-137 are used to calculate the incidence angle i:

$$\cos i = \cos (a_s - a_{wa}) \cos \alpha \sin \beta + \sin \alpha \cos \beta \qquad (6\text{-}136)$$

where

a_s = azimuth of the sun (i.e., the angle of the sun measured in the horizontal plane westward from due south)

a_{wa} = azimuth angle of the normal to the insolated surface, measured in the same way as solar azimuth

$$a_s = \sin^{-1} \left(\frac{\cos \delta_s \sin h_s}{\cos \alpha} \right) \qquad (6\text{-}137)$$

A convenient way to use the results of the preceding equations is by means of sun-path diagrams such as those presented in the *Smithsonian Meteorological Tables*. One of these diagrams is reproduced in Fig. 6-51. An application of these diagrams is given in the following example.

Example 6-18. Find the altitude angle and the azimuth at 10 A.M. solar time on February 23 at 30°N latitude.

Solution: The latitude for each sun-path diagram is given in the figure legend. Fig. 6-51 is for a latitude of 30°N. The time of year can be related to a given sun-path line by reference to the table inset in the diagram. For February 23 the path line labeled "−10°" depicts the motion of the sun. Following this path from the noon position (i.e., due south) to the right to the sun time line for 10 A.M. locates the azimuth angle a_s at 40°E of south. To determine the altitude of the sun (i.e., its elevation above the horizon), note that the intersection of the −10° line and the 10 A.M. line lies at 41°.

The total unattenuated radiation incident during a specified time period Δt on a surface I_{tot}, can be obtained by integrating Eq. 6-134:

$$I_{tot} = \int_0^{\Delta t} I_b(t) \cos i(t) \, dt \qquad (6\text{-}138)$$

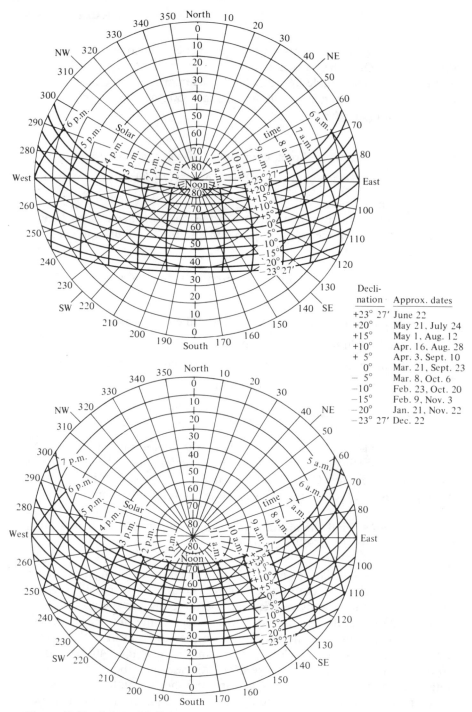

| Decli- | |
nation	Approx. dates
+23° 27'	June 22
+20°	May 21, July 24
+15°	May 1, Aug. 12
+10°	Apr. 16, Aug. 28
+ 5°	Apr. 3, Sept. 10
0°	Mar. 21, Sept. 23
− 5°	Mar. 8, Oct. 6
−10°	Feb. 23, Oct. 20
−15°	Feb. 9, Nov. 3
−20°	Jan. 21, Nov. 22
−23° 27'	Dec. 22

Figure 6-51 Sun-path diagram for 30° north latitude showing altitude and azimuth angles. (From Jan F. Kreider and Frank Kreith, *Solar Heating and Cooling*, rev. 1st ed., McGraw-Hill, New York, 1977.)

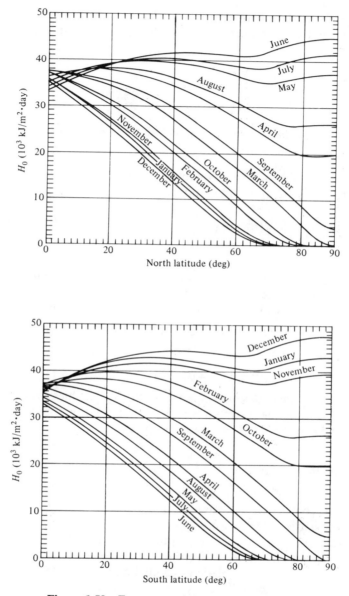

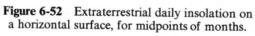

Figure 6-52 Extraterrestrial daily insolation on a horizontal surface, for midpoints of months.

For extraterrestrial or unattenuated radiation $I_b(t)$ is equal to the solar constant times the orbital eccentricity factor e_s from Fig. 6-40:

$$I_b(t) = e_s(t) \cdot I_0 \qquad (6\text{-}139)$$

where

$$e_s(t) = 1 + 0.034 \cos \frac{2\pi n}{365} \qquad (6\text{-}140)$$

$n =$ day number counted from the first day of the year

For a single day, Eq. 6-138 can be integrated for a horizontal surface with the aid of Eqs. 6-132 and 6-140, and we get

$$I_{\text{day},n} = \frac{24}{\pi} I_0 \left[(1 + 0.034) \cos \left(\frac{2\pi n}{365} \right) \right] (\cos L \cos \delta_s \sin h_{ss} + h_{ss} \sin L \sin \delta_s)$$

$$(6\text{-}141)$$

where I_0 is the solar constant (per hour) and h_{ss} is the sunset hour angle at zero altitude in radians on the nth day.

Referring to Eq. 6-132, h_{ss} can be evaluated from the relation

$$\sin \alpha = 0 = \sin L \sin \delta_s + \cos L \cos \delta_s \cos h_{ss} \qquad (\alpha = 0) \qquad (6\text{-}142)$$

or

$$h_{ss}(\alpha = 0) = h_{sr}(\alpha = 0) = \cos^{-1}(-\tan L \tan \delta_s) \qquad (6\text{-}143)$$

(Note that for tilted surfaces sunrise occurs sometimes when $i = 90°$.) Figure 6-52 gives the extraterrestrial daily insolation on a horizontal surface for the midpoint of each month of the year, H_0.

Solar Radiation on Earth

The availability of solar radiation on earth is important for the design of solar energy conversion systems. As mentioned previously, the insolation reaching the earth depends not only on time of year, location, and time of day, but also on the weather. It is practically impossible to predict the insolation for a given time, but from past experience and available data it is possible to predict the insolation expected on the average during a given time period (e.g., a month or a day).

For many years the U.S. National Weather Service (NWS) has operated numerous stations which have recorded the total insolation, $\overline{H}_h$, on a horizontal surface by means of an instrument called a pyranometer. These results have been averaged and have been published in the form of *Solar Maps*, giving lines of constant mean daily solar radiation received during a given month in the continental United States. Figures 6-53 and 6-54 are examples of these solar maps for the months of February and June,

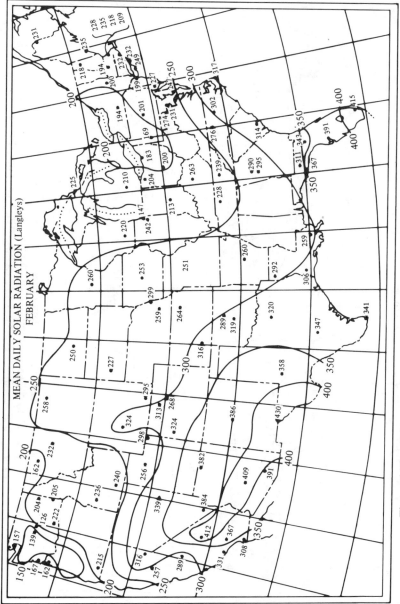

MEAN DAILY SOLAR RADIATION (Langleys)
FEBRUARY

Figure 6-53 Solar map giving constant mean daily solar radiation in the continental United States for February. (From Jan F. Kreider and Frank Kreith, *Solar Heating and Cooling*, rev. 1st ed., McGraw-Hill, New York, 1977.)

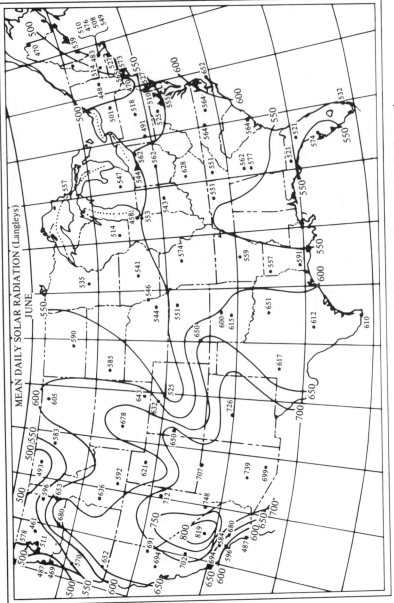

Figure 6-54 Solar map giving constant mean daily solar radiation in the continental United States for June. (From Jan F. Kreider and Frank Kreith, *Solar Heating and Cooling*, rev. 1st ed., McGraw-Hill, New York, 1977.)

respectively. The unit of radiation used by NWS is the *langley* (la):

$$1 \text{ la} = 4.186 \text{ J/cm}^2$$

For the engineering design of solar conversion systems it is necessary to know not only the total insolation, but also how much of the total is beam and how much is diffuse. This is a very complex problem, but Lui and Jordan (Ref. 26) have developed a simple method by which the total insolation measured by the NWS can be decomposed into the beam and diffuse components. The key parameter for this decomposition is the monthly clearness index, $\overline{K}_T$, defined as

$$\overline{K}_T = \frac{\overline{H}_h}{\overline{H}_0} \tag{6-144}$$

where

$\overline{H}_h$ = average monthly horizontal terrestrial insolation

$\overline{H}_0$ = average monthly extraterrestrial horizontal radiation evaluated for middle of month by Eq. 6-141

Available solar data shown in Fig. 6-55 indicate that the ratio of the average monthly diffuse radiation, $\overline{D}_h$, to the average monthly to total radiation, $\overline{H}_h$, is a function of the monthly clearness index. Klein (Ref. 27)

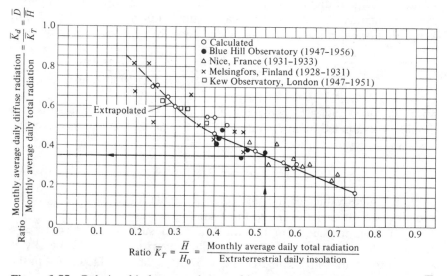

Figure 6-55 Relationship between the monthly average daily diffuse radiation, $\overline{D}$, and the monthly average daily total radiation, $\overline{H}$, for horizontal surfaces. (From Benjamin Liu and Richard C. Jordan, "Availability of Solar Energy for Flat-Plate Solar Heat Collectors," ASHRAE GRP170, 1977.)

found that the function can be expressed empirically as

$$\frac{\overline{D}_h}{\overline{H}_h} = 1.39 - 4.03\,\overline{K}_T + 5.53\,\overline{K}_T^{\,2} - 3.11\,\overline{K}_T^{\,3} \qquad (6\text{-}145)$$

if the solar constant is taken at 1.394 kW/m^2.

Once the diffuse radiation component is known, the average monthly beam component on a horizontal surface, $\overline{B}_h$, can be calculated from the relation

$$\overline{B}_h = \overline{H}_h - \overline{D}_h \qquad (6\text{-}146)$$

For a tilted surface the beam component of radiation $I_{b,c}$ is, from Eq. 6-134 and Eq. 6-142, given by

$$I_{b,c} = I_{b,h}\left(\frac{\cos i}{\cos \alpha}\right) = R_b I_{b,h} \qquad (6\text{-}147)$$

where $I_{b,h}$ is the instantaneous beam component on a horizontal surface and the ratio $(\cos i/\cos\alpha)$, called the *tilt factor*, R_b, is given by the relation

$$R_b = \frac{\cos i}{\cos \alpha} = \frac{\sin \delta_s \sin(L-\beta) + \cos \delta_s \cos(L-\beta)\cos h_s}{\sin L \sin \delta_s + \cos L \cos \delta_s \cos h_s} \qquad (6\text{-}148)$$

R_b is a geometric parameter relating beam radiation on a horizontal surface to beam radiation on a tilted surface. Although it is a continuously varying quantity, for engineering calculations Lui and Jordan (Ref. 26) found that one can use an average value given by

$$\overline{R}_b = \frac{\overline{\cos i}}{\overline{\cos \alpha}} = \frac{\cos(L-\beta)\cos \delta_s \sin h_{ss} + h_{ss}\sin(L-\beta)\sin \delta_s}{\cos L \cos \delta_s \sin_{ss}(\alpha=0) + h_{ss}(\alpha=0)\sin L \sin \delta_s} \qquad (6\text{-}149)$$

where the hour angles $h_{ss}(\alpha=0)$ and h_{ss} must be expressed in radians and should be evaluated at midmonth. Using Eq. 6-149 the monthly average beam radiation on a tilted surface, $\overline{B}_c$, is

$$\overline{B}_c = \overline{R}_b \cdot \overline{B}_h \qquad (6\text{-}150)$$

Table 6-5 gives values of the tilt factor R_b for latitudes of $20°$, $30°$, $40°$, and $50°$, and tilt angles of interest in solar engineering.

In addition to the beam radiation, a tilted surface will also receive diffuse radiation from the portion of the sky within view. If the sky is assumed to be an isotropic source, Eq. 6-57 gives the view factor, $\overline{R}_d$, in the form

$$\overline{R}_d = \tfrac{1}{2}(1 + \cos \beta) \qquad (6\text{-}151)$$

Table 6-5 Solar-Collector Tilt Factor at Various Latitudes

LATITUDE	$L=20°$		$L=30°$		$L=40°$		$L=50°$	
MONTH	$\beta=20°$	$\beta=40°$	$\beta=30°$	$\beta=50°$	$\beta=40°$	$\beta=60°$	$\beta=50°$	$\beta=70°$
Jan	1.36	1.52	1.68	1.88	2.28	2.56	3.56	3.94
Feb	1.22	1.28	1.44	1.52	1.80	1.90	2.49	2.62
Mar	1.08	1.02	1.20	1.15	1.36	1.32	1.65	1.62
Apr	1.00	0.83	1.00	0.84	1.05	0.90	1.16	1.00
May	0.92	0.70	0.87	0.66	0.88	0.66	0.90	0.64
Jun	0.87	0.63	0.81	0.58	0.79	0.60	0.80	0.56
Jul	0.89	0.66	0.83	0.62	0.82	0.64	0.84	0.62
Aug	0.95	0.78	0.93	0.76	0.96	0.78	1.02	0.83
Sep	1.04	0.95	1.11	1.00	1.24	1.12	1.44	1.32
Oct	1.17	1.20	1.36	1.36	1.62	1.64	2.10	2.14
Nov	1.30	1.44	1.60	1.76	2.08	2.24	3.16	3.32
Dec	1.39	1.60	1.76	1.99	2.48	2.80	4.04	4.52

When the surface is near the ground, as for example a flat-plate collector on the roof of a house, it can also receive beam and diffuse radiation reflected from the ground with a view factor, $\bar{R}_r$, approximately equal to

$$\bar{R}_r = \frac{\bar{R}}{\bar{D}_h + \bar{B}_h} = \rho_s \frac{1}{2}(1 - \cos\beta) \qquad (6\text{-}152)$$

where ρ_s is the solar reflectance of the ground surface to the south. No precise measurement of ρ_s have ever been made, but climatological measurements of the earth albedo suggests that one may use $\rho_s = 0.7$ for snow or ice; $\rho_s = 0.2$ for concrete; $\rho_s = 0.1$ for asphalt, water, or dark earth; and $\rho_s = 0.4$ for sand.

Combining Eqs. 6-149, 6-150, 6-151, and 6-152, the total monthly average of the solar radiation incident on a surface on earth, $\bar{I}_c$, is

$$\bar{I}_c = \bar{R}_b \bar{B}_h + \bar{D}_h \frac{1 + \cos\beta}{2} + (\bar{D}_h + \bar{B}_h) \frac{\rho_s(1 - \cos\beta)}{2} \qquad (6\text{-}153)$$

The prediction of average daily or hourly solar radiation incident on a surface can be obtained similarly, and the reader is referred to Reference 25 for details.

Example 6-19. Calculate the total daily average solar radiation incident on a south-facing solar collector, inclined 40°, for the months of February and June in Boulder, Colorado. In February the ground to the south is covered with snow, and June it is a lawn ($\rho_s = 0.3$)

Solution: Boulder is located at 40° latitude. From the solar maps, the average monthly insolation on a horizontal surface $\bar{H}_h$, is 268 la/day ($= 11,218 kJ/m^2 \cdot day$) in February and 525 la/day ($= 21,977 kJ/m^2 \cdot day$)

in June. From the nomagraph in Fig. 6-49, the length of day on Feb. 15 ($\delta_s = 14°$) is 10.5 h and on June 15 ($\delta_s = 22°$) it is 14.5 h. The average clearness index, $\bar{K}_T$, is obtained from Fig. 6-55. First, we calculate or look up the extraterrestrial radiation at midmonth.

From Fig. 6-52 or Eq. 6-141, $\bar{H}_0$ is 500 la/day = 20,930kJ/m² · day in February and 1000 la/day = 41,860kJ/m² · day in June. From Fig. 6-55:

For February

$$\bar{K}_T = \frac{\bar{H}_h}{\bar{H}_0} = 0.536 \quad \text{and} \quad \frac{\bar{D}_h}{\bar{H}_h} = 0.36$$

For June

$$\bar{K}_T = \frac{\bar{H}_h}{\bar{H}_0} = 0.524 \quad \text{and} \quad \frac{\bar{D}_h}{\bar{H}_h} = 0.35$$

Thus the diffuse radiation received is $0.36 \times 11,218 = 4038kJ/m²$ · day in February and $0.35 \times 21,977 = 7692kJ/m²$ · day in June.

From Eq. 6-148 or Table 6-5, $R_b = 1.8$ in February and 0.79 in June. Thus, from Eq. 6-153,

$$\bar{I}_c = 1.8 \times (11,218 - 4038) + 4038 \times \frac{1 + 0.766}{2} + 11,218 \times \frac{0.7(1 - 0.766)}{2}$$

$$= 17,400 \text{ kJ/m}^2 \cdot \text{ day}$$

in February and

$$\bar{I}_c = 0.79 \times (21,977 - 7692) + 7692 \times \left(\frac{1 + 0.766}{2}\right) + 21,977 \times \frac{0.3(1 - 0.766)}{2}$$

$$= 18,850 \text{ kJ/m}^2 \cdot \text{ day}$$

in June.

REFERENCES

1. R. V. Dunkle, "Thermal-Radiation Tables and Applications," *Trans. ASME*, vol. 65, pp. 549–552, 1954.
2. Y. S. Touloukian and D. P. DeWitt, *Thermal Radiative Properties-Metallic Elements and Alloys*, Vol. 7, IFI/Plenum, Publishing Corp., New York, 1970, pp. 10a–11a.
3. E. M. Sparrow, Radiant Emission, Absorption, and Transmission Characteristics of Cavities and Passages, *NASA-SP-55*, pp. 103–114, 1965.
4. W. Truenfels, "Emissivity of Isothermal Cavities," *J. Opt. Soc. Amer.*, vol. 53, No. 10, pp. 1162–68, October 1963.
5. E. M. Sparrow and V. K. Jonsson, "Radiant Emission Characteristics of Diffuse Conical Cavities," *J. Opt. Soc. Amer.*, vol. 53, pp. 816–821, 1963.

6. E. Schmidt and E. Eckert, "Uber die Richtungs-verteilung der Warmestrahlung," *Forsch. Gebeite Ingenieurwesen*, vol. 6, 1935.

7. E. M. Sparrow and R. D. Cess, *Radiation Heat Transfer*, Brooks/Cole Publishing Company, Monterey, Calif., 1966, pp. 56–63.

8. K. E. Torrance and E. M. Sparrow, "Biangular Reflectance of an Electrical Nonconductor as a Function of Wavelength and Surface Roughness," *J. Heat Transfer*, vol. 87, pp. 283–292, 1965.

9. E. R. G. Eckert and E. M. Sparrow, "Radiative Heat Exchange Between Surfaces with Specular Reflection," *Int. J. Heat Mass Transfer*, vol. 3, pp. 42–54, 1961.

10. E. M. Sparrow, E. R. G. Eckert, and V. K. Jonsson, "An Enclosure Theory for Radiative Exchange Between Specular and Diffusely Reflecting Surfaces," *J. Heat Transfer*, vol. 84, pp. 294–299, 1962.

11. D. C. Hamilton and W. R. Morgan, Radiant-Interchange Configuration Factors, *NACA Tech. Note 2836*, 1952.

12. R. Siegel and J. R. Howell, *Thermal Radiation Heat Transfer*, McGraw-Hill Book Company, New York, 1972.

13. H. C. Hottel and A. F. Sarofim, *Radiative Transfer*, McGraw-Hill Book Company, New York, 1967, pp. 31–39.

14. E. Schmidt, *Forsch. Gebiete. Ingenieurs*, vol. 3, no. 57, 1932.

15. S. A. Guerrieri, S.M. thesis in chemical engineering, MIT, Cambridge, Mass., 1932.

16. W. Malkmurs and H. Thompson, *J. Quant. Spectro. Radiative Transfer*, vol. 2, no. 16, 1962.

17. R. H. C. Lee and J. Happel, "Thermal Radiation of Methane Gas," *Ind. Eng. Chem. Fundamentals*, vol. 3, no. 167, 1964.

18. W. Ullrich, Sc.D. thesis in chemical engineering, MIT, Cambridge, Mass., 1953.

19. E. R. G. Eckert, *Forschungsheft*, vol. 387, pp. 1–20, 1937.

20. H. C. Hottel, in W. H. McAdams, *Heat Transmission*, 3rd ed., chap. 4, McGraw-Hill Book Company, New York, 1954.

21. H. C. Hottel and R. B. Egbert, *Trans. AIChE*, vol. 38, pp. 531–565, 1942.

22. H. C. Hottel and A. F. Sarofin, *Radiative Transfer*, McGraw-Hill Book Company, New York, 1967.

23. R. B. Egbert, Sc.D. thesis in chemical engineering, MIT, Cambridge, Mass., 1941.

24. M. Rohsenow and J. P. Hartnett, eds., *Handbook of Heat Transfer*, McGraw-Hill Book Company, New York, 1973.

25. F. Kreith and J. F. Kreider, *Principles of Solar Engineering*, McGraw-Hill Book Company, New York, 1978.

26. B. Y. H. Lui and R. C. Jordon, "Availability of Solar Energy for Flat Plate Solar Collectors," in *Low Temperature Engineering Application of Solar Energy*, chap. 1, ASHRAE, New York, 1967.

27. S. A. Klein, A Design Procedure for Solar Heating Systems, Ph.D. thesis, University of Wisconsin, Madison, 1976.

PROBLEMS

The problems in this chapter are organized in the manner shown in the table. Three problems suggest computer solutions. They are Problems 6-61 through 6-63. No original programs need to be written. All programs needed for solutions are developed in examples within the chapter.

PROBLEM NUMBERS	SECTIONS	SUBJECT
6-1 to 6-9	6-2	Physics of radiation
6-10 to 6-18	6-3	Radiation properties
6-19 to 6-28	6-4	Radiation shape factor
6-29 to 6-35	6-5	Radiative exchange between black surfaces
6-36 to 6-56	6-6	Radiative exchange between gray surfaces
6-57 to 6-63	6-7	Matrix methods
6-64 to 6-73	6-8 and 6-9	Radiation through absorbing media
6-74 to 6-80	6-10	Solar radiation

6-1 Calculate the maximum monochromatic black-body emissive power for surfaces that have a temperature of (a) 100 K, (b) 500 K, (c) 1000 K, and (d) 5000 K.

6-2 Determine the wavelength at which the black-body emissive power is a maximum for surfaces that have a temperature of (a) 100 K, (b) 500 K, (c) 1000 K, and (d) 5000 K.

6-3 Plot the monochromatic blackbody emissive power for surface temperatures of 200 K, 1000 K, 5000 K, and 10,000 K.

6-4 The filament of a light bulb has a temperature of 3200 K and emits as a blackbody. What percentage of the emitted radiation is in (a) the infrared-wavelength range, and (b) the visible-wavelength range?

6-5 Determine the temperature to which a black surface must be heated so that 40 percent of its emitted energy is in the infrared-wavelength range.

6-6 Determine the temperature to which a black surface must be heated so that 20 percent of its emitted energy is in the visible-wavelength range.

6-7 The tungsten filament of a photoflood lamp operates at 3500 K. Assuming that the filament emits as a blackbody, determine the proportions of total energy emitted from the filament in the ultraviolet, visible, and infrared regions. Repeat your calculations for an ordinary tungsten filament that operates at a temperature of 2500 K.

6-8 Derive Wien's law (Eq. 6-3) by differentiating Planck's law.

6-9 Show that the Stefan-Boltzmann constant is given by

$$\sigma = \left(\frac{\pi}{C_2}\right)^4 \frac{C_1}{15}$$

6-10 Suppose that you are asked to select between two materials that are being considered for use as the external surface of a large office building. One material transmits 60% of the incident radiation between 0.3×10^{-6} and 0.6×10^{-6} m and 20% between 0.6×10^{-6} and 40×10^{-6} m. A second material transmits 40% of the incident energy between 0.3×10^{-6} and 2×10^{-6} m and 30% between 2×10^{-6} and 30×10^{-6} m. At other wavelengths both materials are opaque. The building is located where approximately 80% of the energy consumed is used for air-conditioning purposes and only 20% is consumed in heating. Which material would you select and for what reasons?

6-11 A greenhouse is to be constructed of silica glass that is known to transmit 92% of the incident radiant energy between wavelengths 0.35×10^{-6} and 2.7×10^{-6} m. Assume that the glass is completely opaque at shorter and longer wavelengths. Calculate the percentage of solar radiation that will reach the plants if the sun is a blackbody with a temperature of 5550 K. If the average plant temperature is 300 K, and they are assumed to radiate as a blackbody, calculate the percentage of radiant energy emitted by the plants that is transmitted by the glass to the surroundings.

6-12 A surface has a temperature of 500 K and a monochromatic emissivity of

$\epsilon_\lambda = 0$	for $0 \leqslant \lambda \leqslant 5 \times 10^{-6}$ m
$\epsilon_\lambda = 0.002\lambda \times 10^{-6} - 0.01$	for 5×10^{-6} m $< \lambda \leqslant 100 \times 10^{-6}$ m
$\epsilon_\lambda = 0.19$	for 100×10^{-6} m $< \lambda \leqslant 200 \times 10^{-6}$ m
$\epsilon_\lambda = 0$	for 200×10^{-6} m $< \lambda < \infty$

where λ is measured in meters. Determine the total emissivity of the surface.

6-13 Determine the absorptivity of the surface described in Problem 6-12 if the irradiation on the surface is due to a blackbody at (a) 500 K, and (b) 1000 K.

6-14 A surface has a temperature of 1000 K and a monochromatic absorptivity of

$\alpha_\lambda = 0$	for $0 < \lambda \leqslant 0.5 \times 10^{-6}$ m
$\alpha_\lambda = 0.5(\lambda \times 10^{-6} - 0.5)$	for 0.5×10^{-6} m $< \lambda \leqslant 1.5 \times 10^{-6}$ m
$\alpha_\lambda = 0.50$	for 1.5×10^{-6} m $< \lambda \leqslant 2.0 \times 10^{-6}$ m
$\alpha_\lambda = 0$	for 2.0×10^{-6} m $< \lambda < \infty$

where λ is measured in meters. Determine the total absorptivity of the surface if the source of irradiation is a blackbody at (a) 5000 K, and (b) 500 K.

6-15 The monochromatic emissivity of a real surface is approximated by the relationship

$$\epsilon_\lambda = 0 \quad \text{for} \quad 0 \quad \leqslant \lambda < 2 \times 10^{-6} \, \text{m}$$
$$\epsilon_\lambda = 0.2 \quad \text{for} \quad 2 \times 10^{-6} \, \text{m} \leqslant \lambda < 8 \times 10^{-6} \, \text{m}$$
$$\epsilon_\lambda = 0.4 \quad \text{for} \quad 8 \times 10^{-6} \, \text{m} \leqslant \lambda < 25 \times 10^{-6} \, \text{m}$$
$$\epsilon_\lambda = 0 \quad \text{for} \quad 25 \times 10^{-6} \, \text{m} < \lambda < \infty$$

(a) Calculate the total emissivity of the surface when its temperature is 800 K.
(b) Calculate the total absorptivity of the surface if the irradiation on the surface is due to blackbody radiation at 1000 K.

6-16 An earth satellite is to be coated with a special thermal protective paint. Two paints are being considered. The monochromatic absorptivity of the two paints is shown in the figure. Assuming that the sun is a blackbody at 5550 K, calculate the total absorptivity of both paints for irradiation from the sun. Which paint will cause more of the sun's energy to be absorbed?

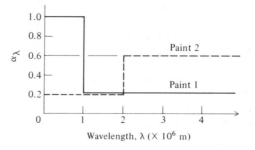

Wavelength, λ ($\times$ 10^6 m)

6-17 A satellite coated with the paints described in Problem 6-16 is tested in a space simulator prior to being placed into orbit. The satellite coated with paint 1 reaches an equilibrium temperature of 550 K and the one coated with paint 2 reaches 450 K. Determine the total emissivity of both paints.

6-18 A surface has a directional emissivity given by $\epsilon(\theta) = 0.60 \cos\theta$, where θ is the angle measured relative to the surface normal. Determine a value for the total emissivity of the surface.

6-19 Use the stretched-string method to determine an expression for the shape factor $F_{1 \to 2}$, where the geometry in the figure is infinite in extent in and out of the

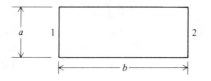

plane of the page. Plot $F_{1\to2}$ as a function of the ratio b/a and compare with the results given in Fig. 6-16.

6-20 Determine a value for the shape factor between the sky and an infinitely long pipe with radius r resting on the ground.

6-21 Determine a value for the shape factor between the sky and an infinitely long pipe with radius r one-half buried in the ground.

6-22 Show that the shape factor $F_{dA_1\to A_2}$ for the rectangular geometry in the figure is

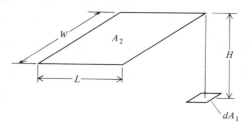

$$F_{dA_1\to A_2} = \frac{1}{2\pi}\left[\frac{X}{\sqrt{1+X^2}}\tan^{-1}\left(\frac{Y}{\sqrt{1+X^2}}\right)\right.$$
$$\left.+ \frac{Y}{\sqrt{1+Y^2}}\tan^{-1}\left(\frac{X}{\sqrt{1+Y^2}}\right)\right]$$

where

$$X=\frac{L}{H}$$

$$Y=\frac{W}{H}$$

6-23 Calculate an expression for the shape factor $F_{dA_1\to A_2}$ for the circular disk geometry shown in the figure.

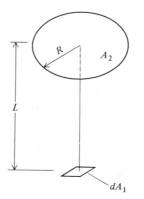

6-24 Calculate a value for $F_{1\to2}$ for each pair of surfaces shown in the figure.

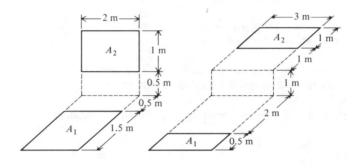

6-25 Determine a value for the shape factor $F_{1\to2}$ for the geometry in the figure.

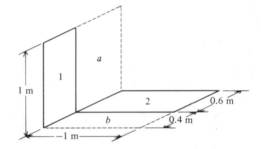

6-26 Determine a value for the shape factor $F_{1\to2}$ for the geometry shown in the figure.

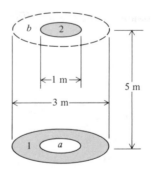

6-27 Determine the shape factors $F_{1\to3}$, $F_{1\to4}$, $F_{1\to5}$, and $F_{1\to6}$, where the cross section of the geometry is shown in the figure. The surfaces are infinite in extent in and out of the plane of the page.

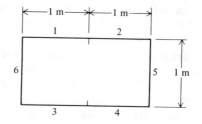

6-28 Determine the shape factors $F_{1\to2}$, $F_{1\to3}$, and $F_{1\to4}$ for the cylindrical geometry shown in the figure. Surfaces 1 and 4 are the end planes of the cylinder and surfaces 2 and 3 are the curved external portion of the cylinder.

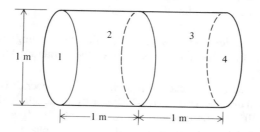

6-29 Two parallel black rectangles 5×10 m are directly opposed from each other and spaced 5 m apart. Assume that the surroundings are black at 0 K. The rectangles are designated as surfaces 1 and 2. The temperatures of the surfaces are $T_1=100$ K and $T_2=2000$ K. Draw a thermal circuit for the system, labeling all potentials, resistances, and currents:

    (a) Determine the net radiant heat transfer between surface 1 and 2.
    (b) Determine the net energy supply to surface 1.
    (c) Determine the net energy supply to surface 2.
    (d) Determine the net radiant heat transfer between surface 1 and the surroundings.
    (e) Determine the net radiant heat transfer between surface 2 and the surroundings.

6-30 Repeat the calculations in Problem 6-29 when the surroundings are replaced by a refractory surface. Calculate the temperature of the refractory surface.

6-31 A black solar collector is placed on the roof of a house. The collector has an area of 50 m². The incident energy from the sun produces a radiant flux of 800 W/m² on the collector. The surroundings may be assumed to be a blackbody at an effective radiative temperature of 30°C. Neglect conduction and convection rates from the collector. Calculate:

    (a) The equilibrium temperature of the collector.
    (b) The net radiative exchange between the collector and the surroundings.

6-32 Rework Problem 6-31 considering convection from the surface of the collector to the ambient air at 30°C when $\bar{h}_c = 60$ W/m²·K.

6-33 Two parallel black disks are 30 cm in diameter and 20 cm apart. One disk has a temperature of 800 K and the other has a temperature of 400 K. The disks are placed in a vacuum and the backs of both surfaces are insulated. The surroundings are black at 300 K:

    (a) Draw a thermal circuit for this problem and label all potentials, resistances, and currents.

    (b) Calculate the net radiative heat transfer between the two disks.

    (c) Calculate the net heat-transfer rate for the two disks and the surroundings. Verify that the three net heat-transfer rates sum to zero.

    (d) Calculate the net radiative-heat-transfer rates between the two disks and the surroundings.

6-34 Rework Problem 6-33 assuming that the surroundings are replaced by a refractory surface. Calculate the temperature of the refractory surface.

6-35 An insulated black rectangular surface on a spacecraft is oriented normal to the sun's rays. The spacecraft is 1.5×10^8 km from the sun and the diameter of the sun is 1.39×10^6 km. Assume that the sun is a blackbody at 5550 K. Incident energy from other parts of the spacecraft and other planets is negligible. The surroundings have an equivalent radiation temperature of 0 K. Calculate the equilibrium temperature of the black surface.

6-36 An asphalt street has an emissivity of 0.6 for long-wavelength-emitted energy and an absorptivity of 0.95 for the short-wavelength energy incident from the sun. On a clear day the incident radiant energy from the sun is 1000 W/m². Estimate the equilibrium temperature of the asphalt if the air temperature is 290 K and the convective-heat-transfer coefficient between the asphalt and air is 10 W/m²·K. Neglect conduction of heat into the ground.

6-37 Rework Problem 6-36 replacing the asphalt with a concrete street that has an emissivity of 0.5 for long-wavelength-emitted energy and an absorptivity 0.65 for the short-wavelength energy incident from the sun.

6-38 A liquid nitrogen dewar is made of two concentric spheres separated by a vacuum. The inner sphere has an outer diameter of 1 m and the outer sphere has an inner diameter of 2 m. Both spheres are gray surfaces with $\epsilon = 0.20$. The saturation temperature for nitrogen at 1 atm pressure is 78 K, and its latent heat of vaporization is 2×10^5 J/kg. Estimate the N_2 boil-off rate when the inner sphere is full of liquid nitrogen and the outer sphere is at 300 K.

6-39 Two large gray plates are placed parallel to each other and they are separated by a small distance. One plate has an emissivity of 0.8 and a temperature of 500 K while the other has an emissivity of 0.2 and a temperature of 400 K. The space between the plates is evacuated and the back surfaces of both plates are

insulated. Calculate:
  (a) The net heat-transfer rate to both surfaces.
  (b) The net radiative-heat-transfer rate between the surfaces.

6-40 A clothing manufacturer has introduced a new line of lightweight jackets that utilize "space-age technology." The jackets are coated with a silver material that is supposed to reduce the radiation loss from the body. The silvered inner surface of the jacket has an emissivity of 0.10. Calculate the radiant loss from a person wearing the jacket assuming that the jacket has a temperature of 270 K and the clothing worn by the person has a temperature of 295 K and an emissivity of 0.70. Since the jacket is worn tightly, assume that the radiant exchange between clothing and jacket can be approximated by a pair of infinite parallel plates. Compare the heat loss by radiation to that by convection when $\bar{h}_c$ over the exterior part of the jacket is 200 W/m²·K and the ambient air temperature is 260 K. Is the radiation loss from the body a significant portion of the total heat loss?

6-41 An enclosure consists of three infinitely long gray surfaces which form an equilateral cross section. Surface 1 has a known temperature T_1; surface 2 has a known net heat flux $(q_2)''_{net}$ but an unknown temperature; surface 3 has a known temperature T_3. Draw the thermal circuit for the problem. Clearly indicate all resistances and evaluate their values. Calculate the radiosity for each surface. Also calculate values for $(q_1)''_{net}$, T_2, and $(q_3)''_{net}$.

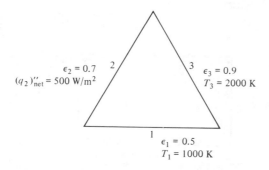

$\epsilon_2 = 0.7$
$(q_2)''_{net} = 500 \text{ W/m}^2$

2

3
$\epsilon_3 = 0.9$
$T_3 = 2000 \text{ K}$

1
$\epsilon_1 = 0.5$
$T_1 = 1000 \text{ K}$

6-42 A long pipe is laid flat on level ground. The pipe is gray, has an emissivity of 0.50, and has a diameter of 1 m. The temperature of the pipe, air, and ground are all 285 K. Calculate the net radiant energy exchange per unit length of pipe between the pipe and sky on a clear night when the effective sky temperature is 150 K.

6-43 Rework Problem 6-42 when the pipe is one-half buried in the ground.

6-44 An oven is used to heat flat, square (1 × 1 m) sheets of plastic that are gray and have an emissivity of 0.40. Heaters are placed above and below the plastic as shown in the figure. The heaters have a temperature of 700 K and an emissivity of 0.90. The backs of both heaters are well insulated. The temperature of the walls is 450 K and $\epsilon = 0.2$. Neglect convection and calculate the equilibrium temperature of the plastic.

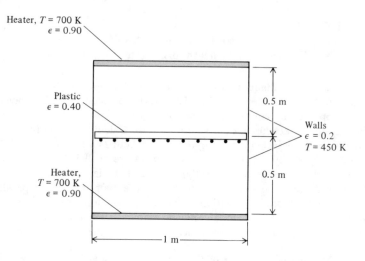

Heater, $T = 700$ K
$\epsilon = 0.90$

Plastic
$\epsilon = 0.40$

0.5 m

Walls
$\epsilon = 0.2$
$T = 450$ K

Heater,
$T = 700$ K
$\epsilon = 0.90$

0.5 m

1 m

6-45 A furnace is used to heat steel ingots. The top of the furnace is covered with electrical radiant heaters with a total power rating of 100 kW. The emissive of the heaters is 0.75. The side walls of the furnace are well insulated and are refractory surfaces. The ingots are gray ($\epsilon = 0.40$) and they are placed on the floor of the furnace as shown in the figure. Initially, the ingots have a temperature of 300

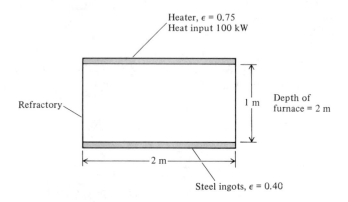

Heater, $\epsilon = 0.75$
Heat input 100 kW

Refractory

1 m

Depth of
furnace = 2 m

2 m

Steel ingots, $\epsilon = 0.40$

K. For this initial temperature, calculate:
  (a) The heater temperature.
  (b) The temperature of the side surfaces of the furnace.
  (c) The net heat-transfer rates to the heater and ingots.
  (d) The net radiative-heat-transfer rate between the heater and ingots.

6-46 Repeat the calculations in Problem 6-45 when the steel ingots reach a temperature of 600 K.

6-47 The filament of a 100 W light is surrounded by a spherical bulb. The bulb is gray with $\tau = 0.88$ and $\alpha = 0.12$. Assume that the filament is turned on and it quickly reaches its equilibrium temperature of 300 K. Determine an expression for the transient temperature of the bulb if it has an inside diameter of 10 cm, an outside diameter of 10.2 cm, density of 2.7×10^3 kg/m^3, and specific heat of 840 J/kg·K. Radiant energy incident on the bulb from the surroundings is negligible.

6-48 A 10×2 m solar collector consists of cooling tubes that circulate a fluid through the base of the collector. The collector is covered with a single pane of glass, and the space between the glass and collector plate is evacuated as shown in the figure. The collector plate is black and the side walls are refractory surfaces. The bottom surface of the collector is well insulated. The glass properties are $\alpha = 0.05$, $\tau = 0.88$ for short-wavelength radiation and $\alpha = 0.90$, $\tau = 0.04$ for long-wavelength radiation. The coolant flow rate is adjusted to maintain the surface of the collector plate at 55°C. When the solar intensity is 950 W/m^2, calculate:

(a) The glass temperature, neglecting conduction and convection effects.
(b) The net radiative energy gain of the collector plate.

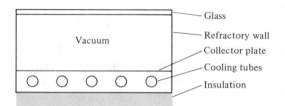

6-49 Repeat Problem 6-48 assuming that the air above the glass has an ambient temperature of 25°C and the convective-heat-transfer coefficient between the air and glass is 25 W/m^2·K.

6-50 The incandescent bulb in a slide projector is rated at 400 W. A reflector placed behind the filament directs 30% of the energy toward the slide. The slide is gray with $\tau = 0.50$, $\epsilon = 0.20$. The area of the slide is 8.9 cm^2. Assume that any radiative energy that leaves the slide never returns. Determine the equilibrium temperature of the slide if:

(a) Convection losses from the slide are negligible.
(b) A fan blows 35°C air over both sides of the slide with $h = 500$ W/m^2·K.

6-51 A small electronics package is placed on the floor of a large room. The surface area of the package is 0.5 m^2 and it has a surface emissivity of 0.70. The power input to the electronics package is 850 W and the convective-heat-transfer coefficient between the package and air is 15 W/m^2·K. The temperature of the room and air is 25°C. Determine the temperature of the package, assuming it is isothermal and is well insulated where it rests on the floor.

6-52 Heat lamps are used to dry paint on a rectangular plate. The painted surface is 4×4 m and has a surface emissivity of 0.25. The heat lamps are formed in a rectangular array in a size equal to the painted surface, and they can be assumed to be diffuse emitters. The power input to the lamps is 50 kW. The paint will blister if its temperature exceeds 450 K. Assume that the back surfaces of the heat lamps and the painted plate are insulated and neglect convection from both surfaces. The drying process takes place in a large room with a temperature of 20°C. Determine the minimum spacing between the heat lamps and the painted surface.

6-53 Rework Problem 6-52 using the data given in the problem but accounting for convection from the lamps and painted surface. Assume that 15% of the energy input into the lamps is dissipated by convection to the surrounding air. The painted surface is oriented vertically and the ambient air temperature is 20°C. Use an appropriate Nusselt number correlation to account for convection from the painted surface. Determine the minimum spacing between the heat lamps and the painted surface for this case.

6-54 Steaks are placed on a charcoal grill 20 cm above the coals. The bed of charcoal has an area of 60×80 cm, an emissivity of 0.8, and a temperature of 1000 K. The steaks are placed parallel to the coals and have an area equal to that of the bed of charcoal. The emissivity of the steaks is estimated to be 0.40. The steaks are cooked outdoors where the temperature is 300 K. The convection rate between the charcoal and steak averages approximately 600 W while the steaks are cooking. Heat transfer from the upper surface of the steaks and lower surface of the charcoal may be neglected:

(a) Estimate the equilibrium temperature of the steaks.
(b) Estimate the rate at which charcoal must be added to maintain steady conditions, assuming that the heating value of the charcoal is 2.8×10^7 J/kg.

6-55 Rework Problem 6-54 assuming that the opening between the grill and charcoal is completely enclosed by a perfectly reflecting foil.

6-56 A furnace with a cylindrical shape is used to heat circular objects placed at the top of the furnace as shown in the figure. The heater is the entire circular bottom of the furnace and it has an input rating of 1.3 kW. The cylindrical side

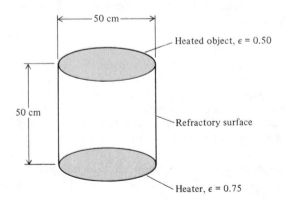

surfaces of the furnace are refractory surfaces. The heater has an emissivity of 0.75 and the object being heated has an emissivity of 0.50. Neglect conduction and convection effects between the surfaces. The temperature of the object being heated is determined to be 500 K. Determine:

    (a)   The temperature of the heater.
    (b)   The temperature of the refractory surface.
    (c)   The irradiation and radiosity of all three surfaces per unit area.
    (d)   The net heat flux to the heated object.

6-57 Consider an enclosure consisting of three gray surfaces with known geometry and emissivities. Surfaces 1 and 2 have known net heat fluxes but unknown temperatures. Surface 3 has a known temperature but unknown net heat flux. Develop three algebraic expressions that can be used to solve for T_1, T_2, and $(q_3)''_{net}$.

6-58 Four gray, diffuse, isothermal, opaque surfaces form an enclosure. Surfaces 1 and 2 have a known net heat flux while surfaces 3 and 4 have known temperatures. Assume that all surface properties and geometry are specified. Derive the equations that could be used to determine T_1, T_2, $(q_3)''_{net}$, and $(q_4)''_{net}$ and put the equations in the standard matrix form.

6-59 Work Example 6-8 by the matrix method.

6-60 Work Example 6-10 by the matrix method.

6-61 A long structural member has a cross section in the shape of a hexagon. All surfaces of the hexagon are gray, opaque, and diffuse. Calculate the net heat-transfer rates per unit area to each of the six surfaces. Neglect convection and conduction between the surfaces. Use the computer program described in Section 6-7 to determine values for $(q_i)''_{net}$. The geometry is shown in the figure. The

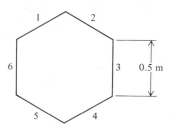

temperatures and emissivities of the surfaces are:

$$T_1 = T_3 = 400 \text{ K} \qquad T_2 = T_4 = 500 \text{ K} \qquad T_5 = T_6 = 300 \text{ K}$$
$$\epsilon_1 = \epsilon_2 = 0.2 \qquad\qquad \epsilon_3 = \epsilon_4 = 0.3 \qquad\qquad \epsilon_5 = \epsilon_6 = 0.4$$

6-62 The member with a hexagon cross section shown in Problem 6-61 has the following given conditions: $\epsilon_1 = 0.1$, $\epsilon_2 = 0.2$, $\epsilon_3 = 0.3$, $\epsilon_4 = 0.4$, $\epsilon_5 = 0.5$, $\epsilon_6 = 0.6$, $T_1 = 300$ K, $T_2 = 350$ K, and $T_3 = 400$ K. Surfaces 4, 5, and 6 are refractory surfaces. Use the computer program described in Section 6-7 and determine values for the temperatures of the three refractory surfaces and the net heat-transfer rates to surfaces 1, 2, and 3 per unit area.

6-63 A new and improved furnace such as the one described in Problem 6-56 is designed to heat an object to higher temperatures. This new design features a bottom heating element plus a heater that occupies one-half of the side cylindrical surface as shown in the figure. The remainder of the cylindrical side surface is a refractory surface. Both heaters have an emissivity of 0.90. When an object with emissivity of 0.5 is placed at the top of the furnace, its equilibrium temperature is 650 K. The power rating of the bottom heater is 1.3 kW and the rating of the side

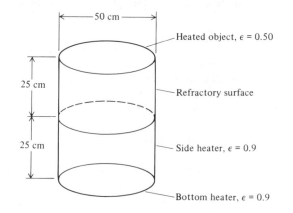

heater is 1.0 kW. Use the computer program described in Section 6-7 to determine:
- (a) The temperatures of both heaters.
- (b) The temperature of the refractory surface.
- (c) The irradiation and radiosity of all four surfaces per unit area.
- (d) The net heat flux to the heated object.

6-64 A mixture of CO_2 and N_2 is heated to 1000 K. The partial pressure of the CO_2 is 0.1 atm and the total pressure of the mixture is 1 atm. The shape of the gas mixture is hemispherical with a radius of 0.8 m. Determine the emissivity of the gas mixture.

6-65 A mixture of dry air and water vapor has a total pressure of 2 atm. The partial pressure of the water vapor is 0.1 atm. The mixture of gases occupies the space between two large parallel plates separated by 1.5 m. Calculate the emissivity of the gas mixture assuming that the temperature is 800 K.

6-66 A mixture of CO_2 and H_2O has a total pressure of 0.5 atm and a temperature of 700 K. The gas mixture is enclosed in a cubical container that is 1 m on a side. Determine the mixture emissivity if the partial pressure of the CO_2 is 0.2 atm.

6-67 A mixture of H_2O and CO_2 at a total pressure of 1.5 atm is placed between two large parallel plates separated by a distance of 0.5 m. The partial pressure of the water vapor is 0.7 atm. The temperature of the gases is 800 K. Determine the emissivity of the gas mixture.

6-68 Determine the absorptivity of the gas mixture in Problem 6-67 for incident radiant energy from a surface at 400 K.

6-69 Consider the given information in Problem 6-67 and assume that the gas is gray. The upper plate is surface 1 and it is gray ($\epsilon = 0.4$) with a temperature of 500 K. The lower plate is surface 2 and it is gray ($\epsilon = 0.8$) and has a temperature of 1000 K. Calculate the net heat-transfer rate to surface 1 $(q_1)''_{net}$ when the gas is present and when it is removed. Determine the temperature of the gas.

6-70 Use the information given in Problem 6-65 and determine the absorptivity of the gas mixture for irradiation from a surface at 900 K.

6-71 A large plate of steel ($\epsilon = 0.5$) leaves a rolling mill at 1000 K. A small instrument is to be placed 1 cm from the steel to monitor its surface properties. The space between the instrument and steel is occupied by a gray gas with an emissivity of 0.15. Calculate the surface heat flux necessary to maintain the instrument package at 500 K if the instrument is a blackbody. Calculate the heat flux if the absorbing gas is not present.

6-72 Calculate the net radiant exchange between two gray ($\epsilon = 0.6$) infinite plates separated by a distance of 4 cm. The space between the plates is filled with CO_2 at 1 atm pressure and the two plates are at 800 K and 600 K. Assume that the CO_2 is a gray gas. Also calculate the net energy supplied to each plate. Calculate the net radiant exchange between the two plates when no gas is present.

6-73 Exhaust gases from a combustion process leave a 20-cm-diameter pipe with a temperature of 2200 K. The gases consist of CO_2 and H_2O with at total pressure 1 atm and partial pressures of 0.05 and 0.1 atm, respectively. Suppose that a black surface placed next to the exhaust gases is to be cooled until its temperature is less than 500 K. What is the minimum cooling capacity per unit area required? Assume that the gas is gray.

6-74 Calculate the fraction of extraterrestrial radiation at wavelengths below 1 μm.

6-75 Determine the solar declination on January 30 and September 1.

6-76 Calculate the angle of incidence of beam radiation at 2 P.M. solar time on February 15 at a latitude of 40°N on surfaces with the following orientation:
    (a) Horizontal.
    (b) Tilted to south with a slope of 40°.
    (c) At a slope of 40°, but facing 30°W of S.
    (d) Vertical, facing S.
    (e) Vertical, facing W.

6-77 Estimate for Denver, Colorado, on March 1, the hourly total radiation from sunrise to sunset on a surface tilted at a slope equal to latitude and facing due South. Assume that the ground is covered with snow, having a reflectivity of 0.7.

6-78 Calculate the equilibrium temperature of a black horizontal plate, insulated at the bottom, in Madison, Wisconsin, at noon on May 11, with wind blowing at 5 m/s, in an air temperature of 10°C. Repeat, with the plate surface having a reflectance of 0.8 below 1 μm, and an emittance of 0.2 above 1 μm. Then estimate the rate at which heat could be withdrawn from this plate if its temperature were lowered by 30°C.

6-79 Repeat the calculations in Problem 6-78, but assume that the plate is covered by a thin plastic film which transmits 95% solar radiation, but only 10% infrared and is placed 1 cm above the plate.

6-80 The sketch below shows a solar system designed to supply electric power. Determine the maximum amount of power that can be produced by this system when the incident radiation is 3,200 kJ/h·m². The characteristics of the collector are: Overall heat loss coefficient $(U_l) = 7.0$ kJ/h·m²·K; Collector efficiency factor $(F') = 0.92$. Mass flow of water through the collector, $\dot{m} = 30$ kg/h·m². Transmittance absorptance product $(\tau\alpha) = 0.80$. Ambient temperature, $T_{amb} = 20$°C. Collector area, $A_c = 50,000$ m². Assume that the heat engine is approximately one-third as efficient as a Carnot engine operating between the ambient temperature as a sink and the collector outlet temperature, $T_{c,out}$.

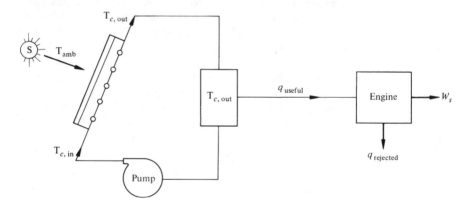

Chapter 7

HEAT EXCHANGERS

7-1 INTRODUCTION

This chapter presents the thermal analysis of various types of heat exchangers. Specifically, two methods of predicting the performance of industrial heat exchangers will be outlined and techniques of estimating the required size and the most suitable type of heat exchangers to accomplish a specified task will be presented. As a part of the thermal performance, the pressure-drop characteristics of heat exchangers and equipment necessary to move the fluid through the exchanger will also be discussed.

In a complete engineering design of thermal equipment, not only the thermal performance characteristics, but also the economics of the system are important. The role of heat exchangers has taken on increasing importance recently as engineers are becoming energy-conscious and want to optimize designs not only in terms of a thermal analysis and economic return on the investment, but also in terms of the energy payback of a system. Thus economics, as well as considerations such as the availability and amount of energy and raw materials necessary to accomplish a given task, should be considered.

Any time a heat exchanger is placed into a thermal transfer system, a temperature drop is required to transfer the heat. The size of this temperature drop can be decreased by utilizing a larger heat exchanger, but this in turn will impose an economic as well as energy investment penalty. Considerations along these lines will play an increasingly important part in

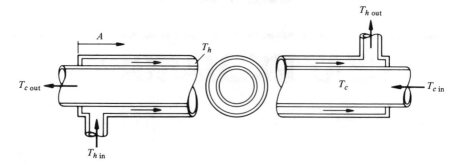

Figure 7-1 Simple tube-within-a-tube counterflow heat exchanger.

years to come, but they are beyond the scope of this chapter. We will, however, consider as a part of the total spectrum of heat-exchanger equipment not only shell-and-tube heat exchangers, but also fixed beds, solar collection heat exchangers, and heat pipes.

7-2 BASIC TYPES OF HEAT EXCHANGERS

The simplest type of heat exchanger consists of a tube within a tube as shown in Fig. 7-1. Such an arrangement can be operated either in counterflow or in parallel flow, with either the hot or cold fluid passing through the annular space and the other fluid passing through the inside of the inner pipe.

A more common type of heat exchanger, widely used in the chemical and process industry, is the shell-and-tube arrangement shown in Fig. 7-2. In this type of heat exchanger one fluid flows inside the tubes while the other fluid is forced through the shell and over the outside of the tubes. The reason for forcing the fluid to flow over the tubes rather than along

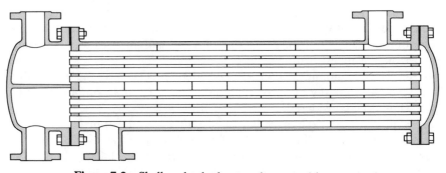

Figure 7-2 Shell-and-tube heat exchanger with segmental baffles: two tube passes, one shell pass.

the tubes is that a higher heat-transfer coefficient can be achieved in cross flow than in flow parallel to the tubes. To achieve cross flow on the shell side, baffles are placed inside the shell as shown in Fig. 7-2. These baffles ensure that in each section the flow passes across the tubes flowing downward in the first, upward in the second section, and so on. Depending upon the header arrangements at the two ends of the heat exchanger, one or more tube passes can be achieved. For a two-tube pass arrangement, the inlet header is split so that the fluid flowing into the tubes passes through one-half of the tubes in one direction, then turns around and returns through the other half of the tubes to where it started, as shown in Fig. 7-2.

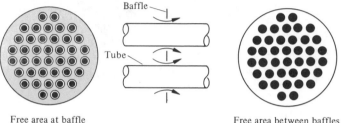

Free area at baffle

Free area between baffles

(a)

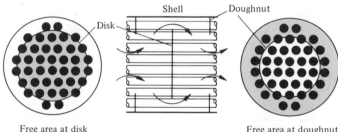

Free area at disk

Free area at doughnut

(b)

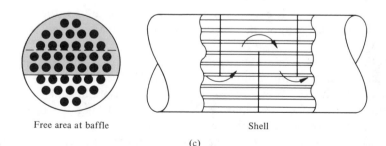

Free area at baffle

Shell

(c)

Figure 7-3 Three types of baffles used in shell-and-tube heat exchangers: (a) orifice baffle; (b) disk-and-doughnut baffle; (c) segmental baffle. (After C. B. Cramer, *Heat Transfer*, 2nd. ed., Intext Publishers Group, New York.)

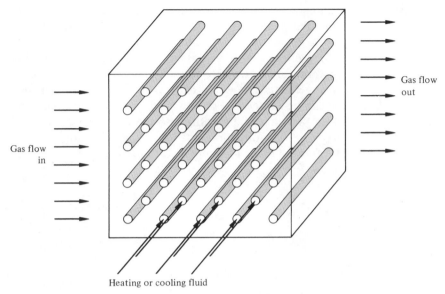

Gas flow out

Gas flow in

Heating or cooling fluid

Figure 7-4 Cross-flow air heater illustrating cross flow with one fluid mixed, the other unmixed.

Three- and four-tube passes can be achieved by rearrangement of the header space. A variety of baffles have been used in industry, but the most common kind is the disk-and-doughnut baffle shown in Fig. 7-3b.

In gas heating or cooling it is often convenient to use a cross-flow heat exchanger such as shown in Fig. 7-4. In such a heat exchanger, one of the fluids passes through the tubes, whereas the gaseous fluid is forced across the tube bundle. The flow of the exterior fluid may be by forced or by free convection. In this type of exchanger the gas flowing across the tube is considered to be *mixed*, whereas the fluid in the tube is considered to be *unmixed*. The exterior gas flow is mixed because it can move about freely between the tubes as it exchanges heat, whereas the fluid within the tubes is confined and cannot mix with any other stream during the heat-exchange process.

Another type of cross-flow heat exchanger, which is widely used in the space and comfort heating industry for homes, is shown in Fig. 7-5. In this arrangement the gas flows across a finned tube bundle and is unmixed because it is confined in separate flow passages during the process. As shown in the temperature profile above the cross-flow heat exchanger, when the fluid is unmixed it will have a temperature gradient both parallel and normal to the flow direction. On the other hand, when the fluid is well mixed, there will be a tendency for the fluid temperature to equalize in the direction normal to the flow, and there will only be a temperature gradient

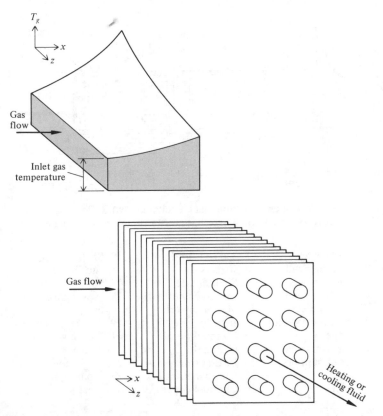

Figure 7-5 Type of cross-flow heat exchanger, widely used in space and comfort heating industry for homes.

in the direction of the flow. In the design of heat exchangers it is important to specify whether or not the fluids are mixed or unmixed, and which of the fluids is mixed. It is also important to balance the temperature drop by obtaining approximately equal heat-transfer coefficients on the exterior and interior of the tubes. If this is not done, one of the thermal resistances may be unduly large and cause an unnecessarily high overall temperature drop for a given rate of heat transfer, which in turn demands larger equipment and results in poor economics.

The shell-and-tube heat exchanger illustrated in Fig. 7-2 has fixed tube plates at each end and the tubes are welded or expanded into the plates. This type of construction has the lowest initial cost but can only be used for small temperature differences between the hot and the cold fluid because no provision is made to prevent thermal stresses due to the differential expansion between the tubes and the shell. Another disadvantage is that the tube bundle cannot be removed for cleaning. These drawbacks can be overcome by the modification of the basic design as

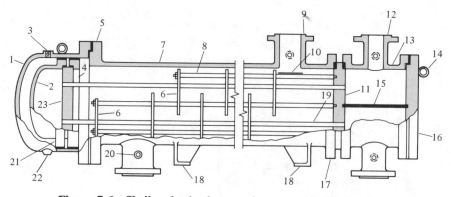

Figure 7-6 Shell-and-tube heat exchanger with floating head. (Courtesy of the Tubular Exchange Manufacturers Association.)

Key:

1. Shell cover
2. Floating head
3. Vent connection
4. Floating-head backing device
5. Shell cover—end flange
6. Transverse baffles or support plates
7. Shell
8. Tie rods and spacers
9. Shell nozzle
10. Impingement baffle
11. Stationary tube sheet
12. Channel nozzle
13. Channel
14. Lifting ring
15. Pass partition
16. Channel cover
17. Shell channel—end flange
18. Support saddles
19. Heat-transfer tube
20. Test connection
21. Floating-head flange
22. Drain connection
23. Floating tube sheet

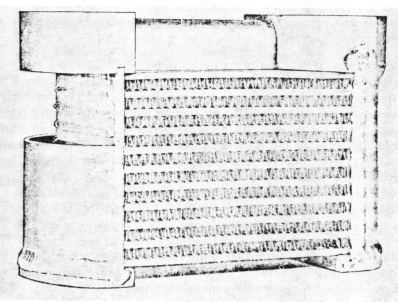

Figure 7-7 Typical compact heat-exchanger section. (Courtesy of the Harrison Radiator Division, General Motors Corp.)

shown in Fig. 7-6. In this arrangement one tube plate is fixed but the other is bolted to a floating-head cover which permits the tube bundle to move relative to the shell. The floating tube sheet is clamped between the floating head and a flange so that it is possible to remove the tube bundle for cleaning. The heat exchanger shown in Fig. 7-6 has one shell pass and two tube passes.

For certain special applications such as regenerators for aircraft or automobile gas turbines, the rate of heat transfer per unit weight and unit volume is the prime consideration. Compact, lightweight heat exchangers for this type of service have been investigated by Kays and London (Ref. 1). A typical design is shown in Fig. 7-7. For a complete description and analysis of compact heat exchangers, especially for the application of fins to increase the effectiveness of such units, the reader is referred to References 2, 3, and 4.

7-3 THE OVERALL HEAT-TRANSFER COEFFICIENT

One of the first tasks in a thermal analysis of a shell-and-tube heat exchanger is to evaluate the overall heat-transfer coefficient between the two fluid streams. It was already shown in Chapter 1 that the overall heat-transfer coefficient between a hot fluid at temperature T_h and a cold fluid at temperature T_c separated by a solid plane wall is defined by the equation

$$q = UA(T_h - T_c) \tag{7-1}$$

where

$$UA = \frac{1}{\sum\limits_{n=1}^{n=3} R_n} = \frac{1}{(1/h_1 A) + (L/kA) + (1/h_2 A)}$$

For a tube-within-a-tube heat exchanger, as shown in Fig. 7-1, the area at the inner surface is $2\pi r_i L$ and the area at the outer surface is $2\pi r_o L$. Thus, if the overall heat-transfer coefficient is *based on the outer area A_o*,

$$U_o = \frac{1}{\dfrac{A_o}{A_i h_i} + \dfrac{A_o \ln(r_o/r_i)}{2\pi k L} + \dfrac{1}{h_o}} \tag{7-2}$$

while *on the basis of the inner area A_i* we get

$$U_i = \frac{1}{\dfrac{1}{h_i} + \dfrac{A_i \ln(r_o/r_i)}{2\pi k L} + \dfrac{A_i}{A_o h_o}} \tag{7-3}$$

Table 7-1   Overall Heat-Transfer Coefficients (W/m². K)

HEAT FLOW→TO: from:	GAS (FREE CONVECTION) $\bar{h}_c = 5-15$	GAS (FLOWING) $\bar{h}_c = 10-100$	LIQUID (FREE CONVECTION) $\bar{h}_c = 50-1000$	LIQUID (FLOWING) WATER $\bar{h}_c = 3000-1000$ OTHER LIQUIDS $\bar{h}_c = 500-2000$	BOILING LIQUID WATER $\bar{h}_c = 3500-60{,}000$ OTHER LIQUIDS $\bar{h}_c = 1000-20{,}000$
Gas (free convection) $\bar{h}_c = 5-15$	Room/outside air through glass $U = 1-2$	Superheaters $U = 3-10$		Combustion chamber $U = 10-40+$ radiation	Steam boiler $U = 10-40+$ radiation
Gas (flowing) $\bar{h}_c = 10-100$		Heat exchangers for gases $U = 10-30$	Gas boiler $U = 10-50$		
Liquid (free convection) $\bar{h}_c = 50-10{,}000$	Radiator central heating		Oil bath for heating $U = 25-500$	Cooling coil $U = 500-1500$ if stirred	
Liquid (flowing) water $\bar{h}_c = 3000-10{,}000$ other liquids $\bar{h}_c = 500-3000$		Gas coolers $U = 10-50$	Heating coil in vessel water/water without stirring $U = 50-250$ with stirring $U = 500-2000$	Heat exchanger water/water $U = 900-2500$ water/other liquids $U = 200-1000$	Evaporators of refrigerators or brine coolers $U = 300-1000$
Condensing vapor water $\bar{h}_c = 5000-30{,}000$ other liquids $\bar{h}_c = 1000-4000$	Steam radiators $U = 5-20$	Air heaters $U = 10-50$	Steam jackets around vessels with stirrers, water $U = 300-1000$ other liquids $U = 150-500$	Condensers steam/water $U = 1000-4000$ other vapor/water $U = 300-1000$	Evaporators steam/water $U = 1500-6000$ steam/other liquids $U = 300-2000$

Source: W. J. Beek and K. M. K. Muttzall, *Transport Phenomena*, John Wiley & Sons, Inc., New York, 1975.

Although for a careful and precise design it is always necessary to calculate the individual heat-transfer coefficients, for preliminary estimates it is often useful to have an approximate value of U, typical of conditions encountered in practice. Table 7-1 lists a few typical values of U for various applications (Ref. 5). It should be noted that in many cases the value of U is almost completely determined by the thermal resistance at one of the fluid/solid interfaces, as for example when one of the fluids is a gas and the other a liquid or if one of the fluids is a boiling liquid with a very large heat-transfer coefficient.

Fouling Factors

The overall heat-transfer coefficient of a heat exchanger under operating conditions, especially in the process industry, can often not be predicted from thermal analysis alone. During operation with most liquids and some gases, a dirt film gradually builds up on the heat-transfer surface. This deposit may be rust, boiler scale, silt, coke, or any number of things. Its effect, which is referred to as *fouling*, is to increase the thermal resistance. The manufacturer cannot usually predict the nature of the dirt deposit or the rate of fouling. Therefore, only the performance of clean exchangers can be guaranteed. The thermal resistance of the deposit can generally be obtained only from actual tests or from experience. If performance tests are made on a clean exchanger and repeated later after the unit has been in service for some time, the thermal resistance of the deposit (or *fouling factor*), R_d, can be determined from the relation

$$R_d = \frac{1}{U_d} - \frac{1}{U} \tag{7-4}$$

where

$$U = \text{unit conductance of clean exchanger}$$
$$U_d = \text{conductance after fouling has occurred}$$
$$R_d = \text{unit thermal resistance of deposit.}$$

A convenient working form of Eq. 7-4 is

$$U_d = \frac{1}{R_d + 1/U}$$

Fouling factors for various applications have been compiled by the Tubular Exchanger Manufacturers Association and are available in their publication (Ref. 6). A few samples are given in Table 7-2. The fouling factors should be applied as indicated in the following equation for the overall

Table 7-2 Normal Fouling Factors

TYPE OF FLUID	FOULING FACTOR, R_d $(m^2 \cdot K/W)$
Seawater	
Below 325 K	0.00009
Above 325 K	0.0002
Treated boiler feedwater above 325 K	0.0002
Fuel oil	0.0009
Quenching oil	0.0007
Alcohol vapors	0.00009
Steam, non-oil-bearing	0.00009
Industrial air	0.0004
Refrigerating liquid	0.0002

Source: *Standards of Tubular Exchanger Manufacturers Association*, 4th ed., 1959.

design heat-transfer coefficient U_d of *unfinned* tubes:

$$U_d = \frac{1}{\dfrac{1}{\bar{h}_o} + R_o + R_k + \dfrac{R_i A_o}{A_i} + \dfrac{A_o}{\bar{h}_i A_i}} \tag{7-5}$$

where

U_d = design overall coefficient of heat transfer, $W/m^2 \cdot K$, based on unit area of outside tube surface

$\bar{h}_o$ = average unit-surface conductance of fluid on outside of tubing, $W/m^2 \cdot K$

$\bar{h}_i$ = average unit-surface conductance of fluid inside tubing, $W/m^2 \cdot K$

R_o = unit fouling resistance on outside of tubing, $m^2 \cdot K/W$

R_i = unit fouling resistance on inside of tubing, $m^2 \cdot K/W$

R_k = unit resistance of tubing in $m^2 \cdot K/W$ based on outside tube surface area

$\dfrac{A_o}{A_i}$ = ratio of outside tube surface to inside tube surface area

7-4 THE LOG-MEAN TEMPERATURE DIFFERENCE

The temperatures of fluids in a heat exchanger are generally not constant, but vary from point to point as heat flows from the hotter to the colder fluid. Even for a constant thermal resistance, the rate of heat flow will therefore vary along the path of the exchangers because its value depends on the temperature difference between the hot and the cold fluid at the section. Figures 7-8, 7-9, 7-10, and 7-11 illustrate the changes in temperature that may occur in either or both fluids in a simple shell-and-tube exchanger (Fig. 7-1). The distances between the solid lines are proportional to the temperature differences ΔT between the two fluids.

Figure 7-8 illustrates the case where a vapor is condensing at a constant temperature while the other fluid is being heated. Figure 7-9 is representative of a case where a liquid is evaporated at constant temperature while heat is flowing from a warmer fluid whose temperature decreases as it passes through the heat exchanger. For both of these cases the direction of flow of either fluid is immaterial and the constant-temperature medium may also be at rest. Figure 7-10 represents conditions in a parallel-flow exchanger, and Fig. 7-11 applies to counterflow. No change of phase

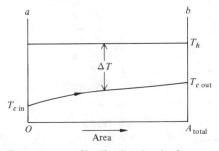

Figure 7-8 Temperature distribution in single-pass condenser.

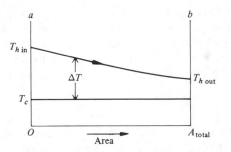

Figure 7-9 Temperature distribution in single-pass evaporator.

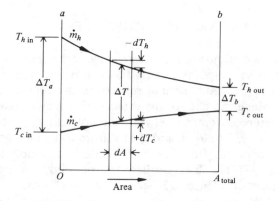

Figure 7-10 Temperature distribution in single-pass parallel-flow heat exchanger.

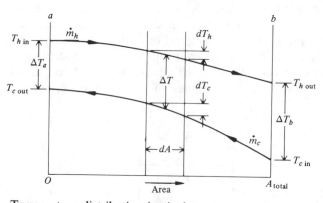

Figure 7-11 Temperature distribution in single-pass counterflow heat exchanger.

occurs in the latter two cases. Inspection of Fig. 7-10 shows that, no matter how long the exchanger is, the final temperature of the colder fluid can never reach the exit temperature of the hotter fluid in parallel flow. For counterflow, on the other hand, the final temperature of the cooler fluid may exceed the outlet temperature of the hotter fluid, since a favorable temperature gradient exists all along the heat exchanger. An additional advantage of the counterflow arrangement is that, for a given rate of heat flow, less surface area is required than in parallel flow.

To determine the rate of heat transfer in any of the aforementioned cases the equation

$$dq = U\,dA\,\Delta T \tag{7-6}$$

must be integrated over the heat-transfer area A along the length of the exchanger. If the overall unit conductance U is constant, if changes in kinetic energy are neglected, and if the shell of the exchanger is insulated, Eq. 7-6 can easily be integrated analytically for parallel or counterflow. An energy balance over a differential area dA yields

$$dq = -\dot{m}_h c_{ph} dT_h = \pm \dot{m}_c c_{pc} dT_c = U dA (T_h - T_c) \qquad (7\text{-}7)$$

where $\dot{m}$ is the mass rate of flow in kg/s, c_p is the specific heat at constant pressure in J/kg·K, and T is the average bulk temperature of the fluid in K. The subscripts h and c refer to the hot and cold fluid, respectively; the plus sign in the third term applies to parallel flow and the minus sign to counterflow. If the specific heats of the fluids do not vary with temperature, we can write a heat balance from the inlet to an arbitrary cross section in the exchanger, or

$$-C_h(T_h - T_{h,\text{in}}) = C_c(T_c - T_{c,\text{in}}) \qquad (7\text{-}8)$$

where

$$C_h = \dot{m}_h c_{ph}, \text{ heat-capacity rate of hotter fluid, W/K}$$

$$C_c = \dot{m}_c c_{pc}, \text{ heat-capacity rate of colder fluid, W/K}$$

Solving Eq. 7-8 for T_h gives

$$T_h = T_{h,\text{in}} - \frac{C_c}{C_h}(T_c - T_{c,\text{in}}) \qquad (7\text{-}9)$$

from which we obtain

$$T_h - T_c = -\left(1 + \frac{C_c}{C_h}\right)T_c + \frac{C_c}{C_h}T_{c,\text{in}} + T_{h,\text{in}} \qquad (7\text{-}10)$$

Substituting Eq. 7-10 for $T_h - T_c$ in Eq. 7-7 yields, after some rearrangement,

$$\frac{dT_c}{-[1 + (C_c/C_h)]T_c + (C_c/C_h)T_{c,\text{in}} + T_{h,\text{in}}} = \frac{U \, dA}{C_c} \qquad (7\text{-}11)$$

Integrating Eq. 7-11 over the entire length of the exchanger (i.e., from $A = 0$ to $A = A_{\text{total}}$) yields

$$\ln\left\{\frac{-[1 + (C_c/C_h)]T_{c,\text{out}} + (C_c/C_h)T_{c,\text{in}} + T_{h,\text{in}}}{-[1 + (C_c/C_h)]T_{c,\text{in}} + (C_c/C_h)T_{c,\text{in}} + T_{h,\text{in}}}\right\} = -\left(\frac{1}{C_c} + \frac{1}{C_h}\right)UA$$

which can be simplified to read

$$\ln\left[\frac{(1 + C_c/C_h)(T_{c,\text{in}} - T_{c,\text{out}}) + T_{h,\text{in}} - T_{c,\text{in}}}{T_{h,\text{in}} - T_{c,\text{in}}}\right] = -\left(\frac{1}{C_c} + \frac{1}{C_h}\right)UA$$

$$(7\text{-}12)$$

From Eq. 7-8 we obtain for the total length of the exchanger

$$\frac{C_c}{C_h} = -\frac{T_{h,\text{out}} - T_{h,\text{in}}}{T_{c,\text{out}} - T_{c,\text{in}}} \qquad (7\text{-}13)$$

which can be used to eliminate the heat capacity rates in Eq. 7-12. After some rearrangement we get

$$\ln\left(\frac{T_{h,\text{out}} - T_{c,\text{out}}}{T_{h,\text{in}} - T_{c,\text{in}}}\right) = \left[(T_{h,\text{out}} - T_{c,\text{out}}) - (T_{h,\text{in}} - T_{c,\text{in}})\right]\frac{UA}{q} \qquad (7\text{-}14)$$

since

$$q = C_c(T_{c,\text{out}} - T_{c,\text{in}}) = C_h(T_{h,\text{in}} - T_{h,\text{out}})$$

Letting $T_h - T_c = \Delta T$, Eq. 7-14 can be written

$$q = UA\,\frac{\Delta T_a - \Delta T_b}{\ln(\Delta T_a / \Delta T_b)} \qquad (7\text{-}15)$$

where the subscripts a and b refer to the respective ends of the exchanger (see Figs. 7-10 and 7-11). In practice it is convenient to use an average effective temperature difference $\overline{\Delta T}$ for the entire heat exchanger, defined by

$$q = UA\,\overline{\Delta T} \qquad (7\text{-}16)$$

Comparing Eqs. 7-15 and 7-16, one finds that, for parallel or counterflow,

$$\overline{\Delta T} = \frac{\Delta T_a - \Delta T_b}{\ln(\Delta T_a / \Delta T_b)} \qquad (7\text{-}17)$$

which is called the *logarithmic-mean temperature difference* often designated by LMTD. The LMTD also applies when the temperature of one of the fluids is constant, as shown in Figs. 7-8 and 7-9. When $\dot{m}_h c_{ph} = \dot{m}_c c_{pc}$, the temperature difference is constant in counterflow and $\overline{\Delta T} = \Delta T_a = \Delta T_b$. If the temperature difference ΔT_a is not more than 50 percent greater than ΔT_b, the arithmetic mean temperature difference will be within 1 percent of the LMTD and may be used to simplify calculations.

The use of the logarithmic mean temperature is only an approximation in practice because U is generally not constant. In design work, however, the overall conductance is usually evaluated at a mean section, halfway between ends, and treated as constant. If U varies considerably, a numerical step-by-step integration of Eq. 7-6 may be necessary.

For more complex heat exchangers such as the shell-and-tube arrangements with several tube or shell passes and with cross-flow exchangers having mixed and unmixed flow, the mathematical derivation of an expression for the mean temperature difference becomes quite complex. The usual procedure is to modify the simple LMTD by correction factors which have been published in chart form by Bowman et al. (Ref. 7) and by the Tubular Exchanger Manufacturer's Association (Ref. 6). Four of these graphs* are shown in Figs. 7-12, 7-13, 7-14, and 7-15. The ordinate of each is the correction factor F. To obtain the true mean temperature for any of these arrangements, the LMTD calculated for *counterflow* must be multi-

---

*Correction factors for several other arrangements are presented in Reference 6.

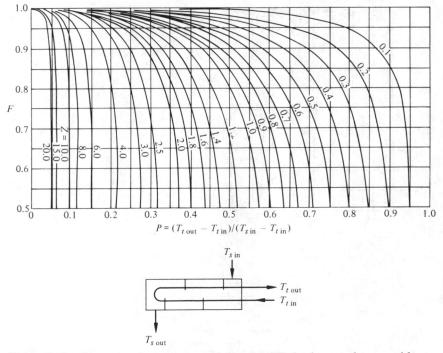

Figure 7-12 Correction factor to counterflow LMTD for heat exchanger with one shell pass and two, or a multiple of two tube passes. (Courtesy of the Tubular Exchange Manufacturers Association.)

plied by the appropriate correction factor, that is,

$$\Delta T_{\text{true mean}} = \text{LMTD} \times F \tag{7-18}$$

The values shown on the abscissa are for the dimensionless temperature-difference ratio

$$P = \frac{T_{t,\text{out}} - T_{t,\text{in}}}{T_{s,\text{in}} - T_{t,\text{in}}}. \tag{7-19}$$

where the subscripts t and s refer to the tube and shell fluid, respectively, and the subscripts in and out refer to the inlet and outlet conditions, respectively. The ratio P is an indication of the heating or cooling effectiveness and can vary from zero for a constant temperature of one of the fluids to unity for the case when inlet temperature of the hotter fluid equals the outlet temperature of the colder fluid. The parameter for each of the curves Z is equal to the ratio of the products of the mass flow rate times the heat capacity of the two fluids $\dot{m}_t c_{pt} / \dot{m}_s c_{ps}$. This ratio is also

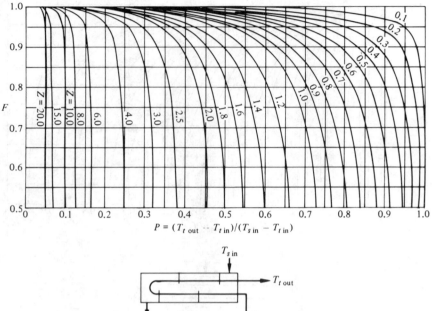

$$P = (T_{t\,out} - T_{t\,in})/(T_{s\,in} - T_{t\,in})$$

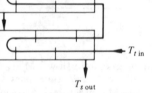

Figure 7-13 Correction factor to counterflow LMTD for heat exchanger with two shell passes and a multiple of two tube passes. (Courtesy of the Tubular Exchange Manufacturers Association.)

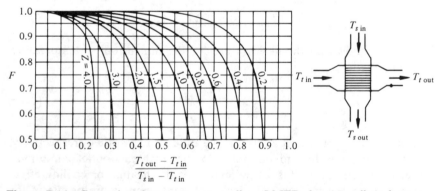

$$\frac{T_{t\,out} - T_{t\,in}}{T_{s\,in} - T_{t\,in}}$$

Figure 7-14 Correction factor to counterflow LMTD for cross-flow heat exchangers, fluid on shell side mixed, other fluid unmixed, one tube pass. (Extracted from R. A. Bowman, A. C. Mueller, and W. M. Nagel, "Mean Temperature Difference in Design," *Trans. ASME*, vol. 62, 1940, with permission of the publishers, The American Society of Mechanical Engineers.)

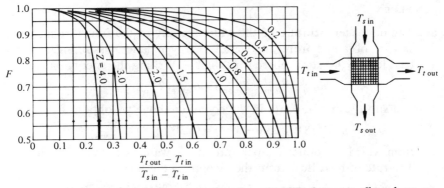

Figure 7-15 Correction factor to counterflow LMTD for cross-flow heat exchanger, both fluids unmixed, one tube pass. (Extracted from R. A. Bowman, A. C. Mueller, and W. M. Nagel, "Mean Temperature Difference in Design," *Trans. ASME*, vol. 62, 1940, with permission of the publishers, The American Society of Mechanical Engineers.)

equal to the temperature change of the shell fluid divided by the temperature change of the fluid in the tubes:

$$Z = \frac{\dot{m}_t c_{pt}}{\dot{m}_s c_{ps}} = \frac{T_{s,\text{in}} - T_{s,\text{out}}}{T_{t,\text{out}} - T_{t,\text{in}}} \tag{7-20}$$

In the application of the correction factors it is immaterial whether the warmer fluid flows through shell or tubes. If the temperature of either of the fluids remains constant, the direction of flow is also immaterial, since F equals 1 and the LMTD applies directly.

Example 7-1. Determine the heat-transfer surface area required for a heat exchanger constructed from a 0.0254-m-o.d tube to cool 6.93 kg/s of a 95% ethyl alcohol solution ($c_p = 3810$ J/kg·K) from 65.6°C to 39.4°C, using 6.30 kg/s of water available at 10°C. Assume that the overall coefficient of heat transfer based on the outer-tube area is 568 W/m²·K and consider each of the following arrangements:

a. Parallel-flow tube and shell.
b. Counterflow tube and shell.
c. Reversed-current exchanger with two shell passes and 72 tube passes, the alcohol flowing through the shell and the water flowing through the tubes.
d. Cross flow, with one tube pass and one shell pass, shell-side fluid mixed.

Solution

a. The outlet temperature of the water for any of the four arrangements can be obtained from an overall energy balance, assuming that the heat loss to the atmosphere is negligible. Writing the energy balance as

$$\dot{m}_h c_{ph}(T_{h,\text{in}} - T_{h,\text{out}}) = \dot{m}_c c_{pc}(T_{c,\text{out}} - T_{c,\text{in}})$$

and substituting the data in the equation above we obtain

$$(6.93)(3810)(65.6 - 39.4) = (6.30)(4187)(T_{c,\text{out}} - 10)$$

from which the outlet temperature of the water is found to be 36.2°C. The rate of heat flow from the alcohol to the water is

$$q = \dot{m}_h c_{ph}(T_{h,\text{in}} - T_{h,\text{out}}) = (6.93)(3810)(65.6 - 39.4)$$
$$= 691,800 \text{ W}$$

From Eq. 7-17 the LMTD for parallel flow is

$$\text{LMTD} = \frac{\Delta T_a - \Delta T_b}{\ln(\Delta T_a / \Delta T_b)} = \frac{55.6 - 3.2}{\ln(55.6/3.2)} = 18.4°\text{C}$$

From Eq. 7-12 the heat-transfer surface area is

$$A = \frac{q}{(U)(\text{LMTD})} = \frac{691,800}{(568)(18.4)} = 66.2 \text{ m}^2$$

The 830-m length of the exchanger for a 0.0254-m -o.d. tube would be too great to be practical.

b. For the counterflow arrangement, the appropriate mean temperature difference is $65.6 - 36.2 = 29.4°\text{C}$, because $\dot{m}_c c_{pc} = \dot{m}_h c_{ph}$. The required area is

$$A = \frac{q}{(U)(\text{LMTD})} = \frac{619,800}{(568)(29.4)} = 41.4 \text{ m}^2$$

which is 40% less than the area necessary for parallel flow.

c. For the reversed-current arrangement, we determine the appropriate mean temperature difference by applying the correction factor found from Fig. 7-13 to the mean temperature for counterflow.

$$P = \frac{T_{c,\text{out}} - T_{c,\text{in}}}{T_{h,\text{in}} - T_{c,\text{in}}} = \frac{36.2 - 10}{65.6 - 10} = 0.47$$

and the heat-capacity-rate ratio is

$$Z = \frac{\dot{m}_c c_{pc}}{\dot{m}_h c_{ph}} = 1$$

From the chart of Fig. 7-13, $F = 0.97$ and the heat-transfer area is

$$A = \frac{41.4}{0.97} = 42.7 \text{ m}^2$$

The length of the exchanger for seventy-two 0.0254-m-od tubes in parallel would be

$$L = \frac{A/72}{\pi D} = \frac{42.7/72}{\pi(0.0254)} = 7.4 \text{ m}$$

This length is not unreasonable, but if it is desirable to shorten the exchanger, more tubes could be used.

d. For the cross-flow arrangement (Fig. 7-3), the correction factor is found from the chart of Fig. 7-14 to be 0.88. The required surface area is thus 48.5 m², about 10% larger than that for the reversed-current exchanger.

7-5 HEAT-EXCHANGER EFFECTIVENESS

In the thermal analysis of the various types of heat exchangers presented in the preceding section, an equation (Eq. 7-16) of the type

$$q = UA\,\Delta T_{\text{mean}}$$

was used. This form will be found convenient when all the terminal temperatures necessary for the evaluation of the appropriate mean temperature are known, and Eq. 7-16 is widely employed in the design of heat exchangers to given specifications. There are, however, numerous occasions when the performance of a heat exchanger (i.e., U) is known, or can at least be estimated, but the temperatures of the fluids leaving the exchanger are not known. This type of problem is encountered in the selection of a heat exchanger or when the unit has been tested at one flow rate, but service conditions require different flow rates for one or both fluids. The outlet temperatures and the rate of heat flow can only be found by a rather tedious trial-and-error procedure if the charts presented in the preceding section are used. In such cases it is desirable to circumvent entirely any reference to the logarithmic or any other mean temperature difference. A method that accomplishes this has been proposed by Nusselt (Ref. 8) and Ten Broeck (Ref. 9).

To obtain an equation for the rate of heat transfer which does not involve any of the outlet temperatures, we introduce the *heat-exchanger effectiveness* $\mathscr{E}$. The heat-exchanger effectiveness is defined as the ratio of the actual rate of heat transfer in a given heat exchanger to the maximum possible rate of heat exchange. The latter would be obtained in a counterflow heat exchanger of infinite heat-transfer area. In this type of unit, if there are no external heat losses, the outlet temperature of the colder fluid equals the inlet temperature of the hotter fluid when $\dot{m}_c c_{pc} < \dot{m}_h c_{ph}$; when $\dot{m}_h c_{ph} < \dot{m}_c c_{pc}$, the outlet temperature of the warmer fluid equals the inlet temperature of the colder one. In other words, the effectiveness compares the actual heat-transfer rate to the maximum rate whose only limit is the second law of thermodynamics. Depending on which of the heat capacity rates is smaller, the effectiveness is

$$\mathcal{E} = \frac{C_h(T_{h,\text{in}} - T_{h,\text{out}})}{C_{\min}(T_{h,\text{in}} - T_{c,\text{in}})} \tag{7-21a}$$

or

$$\mathcal{E} = \frac{C_c(T_{c,\text{out}} - T_{c,\text{in}})}{C_{\min}(T_{h,\text{in}} - T_{c,\text{in}})} \tag{7-21b}$$

where $C_{\min}$ is the smaller of the $\dot{m}_h c_{ph}$ and $\dot{m}_c c_{pc}$ magnitudes.

Once the effectiveness of a heat exchanger is known, the rate of heat transfer can be determined directly from the equation

$$q = \mathcal{E} C_{\min}(T_{h,\text{in}} - T_{c,\text{in}}) \tag{7-22}$$

since

$$\mathcal{E} C_{\min}(T_{h,\text{in}} - T_{c,\text{in}}) = C_h(T_{h,\text{in}} - T_{h,\text{out}}) = C_c(T_{c,\text{out}} - T_{c,\text{in}})$$

Equation 7-22 is the basic relation in this analysis because it expresses the rate of heat transfer in terms of the effectiveness, the smaller heat capacity rate, and the difference between the inlet temperatures. It replaces Eq. 7-16 in the LMTD analysis but does not involve the outlet temperatures. Equation 7-22 is, of course, also suitable for design purposes instead of Eq. 7-16.

We shall illustrate the method of deriving an expression for the effectiveness of a heat exchanger by applying it to a parallel-flow arrangement. The effectiveness can be introduced into Eq. 7-13 by replacing $(T_{c,\text{in}} - T_{c,\text{out}})/(T_{h,\text{in}} - T_{c,\text{in}})$ by the effectiveness relation from Eq. 7-21. We obtain

$$\ln\left[1 - \mathcal{E}\left(\frac{C_{\min}}{C_h} + \frac{C_{\min}}{C_c}\right)\right] = -\left(\frac{1}{C_c} + \frac{1}{C_h}\right)UA$$

or

$$1 - \mathcal{E}\left(\frac{C_{\min}}{C_h} + \frac{C_{\min}}{C_c}\right) = e^{-(1/C_c + 1/C_h)UA}$$

Solving for $\mathcal{E}$ yields

$$\mathcal{E} = \frac{1 - e^{-[1 + (C_h/C_c)]UA/C_h}}{(C_{\min}/C_h) + (C_{\min}/C_c)} \tag{7-23}$$

When C_h is less than C_c, the effectiveness becomes

$$\mathcal{E} = \frac{1 - e^{-[1 + (C_h/C_c)]UA/C_h}}{1 + (C_h/C_c)} \tag{7-24a}$$

and when $C_c < C_h$, we obtain

$$\mathcal{E} = \frac{1 - e^{-[1+(C_c/C_h)]UA/C_c}}{1+(C_c/C_h)} \qquad (7\text{-}24b)$$

The effectiveness for both cases can therefore be written in the form

$$\mathcal{E} = \frac{1 - e^{-[1+(C_{min}/C_{max})]UA/C_{min}}}{1+(C_{min}/C_{max})} \qquad (7\text{-}25)$$

The foregoing derivation illustrates how the effectiveness for a given flow arrangement can be expressed in terms of two dimensionless parameters, the heat-capacity rate-ratio C_{min}/C_{max} and the ratio of the overall conductance to the smaller heat capacity rate, UA/C_{min}. The latter of the two parameters is called the *number of heat-transfer units*, or NTU. The number of heat-transfer units is a measure of the heat-transfer size of the exchanger. The larger the value of NTU, the closer the heat exchanger approaches its thermodynamic limit. By analyses which in principle are similar to the one presented here for parallel flow, effectivenesses may be evaluated for most flow arrangements of practical interest. The results have been put by Kays and London (Ref. 1) into convenient graphs from

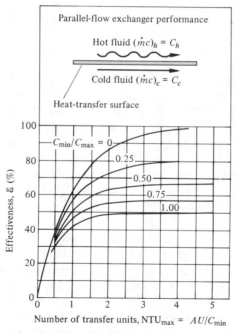

Figure 7-16 Heat-exchanger effectiveness for parallel flow. (By permission from W. M. Kays and A. L. London, *Compact Heat Exchangers*, National Press, Palo Alto, Calif., 1955.)

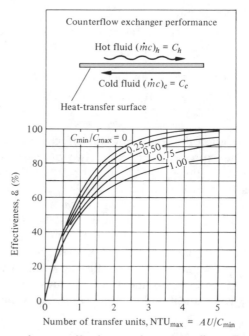

Figure 7-17 Heat-exchanger effectiveness for counterflow. (By permission from W. M. Kays and A. L. London, *Compact Heat Exchangers*, National Press, Palo Alto, Calif., 1955.)

which the effectiveness can be determined for given values of NTU and C_{min}/C_{max}. The effectiveness curves for some common flow arrangements are shown in Figs. 7-16 to 7-20. The abscissas of these figures are the NTUs of the heat exchangers. The constant parameter for each curve is the heat-capacity-rate ratio C_{min}/C_{max}, and the effectiveness is read on the ordinate. Note that, for an evaporator or condenser, $C_{min}/C_{max}=0$, because if one fluid remains at constant temperature throughout the exchanger, its effective specific heat, and thus its heat capacity rate, is by definition equal to infinity.

Example 7-2. From a performance test on a well-baffled single-shell, two-tube-pass heat exchanger, the following data are available: oil ($c_p = 2100$ J/kg·K) in turbulent flow inside the tubes entered at 340 K at the rate of 1.00 kg/s and left at 310 K; water flowing on the shell side entered at 290 K and left at 300 K. A change in service conditions requires the cooling of a similar oil from an initial temperature of 370 K but at three-fourths of the flow rate used in the performance test. Estimate the outlet temperature of the oil for the same water rate and inlet temperature as before.

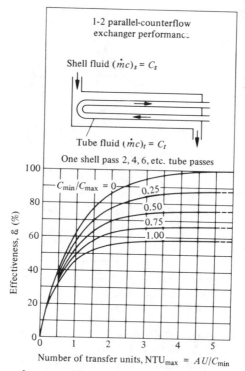

Figure 7-18 Heat-exchanger effectiveness for shell-and-tube heat exchanger with one well-baffled shell pass and two, or a multiple of two, tube passes. (By permission from W. M. Kays and A. L. London, *Compact Heat Exchangers*, National Press, Palo Alto, Calif., 1955.)

Solution: The test data may be used to determine the heat capacity rate of the water and the overall conductance of the exchanger. The heat capacity rate of the water is, from Eq. 7-13,

$$C_c = C_h \frac{T_{h,\text{in}} - T_{h,\text{out}}}{T_{c,\text{out}} - T_{c,\text{in}}} = (1.00)(2100)\frac{340-310}{300-290}$$

$$= 6300 \text{ W/K}$$

and the temperature ratio P is, from Eq. 7-19,

$$P = \frac{T_{t,\text{out}} - T_{t,\text{in}}}{T_{s,\text{in}} - T_{t,\text{in}}} = \frac{340-310}{340-290} = 0.6$$

$$Z = \frac{300-290}{340-310} = 0.33$$

From Fig. 7-12, $F = 0.94$ and the mean temperature difference is

$$\overline{\Delta T} = F \times \text{LMTD} = (0.94)\frac{(340-300)-(310-290)}{\ln\left[(340-300)/(310-290)\right]} = 28.9 \text{ K}$$

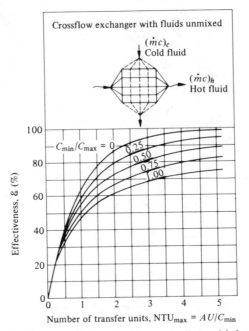

Figure 7-19 Heat-exchanger effectiveness for cross flow with both fluids unmixed. (By permission from W. M. Kays and A. L. London, *Compact Heat Exchangers*, National Press, Palo Alto, Calif., 1955.)

From Eq. 7-16 the overall conductance is

$$UA = \frac{q}{\Delta T} = \frac{(1.00)(2100)(340 - 310)}{28.9} = 2180 \; W/K$$

Since the thermal resistance on the oil side is controlling, a decrease in velocity to 75 percent of the original value will increase the thermal resistance roughly by the velocity ratio raised to the 0.8 power. This can be verified by reference to Eq. 4-88. Under the new conditions the conductance, the NTU, and the heat capacity-rate ratio will therefore be approximately

$$UA \simeq (2180)(0.75)^{0.8} = 1730 \; W/K$$

$$NTU = \frac{UA}{C_{oil}} = \frac{1730}{(0.75)(1.00)(2100)} = 1.10$$

and

$$\frac{C_{oil}}{C_{water}} = \frac{C_{min}}{C_{max}} = \frac{(0.75)(1.00)(2100)}{6300} = 0.25$$

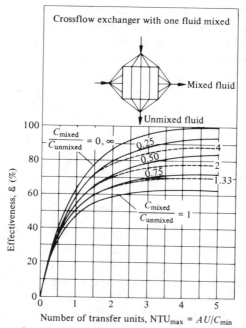

Figure 7-20 Heat-exchanger effectiveness for cross flow with one fluid mixed, the other unmixed. When $C_{mixed}/C_{unmixed} > 1$, NTU_{max} is based on $C_{unmixed}$. (By permission from W. M. Kays and A. L. London, *Compact Heat Exchangers*, National Press, Palo Alto, Calif., 1955.)

From Fig. 7-18 the effectiveness is equal to 0.61. Hence from the definition of $\mathscr{E}$ in Eq. 7-21a, the oil outlet temperature is

$$T_{oil\ out} = T_{oil\ in} - \mathscr{E}\,\Delta T_{max} = 370 - [1.00(370-290)] = 321.2\ K$$

Example 7-3. A flat-plate-type heater (Fig. 7-21) is to be used to heat air with the hot-exhaust gases from a turbine. The required air-flow rate is 0.75 kg/s, entering at 290 K; the hot gases are available at a temperature of 1150 K and at a rate of 0.60 kg/s. Determine the temperature of the air leaving the heat exchanger.

P_a = wetted perimeter on air side—0.703 m

P_g = wetted perimeter on gas side–0.416 m

A_a = cross-sectional area of air passage (per passage)—2.275×10^{-3} m²

A_g = cross-sectional area of gas passage (per passage)—1.600×10^{-3} m²
air passages–19
gas passages—18

Solution: Inspection of Fig. 7-21 shows that the unit is of the cross-flow type, both fluids unmixed. As a first approximation the end effects will be neglected. The flow systems for the air and gas streams are similar to flow in straight ducts having the following dimensions:

$$\text{length of air duct, } L_a = 0.20 \text{ m}$$

$$\text{hydraulic diameter of air duct, } D_{Ha} = \frac{4A_a}{P_a} = 0.0128 \text{ m}$$

$$\text{length of gas duct, } L_g = 0.35 \text{ m}$$

$$\text{hydraulic diameter of gas duct, } D_{Hg} = \frac{4A_g}{P_g} = 0.0154 \text{ m}$$

$$\text{heat-transfer surface area, } A = 2.52 \text{ m}^2$$

The unit conductances may be evaluated from Eq. 5-15 for flow in ducts $(L_a/D_{Ha} = 15.6, L_g/D_{Hg} = 22.7)$. A difficulty arises, however, because the temperatures of both fluids vary along the duct. It is therefore necessary to estimate an average film temperature and refine the calculations after the outlet and wall temperatures have been found. Selecting the average air-side film temperature at 573 K and the average gas-side film temperature at 973 K, the properties at those temperatures are, from Table G-1 (assuming that the properties of the gas are approximated by those of air):

$$\mu_{air} = 3.932 \times 10^{-5} \text{N} \cdot \text{s/m}^2$$

$$\text{Pr}_{air} = 0.71$$

$$k_{air} = 0.0429 \text{ W/m} \cdot \text{K}$$

$$\mu_{gas} = 4.085 \times 10^{-5} \text{ N} \cdot \text{s/m}^2$$

$$\text{Pr}_{gas} = 0.73$$

$$k_{gas} = 0.0623 \text{ W/m} \cdot \text{K}$$

$$c_{p_{air}} = 1047 \text{ J/kg} \cdot \text{K}$$

$$c_{p_{gas}} = 1101 \text{ J/kg} \cdot \text{K}$$

The mass rates per unit area are:

$$(\dot{m}/A)_{air} = \frac{0.75}{(19)(2.275 \times 10^{-3})} = 17.35 \text{ kg/m}^2 \cdot \text{s}$$

$$(\dot{m}/A)_{gas} = \frac{0.60}{(18)(1.600 \times 10^{-3})} = 20.83 \text{ kg/m}^2 \cdot \text{s}$$

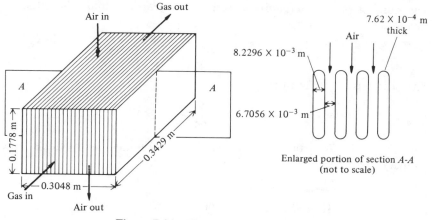

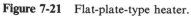

Figure 7-21 Flat-plate-type heater.

The Reynolds numbers are:

$$\text{Re}_{\text{air}} = \frac{(\dot{m}/A)_{\text{air}} D_{Ha}}{\mu_a} = \frac{(17.35)(0.0128)}{3.932 \times 10^{-5}} = 5650$$

$$\text{Re}_{\text{gas}} = \frac{(\dot{m}/A)_{\text{gas}} D_{Hg}}{\mu_g} = \frac{(20.83)(0.0154)}{4.085 \times 10^{-5}} = 7850$$

Using Eq. 5-15, the average unit conductances are:

$$\bar{h}_{\text{air}} = \left(0.036 \frac{k_a}{D_{Ha}} \text{Re}_D^{0.8} \text{Pr}^{0.33}\right)\left(\frac{D_{Ha}}{L_a}\right)^{0.055}$$

$$= \left[(0.036)\frac{0.0429}{0.0128}(5650)^{0.8}(0.71)^{0.33}\right]\left(\frac{1}{15.6}\right)^{0.055}$$

$$= 93.0 \text{ W/m}^2 \cdot \text{K}$$

$$\bar{h}_{\text{gas}} = \left[(0.036)\frac{0.0623}{0.0154}(7850)^{0.8}(0.73)^{0.33}\right]\left(\frac{1}{22.7}\right)^{0.055}$$

$$= 144 \text{ W/m}^2 \cdot \text{K}$$

These unit conductances show that approximately 60% of the overall temperature drop occurs on the air side. If the thermal resistance of the metal wall is neglected, the overall conductance is

$$UA = \cfrac{1}{\cfrac{1}{\bar{h}_a A} + \cfrac{1}{\bar{h}_g A}} = \cfrac{1}{\cfrac{1}{(93.0)(2.52)} + \cfrac{1}{(144)(2.52)}}$$

$$= 142 \text{ W/K}$$

The number of transfer units, based on the warmer fluid which has the smaller heat-capacity rate, are

$$\mathrm{NTU} = \frac{UA}{C_{\min}} = \frac{142}{(0.60)(1101)} = 0.215$$

The heat capacity-rate ratio is

$$\frac{C_g}{C_a} = \frac{(0.60)(1101)}{(0.75)(1047)} = 0.841$$

And from Fig. 7-19 the effectiveness is 0.13. Finally, the average outlet temperatures of the gas and air are

$$T_{\text{gas out}} = T_{\text{gas in}} - \mathcal{E}\,\Delta T_{\max}$$

$$= 1150 - 0.13(1150 - 290) = 1038 \text{ K}$$

$$T_{\text{air out}} = T_{\text{air in}} + \frac{C_g}{C_a}\,\mathcal{E}\,\Delta T_{\max} = 290 + (0.841)(0.13)(1150 - 290)$$

$$= 384 \text{ K}$$

A check on the average air-side and gas-side film temperatures gives values (567 K, 946 K) which are sufficiently close to the assumed values (573 K, 973 K) to make second approximations unnecessary. To appreciate the usefulness of the approach based on the concept of heat exchanger effectiveness, it is suggested that this same problem be worked out by trial and error, using Eq. 7-16 and the chart of Fig. 7-15.

The effectiveness of the heat exchanger in Example 7-3 is very low (13 percent) because the heat-transfer area is too small to utilize the available energy efficiently. The relative gain in heat-transfer performance which can be achieved by increasing the heat-transfer area is well represented on the effectiveness curves. A fivefold increase in area would raise the effectiveness to 60 percent. If, however, a particular design falls near or above the knee of these curves, increasing the surface area will not improve the performance appreciably but may cause an undue increase in the frictional pressure drop.

7-6 SOLAR ENERGY COLLECTORS

A solar energy collector is a heat exchanger capable of using solar radiation to increase the internal energy and temperature of a working fluid. In its simplest form it consists of a tube exposed to solar radiation. The solar insolation is partly absorbed by the tube, the temperature of the tube wall increases, and if a fluid at ambient temperature passes through the tube, heat is transferred from the tube to the fluid and the temperature of the fluid increases until the heat loss from the tube to the surroundings

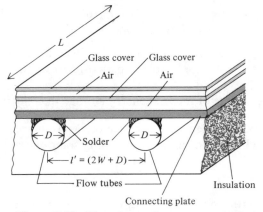

Figure 7-22 Flat-plate-collector cross section.

is equal to the solar energy absorbed. To improve the thermal performance of this simple system, fins can be attached to the tube to increase the area exposed to solar insolation and the heat losses can be reduced by placing one or two layers of glass between the incoming solar energy and the surface absorbing it. Figure 7-22 shows the cross section of a typical solar collector of the flat-plate type. If a fluid such as water passes through the tubes, the useful energy delivered to the working fluid, q_u, is

$$q_u = \dot{m}c_p\left(T_{f,\text{out}} - T_{f,\text{in}}\right) \qquad (7\text{-}26)$$

where

$\dot{m}$ = mass flow rate through collector

c_p = specific heat at constant pressure of working fluid

$T_{f,\text{out}} - T_{f,\text{in}}$ = temperature rise of working fluid passing through collector

Energy Balance for a Flat-Plate Collector

The thermal performance of a solar collector can be evaluated by an energy balance that determines the portion of the incoming radiation delivered as useful energy to the working fluid. For a flat-plate collector of area A_c, this energy balance is

$$I_c A_c \bar{\tau}_s \alpha_{s,c} = q_u + q_{\text{loss}} + \frac{de_c}{dt} \qquad (7\text{-}27)$$

where

I_c = solar irradiation on collector surface

$\bar{\tau}_s$ = effective solar transmittance of collector cover(s)

$\alpha_{s,c}$ = solar absorptance of collector absorber plate surface

q_u = rate of heat transfer from collector absorber plate to working fluid

q_{loss} = rate of heat transfer (or heat loss) from collector absorber plate to surroundings

$\dfrac{de_c}{dt}$ = rate of internal energy storage in collector

The instantaneous efficiency of a collector, η_c, is simply the ratio of the useful energy delivered to the total incoming solar energy:

$$\eta_c = \frac{q_u}{A_c I_c} \tag{7-28}$$

In practice, the efficiency must be measured over a finite time period. In a standard NBS performance test (Ref. 10) the required period is of the order of 15 or 20 min, whereas for design the performance over some longer period t (e.g., a day or a month) is important. Then, the average efficiency is

$$\bar{\eta}_c = \frac{\displaystyle\int_0^t q_u\, dt}{\displaystyle\int_0^t A_c I_c\, dt} \tag{7-29}$$

where t is the duration of the time period over which the performance is averaged.

A detailed and precise analysis of the efficiency of a solar collector is complicated by the nonlinear behavior of radiation heat transfer. However, a simple linearized analysis is usually sufficiently accurate in practice and it illustrates the parameters of significance for a solar collector and how these parameters interact. Although for the design and economic evaluation solar systems the results of standard NBS performance tests are generally used, for a proper analysis and interpretation of these test results an understanding of the thermal analysis is imperative.

Collector-Heat-Loss Conductance

To gain an understanding of the parameters that determine the thermal efficiency of a solar collector, it is important to develop the concept of overall *collector-heat-loss conductance*. Once the collector-heat-loss conductance, U_c, is known, if the collector plate is at an average temperature, T_c, the second right-hand term in Eq. 7-27 can be written for a given ambient temperature, T_a, in the simple form

$$q_{loss} = U_c A_c (T_c - T_a) \tag{7-30}$$

The simplicity of this relation is somewhat misleading because the collec-

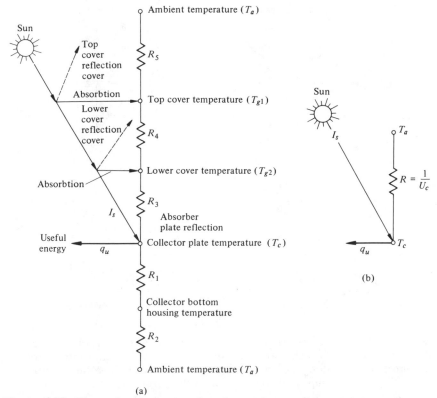

(a)

Figure 7-23 Thermal circuits for flat-plate collector shown in Fig. 7-22, (a) Detailed circuit; (b) approximate, equivalent circuit to (a). In both circuits, the absorber plate absorbs incident energy equal to $\alpha_s I_s$, where $I_s = \bar{\tau}_s I_c$.

tor-heat-loss conductance cannot be specified without a detailed analysis of all the heat losses. Figure 7-22 shows a schematic diagram of a double-glazed collector, while Fig. 7-23(a) shows the thermal circuit with all the elements that must be analyzed before they can be combined into a single conductance element, as shown in Fig. 7-23(b).

Figure 7-24 shows qualitatively the temperature distributions in a flat-plate collector. Radiation impinges on the top of the plate connecting any two adjacent flow tubes. It is absorbed uniformly by the plate and conducted laterally toward the flow tubes, where it is then transferred by convection to the working fluid flowing through the ducts. It is apparent that at any cross section perpendicular to the flow direction the temperature is a maximum at the midpoint between two adjacent flow ducts and decreases along the plate toward the tube, as shown in Fig. 7-24(b). Since heat is transferred to the working fluid, the temperature of the fluid as well

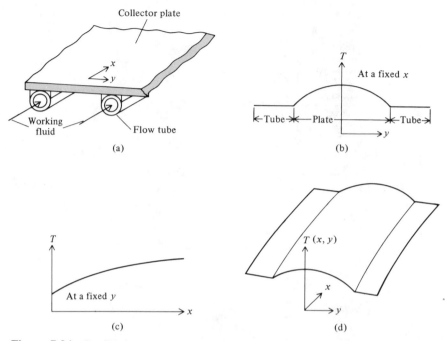

Figure 7-24 Qualitative temperature distribution in the absorber plate of a flat-plate collector. (a) Schematic diagram of absorber. (b) Temperature profile in the direction of the flow of the working fluid. (c) Temperature profile at given y. (d) Temperature distribution in the absorber plate.

as of the entire collector system will increase in the direction of flow. The increase in temperature at the midpoint between the two tubes is shown qualitatively in Fig. 7-24(c). The temperature distribution both in the x and y directions is shown in three-dimensional view in Fig. 7-24(d).

To construct a model suitable for a thermal analysis of a flat-plate collector, the following simplifying assumptions will be made:

1. The collector is thermally in steady state.
2. The temperature drop between the top and bottom of the absorber plate is negligible.
3. Heat flow is one-dimensional through the covers as well as through the back insulation.
4. The headers connecting the tubes cover only a small area of the collector and provide uniform flow to the tubes.
5. The sky can be treated as though it were a blackbody source for infrared radiation at an equivalent sky temperature.
6. The irradiation on the collector plate is uniform.

For a quantitative analysis consider a location at x,y on a typical flat-plate collector as shown in Fig. 7-24(a). Let the plate temperature at this point be $T_c(x,y)$ and assume that solar energy is absorbed at the rate $I_s\alpha_s$. If the lower surface of the collector is well insulated, most of the heat loss occurs from the upper surface. The conductance for the upper surface of the collector can be evaluated by determining the thermal resistances R_3, R_4, and R_5 in Fig. 7-23. Heat is transferred between the cover and the second glass plate and between the two glass plates by convection and radiation in parallel. Except for absorptance of solar energy by the second glass plate, the relations for the rate of heat transfer between T_c and T_{g2} and between T_{g2} and T_{g1} are the same. Thus, the rate of heat transfer per unit surface area of collector between the absorber plate and the second glass cover is

$$q_{\text{top loss}} = A_c \bar{h}_{c2}(T_c - T_{g2}) + \frac{\sigma(T_c^4 - T_{g2}^4)A_c}{1/\epsilon_{p,i} + 1/\epsilon_{g2,i} - 1} \qquad (7\text{-}31)$$

where

$$\bar{h}_{c2} = \text{heat-transfer coefficient between plate and second glass cover}$$

$$\epsilon_{p,i} = \text{infrared emittance of plate}$$

$$\epsilon_{g2,i} = \text{infrared emittance of second cover}$$

As shown previously in Chapter 1, if the radiation term is linearized, Eq. 7-31 becomes

$$q_{\text{top loss}} = (\bar{h}_{c2} + h_{r2})A_c(T_c - T_{g2}) = \frac{T_c - T_{g2}}{R_3} \qquad (7\text{-}32)$$

where

$$h_{r2} = \frac{\sigma(T_c + T_{g2})(T_c^2 + T_{g2}^2)}{(1/\epsilon_{p,i}) + (1/\epsilon_{g2,i}) - 1} \qquad (7\text{-}33)$$

A similar derivation for the rate of heat transfer between the two cover plates gives

$$q_{\text{top loss}} = (\bar{h}_{c1} + h_{r1})A_c(T_{g2} - T_{g1}) = \frac{T_{g2} - T_{g1}}{R_4} \qquad (7\text{-}34)$$

where

$$h_{r1} = \frac{\sigma(T_{g1} + T_{g2})(T_{g1}^2 + T_{g2}^2)}{(1/\epsilon_{g1,i}) + (1/\epsilon_{g2,i}) - 1} \qquad (7\text{-}35)$$

and h_{c1} = heat-transfer coefficient between two transparent covers.

The emittances of the two covers will, of course, be the same if they are made of the same material. However, economic advantages can sometimes be achieved by using a plastic cover between an outer cover of glass and the plate and in such a sandwich construction, the radiative properties of the two covers may not be the same.

The equation for the thermal resistance between the upper surface of the outer collector cover and the ambient air has a form similar to the two preceding relations, but the heat-transfer coefficient at the outer surface must be evaluated differently. If the air is still, free convection relations should be used, but when wind is blowing over the collector, forced convection correlations apply as shown in Chapter 5. Radiation exchange occurs between the top cover and the sky at T_{sky}, whereas convection heat exchange occurs between T_{g1} and the ambient air at T_{air}. For convenience we shall refer both conductances to the air temperature. This gives

$$q_{top\ loss} = (h_{c,\infty} + h_{r,\infty})(T_{g1} - T_{air}) = \frac{T_{g1} - T_{air}}{R_5} \tag{7-36}$$

where

$$h_{r,\infty} = \epsilon_{g1,i}\sigma(T_{g1} + T_{sky})(T_{g1}^2 + T_{sky}^2)\frac{T_{g1} - T_{sky}}{(T_{g1} - T_{air})} \tag{7-37}$$

For a double-glazed flat-plate collector the total heat-loss conductance, $U_{c\ total}$, can then be expressed in the form

$$U_{c,total} = \frac{1}{R_1} + \frac{1}{R_3 + R_4 + R_5} \tag{7-38}$$

where R_1 is the thermal resistance for the lower surface.

The evaluation of the collector-heat-loss conductance defined by Eq. 7-38 requires iterative solution of Eqs. 7-35 and 7-36 because the unit radiation conductances are functions of the cover and plate temperatures, which are not known a priori. An empirical procedure for calculating U_c for collectors with all covers of the same material, which is often sufficiently accurate and more convenient to use, has been suggested by Klein (Ref. 11). For this approach the collector top loss in watts is written in the form

$$q_{top\ loss} = \frac{(T_c - T_a)A_c}{\dfrac{N}{(C/T_p)(T_c - T_a)/(N+f)^{0.33} + \dfrac{1}{h_{c,\infty}}}}$$

$$+ \frac{\sigma(T_c^4 - T_a^4)A_c}{\dfrac{1}{\epsilon_{p,i} + 0.05N(1 + \epsilon_{p,i})} + \dfrac{2N + f - 1}{\epsilon_{g,i}} - N} \tag{7-39}$$

where

$$f = (1 - 0.04 h_{c,\infty} + 0.0005 h_{c,\infty}^2)(1 + 0.091 N)$$

$$C = 365.9(1 - 0.00883\beta + 0.00013\beta^2)$$

N = number of covers

$h_{c,\infty} = 5.7 + 3.8 V$ (V in m/s)

$\epsilon_{g,i}$ = infrared emittance of covers

The values of $q_{\text{top loss}}$ calculated from Eq. 7-39 agreed closely with the values obtained from Eq. 7-38.

To determine the efficiency of a solar collector, the rate of heat transfer to the working fluid must be calculated. If transient effects are neglected, the rate of heat transfer to the fluid flowing through a collector depends only on the temperature of the collector surface from which heat is transferred by convection to the fluid, the temperature of the fluid, and the heat-transfer coefficient between the collector and the fluid. To calculate the rate of heat transfer, consider first the condition at a cross section of the collector shown in Fig. 7-22. Solar radiant energy impinging on the upper face of the collector plate is conducted in a transverse direction toward the flow channels. The temperature is a maximum at any midpoint between adjacent channels and the collector plate acts as a fin attached to the walls of the flow channel. The thermal performance of the fin plate can be expressed in terms of its fin efficiency, η_f, defined as the ratio of the rate of heat flow through the real fin to the rate of heat flow through a fin of infinite thermal conductivity.

If U_c is the overall unit conductance from the collector plate surface to the ambient air, the rate of heat loss from a given segment of the collector plate at x,y in Fig. 7-25 is

$$q(x,y) = U_c[T_c(x,y) - T_a] \, dx \, dy \tag{7-40}$$

where

T_c = local collector plate temperature

T_a = ambient air temperature

U_c = overall unit conductance between plate and ambient air

If conduction in the x direction is negligible, a heat balance at a given distance x_0 for a cross section of the flat-plate collector per unit length in the x direction can be written in the form

$$\alpha_s I_s \, dy - U_c(T_c - T_a) \, dy + \left[\left(-kt\frac{dT_c}{dy}\bigg|_{y,x_0}\right) - \left(-kt\frac{dT_c}{dy}\bigg|_{y+dy,x_0}\right)\right] = 0 \tag{7-41}$$

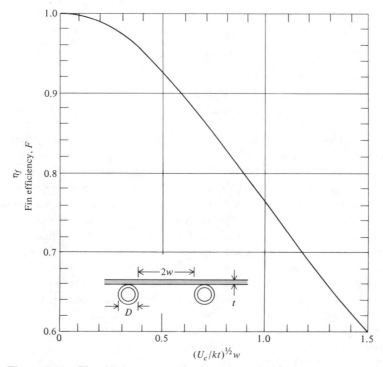

Figure 7-25 Fin efficiency for tube and sheet flat-plate solar collectors.

If the plate thickness t is uniform and the thermal conductivity of the plate is independent of temperature, Eq. 7-41 can be cast into the form of a second-order differential equation:

$$\frac{d^2T_c}{dy^2} = \frac{U_c}{kt}\left[T_c - \left(T_a + \frac{\alpha_s I_s}{U_c}\right)\right]$$ (7-42)

The boundary conditions for the system described above are:

1. At the center between any two ducts, the heat flow is zero, or at $y=0$: $dT_c/dy = 0$
2. At the duct the plate temperature is $T_b(x_0)$, or at $y=w=[(l'-D)/2]$: $T_c = T_b(x_0)$, where $T_b(x_0)$ is the temperature at the fin base.

If we let $m^2 = U_c/kt$ and $\Phi = T_c - (T_a + \alpha_s I_s/U_c)$, Eq. 7-42 becomes

$$\frac{d^2\Phi}{dy^2} = m\Phi$$ (7-43)

subject to the boundary conditions

$$\frac{d\Phi}{dy} = 0 \qquad \text{at} \quad y=0$$

and

$$\Phi = T_b(x_0) - \left(T_a + \frac{\alpha_s I_s}{U_c}\right) \qquad \text{at } y = \frac{l'-D}{2}$$

The general solution of Eq. 7-43 is

$$\Phi = C_1 \sinh my + C_2 \cosh my \qquad (7\text{-}44)$$

The constants C_1 and C_2 can be determined by substituting the two boundary conditions and solving the two resulting equations for C_1 and C_2. This gives

$$\frac{T_c - (T_a + \alpha_s I_s / U_c)}{T_b(x_0) - (T_a + \alpha_s I_s / U_c)} = \frac{\cosh my}{\cosh mw} \qquad (7\text{-}45)$$

From the preceding equation the rate of heat transfer to the conduit from the portion of the plate between two conduits can be determined by evaluating the temperature gradient at the base of the fin per unit width, or

$$q_{\text{fin}} = -kt\frac{dT_c}{dy}\bigg|_{y=w} = \frac{1}{m}\left[\alpha_s I_s - U_c(T_b(x_0) - T_a)\tanh mw\right] \qquad (7\text{-}46)$$

Since the conduit is connected to fins on both sides, the total rate of heat transfer is

$$q_{\text{total}}(x_0) = 2w\left[\alpha_s I_s - U_c(T_b(x_0) - T_a)\right]\frac{\tanh mw}{mw} \qquad (7\text{-}47)$$

If the entire fin were at the temperature $T_b(x)$, a situation corresponding physically to a plate of infinitely large thermal conductivity, the rate of heat transfer would be a maximum, $q_{\text{total, max}}$. As mentioned previously, the ratio of the rate of heat transfer with a real fin to the maximum rate obtainable is the fin efficiency η_f. Using this definition, Eq. 7-47 can be written in the form

$$q_{\text{total}}(x) = 2w\eta_f\left[\alpha_s I_s - U_c(T_b(x_0) - T_a)\right] \qquad (7\text{-}48)$$

where

$$\eta_f \equiv \frac{\tanh mw}{mw}$$

The fin efficiency η_f is plotted as a function of the dimensionless parameter $(U_c/kt)^{1/2}w$ in Fig. 7-25. When the fin efficiency approaches unity, the maximum portion of the radiant energy impinging on the fin becomes available for heating the fluid.

In addition to the heat transferred through the fin, the energy impinging on the portion of the plate above the flow passage also provides useful energy. The rate of useful energy from this region available to heat the working fluid is

$$q_{\text{duct}}(x) = D\left[\alpha_s I_s - U_c(T_b(x_0) - T_a)\right] \qquad (7\text{-}49)$$

Thus the useful energy per unit length in the flow direction becomes

$$q_u(x) = (D + 2w\eta_f) \left[\alpha_s I_s - U_c(T_b(x_0) - T_a) \right] \tag{7-50}$$

The energy $q_u(x)$ must be transferred as heat to the working fluid. If the thermal resistance of the metal wall of the flow duct is negligibly small and there is no contact resistance between the duct and the plate, the rate of heat transfer to the fluid is

$$q_u(x) = (\pi D_i) \bar{h}_{c,i} \left[T_b(x_0) - T_f(x_0) \right] \tag{7-51}$$

Collector Efficiency Factor

To obtain a relation for the useful energy delivered by a collector in terms of known physical parameters, the fluid temperature, and the ambient temperature, the collector temperature must be eliminated from Eqs. 7-50 and 7-51. Solving for $T_b(x_0)$ in Eq. 7-51 and substituting this relation in Eq. 7-50 gives

$$q_u(x) = (D + 2w) F' \left[\alpha_s I_s - U_c(T_f(x_0) - T_a) \right] \tag{7-52}$$

where F' is called the *collector efficiency factor*. It is given by

$$F' = \frac{1/U_c}{(D + 2w) \left[\dfrac{1}{U_c(D + 2w\eta_f)} + \dfrac{1}{\bar{h}_{c,i}(\pi D_i)} \right]} \tag{7-53}$$

Physically, the denominator in Eq. 7-53 is the thermal resistance between the fluid and the environment, whereas the numerator is the thermal resistance between the collector and the ambient air. The collector plate efficiency factor F' depends on U_c, $h_{c,i}$, and η_f. It is only slightly dependent on temperature and can for all practical purposes be treated as a design

Table 7-3 Typical Values for the Parameters That
Determine the Collector Efficiency Factor F'
for a Flat-Plate Collector in Eqs. 7-46 and 7-53

U_c	two glass covers: $4 \, \text{W/m}^2 \cdot \text{K}$
	one glass cover: $8 \, \text{W/m}^2 \cdot \text{K}$
kt	copper plate, 1 mm thick: $0.4 \, \text{W/K}$
	steel plate, 1 mm thick: $0.005 \, \text{W/K}$
$h_{c,i}$	water in laminar flow forced convection: $300 \, \text{W/m}^2 \cdot \text{K}$
	water in turbulent flow forced convection: $1{,}500 \, \text{W/m}^2 \cdot \text{K}$
	air in turbulent forced $\text{W/m}^2 \cdot \text{K}$ convection: $100 \, \text{W/m}^2 \cdot \text{K}$

parameter. Typical values for the factors that determine the value of F' are given in Table 7-3.

The collector efficiency factor increases with increasing plate thickness and plate thermal conductivity but decreases with increasing distance between flow channels. Also, increasing the heat-transfer coefficient between the walls of the flow channel and the working fluid increases F', but an increase in the overall conductance U_c will cause F' to decrease.

Collector-Heat-Removal Factor

Equation 7-52 yields the rate of heat transfer to the working fluid at a given point x along the plate for specified collector and fluid temperatures. However, in a real collector the fluid temperature increases in the direction of flow as heat is transferred to it. An energy balance for a section of flow duct dx can be written in the form

$$- \dot{m}c_p \left(T_f \big|_{x+dx} - T_f \big|_x \right) = q_u(x)\,dx \qquad (7\text{-}54)$$

Substituting Eq. 7-52 for $q_u(x)$ and setting

$$\left[T_f(x) + \frac{dT_f(x)}{dx}\,dx \right] = T_f \bigg|_{x+dx}$$

in Eq. 7-54 gives the differential equation

$$- \dot{m}c_p \frac{dT_f(x)}{dx} = (D+2w)F'\left[\alpha_s I_s - U_c(T_f(x) - T_a) \right] \qquad (7\text{-}55)$$

Separating the variables gives, after some rearranging,

$$\frac{dT_f(x)}{T_f(x) - T_a - (\alpha_s I_s / U_c)} = \frac{(D+2w)F'U_c}{\dot{m}c_p}\,dx \qquad (7\text{-}56)$$

Equation 7-56 can be integrated and solved for the outlet temperature of the fluid, $T_{f,\text{out}}$, for a duct length L, and fluid inlet temperature $T_{f,\text{in}}$, if we assume that F' and U_c are constant, or

$$\frac{T_{f,\text{out}} - T_a - \alpha_s I_s / U_c}{T_{f,\text{in}} - T_a - \alpha_s I_s / U_c} = \exp\left(- \frac{U_c(D+2w)F'L}{\dot{m}c_p} \right) \qquad (7\text{-}57)$$

To compare the performance of a real collector with the thermodynamic optimum, it is convenient to define the heat-removal factor, F_R, as the ratio of the actual rate of heat transfer to the working fluid to the rate of heat transfer at the maximum temperature difference between the absorber and the environment. The thermodynamic limit corresponds to the condition of the working fluid remaining at the inlet temperature throughout the collector. This can be approached when the fluid velocity is very high.

From its definition F_R can be expressed

$$F_R = \frac{G_c c_p (T_{f,\text{out}} - T_{f,\text{in}})}{\alpha_s I_s - U_c (T_{f,\text{in}} - T_a)} \tag{7-58}$$

where G_c is the flow rate per unit surface area of collector $\dot{m}/A_c$. By regrouping the right-hand side of Eq. 7-58 and combining with Eq. 7-57 it can easily be verified that

$$F_R = \frac{G_c c_p}{U_c} \left[1 - \frac{(\alpha_s I_s / U_c) - (T_{f,\text{out}} - T_a)}{(\alpha_s I_s / U_c) - (T_{f,\text{in}} - T_a)} \right]$$

or

$$F_R = \frac{G_c c_p}{U_c} \left[1 - \exp\left(-\frac{U_c F'}{G_c c_p} \right) \right] \tag{7-59}$$

Inspection of the relation above shows that F_R increases with increasing flow rate and approaches as an upper limit F', the collector efficiency factor. Since the numerator of the right-hand side of Eq. 7-58 is q_u, the rate of useful heat transfer can now be expressed in terms of the fluid inlet temperature:

$$q_u = A_c F_R [\alpha_s I_s - U_c (T_{f,\text{in}} - T_a)] \tag{7-60}$$

This is a convenient form for design because the fluid inlet temperature to the collector is usually known or can be specified.

Example 7-4. Calculate the averaged hourly and daily efficiency of a water solar collector on January 15, in Boulder, Colorado. The collector is tilted at an angle of 60° and has an overall conductance of 8.0 W/m²·K on the upper surface. It is made of copper tubes, 1 cm i.d., 0.05 cm thick, connected by a 0.05-cm-thick plate at a center-to-center distance of 15 cm. The heat-transfer coefficient for the water in the tubes is 1500 W/m²·K, the cover transmittance is 0.9, and the solar absorptance of the copper surface is 0.9. The collector is 1 m wide and 2 m long, the water inlet temperature is 330 K, and the water flow rate is 0.02 kg/s. The insolation (total), H_s, and the environmental temperature are tabulated below. Assume that diffuse radiation accounts for 25% of the total insolation.

TIME (h)	H_s (W/m²)	T_{amb} (K)
7–8	12	270
8–9	80	280
9–10	192	283
10–11	320	286
11–12	460	290
12–13	474	290
13–14	395	288
14–15	287	288
15–16	141	284
16–17	32	280

Solution: The total radiation received by the collector is calculated from Eq. 6-153:

$$I_c = I_{c,\text{diffuse}} + I_{c,\text{beam}}$$

$$= H_s \times 0.25\left(\frac{1+\cos 60}{2}\right) + H_s(1-0.25)R_b$$

The tilt factor, R_b, is obtained from its definition, Eq. 6-148:

$$R_b = \frac{\cos i}{\cos \alpha} = \frac{\sin \delta_s \sin(L-\beta) + \cos \delta_s \cos(L-\beta)\cos h_s}{\sin L \sin \delta_s + \cos L \cos \delta_s \cos h_s}$$

where $L = 40°$, $\delta_s = -21.1°$ on January 15 (from Fig. 6-10) and $\beta = 60°$. The hour angle h_s equals 15° for each hour away from noon.

The fin efficiency is obtained from Eq. 7-48:

$$\eta_f = \frac{\tanh mw}{mw}$$

with

$$m = \left(\frac{U_c}{kt}\right)^{1/2} = \left(\frac{8}{390 \times 5 \times 10^{-4}}\right)^{1/2} = 6.4$$

$$\eta_f = \frac{\tanh[6.4\,(0.15-0.01)/2]}{6.4(0.15-0.01)/2} = 0.938$$

The collector efficiency factor F' is from Eq. 7-53:

$$F' = \frac{1/U_c}{(D+2w)\left[\dfrac{1}{U_c(D+2w\eta_f)} + \dfrac{1}{\bar{h}_{c,i}\pi D_i}\right]}$$

$$= \frac{1/8.0}{0.15\left[\dfrac{1}{8.0(0.01+0.14\times0.938)} + \dfrac{1}{1500\pi \times 0.01}\right]} = 0.920$$

The daily average is obtained by summing the useful energy for those hours during which the collector delivers heat and dividing by the total insolation between sunrise and sunset. This yields

$$\bar{\eta}_{b,\text{day}} = \frac{\Sigma q_{u,\text{total}}}{\Sigma A_c I_{c,\text{total}}} = \frac{2099}{2 \times 4837} = 0.217\%$$

Then we obtain the heat-removal factor from Eq. 7-59:

$$F_R = \frac{G_c c_p}{U_c}\left[1 - e^{-(U_c F'/G_c c_p)}\right]$$

$$= \frac{0.01 \times 4184}{8}\left[1 - e^{-(8.0 \times 0.920/0.01 \times 4184)}\right] = 0.844$$

The useful heat-delivery rate is, from Eq. 7-60,

$$q_u = A_c F_R [\alpha_s I_s - U_c (T_{f,\text{in}} - T_a)]$$

In the relation above I_s is the radiation incident on the collector. If the transmittance of the glass is 0.9,

$$I_s = \tau I_{c,\text{total}} = 0.9 I_{c,\text{total}}$$

$$q_u = 2 \times 0.844 [I_{c,t} \times 0.81 - 8.0 (T_{f,\text{in}} - T_a)]$$

The efficiency of the collector is $\eta_b = q_u / A I_{c,t}$ and the hourly averages are calculated in the table below.

HOUR	H_s (W/m^2)	R_b	I_{dif} (W/m^2)	I_{beam} (W/m^2)	I_{tot} (W/m^2)	q_u(W)	T_{amb}(K)	η_c
7–8	12	10.89	2.3	92	94	0	270	0
8–9	80	3.22	15	193	208	0	280	0
9–10	192	2.44	36	351	387	0	283	0
10–11	320	2.18	60	523	583	221	286	0.190
11–12	460	2.08	86	718	804	584	290	0.363
12–13	474	2.08	89	739	828	619	290	0.374
13–14	395	2.18	74	646	720	438	288	0.304
14–15	287	2.44	54	525	579	237	288	0.205
15–16	141	3.22	26	341	367	0	284	0
16–17	32	10.89	6	261	267	0	280	0
					ΣI_{tot} = 4837 W/m^2	Σq_u = 2099 W		

7-7 HEAT TRANSFER AND FLOW IN PACKED BEDS

In addition to conventional heat exchangers in which the hot and cold fluids are separated by a solid wall and operate in steady state, there are systems that make use of a transient type of heat exchanger. These devices consist of a bed of solid particles that can be heated by a working fluid when it is hot. However, the particles can transfer the energy stored in them to the same or another working fluid at a later time when it is colder than the particles and must be heated. A common example of this type of heat-exchanger storage system is the "pebble bed" of solar collector systems that uses air as the working fluid.

In general, two types of particle heat and mass transfer exchange systems are encountered in engineering: the fixed bed and the fluidized bed. In the former, the particles lie on each other and their position is not changed by the fluid flowing through the voids. However, if the fluid is flowing upward and its velocity is increased sufficiently, the net drag force will eventually counteract the gravitational force, the fluid will lift the particles in the bed, and the particles will become suspended. This condition is called a *fluidized bed*, characterized by thorough mixing of particles, which largely eliminates the temperature gradient in the bed. This type of operation is often sought in the chemical industry, especially for processes in which heat and mass transfer occur simultaneously.

To analyze the flow and convection heat transfer in a fixed bed of pebbles, the particles are treated as though they were approximately spherical in shape. For such a fixed packed bed Löf and Hawley (Ref. 12) investigated the heat transfer and recommended the following empirical relation to evaluate the heat-transfer coefficient:

$$h_v = 650 \left(\frac{\dot{m}_b}{A_b \cdot D_s} \right)^{0.7} \tag{7-61}$$

where

h_v = heat-transfer coefficient per unit volume, $W/m^3 \cdot K$

$\dot{m}_b$ = mass rate of flow, kg/s

A_b = cross-sectional area of pebble bed, m^2

D_s = equivalent spherical diameter of particles in m given by

$$D_s = \left(\frac{6}{\pi} \times \frac{\text{net volume of particles}}{\text{number of particles}} \right)^{1/3} \tag{7-62}$$

Experimental verification of the range of variables over which this relation is applicable extends specifically over entering air temperatures between 300 and 380 K, particle sizes between 0.5 to 3.8 cm, air flow rates between 0.66 and 3.6 m^3/s per square meter of empty bed cross section and voids obtained by normal filling.

For a bed of approximately spherical particles having a total particle surface area, A_p, given by the relation

$$A_p = \frac{6(1 - \epsilon_v) A_b L}{D_s} \tag{7-63}$$

where

ϵ_v = void fraction (or porosity) of the bed

L = length in the direction of flow

Handley and Heggs (Ref. 13) measured the heat transfer and frictional pressure drop in flow through packed beds and correlated their results with those of previous investigators. They found that the heat-transfer coefficients for a bed of approximately spherical particles can be correlated by the dimensionless equation

$$j_b = \frac{0.255}{\epsilon_v \text{Re}_{D_s}^{0.33}} \tag{7-64}$$

where

$$j_b = \text{Stanton number } \left(\text{Nu}_{D_s}/\text{Re}_{D_s}\text{Pr}^{1/3}\right)$$

$$\text{Re}_{D_s} = \text{Reynolds number } (GD_s/\mu)$$

$$D_s = \text{equivalent sphere diameter}$$

$$\text{Nu}_{D_s} = \text{Nusselt number } \left(\bar{h}_c D_s/k_{\text{fluid}}\right)$$

$G =$ superficial mass velocity of gas, kg/s·m², based on cross-sectional area of bed

$\bar{h}_c =$ heat-transfer coefficient based on surface area of particles in bed as given by Eq. 7-63

For air (Pr = 0.71) this relation becomes

$$\text{Nu}_{D_s} = \frac{0.23\text{Re}_{D_s}^{2/3}}{\epsilon_v} \tag{7-65}$$

The pressure drop in a bed of length L can be determined from the relation

$$\frac{\Delta p}{\rho V_s^2} = \frac{L}{D_s}\frac{1-\epsilon_v}{\epsilon_v^3}\left(170\frac{1-\epsilon_v}{\text{Re}_{\text{bed}}} + 1.75\right) \tag{7-66}$$

where

$$\text{Re}_{\text{bed}} = \frac{2}{3}\frac{\rho V_s D_s}{(1-\epsilon_v)\mu} \tag{7-67}$$

with $V_s = \dot{m}/A_b \cdot \rho$, the superficial velocity. For a pebble bed, ϵ_v is between 0.35 and 0.45.

More detailed analysis of the transient heat-transfer and data on the effect of particle shape on the heat-transfer coefficients are presented in Reference 14. Information about the application of fixed beds for heat storage in solar air collectors is given in Reference 15.

7-8 HEAT PIPES

One of the main objectives of energy conversion systems is to transfer energy from a receiver to some other location where it can be used to heat a working fluid. A relatively novel device that can transfer large quantities of heat through small surface areas is the heat pipe. The method of operation of a heat pipe is shown schematically in Fig. 7-26. The device consists of a circular pipe with an annular layer of wicking material covering the inside. The core of the system is hollow in the center to permit the working fluid to pass freely from the heat addition end on the left, to the heat rejection end on the right. The heat addition end is equivalent to an evaporator, whereas the heat rejection end corresponds to a condensor.

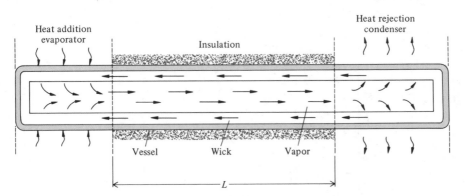

Figure 7-26 Schematic diagram of heat pipe.

The condensor and the evaporator are connected by an insulated section of length L. The liquid permeates the wicking material by capillary action, and when heat is added to the evaporator end of the heat pipe, liquid is vaporized in the wick and moves through the central core to the condensor end where heat is removed. Then the vapor condenses back into the wick and the cycle repeats.

A large variety of fluid and pipe material combinations have been used for heat pipes, and some typical working fluid and material combinations, as well as the temperature ranges over which they can operate, are presented in Table 7-4. In the fourth and fifth columns of the table measured axial heat fluxes and measured surface heat fluxes are listed and it is apparent that very high heat fluxes can be obtained.

A widely used correlation between the maximum achievable power transfer by a heat pipe and its dominant dimensions and operating parameters is (Ref. 16)

$$q = 2A_w g H_e \rho_l^{3/2} \left(\frac{\rho_v \rho_l}{\mu_l \mu_v} \right)^{1/2} \frac{l_m K_1}{L} \tag{7-68}$$

where

A_w = wick area

g = gravitational acceleration

H_e = heat of evaporization

ρ_l = liquid density

ρ_v = vapor density

μ_l = liquid viscosity

μ_v = vapor viscosity

l_m = wicking height of fluid in wick

K_1 = wick factor

L = heat pipe length

Table 7-4 Some Typical Operating Characteristics of Heat Pipes

TEMPERATURE RANGE (K)	WORKING FLUID	VESSEL MATERIAL	MEASURED AXIAL HEAT FLUX[a] (W/cm^2)	MEASURED SURFACE HEAT FLUX[a] (W/cm^2)	COMMENTS
230–400	Methanol	Copper, nickel, stainless steel	0.45 at 373 K	75.5 at 373 K	Using threaded artery wick
280–500	Water	Copper, nickel	0.67 at 473 K	146 at 443 K	
360–850	Mercury +0.02%, magnesium +0.001%	Stainless steel	25.1 at 533 K	181 at 533 K	Based on sonic limit in heat pipe
673–1073	Potassium	Nickel, stainless steel	5.6 at 1023 K	181 at 1023 K	
773–1173	Sodium	Nickel, stainless steel	9.3 at 1123 K	224 at 1033 K	

[a]Varies with temperature.
Source: Abstracted from C. H. Dutcher and M. R. Burke, "Heat Pipes: A Cool Way to Cool Circuits," *Electronics*, pp. 93–100, February 16 1970.

The wicking height is given by

$$l_m = \frac{2\gamma}{r_c \rho_l g} \tag{7-69}$$

where

$$\gamma = \text{surface tension}$$
$$r_c = \text{effective pore radius}$$

The maximum wicking height with sodium as the working fluid is about 38.5 cm, which is calculated by assuming an effective pore diameter of 8.6×10^{-3} cm. This is typical for a screen made with eight 4.1×10^{-3}-cm-diameter wires per millimeter. Inherent elevation differences are those typically associated with the heat-pipe diameter and extrusions in the heat pipe. Operational elevation differences occur from tilting of the heat pipe or effective bends in the vertical plane of testing. If the heat pipe is to be operated with an elevation differential, l_e, the driving force is reduced by this height and the wicking height, l_m, in Eq. 7-68 must be replaced by

$$l_w = l_m - l_e$$

From the equation above it is seen that the heat pipe will lose its entire power-transfer capacity when the total elevation differential in the heat pipe exceeds the available suction head produced by the surface tension.

The most dominant parameters affecting the total power-transfer capacity are the wick area, A_w, effective wicking height, l_w, and heat pipe length. For any effective wicking height, l_w, a wicking area, A_w, can be selected to achieve the desired total power transfer if the operating temperature as well as the temperature drops at the evaporator section and the condenser section can be freely selected. However, when a limit to the upper operating temperature as well as to the temperature of the heat pipe at the condenser section exists, the wicking thickness might be determined by these temperature considerations. In general, the temperature drops and the operating temperature increases with increasing wick thickness. If the wick thickness is based on temperature and temperature-drop considerations, the maximum heat-pipe length for a given power transfer is determined.

Equation 7-68 expresses the correlation between the total power transfer and the heat-pipe configuration assuming constant wick flow area, A_w. To decrease the temperature drop at the evaporator and condenser, the wick thickness can be decreased selectively in these two areas of the heat pipe. For this case, Eq. 7-68 would need to be modified to take the nonuniformity of the wick flow area into account, as indicated in the following equation:

$$q = 2gH_e\rho_l \left(\frac{\rho_v \rho_l}{\mu_l \mu_v} \right)^{1/2} (l_m - l_c) K_1' \left(\frac{L_e}{A_{we}} + \frac{L_a}{A_{wa}} + \frac{L_c}{A_{wc}} \right) \tag{7-70}$$

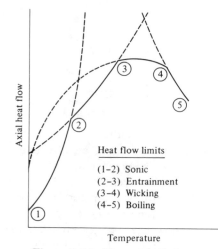

Figure 7-27 Heat-pipe limitations.

where the subscripts e, a, and c pertain to the evaporator, the adiabatic, and the condenser section of the heat pipe, respectively.

Although a heat pipe behaves like a structure of very high thermal conductance, it possesses heat-transfer limitations which are governed by certain principles of fluid mechanics. The possible effects of these limitations on the capability of a heat pipe with a liquid-metal working fluid are shown in Fig. 7-27. Individual limitations indicated in the figure are discussed below.

For a more complete treatment of the heat-pipe theory and practice, the reader is referred to Reference 17.

Sonic Limitation

When heat is transferred from the evaporator section of a heat pipe to the condenser section, the rate of heat transfer, q, between the two sections is given by

$$q = \dot{m}_v h_{fg} \tag{7-71}$$

where $\dot{m}_v$ is the rate of mass flow of vapor at evaporator exit and h_{fg} the latent heat of the fluid. Because the latent energy of the working fluid is used instead of its heat capacity, large heat-transfer rates can be achieved with a relatively small mass flow. Furthermore, if the heat is transferred by high density/low velocity vapor, the transfer is nearly isothermal because only small pressure gradients are necessary to move the vapor.

To show the effect of vapor density and velocity on heat transfer, Eq. 7-71 can be modified by using the continuity equation

$$\dot{m}_v = \bar{\rho}_v \bar{V} A \tag{7-72}$$

where $\bar{\rho}_v$ is the radial average vapor density at evaporator exit and A the cross-sectional area of vapor passage. By combining Eqs. 7-71 and 7-72 and rearranging, the result is

$$\frac{q}{A} = \bar{\rho}_v \overline{V} h_{fg} \qquad (7\text{-}73)$$

where q/A is the axial heat flux based on the cross-sectional area of the vapor passage.

Equation 7-73 shows that the axial heat flux in a heat pipe can be held constant and the condenser environment adjusted to lower the pressure, temperature, and density of the vapor until the flow at the evaporator exit becomes sonic. Once this occurs, pressure changes in the condenser will not be transmitted to the evaporator. This sonic limiting condition is represented in Fig. 7-27 by the solid curve between points 1 and 2. Some values for sonic heat-flux limits as a function of evaporator exit temperature are given in Table 7-5 for Cs, K, Na, and Li.

Table 7-5   Sonic Limitations of Heat-Pipe Working Fluids

EVAPORATOR EXIT TEMPERATURE (°C)	HEAT-FLUX LIMITS (kW/cm²)			
	Cs	K	Na	Li
400	1.0	0.5	—	—
500	4.6	2.9	0.6	—
600	14.9	12.1	3.5	—
700	37.3	36.6	13.2	—
800	—	—	38.9	1.0
900	—	—	94.2	3.9
1000	—	—	—	12.0
1100	—	—	—	31.1
1200	—	—	—	71.0
1300	—	—	—	143.8

Although heat pipes are normally not operated at sonic flow, such conditions have been encountered during startup with the working fluids listed in Table 7-5. Temperatures during such startups are always higher at the beginning of the heat-pipe evaporator than at the evaporator exit. The difference in temperature can be predicted by using the momentum equation for a system with mass addition

$$P_1 = P_2 + \rho_v V^2 \qquad (7\text{-}74)$$

where P_1 is the static pressure at beginning of evaporator, P_2 the static pressure at evaporator exit, and $\rho_v V^2$ the dynamic pressure at evaporator exit. If the velocity is expressed in terms of a Mach number, use of the ideal gas relationships gives

$$\frac{P_1}{P_2} = 1 + M^2 \gamma \qquad (7\text{-}75)$$

where M is the Mach number at exit and γ the ratio of specific heats of vapor. The use of Eq. 7-75 and of appropriate vapor pressure/temperature curves allows the evaporator temperature gradients to be determined at sonic and subsonic conditions.

Entrainment Limitation

Ordinarily, the sonic limitations just discussed do not cause dryout of the wick with attendant overheating of the evaporator. In fact, they often prevent the attainment of other limitations during startup. However, if the vapor density is allowed to increase without an accompanying decrease in velocity, some liquid from the wick-return system may be entrained. The onset of entrainment can be expressed in terms of a Weber number,

$$\frac{\rho_v V^2 L_c}{2\pi\sigma_s} = 1 \qquad (7\text{-}76)$$

where L_c is a characteristic length and σ_s the liquid surface tension. Equation 7-76 simply expresses the ratio of vapor inertial forces to liquid surface tension forces. When this ratio exceeds unity, a condition develops which is very similar to that of a body of water agitated by high-velocity winds into waves which propagate until liquid is torn from their crests. Once entrainment begins in a heat pipe, fluid circulation increases until the liquid-return path cannot accommodate the increased flow. This causes dryout and overheating of the evaporator.

Because the wavelength of the perturbations at the liquid/vapor interface in a heat pipe is determined by the wick structure, the entrainment limit can be estimated by combining Eqs. 7-73 and 7-76 to give

$$\frac{q}{A} = \frac{2\pi\rho_v \sigma h_{fg}}{L_c} \qquad (7\text{-}77)$$

Equation 7-77 can then be used to obtain the type of curve represented by the solid line between points 2 and 3 in Fig. 7-27.

Wicking Limitation

Fluid circulation in a heat pipe is maintained by capillary forces that develop in the wick structure at the liquid/vapor interface. These forces balance the pressure losses due to the flow in the liquid and vapor phases; they are manifest as many menisci which allow the pressure in the vapor to be higher than the pressure in the adjacent liquid in all parts of the system. When a typical meniscus is characterized by two principal radii of curvature (r_1 and r_2) the pressure drop, ΔP_c, across the liquid surface is given by

$$\Delta P_c = \sigma\left(\frac{1}{r_1} + \frac{1}{r_2}\right) \qquad (7\text{-}78)$$

Figure 7-28 Cross sections of various wick structures: (a) artery; (b) channels; (c) screen; (d) concentric annulus; (e) crescent annulus.

These radii, which are smallest at the evaporator end of the heat pipe, become even smaller as the heat-transfer rate is increased. If the liquid wets the wick perfectly, the radii will be defined exactly by the pore size of the wick when a heat-transfer limit is reached. Any further increase in heat transfer will cause the liquid to retreat into the wick, and drying and overheating will occur at the evaporator end of the system.

As indicated by Eq. 7-78, the capillary force in a heat pipe can be increased by decreasing the size of the wick pores that are exposed to vapor flow. However, if the pore size is decreased also in the remainder of the wick, the wicking limit might actually be reduced because of the increased pressure drop in the liquid phase. This is shown by *Poiseuille's equation* for the pressure drop through a capillary tube,

$$\Delta P_e = \frac{8 \mu \dot{m}_e L}{\pi r^4 \rho} \tag{7-79}$$

where μ is the liquid viscosity, $\dot{m}_e$ the rate of mass flow of liquid, r the tube radius, ρ the liquid density, and L the tube length.

Equation 7-79 can be modified to obtain the liquid-pressure drop at a particular heat-transfer rate, q, for various wick structures. The equations given below are for the examples shown in Fig. 7-28.

(a) *Artery*

$$\Delta P_L = \frac{8 \mu q Z_e}{\pi r^4 \rho h_{fg}} \tag{7-80}$$

(b) *Channels*

$$\Delta P_L = \frac{8 \mu q Z_e}{\pi r_e^4 N \rho h_{fg}} \tag{7-81}$$

(c) *Screen*

$$\Delta P_L = \frac{b \mu q Z_e}{\pi (R_W^2 - R^2) \epsilon r_c^2 \rho h_{fg}} \tag{7-82}$$

(d) *Concentric Annulus*

$$\Delta P_L = \frac{12 \mu q Z_e}{\pi D w^3 \rho h_{fg}} \tag{7-83}$$

(e) *Crescent*
   Annulus

$$\Delta P_L = \frac{4.8\mu q Z_e}{\pi D w^3 \rho h_{fg}} \tag{7-84}$$

The quantities are defined as follows:

Z_e = effective length of heat pipe

r_e = effective channel radius

N = number of channels

b = screen tortuosity factor

R_W = outer radius of screen structure

R = radius of vapor passage

ϵ = screen void fraction

r_c = effective radius of screen openings

D = mean diameter of annulus

w = width of annulus

Equations 7-81 through 7-84, except Eq. 7-82, are simple modifications of Eq. 7-79. In all cases q/L is substituted for $\dot{m}_L$ and Z_e for Z. The first substitution comes from Eq. 7-71 because $\dot{m}_L = \dot{m}_V$ in a heat pipe. The second substitution comes from the relationship

$$Z_e = \frac{Z_h + Z_c}{2} + Z_i \tag{7-85}$$

where Z_h is the heated length, Z_c the cooled length, and Z_i the insulated length between heated and cooled sections. When the middle of a heat pipe is heated and the remainder is cooled, the effective length is

$$Z_e = \frac{Z_h + Z_c}{4} \tag{7-86}$$

Although the artery wick system appears ideal, it requires an additional capillary network to distribute the liquid over surfaces which are used for heat addition and removal. Because of this complication, arteries are usually reserved for systems where boiling is likely to occur within the wick if the bulk of the liquid-return network is located in the path of the incoming heat. (The consequences of such boiling will be discussed later.)

Equation 7-81 is essentially the same as Eq. 7-82, except that it involves a number of channels, N, and an effective channel radius, r_e, which is obtained by the hydraulic-radius method:

$$r_e = 2\left(\frac{\text{flow area}}{\text{wetted perimeter}}\right) \tag{7-87}$$

Although open channels are subject to an interaction of vapor and liquid, which causes waves but no liquid entrainment, the interaction can be suppressed by covering the channels with a layer of fine-mesh screen. Because the screen is located at the interface of liquid and vapor, the fine pores of the screen provide large capillary forces for fluid circulation, while the channels provide a less restrictive flow path for liquid return. This general type of structure is called a *composite wick*.

All-screen composite wicks can be made by wrapping a layer of a fine screen around a mandrel followed by a second layer of coarse screen. The assembly can be placed in a container tube, the diameter of which is then drawn down until the inner wall makes contact with the coarse screen. The quantity $(b/\epsilon r_c^2)$ in Eq. 7-82 can next be determined by liquid-flow measurements through the screen before the mandrel is removed.

An ideal wick system for liquid-metal working fluids consists of an inner porous tube separated from an outer container tube by a gap that provides an unobstructed annulus for liquid return. The pressure drop in a concentric annulus (Eq. 7-83) is obtained by deriving Poiseuille's equation for flow between two parallel plates. Although not as precise as the equation for flow between concentric cylinders, it is easier to handle and is fairly accurate provided that the width of the annulus is small compared to its mean diameter. Equation 7-84 for a crescent annulus is obtained by assuming the displacement obeys a cosine function—the width of the annulus doubles at the top of the tube, becomes zero at the bottom, and remains unchanged on the sides.

In Fig. 7-27, the wicking limitation is represented by the solid line between points 3 and 4. Although this limitation is shown to occur at temperatures where essentially all the pressure drop is in the liquid phase, the effect of a significant vapor-pressure drop is indicated by the dotted extension line at lower temperatures.

Boiling Limitations

In most two-phase flow systems the formation of vapor bubbles in the liquid phase (boiling) enhances convection, which is required for heat transfer. Such boiling is often difficult to produce in liquid-metal systems because the liquid tends to fill the nucleation sites necessary for bubble formation. In a heat pipe, convection in the liquid is not required because heat enters the pipe by conduction through a thin saturated wick. Furthermore, the formation of vapor bubbles is undesirable because they could cause hot spots and destroy the action of the wick. Therefore, heat pipes are usually heated isothermally before being used to allow the liquid to wet the inner heat-pipe wall and to fill all but the smallest nucleation sites.

Boiling may occur at high input heat fluxes and high operating temperatures. The curve between points 4 and 5 in Fig. 7-27 is based on the equations

$$P_i - P_L = \frac{2\sigma}{r} \tag{7-88}$$

$$\frac{q}{S} = \frac{k(T_w - T_v)}{t} \tag{7-89}$$

where P_i is the vapor pressure inside the bubble, P_L the pressure in adjacent liquid, r the radius of largest nucleation site, S the heat input area, k the effective thermal conductivity of saturated wick, T_w the temperature at inside wall, T_v the temperature at the liquid/vapor interface, and t the wick thickness.

Since the sizes of nucleation sites in any system are usually unknown, it is not possible to predict when boiling will occur. However, Eqs. 7-88 and 7-89 show how various factors influence boiling. For example, if nucleation sites are small, a large pressure difference will be required for bubbles to grow. For a given heat-input flux, this pressure difference will depend on the thickness and thermal conductivity of the wick, on the saturation temperature of the vapor, and on the pressure drop in the vapor and liquid phases. This pressure drop is often overlooked because it is not a factor in the ordinary treatment of boiling.

Boiling is not a limitation with liquid metals, but when water is used as the working fluid, boiling may be a major heat-transfer limitation because the thermal conductivity of the fluid is low and because it does not readily fill nucleation sites. Unfortunately, little experimental information is available concerning this limitation.

REFERENCES

1. W. M. Kays and A. L. London, *Compact Heat Exchangers*, 2nd ed., McGraw-Hill Book Company, New York, 1964.
2. W. M. Kays and A. L. London, "Remarks on the Behavior and Application of Compact High-Performance Heat Transfer Surfaces," Inst. Mech. Eng. and ASME, *Proc. General Discussion on Heat Transfer*, 1951, pp. 127–132.
3. L. M. K. Boelter, R. C. Martinelli, F. E. Romie, and E. H. Morrin, "An Investigation of Aircraft Heaters XVIII—A Design Manual for Exhaust Gas and Air Heat Exchangers," *NACA Wartime Report*, ARR5AO6, August, 1945.
4. A. L. London and W. M. Kays, "The Gas Turbine Regenerator—the Use of Compact Heat Transfer Surfaces," *Trans. ASME*, vol. 72 (1950), p. 611.
5. F. Kreith, *Principles of Heat Transfer*, 3rd ed., Intext Publishers Group, New York, 1973.
6. *Standards of Tubular Exchanger Manufacturers Association*, 4th ed., 1959.
7. R. A. Bowman, A. C. Mueller, and W. M. Nagle, "Mean Temperature Difference in Design," *Trans. ASME*, Vol. 62 (1940), pp. 283–294.
8. W. Nusselt, "A New Heat Transfer Formula for Cross-Flow," *Technische Mechanik and Thermodynamik*, Vol. 12 (1930).

9. H. Ten Broeck, "Multipass Exchanger Calculations," *Ind. Eng. Chem.*, Vol. 30 (1938), pp. 1041–1042.

10. J. E. Hill and T. Kusuda, Methods of Testing for Rating Solar Collectors Based on Thermal Performance, NBS, U.S. Department of Commerce, Washington, D.C., *Interim Rept. NBSIR* 74-635, December 1974.

11. S. A. Klein, The Effects of Thermal Capacitance upon the Performance of Flat Plate Collectors, M. S. thesis, University of Wisconsin, Madison, 1973.

12. G. O. G. Löf and R. W. Hawley, "Unsteady Heat Transfer Between Air and Loose Solids," *Ind. Eng. Chem.*, vol. 40, pp. 1061–1066 1948.

13. D. Handley and P. J. Heggs, "Momentum and Heat Transfer Mechanisms in Regularly Shaped Packings," *Trans. Inst. Chem. Eng.*, vol. 46, pp. 251–264, 1968.

14. Handley and Heggs, "The Effect of Thermal Conductivity of the Packing Material on the Transient Heat Transfer in a Fixed Bed," *Int. J. Heat Mass Transfer*, vol. 12, pp. 549–570, 1969.

15. F. Kreith and J. Kreider, *Principles of Solar Engineering*, McGraw-Hill Book Company, New York, 1978.

16. R. Richter, Solar Collector Thermal Power System, vol. I, *Rept. AFAPL-TR-74-89-I*, Xerox Corp., Pasadena, Calif.; *NTIS AD/A*-000-940, 1974.

17. S. W. Chi, *Heat Pipe Theory and Practice*, Hemisphere Publishing Corporation, Washington, D.C., 1976.

PROBLEMS

The problems in this chapter are organized in the manner shown in the table.

PROBLEM NUMBERS	SECTION	SUBJECT
7-1 to 7-6	7-3	Overall Heat Transfer Coefficient
7-7 to 7-14	7-4	Log-Mean-Temperature Difference
7-15 to 7-24	7-5	Heat-Exchanger Effectiveness

7-1   A heat exchanger consists of air flow over brass tubes that contain steam. The convective heat-transfer coefficients on the air and steam sides of the tubes are 70 W/m²·K and 210 W/m²·K, respectively. All tubes have an i.d. of 1.8 cm and an o.d. of 2.1 cm. Calculate the overall heat-transfer coefficient for the heat-exchanger (a) based on the inner tube area, (b) based on the outer tube area.

7-2   A refrigerant flows through a copper tube with an i.d. of 2.6 cm and an o.d. of 3.2 cm. Air is forced over the exterior of the tube. The convective heat-transfer coefficient for the refrigerant is 120 W/m²·K and for the air is 35 W/m²·K. Calculate the overall heat-transfer coefficient based on the outside area of the tube (a) considering the thermal resistance of the tube, (b) neglecting the resistance of the tube.

7-3   Work Problem 7-1 assuming the fouling factor on the inside of the tube 0.00018 m²·K/W.

7-4   Work Problem 7-2 assuming the fouling factor is 0.00023 m²·K/W on the inside of the tube and 0.00011 m²·K/W on the outside of the tube.

7-5   The convective heat-transfer coefficient is 560 W/m²·K on the inside of a brass tube (1.9 cm i.d., 2.4 cm o.d.) and it is 105 W/m²·K on the outside of the tube. A deposit has built up inside the tube producing a fouling factor equal to 0.00065 m²·K/W. Estimate the percent increase in heat transfer that could be achieved by removal of the deposit assuming the presence of the deposit causes negligible changes in the temperatures of the fluids inside and outside the tube.

7-6   A long aluminum tube contains water that is to be heated by air flowing over the exterior of the tube perpendicular to the tube axis. The i.d. of the tube is 1.85 cm and its o.d. is 2.3 cm. The mass flow rate of the water through the tube is 0.65 kg/s and the temperature of the water in the tube averages 30°C. The free stream velocity and ambient temperature of the air are 10 m/s and 120°C, respectively. Using heat-transfer correlations given in Chapters 4 and 5, estimate the overall heat-transfer coefficient for the heat exchanger.

7-7   Oil is used to heat water in a counter flow heat exchanger from 20°C to 40°C. The temperature of the oil changes from 95°C to 60°C. Calculate the log mean temperature difference for this situation.

7-8   For the conditions given in Problem 7-7, calculate the heat-transfer rate between the water and oil if the overall heat-transfer coefficient is 325 W/m$^2 \cdot$K and the surface area of the heat exchanger is 1.6 m^2.

7-9   The overall heat-transfer coefficient of a new heat exchanger is 230 W/m$^2 \cdot$K based on an area of 14.0 m^2. The heat exchanger is a shell and tube counterflow design. The hot fluid enters at 150°C and leaves at 90°C while the cold fluid enters at 20°C and leaves at 75°C. Determine the heat-transfer rate between the fluids. After a few years of operation the heat transfer is reduced due to a deposit of rust on the inside of the tube. The fouling factor of the deposit is 0.0003 m$^2 \cdot$K/W. Determine the heat-transfer rate for this condition and calculate the percent decrease in the heat-transfer rate due to the presence of the rust.

7-10   A shell and tube heat exchanger has one shell pass and four tube passes. The tube fluid enters at 250°C and leaves at 110°C. The temperature of the fluid entering the shell is 10°C and it is 90°C as it leaves the shell. The overall heat-transfer coefficient based on a surface area of 12 m^2 is 320 W/m$^2 \cdot$K. Calculate the heat-transfer rate between the fluids.

7-11   Exhaust gases from a power plant are used to preheat air in a cross-flow heat exchanger. The exhaust gases enter the heat exchanger at 450°C and leave at 200°C. The air enters the heat exchanger at 70°C, leaves at 250°C and has a mass flow rate of 10 kg/s. Assume the properties of the exhaust gases can be approximated by those of air. The overall heat-transfer coefficient of the heat exchanger is 154 W/m$^2 \cdot$K. Calculate the heat-exchanger surface area required if (a) the air is unmixed and the exhaust gases are mixed and (b) both fluids are unmixed.

7-12   A shell and tube heat exchanger is used to heat water by condensing steam in the shell. The flow rate of the water 15 kg/s and it is heated from 60° to 80°C. The steam condenses at 140°C and the overall heat-transfer coefficient of the heat exchanger is 820 W/m$^2 \cdot$K. There are two tube passes and a single shell pass. There are 45 tubes with an o.d. of 2.75 cm. Calculate the length of tubes required.

7-13   The heating requirements for a room are 5 kW. The heat is to be supplied by hot water that flows at a rate of 0.8 kg/s through a 2.9-cm o.d. copper pipe and enters the pipe at 110°C. Estimate (a) the length of pipe needed assuming the convective heat-transfer coefficient over the exterior of the pipe is 150 W/m²·K and the room temperature is 25°C, (b) the water temperature as it leaves the pipe, and (c) the average temperature of the surface of the pipe.

7-14 Calculate the heat-transfer rate in a shell and tube heat exchanger for the following conditions: six tube passes, one shell pass, hot fluid in at 500°C, hot fluid out at 250°C, cold fluid in at 100°C, cold fluid out at 300°C, $U = 175$ W/m²·K based on a surface area of 14.5 m².

7-15 Water is heated by hot air in a heat exchanger. The flow rate of the water is 12 kg/s and of the air is 2 kg/s. The water enters at 40°C and the air enters at 460°C. The overall heat-transfer coefficient of the heat exchanger is 275 W/m²·K based on a surface area of 14 m². Determine the effectiveness of the heat exchanger if it is (a) parallel-flow type or a (b) cross-flow type.

7-16 Calculate the heat transfer rate for the two types of heat exchangers described in Problem 7-15.

7-17 Calculate the outlet temperatures of the hot and cold fluids for the conditions given in Problem 7-15.

7-18 Oil ($c_p = 2.1$ kJ/kg·K) is used to heat water in a shell and tube heat exchanger designed with a single shell and two tube passes. The overall heat-transfer coefficient is 525 W/m²·K. The mass flow rates are 7 kg/s for the oil and 10 kg/s for the water. The oil and water enter the heat exchanger at 240°C and 20°C, respectively. The heat exchanger is to be designed so that the water leaves the heat exchanger with a minimum temperature of 80°C. Calculate the heat-transfer surface area required to provide this temperature.

7-19 Steam at 125°C is condensed on the outside of the tubes in a shell and tube heat exchanger. Water enters the tubes at 35°C with a mass flow rate of 1.3 kg/s. The mass flow rate of the steam is 2.5 kg/s and the overall heat-transfer coefficient is 1650 W/m²·K based on a surface area of 3.7 m². Calculate (a) the effectiveness of the heat exchanger, (b) the actual heat transfer rate to the water, and (c) the outlet water temperature.

7-20 A cross-flow heat exchanger ($U = 45$ W/m²·K based on an area of 53 m²) in a power plant preheats inlet air by exchanging heat from exhaust gases. The flow rate of the exhaust gases is 2.7 kg/s and the inlet air has a flow rate of 2.0 kg/s. The inlet temperature of the exhaust gases and air are 350°C and 35°C, respectively. Assuming the properties of the exhaust gases can be approximated by those of air, calculate the heat exchanger effectiveness if (a) both fluids are unmixed and (b) the exhaust gases are unmixed while the air is mixed.

7-21 Calculate the heat transfer rate to the air for both heat exchangers described in Problem 7-20.

7-22 Calculate the outlet temperatures of both fluids for both heat exchanger designs described in Problem 7-20.

7-23 Water flowing at a rate of 10 kg/s through 50 tubes in a double pass shell and tube heat exchanger heats air that flows through the shell side. The developed length of the brass tubes is 6.7 m and they have an outside diameter of 2.6 cm. The convective heat-transfer coefficients of the water and air are 470 W/m²·K and 210 W/m²·K, respectively. The air enters the shell at a temperature of 15°C and a flow rate of 1.6 kg/s. The temperature of the water as it enters the tubes is 75°C. Calculate (a) the heat exchanger effectiveness, (b) the heat transfer rate to the air, and (c) the outlet temperature of the air and water.

7-24 After several years of operation, a scale has built up in the tubes of the heat exchanger described in Problem 7-23. The fouling factor of the scale is 0.0006 m²·K/W. Rework Problem 7-23 for this new condition.

Chapter 8

CONDENSATION, BOILING, AND MASS TRANSFER

8-1 INTRODUCTION

This chapter deals with three specialized types of transport processes: heat transfer by condensation, boiling heat transfer, and mass transfer. All three of these processes are quite complex, and entire books have been written about each of them. For an extensive discussion of boiling and condensation the reader is referred to *Convective Boiling and Condensation* by Collier (Ref. 16), while for a detailed discussion of mass transfer, the treatise by Treybal, *Mass Transfer Operation* (Ref. 47), is recommended. Here only the salient features of each of these processes will be discussed, and some of the correlation equations for making estimates of the heat- or mass-transfer coefficients will be presented.

8-2 CONDENSATION HEAT TRANSFER

When a saturated vapor comes in contact with a surface at a lower temperature, condensation occurs. Under normal conditions a continuous flow of liquid is formed over the surface and the condensate flows downward under the influence of gravity. Unless the velocity of the vapor is very high or the liquid film very thick, the motion of the condensate is laminar and heat is transferred from the vapor/liquid interface to the surface merely by conduction. The rate of heat flow depends, therefore, primarily on the thickness of the condensate film, which in turn depends

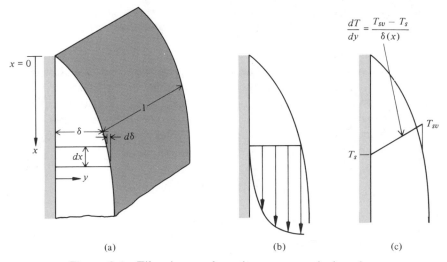

Figure 8-1 Filmwise condensation on a vertical surface: film growth, velocity profile, and temperature distribution.

on the rate at which vapor is condensed and the rate at which the condensate is removed. On a vertical surface the film thickness increases continuously from top to bottom, as shown in Fig. 8-1. As the plate is inclined from the vertical position, the drainage rate decreases and the liquid film becomes thicker. This, of course, causes a decrease in the rate of heat transfer.

Filmwise Condensation

Theoretical relations for calculating the heat-transfer coefficients for filmwise condensation of pure vapors on tubes and plates were first obtained by Nusselt (Ref. 1), in 1916. To illustrate the classical Nusselt approach we shall consider a plane vertical surface at a constant temperature T_s on which a pure vapor at saturation T_{sv} is condensing. As shown in Fig. 8-1, a continuous film of liquid flows downward under the action of gravity, and its thickness increases as more and more vapor condenses at the liquid/vapor interface. At a distance x from the top of the plate the thickness of the film is δ. If the flow of the liquid is laminar and is caused by gravity alone, we can estimate the velocity of the liquid by means of a force balance on the element $dx\ \delta\ 1$. The downward force acting on the liquid at a distance greater than y from the surface is

$$(\delta - y)\,dx\rho_l g$$

Assuming that the vapor outside the condensate layer is in hydrostatic balance

$$dp/dx = \rho_v g$$

a partially balancing force equal to

$$(\delta - y)\, dx \rho_v g$$

will be present as a result of the pressure difference between the upper and lower faces of the element. The other forces retarding the downward motions consist of the drag at the inner boundary of the element. Unless the vapor flows at a very high velocity, the shear at the free surface is quite small and may be neglected. The remaining force will then simply be the viscous shear $\mu_l(du/dy)dx$ at the vertical plane y. Under steady-state conditions the upward and downward forces are equal:

$$(\delta - y)(\rho_l - \rho_v)g = \mu_l \frac{du}{dy} \tag{8-1}$$

where

$$\delta = \text{thickness of the condensate layer}$$
$$\rho_l = \text{liquid density}$$
$$\rho_v = \text{vapor density}$$
$$g = \text{acceleration of gravity}$$
$$u = \text{velocity at } y$$
$$\mu_l = \text{liquid viscosity}$$

The velocity u at y is obtained by separating the variables and integrating. This yields the expression

$$u(y) = \frac{(\rho_l - \rho_v)g}{\mu_l}\left(\delta y - \tfrac{1}{2}y^2\right) + \text{constant} \tag{8-2}$$

The constant of integration is zero because the velocity u is zero at the surface (i.e., $u = 0$ at $y = 0$).

The mass rate of flow of condensate per unit breadth Γ_c is obtained by integrating the local mass flow rate at the elevation x, $\rho u(y)$, between the limits $y = 0$ and $y = \delta$:

$$\Gamma_c = \int_0^\delta \frac{\rho_l(\rho_l - \rho_v)g}{\mu_l}\left(\delta y - \tfrac{1}{2}y^2\right)dy = \frac{\rho_l(\rho_l - \rho_v)\delta^3 g}{3\mu_l} \tag{8-3}$$

The change in condensate flow rate Γ_c with the thickness of the condensate layer δ is

$$\frac{d\Gamma_c}{d\delta} = \frac{g\rho_l(\rho_l - \rho_v)}{\mu_l}\delta^2 \tag{8-4}$$

Heat is transferred through the condensate layer solely by conduction. Assuming that the temperature gradient is linear, the average enthalpy change of the vapor in condensing to liquid and subcooling to the average liquid temperature of the condensate film is

$$h_{fg} + \frac{1}{\Gamma_c}\int_0^\delta \rho_l u c_{pl}(T_{sv} - T)\, dy = h_{fg} + \tfrac{3}{8}c_{pl}(T_{sv} - T_s) \tag{8-5}$$

and the rate of heat transfer to the wall is $(k/\delta)(T_{sv} - T_s)$, where k is the thermal conductivity of the condensate. In the steady state the rate of enthalpy change of the condensing vapor must equal the rate of heat flow to the wall:

$$\frac{q}{A} = k\frac{T_{sv} - T_s}{\delta} = \left[h_{fg} + \tfrac{3}{8}c_{pl}(T_{sv} - T_s) \right]\frac{d\Gamma_c}{dx} \tag{8-6}$$

Equating the expressions for $d\Gamma_c$ from Eqs. 8-4 and 8-6 gives

$$\delta^3 d\delta = \frac{k\mu_l(T_{sv} - T_s)}{g\rho_l(\rho_l - \rho_v)h'_{fg}} dx$$

where $h'_{fg} = h_{fg} + \tfrac{3}{8}c_{pl}(T_{sv} - T_s)$. Integrating between the limits $\delta = 0$ at $x = 0$ and $\delta = \delta$ at $x = x$ and solving for $\delta(x)$ yields

$$\delta(x) = \left[\frac{4\mu_l kx(T_{sv} - T_s)}{g\rho_l(\rho_l - \rho_v)h'_{fg}} \right]^{1/4} \tag{8-7}$$

The local heat-transfer coefficient $h(x)$ is $k/\delta(x)$. Substituting in the expression for δ from Eq. 8-7 gives the unit-surface conductance as

$$h(x) = \left[\frac{\rho_l(\rho_l - \rho_v)gh'_{fg}k^3}{4\mu_l x(T_{sv} - T_s)} \right]^{1/4} \tag{8-8}$$

and the dimensionless local Nusselt number at x is

$$\mathrm{Nu}_x = \frac{h_x x}{k} = \left[\frac{\rho_l(\rho_l - \rho_v)gh'_{fg}x^3}{4\mu_l k(T_{sv} - T_s)} \right]^{1/4} \tag{8-9}$$

Inspection of Eq. 8-8 shows that the unit conductance for condensation decreases with increasing distance from the top as the film thickens. The thickening of the condensate film is similar to the growth of a boundary layer over a flat plate in convection. At the same time it is also interesting to observe that an increase in the temperature difference $(T_{sv} - T_s)$ causes a decrease in the surface conductance. This is caused by the increase in the film thickness as a result of the increased rate of condensation. No comparable phenomenon occurs in simple convection.

The average value of the conductance $\bar{h}$ for a vapor condensing on a plate of height L is obtained by integrating the local value h_x over the plate and dividing by the area. For a vertical plate of unit width and height L, we obtain by this operation

$$\bar{h}_c = \frac{1}{L}\int_0^L h_x\,dx = \frac{4}{3}h_{x=L} \tag{8-10}$$

or

$$\bar{h}_c = 0.943\left[\frac{\rho_l(\rho_l - \rho_v)gh'_{fg}k^3}{\mu_l L(T_{sv} - T_s)} \right]^{1/4} \tag{8-11}$$

It can easily be shown that, for a surface inclined by an angle ψ with the horizontal, the average conductance is

$$\bar{h}_c = 0.943 \left[\frac{\rho_l(\rho_l - \rho_v)gh'_{fg}k^3\sin\psi}{\mu_l L(T_{sv} - T_s)} \right]^{1/4} \tag{8-12}$$

A modified integral analysis for this problem by Rohsenow (Ref. 2), which is in better agreement with experimental data if $Pr > 0.5$ and $c_{pl}(T_{sv} - T_s)/h'_{fg} < 1.0$, yields results identical to Eqs. 8-8 through 8-12 except that h'_{fg} is replaced by $[h_{fg} + 0.68c_{pl}(T_{sv} - T_s)]$. The effect of vapor shear stress on laminar film condensation is usually small, but it can be taken into account in the preceding analysis as shown by Rohsenow and Choi (Ref. 3).

Although the foregoing analysis was made specifically for a vertical flat plate, the development is also valid for the inside and outside surfaces of vertical tubes if the tubes are large in diameter, compared with the film thickness. These results cannot be extended to inclined tubes, however. In such cases the film flow would not be parallel to the axis of the tube and the effective angle of inclination would vary with x.

The average unit conductance of a pure saturated vapor condensing on the outside of a horizontal tube can be evaluated by the same method which was used to obtain Eq. 8-12. For a tube of diameter D it leads to the equation (Ref. 4)

$$\bar{h}_c = 0.725 \left[\frac{\rho_l(\rho_l - \rho_v)gh'_{fg}k^3}{D\mu_l(T_{sv} - T_s)} \right]^{1/4} \tag{8-13}$$

If condensation occurs on N horizontal tubes so arranged that condensate from one tube flows directly onto the tube below, the average unit-surface conductance for the system can be estimated by replacing the tube diameter D in Eq. 8-13 by (DN). This method will in general yield conservative results because a certain amount of turbulence is unavoidable in this type of system (Ref. 8).

An analysis that agrees better with experimental data was made by Chen (Ref. 5) who suggested that, since the liquid film is subcooled, additional condensation occurs on the liquid layer between tubes. Assuming that all the subcooling is used for additional condensation, Chen's analysis yields for N horizontal tubes the equation

$$\bar{h}_c = 0.728 \left[1 + 0.2\frac{c_p(T_{sv} - T_s)}{h_{fg}}(N-1) \right] \left[\frac{g\rho_l(\rho_l - \rho_v)k^3h'_{fg}}{ND\mu_l(T_{sv} - T_s)} \right]^{1/4} \tag{8-14}$$

which is in reasonably good agreement with experimental results, provided that $[(N-1)c_p(T_{sv} - T_s)/h_{fg}] < 2$.

In the preceding equations the unit-surface conductance will be in $W/m^2 \cdot K$ if the other quantities are evaluated in the units listed below:

c_{pl} = specific heat of liquid, $J/kg \cdot K$

k_l = thermal conductivity of liquid, $W/m \cdot K$

ρ_l = density of liquid, kg/m^3

ρ_v = density of the vapor, kg/m^3

g = gravitational acceleration, m/s^2

h_{fg} = latent heat of condensation or vaporization, J/kg

$h'_{fg} = h_{fg} + \frac{3}{8} c_{pl}(T_{sv} - T_s)$

μ_l = viscosity of liquid, $N \cdot s/m^2$

D = tube diameter, m

L = vertical length of plane surface, m

T_{sv} = temperature of saturated vapor, K

T_s = wall-surface temperature, K

The physical properties of the liquid film in Eqs. 8-7 to 8-14 should be evaluated at the arithmetic average of the vapor and wall temperature. When used in this manner, Nusselt's equations are satisfactory for estimating surface conductances for condensing vapors. Experimental data are in general agreement with Nusselt's theory when the physical conditions comply with the assumptions inherent in the analysis. Deviations from Nusselt's film theory occur when the condensate flow becomes turbulent, when the vapor velocity is very high (Ref. 6), or when a special effort is made to render the surface nonwettable. All these factors tend to increase the surface conductance, and the Nusselt film theory will therefore always yield conservative results.

Example 8-1. A 0.013-m-o.d., 1.5-m-long tube is to be used to condense steam at 40,000 N/m^2, $T_{sv} = 349$ K. Estimate the unit-surface conductances for this tube in (a) the horizontal position, and (b) the vertical positions. Assume that the average tube-wall temperature is 325 K.

Solution

a. At the average temperature of the condensate film [$T_f = (349 + 325)/2 = 337$ K], the physical-property values pertinent to the problem are

$$k_l = 0.661 \ W/m \cdot K$$

$$\rho_l = 980.9 \ kg/m^3$$

$$h_{fg} = 2.349 \times 10^6 \ J/kg$$

$$\mu_l = 4.48 \times 10^{-4} \ N \cdot s/m^2$$

$$c_{pl} = 4184 \ J/kg \cdot K$$

For the tube in the horizontal position Eq. 8-13 applies and the unit-surface conductance is

$$\bar{h}_c = 0.725 \left[\frac{(980.9)(980.6)(9.81)(2.387 \times 10^6)(0.661)^3}{(0.013)(4.48 \times 10^{-4})(349 - 325)} \right]^{1/4}$$

$$= 10{,}600 \text{ W/m}^2 \cdot \text{K}$$

b. In the vertical position the tube may be treated as a vertical plate of area $\pi D L$ and, according to Eq. 8-11, the average unit-surface conductance is

$$\bar{h}_c = 0.943 \left[\frac{(980.9)(980.6)(9.81)(2.387 \times 10^6)(0.661)^3}{(4.48 \times 10^{-4})(1.5)(349 - 325)} \right]^{1/4}$$

$$= 4230 \text{ W/m}^2 \cdot \text{K}$$

Effect of Turbulence in the Film

The results of the preceding calculations show that, for a given temperature difference, the average unit conductance is considerably larger when the tube is placed in a horizontal position, where the path of the condensate is shorter and the film thinner, than in the vertical position, where the path is longer and the film thicker. This conclusion is generally valid when the length of the vertical tube is larger than 2.87 times the outer diameter, as can be seen by a comparison of Eqs. 8-11 and 8-13. However, both these equations are based on the assumption that the flow of the condensate film is laminar and consequently do not apply when the flow of the condensate is turbulent. Turbulent flow is hardly ever reached on a horizontal tube but may be established over the lower portion of a vertical surface. When this occurs, the average heat-transfer coefficient becomes larger as the length of the condensing surface is increased because the condensate no longer offers as high a thermal resistance as it does in laminar flow. This phenomenon is somewhat analogous to the behavior of a boundary layer.

Just as a fluid flowing over a surface undergoes a transition from laminar to turbulent flow, so the motion of the condensate becomes turbulent when its Reynolds number exceeds a critical value of about 2000. The Reynolds number of the condensate film Re_δ, when based on the hydraulic diameter, can be written as $\text{Re}_\delta = (4A/P)\Gamma_c/\delta \mu_l$, where P is the wetted perimeter equal to πD for a vertical tube and A is the flow cross-sectional area equal to $P\delta$. According to an analysis by Colburn (Ref. 7), the local heat-transfer coefficient for turbulent flow of the condensate can be evaluated from the equation

$$h_x = 0.056 \left(\frac{4\Gamma_c}{\mu_l} \right)^{0.2} \left(\frac{k_l^3 \rho_l^2 g}{\mu_l^2} \right)^{1/3} \text{Pr}_l^{1/2} \tag{8-15}$$

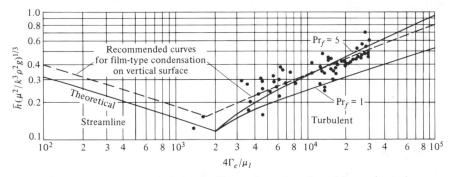

Figure 8-2 Effect of turbulence in film or heat transfer with condensation.

To obtain average values of the conductance, integration of h_x over the surface by means of Eq. 8-8 for values of $(4\Gamma_c/\mu_l)$ less than 2000 and Eq. 8-15 for values larger than 2000 is necessary. The results of such calculations for two values of the Prandtl number are plotted as solid lines in Fig. 8-2, where some experimental data obtained with diphenyl in turbulent flow are also shown (Ref. 8). The heavy dashed line shown on the same graph is an empirical curve recommended by McAdams (Ref. 9) for evaluating the average unit-surface conductance of single vapors condensing on vertical surfaces.

Example 8-2. Determine whether or not the flow of the condensate in Example 8-1 is laminar or turbulent at the lower end of the tube.

Solution: The Reynolds number of the condensate at the lower end of the tube can be written with aid of Eq. 8-3 as

$$\text{Re}_\delta = \frac{4\Gamma_c}{\mu_l} = \frac{4\rho_l^2 g \delta^3}{3\mu_l^2}$$

Substituting Eq. 8-7 for δ yields

$$\text{Re}_\delta = \frac{4\rho_l^2 g}{3\mu_l^2} \left[\frac{4\mu_l k_l L(T_{sv} - T_s)}{g h_{fg} \rho_l^2} \right]^{3/4}$$

$$= \frac{4}{3} \left[\frac{4k_l L(T_{sv} - T_s)\rho_l^{2/3} g^{1/3}}{\mu_l^{5/3} h_{fg}} \right]^{3/4}$$

Inserting in the expression above the numerical values for the problem yields

$$\text{Re}_\delta = \frac{4}{3} \left[\frac{4(0.661)(1.5)(349 - 325)(980.9)^{2/3}(9.81)^{1/3}}{(4.48 \times 10^{-4})^{5/3}(2.349 \times 10^6)} \right]^{3/4}$$

$$= 576$$

Since the Reynolds number at the lower edge of the tube is below 2000, the flow of the condensate is laminar and the result obtained from Eq. 8-11 is valid.

Effect of High Vapor Velocity

One of the approximations made in Nusselt's film theory is that the frictional drag between the condensate and the vapor is negligible. This approximation ceases to be valid when the velocity of the uncondensed vapor is substantial compared with the velocity of the liquid at the vapor-condensate interface. When the vapor flows upward, it adds a retarding force to the viscous shear and causes the film thickness to increase. With downward flow of vapor, the film thickness decreases, and surface conductances substantially larger than those predicted from Eq. 8-11 can be obtained. In addition, the transition from laminar to turbulent flow occurs at condensate Reynolds numbers of the order of 300 when the vapor velocity is high. Carpenter and Colburn (Ref. 10) determined the heat-transfer coefficients for condensation of pure vapors of steam and several hydrocarbons in a vertical tube, 2.44 m long and of 1.27 cm i.d., with inlet vapor velocities at the top up to 152 m/s. Their data are correlated reasonably well by the equation

$$\frac{\bar{h}_c}{c_{pl}G_m} \Pr_l^{1/2} = 0.046\sqrt{\frac{\rho_l}{\rho_v}}f \qquad (8\text{-}16)$$

where
$\Pr_l$ = Prandtl number of liquid

ρ_l = density of liquid, kg/m^3

ρ_v = density of vapor, kg/m^3

c_{pl} = specific heat of liquid, J/kg·K

$\bar{h}_c$ = average unit conductance, W/m^2·K

f = Fanning pipe friction coefficient
evaluated at the average vapor velocity

$$G_m = \sqrt{(G_1^2 + G_1G_2 + G_2^2)/3}$$

= mean value of the mass velocity of the vapor, kg/s·m^2

G_1 = mass velocity at top of the tube

G_2 = mass velocity at bottom of tube

All physical properties of the liquid in Eq. 8-16 are to be evaluated at a reference temperature equal to $0.25T_{sv} + 0.75T_s$. These results have not been verified on other systems but may be used generally as an indication of the influence of vapor velocity on the heat-transfer coefficient of condensing vapors when the vapor and the condensate flow in the same direction.

Condensation of Superheated Vapor

Although all the preceding equations strictly apply only to saturated vapors, they can also be used with reasonable accuracy for condensation of superheated vapors. The rate of heat transfer from a superheated vapor to a wall at T_s will therefore be

$$q = A\bar{h}(T_{sv} - T_s) \tag{8-17}$$

where

$\bar{h}$ = average value of unit conductance determined from the equation appropriate to the geometrical configuration with the same vapor at saturation conditions

T_{sv} = *saturation temperature* corresponding to the prevailing system pressure

Dropwise Condensation

When a condensing surface is contaminated with a substance that prevents the condensate from wetting the surface, the vapor will condense in drops rather than as a continuous film (Ref. 11). This is known as *dropwise condensation*. A large part of the surface is not covered by an insulating film under these conditions, and the heat-transfer coefficients are four to eight times as high as in filmwise condensation. So far, dropwise condensation has been reliably obtained only with steam. For the purpose of calculating the unit conductance in practice, it is recommended that filmwise condensation be assumed because, even with steam, dropwise

Table 8-1   Approximate Values of Unit-Surface Conductances for Condensation of Pure Vapors

VAPOR	SYSTEM	APPROXIMATE RANGE OF $T_{sv} - T_s$(K)	APPROXIMATE RANGE OF AVERAGE UNIT CONDUCTANCE W/m²·K
Steam	Horizontal tubes, 25–75 mm o.d.	3–20	11,000–23,000
Steam	Vertical surface 3 m high	3–20	5700–11,000
Ethanol	Vertical surface 0.15 m high	10–55	1100–1900
Benzene	Horizontal tube, 25 mm o.d.	15–45	1400–2000
Ethanol	Horizontal tube, 50 mm o.d.	5–20	1700–2600
Ammonia	Horizontal 50 to 75 mm annulus	1–4	1400–2600[a]

[a]Overall heat-transfer coefficient U for water velocities between 1.2 and 2.4 m/s (Ref. 14) inside the tube.

condensation can be expected only under carefully controlled conditions which cannot always be maintained in practice (Refs. 12 and 13). Dropwise condensation of steam may, however, be a useful method in some industrial applications.

Table 8-1 gives some approximate values of the unit surface conductance in condensation on horizontal tubes and vertical surfaces.

Mixtures of Vapors and Noncondensable Gases

The analysis of a condensing system containing a mixture of vapors, or a pure vapor mixed with noncondensable gas, is considerably more complicated than the analysis of a pure-vapor system. The presence of appreciable quantities of a noncondensable gas will in general reduce the rate of heat transfer. If high rates of heat transfer are desired, it is considered good practice to vent the noncondensable gas, which otherwise will blanket the cooling surface and add considerably to the thermal resistance. Noncondensable gases inhibit mass transfer by offering a diffusional resistance. A complete treatment of problems involving condensation of mixtures is beyond the scope of this text, and the reader is referred to References 8 and 15 for a comprehensive summary of available information on these topics.

8-3 BOILING HEAT TRANSFER

Boiling may occur when a surface in contact with a liquid is maintained at a temperature above the saturation temperature of the liquid. In contrast to convection heat transfer where the heat flux depends on the temperature difference between the surface and the bulk of the liquid, in boiling heat transfer the temperature difference between the surface and the saturation temperature at the pressure of the liquid is the driving potential for the heat flow. When the heated surface is submerged in a quiescent liquid the process is referred to as *pool boiling*. When the temperature of the liquid is below the saturation temperature and boiling takes place, the process is called *subcooled boiling*. If, on the other hand, the liquid is at the saturation temperature, the process is known as *saturated* or *bulk boiling*.

The different regimes of pool boiling are illustrated schematically in Fig. 8-3. In Fig. 8-3 the heat flux is plotted as a function of the temperature difference between the surface and the saturation temperature, called the *excess temperature*, ΔT_x for short, for the case of a heated platinum wire submerged in water at atmospheric pressure (Ref. 17). In region 1, heat transfer occurs by free convection. In this regime the liquid in the vicinity of the wire is superheated, but the degree of superheat is insufficient to form vapor bubbles. As the temperature of the heating surface is increased, the energy level of liquid adjacent to the surface becomes sufficiently high

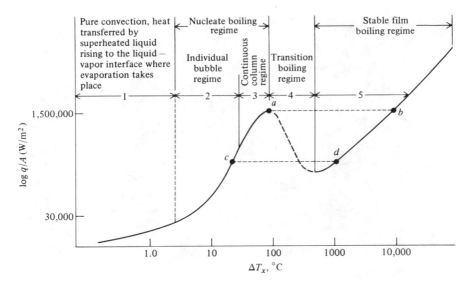

Figure 8-3 Typical boiling curves for a wire, tube, or horizontal surface in a pool of water at atmospheric pressure.

for some of the molecules to break away from surrounding molecules and form vapor nuclei which finally grow into vapor bubbles. This process occurs simultaneously at a number of favored spots on the heating surface (Ref. 18). Initially, the vapor bubbles are small and condense before reaching the surface; but as the temperature is raised further, the bubbles become more numerous and larger. Finally, they reach the free surface. As the temperature excess is increased further, bubbles form more rapidly, and in region 3 they coalesce into more-or-less continuous vapor columns rising to the surface. Both regions 2 and 3 are called the *nucleate boiling* regime.

As the temperature excess is further increased, bubbles form so rapidly that they coalesce on the wire and eventually form a vapor film that covers the surface. Under those conditions heat must be transferred through this film by conduction. Since the thermal conductivity of the vapor is small, the thermal resistance of this film is large and actually causes a reduction in heat flux with increasing excess temperature, as shown by the dotted curve in region 4. This region represents a transition from nucleate to film boiling and is unstable. A stable film-boiling condition is established in region 5 when the heat flux begins again to increase with increasing excess temperature. Since the temperature in this stable film-boiling regime is quite high, a significant part of the total heat is transferred by radiation.

The maximum heat flux of the nucleate boiling regime, indicated by point *a* in Fig. 8-3, bears special attention. The excess temperature corresponding to point *a* is known as the *critical excess temperature*. If the wire

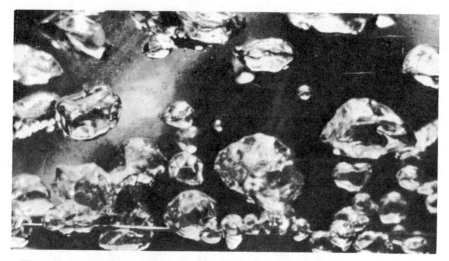

Figure 8-4 Nucleate boiling on a wire in water. (Courtesy of J. T. Castles.)

or the surface from which the heat is dissipated is heated electrically, an increase in the heat flux beyond point *a* requires a transition to point *b* in order to dissipate the heat. Since in many cases the excess temperature at point *b* is so large that the temperature of the surface exceeds the melting point, the wire melts, and a *burnout* results (Ref. 19). Only if the heat input is reduced rapidly below the critical heat flux when the system attains point *a* is it possible to operate in the transition boiling regime and

Figure 8-5 Film boiling on a wire in water. (Courtesy of J. T. Castles.)

finally attain a stable condition in the film-boiling regime (number 5 in Fig. 8-3). The photographs in Figs. 8-4 and 8-5 illustrate the nucleate and film-boiling mechanisms, respectively.

Correlation of Boiling Heat-Transfer Data

In nucleate boiling the rate of heat transfer depends largely on the turbulence generated by the bubbles in the vicinity of the surface. Bubbles are formed by expansion of entrapped gas or vapor in small cavities or crevasses at the surface. Their behavior depends on the temperature and pressure of the liquid (Ref. 20) as well as on the surface tension at the liquid/vapor interface.

To correlate experimental data in nucleate boiling, a dimensionless parameter indicative of the turbulence and mixing motion for the boiling process is used. This parameter is a type of Reynolds number, Re_b, which is obtained by combining the average bubble diameter, D_b, the mass velocity of the bubbles, G_b, and the liquid viscosity, μ_l, in the form

$$Re_b = \frac{D_b G_b}{\mu_l} \tag{8-18}$$

This parameter may be referred to as the *Bubble Reynolds number*, and can be used to calculate the Nusselt number, Nu_b, from an equation of the form

$$Nu_b = \frac{\bar{h}_b D_b}{k_l} = \phi(Re_b)\psi(Pr_l) \tag{8-19}$$

where Pr_l is the Prandtl number of the saturated liquid and $\bar{h}_b$ is the nucleate boiling heat-transfer coefficient defined as

$$\bar{h}_b = \frac{q}{A \Delta T_x} \tag{8-20}$$

Since in nucleate boiling the excess temperature is the physically significant temperature potential, it replaces the temperature difference between the surface and the bulk of the liquid used in the definition of the ordinary convection-heat-transfer coefficient.

Pool Boiling

Using experimental data as a guide, Rohsenow (Ref. 21) modified Eq. 8-19 by means of simplifying assumption to obtain the following dimensionless correlation equation for nucleate pool boiling:

$$\frac{c_l \Delta T_x}{h_{fg} Pr_l^{1.7}} = C_{sf} \left[\frac{q/A}{\mu_l h_{fg}} \sqrt{\frac{\sigma}{g(\rho_l - \rho_v)}} \right]^{0.33} \tag{8-21}$$

where

c_l = specific heat of saturated liquid, J/kg·K

ΔT_x = temperature excess = $T_s - T_{sat}$, K

q/A = heat flux, W/m²

h_{fg} = latent heat of vaporization, J/kg

g = gravitational acceleration, m/s²

ρ_l = density of saturated liquid, kg/m³

ρ_v = density of saturated vapor, kg/m³

σ = surface tension of liquid/vapor interface, N/m

μ_l = viscosity of liquid, N·s/m²

C_{sf} = empirical constant that depends upon the

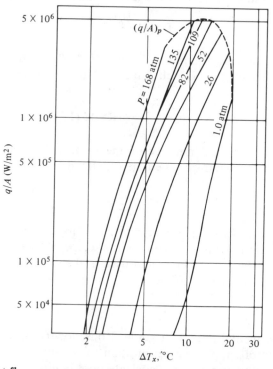

Figure 8-6 Heat flux versus excess temperature for nucleate boiling of water on a 0.024-in-diameter electrically heated platinum wire. (Extracted from W. M. Rohsenow, "A Method of Correlating Heat-Transfer Data from Surface Boiling Liquids," *Trans. ASME*, vol. 75, 1955, with permission of the publishers, The American Society of Mechanical Engineers.)

nature of the heating surface/fluid combination and whose numerical value varies from system to system; dimensionless

Pr_l = Prandtl number of liquid; dimensionless

The empirical constant C_{sf} depends on the surface roughness of the boiling surface and the wettability of a surface with a given fluid. Its value must be determined empirically with each surface/fluid combination. Figure 8-6 shows experimental data obtained by Addoms for pool boiling of water on a small platinum wire at various saturation pressures (Ref. 22). A correlation of these data by the Rohsenow equation is shown in Fig. 8-7 using the parameters in the square bracket of the right-hand side of Eq. 8-21,

$$\frac{q/A}{\mu_l h_{fg}} \sqrt{\frac{\sigma}{g(\rho_l - \rho_v)}}$$

as the ordinate and the combination of parameters on the left-hand side of

Table 8-2   Values of the Coefficient C_{sf} in Eq. 8-21
for Various Liquid-Surface Combinations

FLUID-HEATING SURFACE COMBINATION	C_{sf}
Water–copper (23)[a]	0.0130
Carbon tetrachloride–copper (23)	0.0130
35% K_2CO_3–copper (23)	0.0054
n-Butyl alcohol–copper (19)	0.0030
50% K_2CO_3–copper (23)	0.0027
Isopropyl alcohol–copper (19)	0.0025
n-Pentane–chromium (24)	0.0150
Water–platinum (22)	0.0130
Benzene–chromium (24)	0.0100
Water–brass (25)	0.0060
Ethyl alcohol–chromium (24)	0.0027
n-Pentane on Emery Polished Copper (26)	0.0154
n-Pentane on Emery Polished Nickel (26)	0.0127
Water on Emery Polished Copper (26)	0.0128
Carbon Tetrachloride on Emery Polished Copper (26)	0.0070
Water on Emery Polished, Paraffin-treated Copper (26)	0.0147
n-Pentane on Lapped Copper (26)	0.0049
n-Pentane on Emery Rubbed Copper (26)	0.0074
Water on Scored Copper (26)	0.0068
Water on Ground and Polished Stainless Steel (26)	0.0080
Water on Teflon Pitted Stainless Steel (26)	0.0058
Water on Chemically Etched Stainless Steel (26)	0.0133
Water on Mechanically Polished Stainless Steel (26)	0.0132

[a]Numbers in parentheses are those of references listed at end of chapter.

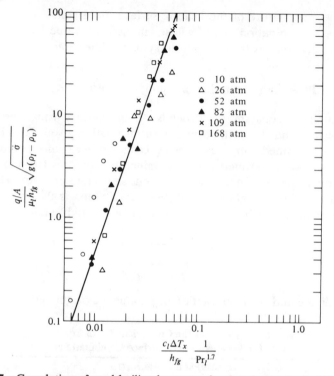

Figure 8-7 Correlation of pool-boiling heat-transfer data by method of Rohsenow. (Extracted from W. M. Rohsenow, "A Method of Correlating Heat-Transfer Data from Surface Boiling Liquids," *Trans. ASME*, vol. 75, 1955, with permission of the publishers, The American Society of Mechanical Engineers.)

Table 8-3 Vapor-Liquid Surface Tension for Water

SURFACE TENSION $\sigma \times 10^3$, N/m	SATURATION TEMPERATURE °C
75.6	0
72.6	20
69.4	40
66.0	60
62.5	80
58.8	100
48.2	150
37.6	200
26.4	250
14.7	300
3.7	350
0.0	374.1

Eq. 8-21,

$$\frac{c_l \Delta T_x}{h_{fg} Pr_l^{1.7}}$$

as the abscissa. The slope of a line fitted through the experimental points is 0.33 and C_{sf} is 0.013. With these two empirical constants, experimental data covering a pressure range between 1 and 150 atm are correlated quite satisfactorily. Values of the coefficient and other fluid heat-transfer surface combinations are listed in Table 8-2. Selected values of the vapor/liquid surface tension for water at various temperatures are given in Table 8-3.

The principal advantage of the Rohsenow correlation is that the heat-transfer performance of any fluid/surface combination in nucleate boiling at any pressure and heat flux can be predicted from a single test. One value of the heat flux and its corresponding value of excess temperature determine the value of C_{sf} in Eq. 8-21. It has been noted, however, that Eq. 8-21 strictly applies only to clean surfaces. For contaminated surfaces the exponent of Pr_l has been found to vary between 0.8 and 2, although contamination apparently does not influence the other exponent in Eq. 8-21. The geometrical shape of the heating surface does not affect the nucleate boiling mechanism appreciably, because the influence of the bubble motion on the fluid condition is limited to a region very near the surface (Refs. 27 and 28).

Example 8-3. Calculate the heat-transfer coefficient for water boiling at atmospheric pressure on a copper surface at 120°C.

Solution: The heat flux can be determined from the Rohsenow equation, using physical properties of water from Table F-1 and C_{sf} from Table 8-2.

$$\frac{4211 \times 20}{2.257 \times 10^6 \times (1.74)^{1.7}} =$$

$$0.013 \left[\frac{q/A}{278 \times 10^{-6} \times 2.257 \times 10^{-6}} \sqrt{\frac{0.0587}{9.81(957 - 0.6)}} \right]^{0.33}$$

Solving for the heat flux gives

$$\frac{q}{A} = 358 \text{ kW/m}^2$$

and the heat-transfer coefficient is

$$\bar{h}_b = \frac{q/A}{\Delta T_x} = \frac{358}{20} = 17.9 \text{ kW/m}^2\cdot\text{K}$$

Nucleate Boiling with Forced Convection

When a liquid flows over a surface which is maintained at a temperature above the saturation temperature of the liquid or when a liquid passes

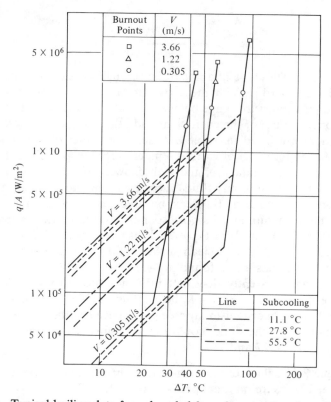

Figure 8-8 Typical boiling data for subcooled forced convection: heat flux versus temperature difference between surface and fluid bulk. (By permission from W. M. McAdams, W. E. Kennel, C. S. Minden, R. Carl, P. M. Picarnell, and J. E. Drew, "Heat Transfer at High Rates to Water with Surface Boiling," *Ind. Eng. Chem.*, vol. 41, 1945.)

through a channel whose walls are above the saturation temperature, forced convection boiling may occur. Figure 8-8 shows experimental results typical of subcooled forced convection in tubes or ducts. The ordinate is the heat flux and the abscissa is the temperature difference between the heating surface and the bulk of the fluid. The dotted lines represent forced convection conditions at various velocities and various degrees of subcooling while the solid lines indicate the deviations from forced convection caused by surface boiling. After the onset of boiling the wall temperature is practically independent of the fluid velocity, which indicates that the agitation caused by the bubbles is much more effective than the turbulence in forced convection without boiling. The heat-flux data with surface boiling from Fig. 8-8 are replotted in Fig. 8-9 as a function of the excess temperature. It can be seen that the resulting curve is similar to Fig. 8-6 for nucleate boiling in a saturated pool. This emphasizes the similarity of the boiling processes with and without convec-

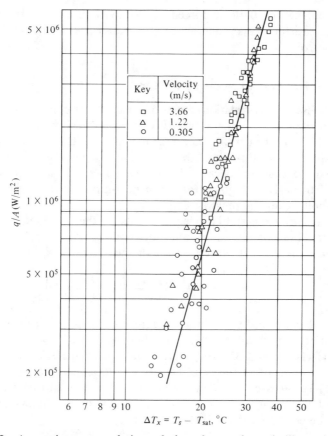

Figure 8-9 Approximate correlation of data for nucleate boiling with forced convection obtained by plotting heat flux versus excess temperature. (By permission from W. M. McAdams, W. E. Kennel, C. S. Minden, R. Carl, P. M. Picarnell, and J. E. Drew, "Heat Transfer at High Rates to Water with Surface Boiling," *Ind. Eng. Chem.*, vol. 41, 1945.)

tion; in both of them the heat flux depends on the excess temperature rather than on the conventional surface/bulk temperature difference. The total heat flux in forced convection boiling can be obtained by adding the boiling heat flux calculated from Eq. 8-21 to the forced convection heat flux calculated from Eq. 8-22:

$$\left(\frac{q}{A}\right)_{\text{total}} = \left(\frac{q}{A}\right)_{\text{boiling}} + \left(\frac{q}{A}\right)_{\text{forced convection}} \tag{8-22}$$

However, to calculate the forced-convection heat flux the temperature difference between the wall and the liquid bulk rather than the excess temperature difference should be used, and Rohsenow and Griffith (Ref. 32) recommend that for flow through tubes the coefficient 0.023 in Eq.

5-12 be replaced by 0.019 in forced-convection boiling. The superposition method is satisfactory as long as no substantial portion of the liquid has been vaporized.

Maximum Heat Flux In Nucleate Boiling

The Rohsenow correlation equation (Eq. 8-21) relates heat flux in boiling to the excess temperature but does not reveal the excess temperature at which the heat flux reaches a maximum and burnout may occur. The maximum heat flux attainable with nucleate boiling is often of great interest to the designer because for efficient heat transfer and operating safety, particularly in high-performance heat-input systems, operations in the film-boiling regime must be avoided.

By considering the stability requirements of the interface between the vapor film and the liquid in nucleate boiling, Zuber et al. (Ref. 31) developed an analytical expression for the peak heat flux,

$$\left(\frac{q}{A}\right)_{max} = \frac{\pi}{24} \rho_v h_{fg} \left[\frac{\sigma(\rho_l - \rho_v)g}{\rho_v^2}\right]^{1/4} \left(\frac{\rho_l + \rho_v}{\rho_l}\right)^{1/2} \qquad (8\text{-}23)$$

which has been found to be in good agreement with experimental results. A simplified version of Eq. 8-23 has been proposed by Rohsenow and Griffith (Refs. 32 and 33). In general, the type of material of the boiling surface does not affect the peak heat flux, although it has been observed that dirty surfaces can actually increase the peak heat-flux value by as much as 15%. When the bulk of the liquid is subcooled, the maximum heat flux can be estimated (Ref. 31) from the relation

$$\left(\frac{q}{A}\right)_{max} = \left(\frac{q}{A}\right)_{max,\,sat} \left\{1 + \left[\frac{2k_l(T_{sat} - T)}{\sqrt{\pi\alpha_l\tau}}\right]\frac{24}{\pi h_{fg}\rho_v}\left[\frac{\rho_v^2}{\sigma g(\rho_l - \rho_v)}\right]\right\}^{1/4}$$

$$(8\text{-}24)$$

where

$$\tau = \frac{\pi}{3}\sqrt{2\pi}\left[\frac{\sigma}{g(\rho_l - \rho_v)}\right]^{1/2}\left[\frac{\rho_v^2}{\sigma g(\rho_l - \rho_v)}\right]^{1/4}$$

and $(q/A)_{max,\,sat}$ is determined from Eq. 8-21. Noncondensible gases and nonwetting surfaces may reduce the peak heat flux at a given bulk temperature.

Boiling and Vaporization In Forced Convection

Forced-convection-vaporization characteristics have been investigated extensively, but because of the large number of variables that can influence this process, only a qualitative description is given (Refs. 34 and 35).

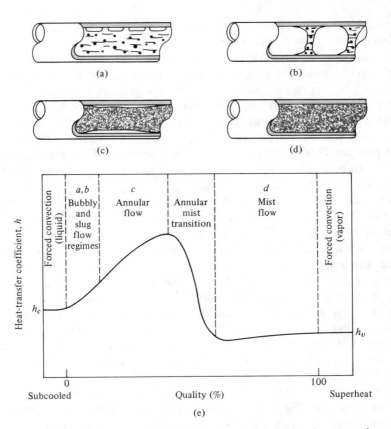

Figure 8-10 Characteristics of forced-convection vaporization-heat-transfer coefficients versus quality and types of flow regimes: (a) bubble; (b) slug; (c) annular; (d) mist; (e) flow regimes.

Suppose that a fluid below its boiling temperature enters a duct in which it is heated so that progressive vaporization occurs. Figure 8-10 shows schematically what happens in a duct in which a fluid is vaporized. Part (e) of the figure is a qualitative graph on which the heat-transfer coefficient at a specific location is plotted as a function of the local quality. As the fluid passes through the duct, the fluid bulk temperature increases toward its saturation point. This occurs usually only a short distance from the inlet in a system designed to vaporize the fluid. Then bubbles begin to form at nucleation sites and are carried into the main stream, as in nucleate pool boiling. The regime, known as the *bubbly-flow regime*, is shown schematically in Fig. 8-10(a). Bubbly flow occurs only at a very low quality and consists of individual bubbles of vapor entrained in the main flow. In the very narrow quality range over which bubbly flow exists, the heat-transfer coefficient can be predicted by superimposing liquid-forced-convection and nucleate-pool-boiling equations as long as the wall temperature is not so large as to produce film boiling.

As the vapor volume fraction increases, the individual bubbles begin to agglomerate and form plugs or slugs of vapor as shown in Fig. 8-10(b). Although in this regime, known as slug-flow regime, the mass fraction of vapor is generally much less than 1 percent, as much as 50% of the volume fraction may be vapor and the fluid velocity in the slug-flow regime may increase appreciably. The plugs of vapor are compressible volumes which also produce flow oscillations within the duct even if the entering flow is steady. Bubbles may continue to nucleate at the wall, and it is probably that the heat-transfer mechanism in plug flow is the same as in the bubbly regime: a superposition of forced convection to a liquid and nucleate pool boiling. Owing to the increased flow velocity, the heat-transfer coefficient rises, as can be seen in Fig. 8-10(c).

As the fluid flows farther along in the tube and the quality increases, a third flow regime, commonly known as the *annular-flow regime*, develops. In this regime the wall of the tube is covered by a thin film of liquid and heat is transferred through this liquid film. In the center of the tube, vapor is flowing at a faster velocity and although there may be a number of active bubble nucleation sites at the wall, vapor is generated primarily by vaporization from the liquid/vapor interface inside the tube and not by the formation of bubbles inside the liquid annulus. In addition to the liquid in the annulus at the wall, there is a significant amount of liquid dispersed throughout the vapor core as droplets. The quality range for this type of flow is strongly affected by fluid properties and geometry, but it is generally believed that transition to the next flow regime, known as the *mist-flow regime*, occurs at qualities of about 25% or higher.

The transition from annular to mist flow is of great interest since this is presumably the point at which the heat-transfer coefficient experiences a sharp decrease, as shown in Fig. 8-10(c). Therefore, this transition point can be the cause of a burnout in forced-convection vaporization unless the heat flux is reduced appropriately before this condition is encountered. An important change takes place in the transition between annular and mist flow: In the former the wall is covered by a relatively high conductivity liquid, whereas in the latter the wall is covered by a low-conductivity vapor.

Most of the heat transfer in mist flow is from the hot wall to the vapor, and after the heat has been transferred into the vapor core it is transferred to the liquid droplets there. Vaporization in mist flow actually takes place in the interior of the duct, not at the wall. For this reason the temperature of the vapor in the mist-flow regime can be greater than the saturation temperature, and thermal equilibrium may not exist in the duct. While the volume fraction of the liquid droplets is small, they account for a substantial mass fraction because of the high ratio of liquid to vapor density.

For the flow of vapor/liquid mixtures through tubes, Davis and David (Ref. 42) found that as long as liquid wets the wall, the empirical equation

$$\left(\frac{\bar{h}D}{k_l}\right) = 0.06 \left(\frac{\rho_l}{\rho_v}\right)^{0.28} \left(\frac{DG\chi}{\mu_l}\right)^{0.87} Pr_l^{0.4} \qquad (8\text{-}25)$$

where χ is the vapor mass fraction or quality. This equation correlates the results of several investigations within about 20%.

Mist flow persists until the quality reaches 100%. Once this condition is reached, the heat-transfer coefficient can again be predicted by equations appropriate for forced convection of a vapor in a tube or a duct.

Griffith (Ref. 36) developed an empirical burnout correlation for forced convection covering a wide range of conditions. He correlated burnout data for water, benzene, n-heptane, n-pentane, and ethanol at pressures varying from 0.5 to 96% of critical pressure, at velocities from 0 to 30 m/s, at subcooling from 255 to 535 K and at qualities ranging from 0 to 70%. The data used in this correlation were obtained in round tubes and rectangular channels. Figure 8-11 shows the correlated data and an inspection of this figure suggests that the burnout can apparently be predicted to within ±33% for the conditions used in this study. In Fig. 8-11, h_g is the saturated vapor enthalpy and h_b is the bulk enthalpy of the fluid, which

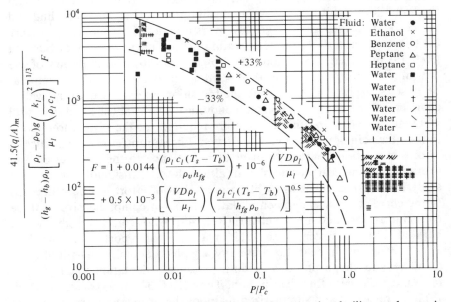

Figure 8-11 Peak-heat-flux correlation for forced-convection boiling and vaporization. (Courtesy of P. Griffith and the American Society of Mechanical Engineers.)

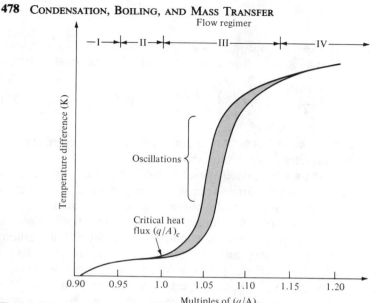

Figure 8-12 Characteristic temperature history near critical heat flux in forced convection boiling. I—forced convection; II—nucliate boiling; III—transition; IV —film boiling.

may be subcooled liquid, saturated liquid, or a two-phase flow mixture at some quality less than 70%.

It should be noted, however, that our ability to predict the critical heat flux in boiling with forced convection is inferior to that in pool boiling and the correlation in Fig. 8-11 grossly oversimplifies the complexity. Figure 8-12 shows qualitatively the temperature history in the vicinity of burnout (Ref. 19) in forced convection. The abscissa is shown in multiples of the critical heat flux. When the critical point is exceeded, the surface temperature rises sharply and oscillates in the transition regime III until the wall is completely dry and film boiling conditions are established, if the wall material can withstand the temperature required for the system to operate in the film boiling regime IV. The flow and heat transfer characteristics in forced convection boiling are of great importance to the safe design and operation of nuclear reactors and although no satisfactory theoretical analysis of all regimes is available, a great deal of effort has been devoted to the correlation of experimental data. The most comprehensive of these correlations are due to Macbeth (Ref. 48) who correlated the world burnout data for water up to 1963 in a series of reports. The variables affecting the critical heat flux are many, e.g., inlet enthalpy, length of duct, diameter and shape of flow conduit, pressure, orientation, quality, flow rate, and axial heat flux distribution. Tong (Ref. 34) and Rohsenow (Ref. 19) have reviewed and summarized many of the proposed correlations and the reader is referred to these references for more detailed information on this difficult and complex topic.

Transition Boiling and Film Boiling

The transition region between the nucleate- and film-boiling regimes is difficult to characterize in a quantitative manner (Ref. 37). Within this *transition-boiling* region, the amount of vapor generated is not enough to support a stable vapor film, but it is too large to allow sufficient liquid to reach the surface to support nucleate boiling. Berenson (Ref. 38) suggests, therefore, that nucleate and film boiling occur alternately at a given location. The process is unstable and photographs show that liquid surges sometimes toward the heating surface and sometimes away from it. At times, this turbulent liquid becomes so highly superheated that it explodes into vapor (Ref. 39). From an industrial viewpoint, the transition-boiling regime is of little interest; equipment designed to operate in the nucleate-boiling region may be sized with more assurance and operate with more reproducible results.

Film boiling is characterized by a vapor film covering the heating surface. Since the vapor has a low thermal conductivity relative to the liquid, very large temperature differences are necessary to transfer heat at a rate approaching the nucleate-boiling regime. Film boiling is, therefore, used industrially only if circumstances make it unavoidable. Examples of such situations are when liquefied gases such as oxygen or hydrogen are boiling at ordinary temperatures. Film boiling may also occur in cryogenic fluids.

Film boiling requires a large temperature difference between the solid surface and the liquid, but it is not possible to predict exactly what the minimum excess temperature difference must be to sustain a stable film. For most organic liquids at atmospheric pressure the value is at least 100 K, but it is known that the lower limit is strongly influenced by pressure. There appears to be no upper limit to the temperature difference which will sustain a stable film, but at very high temperatures an appreciable amount of the heat transfer is due to radiation superimposed on the boiling.

The film-boiling process is classified according to whether the vapor film moves in viscous or turbulent flow. The flow is viscous if the flow path is short (e.g., on horizontal wires, small horizontal tubes, or short vertical surfaces). Figure 8-5 illustrates viscous film boiling on a wire.

When the flow is viscous it is possible to predict the thickness of the vapor film and to calculate the heat flux quite accurately. Bromley (Refs. 40 and 41) has studied stable film boiling on the outside of horizontal tubes experimentally and analytically and his experimental results can be correlated with satisfactory accuracy by the equation

$$\bar{h}_b = 0.62 \left[\frac{k_v{}^3 \rho_v (\rho_l - \rho_v) g \lambda'}{D_o \mu_v \Delta T_x} \right]^{1/4} \tag{8-26}$$

where

k_v = thermal conductivity of saturated vapor, $W/m \cdot K$

D_o = outside diameter of tube, m

μ_v = viscosity of saturated vapor, $N \cdot s/m^2$

while, except for λ', the other symbols are the same as those used in Eq. 8-21. The symbol λ' is defined as

$$\lambda' = h_{fg}\left(1 + \frac{0.4c_{pv}\Delta T_x}{h_{fg}}\right) \tag{8-27}$$

where c_{pv} is the specific heat of the saturated vapor, $J/kg \cdot K$.

The average unit-surface conductance $\bar{h}_b$ in Eq. 8-26 accounts only for the heat which is transferred by conduction through the vapor film and by boiling convection from the surface of the film to the surrounding liquid. Superimposed on this heat-flow path is the contribution of radiation to the total heat transfer. Since the heat transfer by radiation causes an increase in the thickness of the film, the coefficient h_b for conduction and convection in the presence of appreciable radiation is less than in the absence of radiation. The total surface conductance when radiation is appreciable can be estimated from the empirical relation

$$\bar{h} = \bar{h}_b\left(\frac{\bar{h}_b}{\bar{h}}\right)^{1/3} + \bar{h}_r \tag{8-28}$$

assuming an emissivity of unity for the liquid by trial and error. To determine the heat-transfer coefficient when the liquid is flowing past the surface of the tube, Bromley et al. (Ref. 41) suggests the equation

$$\bar{h}_b = 2.7\sqrt{\frac{V_\infty k_v \rho_v \lambda'}{D_o \Delta T_x}} \tag{8-29}$$

if the velocity V_∞ is larger than $2\sqrt{gD_o}$. The total conductance, including radiation, is then

$$\bar{h} = \bar{h}_b + \tfrac{7}{8}h \tag{8-30}$$

under these conditions. At velocities less than $2\sqrt{gD_o}$, the flow is not fully developed turbulent, and the conductance may be evaluated from data in Reference 41.

8-4 MASS TRANSFER

Mass transfer is the transport of a component of a mixture from a region of high concentration of that component to a region of lower concentration.

Our discussion of mass transfer will be kept on a relatively elementary level. Those interested in a more in-depth coverage of the subject of mass transfer are referred to References 43 and 47.

We will consider two broad categories of mass transfer in the next two sections. They are diffusion mass transfer and convective mass transfer. *Diffusion mass transfer* (also frequently called *molecular mass transfer*) occurs in a material predominantly by molecular motion. Diffusion of one component in an essentially stationary mixture in the direction of decreasing concentration of that component is analogous to the transfer of heat by conduction in the direction of decreasing temperature. A familiar example of diffusion mass transfer would be the transport of an aerosol spray into the still air in a room. The spray diffuses from its point of introduction until it permeates the room. Similarly, a wet cloth placed in a room will eventually dry because the high concentration of water vapor surrounding the cloth diffuses into the dryer air. There are numerous other examples of the diffusion mass transfer process.

Convective mass transfer involves the transport of a component due to the motion of the bulk fluid. The convective-mass-transfer process is analogous to the convective-heat-transfer process. Mass transfer by convection can be classified as either forced or free. If the fluid motion is caused by density differences, the process is free convective mass transfer, and if an external device such as a fan or pump causes the fluid motion, the process is classified as forced convective mass transfer. There are numerous examples of the convective-mass-transfer processes, including humidification, distillation, liquid extraction, gas absorption, and leaching.

Our coverage of mass transfer relies heavily on the analogy with the corresponding heat-transfer mode. We will discuss a flux equation for the diffusion of mass which is analogous to the Fourier law for heat conduction. Also, since the conservation equations for both convective heat and mass transfer are similar, we will find that many of the dimensionless convective-heat-transfer correlations give acceptable results when applied to convective-mass-transfer problems.

Mass Transfer by Diffusion

The rate of diffusion of species A in a mixture of A and B (often called *binary diffusion* because only two species are present) is governed by *Fick's first law of diffusion*. Fick's first law can be written on a mass basis, in which the mass transfer rate is measured by kg/s, or on a molar basis, in which the mass transfer rate is measured in mol/s.

On a mass basis the one-dimensional form of Fick's first law is

$$\dot{m}_A'' = \omega_A(\dot{m}_A'' + \dot{m}''_B) - \rho D_{AB} \frac{d\omega_A}{dx} \tag{8-31}$$

where

$\dot{m}_A'', \dot{m}_B'' =$ mass-transfer rate in x direction per unit area of species A and species B, respectively, kg/m$^2 \cdot$s

$\omega_A =$ mass fraction of species A, which is a dimensionless quantity equal to mass of species A divided by mass of mixture of A and B

$\rho =$ density of mixture of A and B, kg/m^3

$D_{AB} =$ mass diffusivity of species A in species B, m^2/s

$x =$ direction in which mass-transfer rate is measured, m

The first term on the right-hand side of Eq. 8-31 physically represents transport of species A and B due to bulk motion of both species.

The second term on the right-hand side of Eq. 8-31 is the diffusion of species A due to the gradient of concentration of species A within the mixture. This term is analogous in form to the Fourier law (Eq. 1-2) of heat conduction. The negative sign preceding this term indicates that mass will be transferred in the $+x$ direction if the concentration of species A decreases in the x direction.

Equation 8-31 can be greatly simplified if the mass-transfer rate of species A per unit area is exactly equal to the mass-transfer rate of species B per unit area in the opposite direction. This condition is called *equimass counterdiffusion*. If conditions are such that equimass counterdiffusion exists, then

$$\dot{m}_A'' = -\dot{m}_B''$$

and the first term on the right side of Eq. 8-31 is eliminated resulting in a simplified version of Fick's first law,

$$\dot{m}_A'' = -\rho D_{AB} \frac{d\omega_A}{dx} \tag{8-32}$$

Equation 8-32 is also valid for one other condition. If the mass fraction of species A is small (i.e., $\omega_A \ll 1.0$), the term $\omega_A(\dot{m}_A'' + \dot{m}_B'')$ is small compared to the mass transfer due to a concentration gradient and Eq. 8-32 will apply even if conditions of equimass counter diffusion are not valid.

Fick's law can also be written on a molar basis, in which the mass-transfer rates per unit area are measured in mol/m$^2 \cdot$s. The mole basis form of Fick's first law analogous to Eq. 8-31 is

$$\dot{N}_A'' = X_A(\dot{N}_A'' + \dot{N}_B'') - CD_{AB} \frac{dX_A}{dx} \tag{8-33}$$

where

$$\dot{N}_A'', \dot{N}_B'' = \text{mass transfer rate in the } x \text{ direction per unit area of species } A \text{ and species } B, \text{ respectively, mol/m}^2 \cdot \text{s}$$

$$X_A = \text{mole fraction of species } A, \text{ which is a dimensionless quantity equal to number of moles of species } A \text{ divided by number of moles of mixture of } A \text{ and } B$$

$$C = \text{total molar concentration of mixture of } A \text{ and } B, \text{ mol/m}^3$$

If conditions exist so that for every mole of A being transported in one direction there is an equal number of moles of B transferred in the opposite direction, we have a condition called *equimolar counterdiffusion*. For equimolar counterdiffusion $\dot{N}_A'' = -\dot{N}_B''$ and Eq. 8-33 reduces to

$$\dot{N}_A'' = -CD_{AB}\frac{dX_A}{dx} \tag{8-34}$$

Equation 8-34 is also applicable when the mole fraction of A is small $(X_A \ll 1.0)$ even if conditions of equimolar counterdiffusion do not exist.

Like the thermal conductivity in the Fourier law, the mass diffusivity D_{AB} is a measurable property. The double subscript is used because its value is a function of both species that constitute the binary mixture. The values for D_{AB} not only depend upon temperature, pressure, and the concentration of both species but also the phase of each species. Measured values for binary systems where both species are gases are given in Table J-1. Several liquid phase mass diffusivity values are listed in Table J-2. The values for mass diffusivity in the liquid phase are several orders of magnitude smaller than in the gas phase because the mobility of the liquid molecules are a great deal less than the gas molecules. A few of the available values for solid phase mass diffusivity are given in Table J-3.

To determine the mass-transfer rate by diffusion we must integrate Fick's first law. Assuming equimass counterdiffusion, constant mass-transfer rate, constant density, and constant mass diffusivity, integration of Eq. 8-32 for steady, one-dimensional diffusion results in

$$\dot{m}_A = \rho D_{AB} A \frac{\omega_A(L) - \omega_A(0)}{L} \tag{8-35}$$

where L is the width of the diffusing layer on whose surfaces the mass fraction of species A are $\omega_A(L)$ and $\omega_A(0)$.

We could derive a similar equation on a molar basis by integrating Eq. 8-34. Assuming constant D_{AB}, constant C, constant mass-transfer rate in one dimension, and equimolar counterdiffusion, the integration would result in

$$\dot{N}_A = CD_{AB}A \frac{X_A(L) - X_A(0)}{L} \tag{8-36}$$

Using the resistance concept, the molar mass-transfer rate by diffusion can be expressed as

$$\dot{N}_A = \frac{X_A(L) - X_A(0)}{R_D} \tag{8-37}$$

where R_D is the resistance to diffusion mass transfer. We can see that the resistance to equimolar counter diffusion in rectangular coordinates is given by

$$R_D = \frac{L}{CD_{AB}A} \tag{8-38}$$

where A is the cross-sectional area through which the mass is diffusing. Notice that the resistance to diffusion is inversely proportional to the mass diffusivity.

It is easy to extend the resistance concept to other coordinate systems. For example, we could show that the resistance for equimolar counterdiffusion through a cylindrical tube with radii r_i and r_o and length L would be

$$R_D = \frac{\ln(r_o/r_i)}{2\pi CD_{AB}L} \tag{8-39}$$

and the resistance for equimolar counterdiffusion through a spherical shell is

$$R_D = \frac{r_o - r_i}{4\pi CD_{AB}r_o r_i} \tag{8-40}$$

Equation 8-40 is similar to Eq. 2-40 and Eq. 8-39 is similar to Eq. 2-35.

When both diffusing species are assumed to be ideal gases, it is often more convenient to express Fick's first law or its integrated form in terms of temperatures and partial pressures instead of mass or mole fractions.

The ideal equation of state for a mixture can be written as

$$P = \frac{\rho R_u T}{M} \tag{8-41}$$

where P is the total pressure of the mixture, R_u the universal gas constant ($R_u = 8.314 \times 10^3$, J/kmol·K), and M the molecular weight of the mixture. The ideal equation of state can also be written for a single species that makes up the ideal gas mixture. For example, for species A, the equation of state is

$$P_A = \frac{\rho_A R_u T}{M_A} \tag{8-42}$$

In this equation P_A is the partial pressure of species A. The mass fraction of species A in an ideal gas can therefore be expressed as

$$\omega_A = \frac{\rho_A}{\rho} = \frac{P_A M_A}{PM} \tag{8-43}$$

Assuming that an ideal gas mixture is isothermal and at a constant total pressure, Eq. 8-35 for equimass counterdiffusion can be converted to partial pressures by substituting Eqs. 8-41, 8-42, and 8-43. The result is

$$\dot{m}_A = \frac{M_A}{R_u T} \frac{P_A(L) - P_A(0)}{L/D_{AB}A} \tag{8-44}$$

If steady equimass counterdiffusion at constant temperature and constant total pressure takes place radially through a hollow cylinder, the mass-transfer rate would be

$$\dot{m}_A = \frac{M_A}{R_u T} \frac{P_A(r_i) - P_A(r_o)}{\ln(r_o/r_i)/2\pi D_{AB}L} \tag{8-45}$$

The corresponding equations to Eqs. 8-44 and 8-45 for equimolar mass transfer are

$$\dot{N}_A = \frac{1}{R_u T} \frac{P_A(L) - P_A(0)}{L/D_{AB}A} \tag{8-46}$$

for diffusion of ideal gases at constant temperature and total pressure in rectangular coordinates and

$$\dot{N}_A = \frac{1}{R_u T} \frac{P_A(r_i) - P_A(r_o)}{\ln(r_o/r_i)/2\pi D_{AB}L} \tag{8-47}$$

for similar conditions in cylindrical coordinates.

Example 8-4. Two large tanks are separated by a 0.75-m-long tube that has an internal diameter of 2 cm. One tank contains pure CO_2 at 0°C and 1 atm (1.0133×10^5 N/m^2) while the other holds pure H_2 at 0°C and 1 atm pressure. Estimate the initial rate at which the CO_2 diffuses into the tank of H_2.

Solution: From Table J-1,

$$D_{AB} = 5.50 \times 10^{-5} \text{m}^2/\text{s}$$

and Eq. 8-46 gives the mass-transfer rate assuming ideal gas behavior and conditions of equimolar counterdiffusion:

$$\dot{N}_A = \frac{AD_{AB}}{R_u TL} [P_A(L) - P_A(0)]$$

$$= \frac{\pi(0.01)^2 \times (5.5 \times 10^{-5})}{(8.314 \times 10^3) \times 273 \times 0.75} (1.0133 \times 10^5)$$

$$= 1.029 \times 10^{-9} \text{kmol/s} = 3.7 \times 10^{-3} \text{mol/h}$$

We can justify the assumption of equimolar counter diffusion by calculating the molar diffusion rate of H_2 from Eq. 8-46:

$$\dot{N}_B = \frac{AD_{AB}}{R_u TL}[P_B(L) - P_B(0)] = \frac{AD_{AB}}{R_u TL}[P_A(0) - P_A(L)]$$

$$= -1.029 \times 10^{-9} \text{kmol/s}$$

Therefore, the assumption of equimolar counterdiffusion is valid.

Example 8-5. Pure helium at 10 atm pressure and 20°C is contained in a Pyrex tube that has an o.d. of 5 cm and a wall thickness of 4 mm. Estimate the leak rate of helium through the tube wall in kg/s for a 1-m length of tube.

Solution: The properties of helium are

$$M_A = 4.0 \text{ kg/mol}$$

$$P_A(r_i) = 10 \text{ atm} = 10 \times 1.0133 \times 10^5 = 1.0133 \times 10^6 \text{ N/m}^2$$

and the D_{AB} value for the helium–Pyrex mixture is from Table J-3,

$$D_{AB} = 4.49 \times 10^{-15} \text{m}^2/\text{s}$$

Assuming that the concentration of helium on the outside of the tube is essentially zero,

$$P_A(r_o) = 0$$

If we assume that the mass fraction of the helium in the Pyrex is very small ($X_A \ll 1.0$), Eq. 8-45 will accurately predict the mass-transfer rate by diffusion through the tube wall.

$$\dot{m}_A = \frac{M_A}{R_u T} \frac{P_A(r_i) - P_A(r_o)}{\ln(r_o/r_i)/2\pi D_{AB}L}$$

$$= \frac{4}{(8.314 \times 10^3) \times 293} \frac{1.0133 \times 10^6}{\ln(2.5/2.1)/(2 \times \pi \times (4.49 \times 10^{-15}) \times 1)}$$

$$= 2.69 \times 10^{-13} \text{ kg/s}$$

To appreciate the size of this mass-transfer rate, consider that it would take approximately 120 years for 1 g of helium to diffuse through the tube wall.

Mass Transfer by Convection

Suppose we wish to determine the evaporation rate of water from a lake when dry air is blown over the surface. Since the mass transfer takes place by a convection process, it is convenient to define the mass transfer rate as proportional to the difference between the mass concentration at the surface and in the ambient fluid:

$$\dot{m}_A = h_m A \left(C_{A_s} - C_{A_\infty} \right) \tag{8-48}$$

Equation 8-48 is the basic rate equation for convective mass transfer assuming low mass-transfer rates and it defines the convective-mass-transfer coefficient h_m. Equation 8-48 is analogous to Newton's law of cooling (Eq. 1-16).

A sketch of the physical problem of water evaporating from a lake is shown in Fig. 8-13. The problem is similar to heat being transferred from a horizontal plane in which a thermal boundary layer develops over the plate. In a similar fashion, a concentration boundary layer exists inside which the concentration varies with distance perpendicular to the horizontal surface of the lake. Outside the boundary layer the concentration of water vapor remains constant at the ambient value.

The example of water evaporating from a lake further illustrates the similarity between the convective-heat and mass-transfer processes. In fact, if the conservation equations are derived for both convection of mass and heat, the equations are similar, with the mass concentration C_A analogous to the temperature T and the mass diffusivity D_{AB} analogous to the thermal diffusivity α.

This analogy suggests that a simple way to predict the value for the mass-transfer coefficient is to use the corresponding dimensionless correlation for convective heat transfer after substituting for the appropriate mass-transfer dimensionless groups. We know that the dimensionless group in heat transfer that contains the heat transfer coefficient is the Nusselt number,

$$\text{Nu} = \frac{h_c L}{k}$$

The analogous mass-transfer dimensionless group is called the *Sherwood number*, which is defined as

$$\text{Sh} = \frac{h_m L}{D_{AB}} \tag{8-49}$$

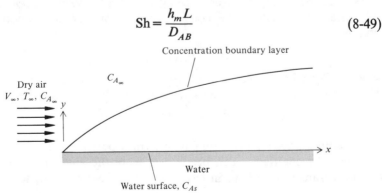

Figure 8-13 Concentration boundary layer on a flat plate.

In heat transfer the dimensionless group that specified the ratio of momentum to thermal diffusion is the Prandtl number,

$$\text{Pr} = \frac{\nu}{\alpha} \tag{8-50}$$

In mass transfer the mass diffusivity replaces the thermal diffusivity and the new dimensionless group is called the *Schmidt number*,

$$\text{Sc} = \frac{\nu}{D_{AB}} \tag{8-51}$$

The Schmidt number is the ratio of momentum to mass diffusion.

From our previous work in forced convection heat transfer, we have seen that the Nusselt number is a function of Reynolds and Prandtl numbers:

$$\text{Nu} = f(\text{Re}, \text{Pr})$$

Utilizing the similarity between convective heat and mass transfer, we would expect the Sherwood number to be a similar function of the Reynolds and Schmidt numbers:

$$\text{Sh} = f(\text{Re}, \text{Sc}) \tag{8-52}$$

For example, we have already seen for turbulent flow inside a tube that the dimensionless heat-transfer correlation is given by Eq. 5-12:

$$\text{Nu}_D = 0.023 \text{Re}_D^{0.8} \text{Pr}^{0.33} \tag{8-53}$$

Using this correlation we could approximate the mass-transfer rate from a liquid that completely wets the inside of a tube to a turbulent gas that is flowing down the axis of the tube by the equation

$$\text{Sh}_D = 0.023 \text{Re}_D^{0.8} \text{Sc}^{0.33} \tag{8-54}$$

For this situation the liquid evaporates into the gas phase and Eq. 8-54 can be used to predict the liquid evaporation rate.

As a second example, let us once again consider water evaporating from the surface of a lake. The convective-heat-transfer correlation for this situation is

$$\overline{\text{Nu}}_L = 0.664 \text{Re}_L^{1/2} \text{Pr}^{1/3} \tag{8-55}$$

which is Eq. 5-22 for heat transfer from a flat plate assuming laminar flow. The corresponding laminar mass transfer correlation would be

$$\overline{\text{Sh}}_L = 0.664 \text{Re}_L^{1/2} \text{Sc}^{1/3} \tag{8-56}$$

If the mass-transfer problem involves transport by free convection, the mass-transfer coefficient can be derived from the analogous free convection-heat-transfer problem. In free convection heat transfer, we know that

$$\text{Nu} = f(\text{Gr}, \text{Pr}) \tag{8-57}$$

The Grashof number for mass transfer is defined as

$$\mathrm{Gr}_{AB} = \frac{g\xi L^3 \Delta C_A}{\nu^2} \tag{8-58}$$

where ξ is defined as

$$\xi = -\frac{1}{\rho} \frac{\partial \rho}{\partial C_A} \bigg]_T$$

For free convection mass transfer, we would expect a mass-transfer correlation of the form

$$\mathrm{Sh} = f(\mathrm{Gr}_{AB}, \mathrm{Sc})$$

Consult Chapter 5 for the form of functional relationships to be used for each problem.

The Reynolds analogy, which relates the heat-transfer rate to drag on a surface, can be extended to mass transfer. The Reynolds analogy given in Eq. 4-50 is

$$\frac{\mathrm{Nu}}{\mathrm{RePr}} = \frac{C_f}{2} \tag{8-59}$$

Reynold's analogy for turbulent mass transfer in the absence of form drag is then

$$\frac{\mathrm{Sh}}{\mathrm{ReSc}} = \frac{C_f}{2} \tag{8-60}$$

Using Eq. 8-60, we can estimate mass-transfer coefficients from a knowledge of the friction coefficient.

Example 8-6. Estimate the evaporation rate of water from a lake that is roughly 500×500 m in size. The wind velocity is 5 m/s. Both the lake and air have a temperature of 25°C. Estimate the evaporation rate when the ambient air has a relative humidity of (a) 10%, and (b) 80%.

Solution: This problem involves forced convective mass transfer from a flat plate. Before we can select the proper dimensionless correlation for the Sherwood number, we must determine whether the flow process is laminar or turbulent. The Reynolds number at the end of the lake is

$$\mathrm{Re}_L = \frac{\rho V_\infty L}{\mu} = \frac{1.146 \times 5 \times 500}{18.46 \times 10^{-6}} = 1.55 \times 10^8$$

The flow of the air is completely turbulent, so the proper dimensionless correlation is analogous to Eq. 4-56:

$$\overline{\mathrm{Nu}}_L = 0.036 \mathrm{Pr}^{1/3} \mathrm{Re}_L^{0.8}$$

which is used for turbulent heat transfer over a flat plate. The proper mass-transfer correlation is therefore

$$\overline{Sh}_L = 0.036 Sc^{1/3} Re_L^{0.8}$$

Using properties from Table G-1 and J-1,

$$D_{AB} = 2.6 \times 10^{-5} \ m^2/s$$

$$Sc = \frac{\nu}{D_{AB}} = \frac{16.18 \times 10^{-6}}{2.6 \times 10^{-5}} = 0.6223$$

The Sherwood number is

$$\overline{Sh}_L = 0.036(0.6223)^{1/3}(1.55 \times 10^8)^{0.8}$$

$$= 1.096 \times 10^5$$

The convective-mass-transfer coefficient is

$$\overline{h}_m = \frac{\overline{Sh}_L D_{AB}}{L} = \frac{1.096 \times 10^5 \times (2.6 \times 10^{-5})}{500} = 5.7 \times 10^{-3} \ m/s$$

Next, the concentration of the water at the surface of the lake and in the ambient air must be determined. At the surface the air is saturated and the relative humidity is 100%. The relationship among the water partial pressure, the relative humidity, and saturation pressure is

$$P_A = \phi P_{sat}$$

From the steam tables, the saturation pressure at 25°C is

$$P_{sat} = 3098 \ N/m^2$$

The partial pressure at the surface of the lake is

$$P_{w_s} = P_{sat} = 3098 \ N/m^2$$

The concentration of the water vapor at the surface of the lake, assuming that the water vapor in an ideal gas, is

$$C_{A_s} = \frac{M_w P_{w_s}}{R_u T} = \frac{18 \times 3098}{8.314 \times 10^3 \times 298} = 0.02251 \ kg/m^3$$

a. For an ambient relative humidity of 10%, the ambient water-vapor concentration is

$$C_{A_\infty} = \frac{M_w \phi_\infty P_{sat}}{R_u T} = \frac{18 \times 0.10 \times 3098}{8.314 \times 10^3 \times 298} = 0.00225 \ kg/m^3$$

The water evaporation transfer rate is

$$\dot{m}_A = \overline{h}_m A (C_{A_s} - C_{A_\infty}) = (5.7 \times 10^{-3})(25 \times 10^4)(0.02251 - 0.00225)$$

$$= 28.9 \ kg/s$$

b. When the ambient relative humidity is 80%, the surface and ambient concentrations are

$$C_{A_s} = 0.02251 \text{ kg/m}^3$$

$$C_{A_\infty} = 0.0180 \text{ kg/m}^3$$

and the evaporation rate is

$$\dot{m}_A = (5.7 \times 10^{-3})(25 \times 10^4)(0.02251 - 0.0180)$$
$$= 6.43 \text{ kg/s}$$

or 78% less than the rate when the ambient relative humidity is 10%.

REFERENCES

1. W. Nusselt, "Die Oberflachenkondensation des Wasserdampfes," *Z. Ver. Deutsch. Ing.*, vol. 60, pp. 541–569, 1916.
2. W. M. Rohsenow, "Heat Transfer and Temperature Distribution in Laminar-Film Condensation," *Trans. ASME*, vol. 78, pp. 1645–1648, 1956.
3. W. M. Rohsenow and H. Choi, *Heat, Mass, and Momentum Transfer*, Prentice-Hall, Inc., Englewood Cliffs, N. J., 1961.
4. F. Kreith, *Principles of Heat Transfer*, Harper & Row, Publishers, New York, 1973.
5. M. M. Chen, "An Analytical Study of Laminar Film Condensation: Part 1, Flat Plates, and Part 2, Single and Multiple Horizontal Tubes," *Trans. ASME*, sec. C, vol. 83, pp. 48–60, 1961.
6. W. M. Rohsenow, J. M. Weber, and A. T. Ling, "Effect of Vapor Velocity on Laminar and Turbulent Film Condensation," *Trans. ASME*, vol. 78, pp. 1637–1744, 1956.
7. A. P. Colburn, "The Calculation of Condensation Where a Portion of the Condensate Layer is in Turbulent Flow," *Trans. AIChE*, vol. 30, p. 187, 1933.
8. C. G. Kirkbridge, "Heat Transfer by Condensing Vapors on Vertical Tubes," *Trans. AIChE*, vol. 30, p. 170, 1933.
9. W. H. McAdams, *Heat Transmission*, 3rd ed., McGraw-Hill Book Company, New York, 1954.
10. E. F. Carpenter and A. P. Colburn, "The Effect of Vapor Velocity on Condensation-Inside Tubes," *Inst. Mech. Eng. ASME, Proc. General Discussion on Heat Transfer*, pp. 20–26, 1951.
11. T. B. Drew, W. M. Nagle, and W. Q. Smith, "The Conditions for Dropwise Condensation of Steam," *Trans. AIChE*, vol. 31, pp. 605–621, 1935.
12. J. W. Rose, "On the Mechanism of Dropwise Condensation," *Int. J. Heat Mass Transfer*, vol. 10, pp. 755–762, 1967.
13. P. Griffith and M. S. Lee, "The Effect of Surface Thermal Properties and Finish on Dropwise Condensation," *Int. J. Heat Mass Transfer*, vol. 10, pp. 697–707, 1967.
14. A. P. Katz, H. J. Macintire, and R. E. Gould, "Heat Transfer in Ammonia Condensers," *Univ. Illinois Eng. Expt. Sta. Bull. 209*, 1930.

15. W. H. McAdams, *Heat Transmission*, 3rd ed., McGraw-Hill Book Company, New York, 1954.

16. J. C. Collier, *Convective Boiling and Condensation*, McGraw-Hill Book Company, New York, 1972.

17. E. A. Farber and R. L. Scorah, "Heat Transfer to Water Boiling Under Pressure," *Trans. ASME*, vol. 70, pp. 369–384, 1948.

18. F. C. Gunther and F. Kreith, Photographic Study of Bubble Formation in Heat Transfer to Subcooled Water, Jet Propulsion Laboratory, California Institute of Technology, *Prog. Rept. 4-120*, March 1950.

19. W. M. Rohsenow, *Boiling*, Section 13, in *Handbook of Heat Transfer*, W. M. Rohsenow and J. P. Hartnett, eds. McGraw-Hill Book Company, 1973, pp. 13-1 to 13-75.

20. H. S. Fath and R. L. Judd, "Influence of System Pressure on Microlayer Evaporation Heat Transfer," *J. Heat Trans.*, vol. 100, pp. 49–55, 1978.

21. W. M. Rohsenow, "A Method of Correlating Heat-Transfer Data for Surface Boiling Liquids," *Trans. ASME*, vol. 74, pp. 969–975, 1952.

22. J. N. Addoms, Heat Transfer at High Rates to Water Boiling Outside Cylinders, D. Sc. thesis, in chemical engineering, MIT, Cambridge, Mass., 1948.

23. E. L. Piret and H. S. Isbin, "Natural Circulation Evaporation Two-Phase Heat Transfer," *Chem. Eng. Prog.*, vol. 50, p. 305, 1954.

24. M. T. Cichelli and C. F. Bonilla, "Heat Transfer to Liquids Boiling Under Pressure, *Trans. AIHE*, vol. 41, pp. 755–787, 1945.

25. D. S. Cryder and A. C. Finalbargo, "Heat Transmission from Metal Surfaces to Boiling Liquids: Effect of Temperature of the Liquid on Film Coefficient," *Trans. AIChE*, vol. 33, pp. 346–362, 1937.

26. R. I. Vachon, G. H. Nix, and G. E. Tanger, "Evaluation of Constants for the Rohsenow Pool-Boiling Correlation," *J. Heat Transfer*, vol. 90, pp. 239–247, 1968.

27. W. H. McAdams et al., "Heat Transfer from Single Horizontal Wires to Boiling Water," *Chem. Eng. Prog.*, vol. 44, pp. 639–646, 1948.

28. W. H. McAdams, *Heat Transmission*, 3rd ed., McGraw-Hill Book Company, New York, 1954.

29. F. Kreith and A. S. Foust, Remarks on the Stability and Mechanism of Surface Boiling Heat Transfer, *ASME Paper 54-A-16*, August 1954.

30. W. H. McAdams, W. E. Kennel, C. S. Minden, R. Carl, P. M. Picornell, and J. E. Dew, "Heat Transfer at High Rates to Water with Surface Boiling," *Ind. Eng. Chem.*, vol. 41, pp. 1945–1953, 1944.

31. N. Zuber, M. Tribus, and J. W. Westwater, "The Hydrodynamic Crisis in Pool Boiling of Saturated and Subcooled Liquids," *Proceedings of the International Conference on Developments in Heat Transfer, ASME*, New York, pp. 230–236, 1962.

32. W. Rohsenow and P. Griffith, "Correlation of Maximum Heat Flux Data for Boiling of Saturated Liquids," Reprint, *Heat Transfer Symposium, AIChE*, Louisville, Ky., March 1955.

33. W. M. Rohsenow and H. Choi, *Heat, Mass, and Momentum Transfer*, Prentice-Hall, Inc., Englewood Cliffs, N. J., 1961.

34. L. S. Tong, *Boiling and Two-Phase Flow*, John Wiley and Sons, New York, 1965.

35. K. Konmutsos, R. Moissis, and A. Spyridonos, "A Study of Bubble Departure in Forced Convection Boiling," *J. Heat Transfer*, vol. 90, pp. 223–230, 1968.

36. P. Griffith, "Correlation of Nucleate-Boiling Burnout Data," *ASME Paper 57-HT-21*.

37. D. P. Jordan, "Film and Transition Boiling," in *Advanced in Heat Transfer*, vol. 5, T. F. Irvine, Jr., and J. P. Hartnett, eds, Academic Press, New York, pp. 55–125, 1968.

38. P. J. Berenson, "Experiments on Pool-Boiling Heat Transfer," *Int. J. Heat Mass Transfer*, vol. 5, pp. 985–999, 1962.

39. J. W. Westwater, "Boiling Heat Transfer," *Amer. Scientist*, vol. 47, no. 3, pp. 427–446, 1959.

40. L. A. Bromley, "Heat Transfer in Stable Film Boiling," *Chem. Eng. Prog.*, vol. 46, pp. 221–227, 1950.

41. L. A. Bromley et al., "Heat Transfer in Forced Convection Film Boiling," *Ind. Eng. Chem.*, vol. 45, pp. 2639–2646, 1953.

42. E. J. Davis and M. M. David, "Two-Phase Gas-Liquid Convection Heat Transfer," *I and E C Fundamentals*, vol. 3 pp. 111–118, 1964.

43. R. Bird, W. E. Stewart, and E. N. Lightfoot, *Transport Phenomena*, John Wiley & Sons, Inc., New York, 1960.

44. J. Crank, *The Mathematics of Diffusion*, 2nd ed., Oxford University Press, Inc., New York, 1975.

45. W. Jost, *Diffusion in Solids, Liquids and Gases*, rev. ed., Academic Press, New York, 1960.

46. T. K. Sherwood, R. C. Pigford, and C. R. Wilke, *Mass Transfer*, McGraw-Hill Book Company, New York, 1975.

47. R. D. Treybal, *Mass Transfer Operations*, 2nd ed., McGraw-Hill Book Company, New York, 1968.

48. R. V. Macbeth, "Forced Convection Burnout in Simple Uniformly Heated Channels: A Detailed Analysis of World Data," European Atomic Energy Soc. Symp. on Two-Phase Flow, Steady State Burnout and Hydrodynamic Instability, Sweden, 1963.

PROBLEMS

The problems in this chapter are organized in the manner shown in the table.

PROBLEM NUMBERS	SECTION	SUBJECT
8-1 to 8-6	8-2	Condensation
8-7 to 8-14	8-3	Boiling
8-14 to 8-26	8-4	Mass Transfer

8-1 Develop the Nusselt film-condensation relation for condensation inside small vertical tubes where the film builds up an annulus.

8-2 Consider a 1-cm i.d. vertical tube at a surface temperature of 90°C with atmospheric saturated steam inside. Determine the tube length at which the condensate fills the tube and chokes the flow.

8-3 Calculate the average heat-transfer coefficient for film-type condensation of water at pressures of 0.05 and 1.0 atm for (a) a vertical surface 2 m high; (b) the outside surface of a 1-cm vertical tube 2 m long; (c) the outside surface of a 1-cm horizontal tube 2 m long; and (d) a 10-tube vertical bank of 1-cm horizontal tubes 2 m long. In all cases, assume that the vapor velocity is negligible and that the surface temperatures are constant at 10°C below saturation temperature.

8-4 A thin-walled horizontal copper tube of 9-mm o.d. is placed in a pool of water at atmospheric pressure and 100°C. Inside the tube an organic vapor is condensing and the outside surface temperature of the tube is uniform at 220°C. Calculate the average unit-surface conductance at the outside of the tube.

8-5 Estimate (a) the heat-transfer surface area required and (b) suggest a suitable arrangement for the condenser of a 70-kW refrigeration machine. The working fluid is ammonia condensing on the outside of horizontal pipes at a pressure of 12 atm. The condenser is to be constructed with 3-cm steel pipes (3-cm o.d. 2.5-cm i.d.), cooling water is available at 25°C, and the average water velocity in the pipes is not to exceed 2 m/s.

8-6 The inside surface of a 1-m long vertical 1-cm i.d. tube is maintained at 120°C. For saturated steam at 3.4 atm condensing inside estimate the average unit-surface conductance and the condensation rate assuming the steam velocity is small.

8-7 Predict the nucleate-boiling heat-transfer coefficient for water boiling at atmospheric pressure on the outside surface of a 1-cm o.d. verticle tube 2m long. Assume the tube-surface temperature is constant at $10°C$ above the saturation temperature.

8-8 Estimate the maximum heat flux obtainable with nucleate pool boiling on a clean surface for (a) water at 1 atm on brass, (b) water at 10 atm on brass, and (c) n-butyl alcohol at 3 atm on copper.

8-9 Determine the excess temperature at one-half of the maximum heat flux for the fluid-surface combinations in Problem 8.8.

8-10 Estimate the time required to freeze a 2-cm thickness of water due to nocturnal radiation with ambient air and initial water temperatures at $5°C$. Neglect evaporation effect.

8-11 For saturated pool boiling of water on a horizontal plate calculate the peak heat flux at pressures of 10, 20, 40, 60, and 80% of the critical pressure p_c and plot your results as q_{max} vs p/p_c. The surface tension of water may be taken as $\sigma = 0.242 - 6.3 \times 10^{-4} T$ where σ is in N/m and T in $°C$.

8-12 A 5-mm-thick flat plate of stainless steel, 1.5 cm high and 35 cm long, is immersed vertically at an initial temperature of $1000°C$ in a large water bath at $100°C$ and at atmospheric pressure. Determine how long it will take this plate to cool to $600°C$.

8-13 Calculate the maximum safe heat flux in the nucleate-boiling regime for water flowing at a velocity of 15 m/s through a 1-cm i.d. tube 0.5 m long if the water enters at 1 atm pressure and $100°C$ and the heat flux in the tube is uniform at a rate of 16×10^3 kW/m^2.

8-14 Water at atmospheric pressure is boiling in a pot with a flat copper bottom on an electric range which maintains the surface temperature at $110°C$. Calculate the boiling heat-transfer coefficient.

8-15 A spherical rubber balloon 1 m in diameter contains helium at a gauge pressure of 3 atm and a temperature of $30°C$. The thickness of the balloon material is 0.1 mm. Calculate the rate of helium diffusion through the balloon material if the mass diffusivity of helium in rubber is 3×10^{-11} m^2/s.

8-16 A 3-m long, 1-cm i.d. tube connects two tanks both containing gases at 1 atm and $0°C$. Calculate the diffusion rate between the two gases assuming equimolar counterdiffusion. The two gases are air and hydrogen.

8-17 Rework Problem 8-16 assuming the gases are carbon dioxide and hydrogen.

8-18 The end of a 2-cm i.d. metal pipe is sealed with a diaphram that has a thickness of 3×10^{-2} mm. The pipe contains hydrogen at a temperature of 20°C and a pressure of 5 atm. The mass diffusivity of hydrogen in the diaphram material is estimated to be 6×10^{-12} m^2/s. Estimate the rate of hydrogen diffusion through the diaphram.

8-19 Estimate the diffusion rate per unit area of amonia gas at 1 atm 0°C across a 25-cm layer of stagnant air at the same conditions.

8-20 Estimate the rate at which a layer of nitrogen with a surface area of 0.5 m^2 diffuses across a 2-m-thick layer of carbon dioxide at a total pressure of 1 atm and a temperature of 25°C.

8-21 Verify that the Sherwood, Schmidt and the Grashof number for mass transfer are all dimensionless.

8-22 Estimate the evaporation rate from a 100×100 m lake to dry air. The ambient air velocity is 20 m/s and the temperature of the lake is 25°C.

8-23 Rework Problem 8-22 when the ambient air has a relative humidity of 50% and a temperature of 25°C.

8-24 A 30×30 cm pan of toluene is placed in a 25°C room. Air is circulated across the surface of the toluene with a velocity of 10 m/s. Estimate the evaporation rate of the toluene into the air.

8-25 A 3-cm o.d. horizontal cylinder is made of a porous material that is saturated with water. Dry air at 25°C is forced across the cylinder with velocity of 100 m/s normal to the axis of the cylinder. Estimate the evaporation rate of the water from the surface per unit length of cylinder.

8-26 Estimate the evaporatiion rate from a single 2-mm diameter droplet of water at 25°C into air with a relative humidity of 50% if the droplet falls through still air at a velocity of 170 m/s.

Appendix A

THE INTERNATIONAL SYSTEM OF UNITS

The International System of Units (SI) has evolved from the MKS system, in which the meter is the unit of length, the kilogram is the unit of mass, and the second is the unit of time. The SI unit system is rapidly becoming the standard unit system throughout the industrialized world.

The SI system is based on seven units. Other derived units may be related to these seven base units through governing equations. The base units are listed in Table A-1 along with the recommended symbols. The derived units of interest in heat transfer and fluid flow are given in Table A-2. Several defined units are listed in Table A-3.

Table A-1   SI Base Units

QUANTITY	NAME OF UNIT	SYMBOL
Length	meter	m
Mass	kilogram	kg
Time	second	s
Electrical current	ampere	A
Thermodynamic temperature	kelvin	K
Luminous intensity	candela	cd
Amount of a substance	mole	mol

Table A-2 SI Derived Units

QUANTITY	NAME OF UNIT	SYMBOL
Acceleration	meters per second squared	m/s^2
Area	square meters	m^2
Capacitance	farad	F
Density	kilogram per cubic meter	kg/m^3
Dynamic viscosity	newton-second per square meter	$N \cdot s/m^2$
Electrical resistance	ohm	Ω
Force	newton	N
Frequency	hertz	Hz
Kinematic viscosity	square meter per second	m^2/s
Plane angle	radian	rad
Potential difference	volt	V
Power	watt	W
Pressure	pascal	Pa
Radiant intensity	watts per steradian	W/sr
Solid angle	steradian	sr
Specific heat	joules per kilogram · kelvin	$J/kg \cdot K$
Thermal conductivity	watts per meter · kelvin	$W/m \cdot K$
Velocity	meters per second	m/s
Volume	cubic meter	m^3
Work, energy, heat	joule	J

Table A-3 SI Defined Units

QUANTITY	UNIT	DEFINING EQUATION
Capacitance	farad, F	$1\ F = 1\ A \cdot s/V$
Electrical resistance	ohm, Ω	$1\ \Omega = 1\ V/A$
Force	newton, N	$1\ N = 1\ kg \cdot m/s^2$
Potential difference	volt, V	$1\ V = 1\ W/A$
Power	watt, W	$1\ W = 1\ J/s$
Pressure	pascal, Pa	$1\ Pa = 1\ N/m^2$
Temperature	kelvin, K	$K = °C + 273.15$
Work, heat, energy	joule, J	$1\ J = 1\ N \cdot m$

Standard prefixes can be used in the SI system to designate multiples of the basic units and thereby conserve space. The standard prefixes are listed in Table A-4.

Table A-4 SI Prefixes

MULTIPLIER	SYMBOL	PREFIX
10^{12}	T	tera
10^9	G	giga
10^6	M	mega
10^3	k	kilo
10^2	h	hecto
10^1	da	deka
10^{-1}	d	deci
10^{-2}	c	centi
10^{-3}	m	milli
10^{-6}	μ	micro
10^{-9}	n	nano
10^{-12}	p	pico
10^{-15}	f	femto
10^{-18}	a	atto

Table A-5 contains an alphabetical listing of physical constants that are frequently used in heat-transfer and fluid-flow problems, along with their values in the SI system of units.

Table A-5 Physical Constants in SI Units

QUANTITY	SYMBOL	VALUE
–	e	2.718281828
–	π	3.141592653
–	g_c	$1.00000 \text{ kg} \cdot \text{m N}^{-1} \text{ s}^{-2}$
Avogadro constant	N_A	$6.022169 \times 10^{26} \text{ kmol}^{-1}$
Boltzmann constant	k	$1.380622 \times 10^{-23} \text{ J K}^{-1}$
First radiation constant	$C_1 = 2\pi hc^2$	$3.741844 \times 10^{-16} \text{ W} \cdot \text{m}^2$
Gas constant	R_u	$8.31434 \times 10^3 \text{ J kmol}^{-1} \text{ K}^{-1}$
Gravitational constant	G	$6.6732 \times 10^{-11} \text{ N} \cdot \text{m}^2 \text{ kg}^{-2}$
Planck constant	h	$6.626196 \times 10^{-34} \text{ J} \cdot \text{s}$
Second radiation constant	$C_2 = hc/k$	$1.438833 \times 10^{-2} \text{ m} \cdot \text{K}$
Speed of light in a vacuum	c	$2.997925 \times 10^8 \text{ m s}^{-1}$
Stefan-Boltzmann constant	σ	$5.66961 \times 10^{-8} \text{ W m}^{-2} \text{ K}^{-4}$

Conversion factors between the SI and engineering systems for commonly used heat-transfer quantities are given in Table A-6. The conversion factors are listed in alphabetical order of the physical quantities.

Table A-6   Conversion Factors

Physical Quantity	Symbol	Conversion Factor
Area	A	1 ft^2 = 0.0929 m^2 1 in.^{2} = 6.452 × 10^{-4} m^2
Density	ρ	1 lb$_m$/ft^3 = 16.018 kg/m^3 1 slug/ft^3 = 515.379 kg/m^3
Energy	Q or W	1 Btu = 1055.1 J 1 cal = 4.186 J 1 (ft)(lb$_f$) = 1.3558 J 1 (hp)(hr) = 2.685 × 10^6 J
Force	F	1 lb$_f$ = 4.448 N
Heat-flow rate	q	1 Btu/hr = 0.2931 W 1 Btu/sec = 1055.1 W
Heat flux	q''	1 Btu/(hr)(ft^2) = 3.1525 W/m^2
Heat generation per unit volume	q'''_G	1 Btu/(hr)(ft^3) = 10.343 W/m^3
Heat-transfer coefficient	h_c	1 Btu/(hr)(ft^2)(°F) = 5.678 W/m^2·K
Length	L	1 ft = 0.3048 m 1 in. = 2.54 cm = 0.0254 m 1 mile = 1.6093 km = 1609.3 m
Mass	m	1 lb$_m$ = 0.4536 kg 1 slug = 14.594 kg
Mass flow rate	$\dot{m}$	1 lb$_m$/hr = 0.000126 kg/s 1 lb$_m$/sec = 0.4536 kg/s
Power	$\dot{W}$	1 hp = 745.7 W 1 (ft)(lb$_f$)/sec = 1.3558 W 1 Btu/sec = 1055.1 W 1 Btu/hr = 0.293 W
Pressure	P	1 lb$_f$/in.^{2} = 6894.8 N/m^2 1 lb$_f$/ft^2 = 47.88 N/m^2 1 atm = 101,325 N/m^2
Specific energy	Q/m	1 Btu/lb$_m$ = 2326.1 J/kg
Specific heat capacity	c	1 Btu/(lb$_m$)(°F) = 4187 J/kg·K
Temperature	T	$T(°R) = (9/5)T(K)$ $T(°F) = [T(°C)](9/5) + 32$ $T(°F) = [T(K) - 273.15](9/5) + 32$
Thermal conductivity	k	1 Btu/(hr)(ft)(°F) = 1.731 W/m·K
Thermal diffusivity	α	1 ft^2/sec = 0.0929 m^2/s 1 ft^2/hr = 2.581 × 10^{-5} m^2/s
Thermal resistance	R_t	1 (hr)(°F)/Btu = 1.8958 K/W

Table A-6. Conversion Factors (Continued)

PHYSICAL QUANTITY	SYMBOL	CONVERSION FACTOR
Velocity	V	1 ft/sec = 0.3048 m/s
		1 mph = 0.44703 m/s
Viscosity, dynamic	μ	1 lb_m/(ft)(sec) = 1.488 N·s/m^2
		1 centipoise = 0.00100 N·s/m^2
Viscosity, kinematic	ν	1 ft^2/sec = 0.0929 m^2/s
		1 ft^2/hr = 2.581 × 10^{-5} m^2/s
Volume	V	1 ft^3 = 0.02832 m^3
		1 in^3 = 1.6387 × 10^{-5} m^3
		1 gal (U.S. liq.) = 0.003785 m^3

Appendix B

VECTOR OPERATIONS

Laplacian

It is often convenient to apply the principles of vector calculus to the field of heat transfer. Vector operations such as the Laplacian and gradient offer a very convenient and compact way to express several of the basic equations used in conduction heat transfer.

The general form of the conduction equation (Eq. 2-6) can be written in terms of the *Laplacian operator*,

$$\nabla^2 T + \frac{q_G'''}{k} = \frac{1}{\alpha} \frac{\partial T}{\partial t} \tag{B-1}$$

The form of the Laplacian operator designated by the symbol ∇^2 varies depending upon the coordinate system being used. The form of the Laplacian in the rectangular, cylindrical, and spherical coordinate systems is given in Table B-1.

Table B-1   Form of the Laplacian

COORDINATE SYSTEM	LAPLACIAN $\nabla^2 T$
Rectangular, $T = T(x,y,z)$	$\dfrac{\partial^2 T}{\partial x^2} + \dfrac{\partial^2 T}{\partial y^2} + \dfrac{\partial^2 T}{\partial z^2}$
Cylindrical, $T = T(r,\phi,z)$	$\dfrac{1}{r} \dfrac{\partial}{\partial r}\left(r \dfrac{\partial T}{\partial r} \right) + \dfrac{1}{r^2} \dfrac{\partial^2 T}{\partial \phi^2} + \dfrac{\partial^2 T}{\partial z^2}$
Spherical, $T = T(r,\theta,\phi)$	$\dfrac{1}{r^2} \dfrac{\partial}{\partial r}\left(r^2 \dfrac{\partial T}{\partial r} \right) + \dfrac{1}{r^2 \sin\theta} \dfrac{\partial}{\partial \theta}\left(\sin\theta \dfrac{\partial T}{\partial \theta} \right)$ $+ \dfrac{1}{r^2 \sin^2\theta} \dfrac{\partial^2 T}{\partial \phi^2}$

The conduction equation written in the form of Eq. B-1 is applicable to all three coordinate systems. The only changes that are necessary when converting from one coordinate system to another is the substitution of the appropriate form of the Laplacian from Table B-1.

The Laplacian operates on a scalar quantity, in this case the temperature, and the result is a scalar quantity. The Laplacian of the temperature when multiplied by the thermal conductivity physically represents the net energy per unit volume conducted into an elemental volume of the solid material.

Gradient

The *gradient* is another vector operation that is frequently used in heat transfer. The Fourier law of conduction may be written in a vector form in terms of the gradient as

$$\bar{q}'' = -k\,\nabla T \qquad\qquad (B-2)$$

where the gradient of the temperature, ∇T, is a vector quantity. An important property of the gradient is the fact that it is a vector which points in the direction of maximum change in temperature. That is, the heat flows in the direction of maximum change in temperature.

Normally, the heat flux is not a vector quantity. But when the Fourier law of heat conduction is written in the form of Eq. B-2, the heat flux can be considered to be a vector consisting of three components, one in each of the three coordinate directions. The heat flux vector in rectangular coordinates would be, for example, $\bar{q}'' = q_x''\hat{i} + q_y''\hat{j} + q_z''\hat{k}$, where $\hat{i}, \hat{j}$, and $\hat{k}$ are unit vectors in the x,y, and z directions, respectively. Similar expressions can be written for the heat-flux vector in the cylindrical and spherical coordinate systems.

The form of the gradient operation in the three coordinate systems is given in Table B-2.

Table B-2   Form of the Gradient

COORDINATE SYSTEM	GRADIENT ∇T
Rectangular	$\dfrac{\partial T}{\partial x}\hat{i} + \dfrac{\partial T}{\partial y}\hat{j} + \dfrac{\partial T}{\partial z}\hat{k}$
Cylindrical	$\dfrac{\partial T}{\partial r}\hat{i}_r + \dfrac{1}{r}\dfrac{\partial T}{\partial \phi}\hat{i}_\phi + \dfrac{\partial T}{\partial z}\hat{i}_z$
Spherical	$\dfrac{\partial T}{\partial r}\hat{i}_r + \dfrac{1}{r}\dfrac{\partial T}{\partial \theta}\hat{i}_\theta + \dfrac{1}{r\sin\theta}\dfrac{\partial T}{\partial \phi}\hat{i}_\phi$

By equating coefficients of Eq. B-2 to the corresponding terms in the gradient, the vector components of the heat-flux vector in the rectangular coordinate system are

$$q_x'' = -k\frac{\partial T}{\partial x}, \qquad q_y'' = -k\frac{\partial T}{\partial y}, \qquad q_z'' = -k\frac{\partial T}{\partial z}$$

and for the cylindrical coordinate system they are

$$q_r'' = -k\frac{\partial T}{\partial r}, \qquad q_\phi'' = -\frac{k}{r}\frac{\partial T}{\partial \phi}, \qquad q_z'' = -k\frac{\partial T}{\partial z}$$

Appendix C

HYPERBOLIC FUNCTIONS

Certain combinations of exponential functions are given special names. Because they are related to the hyperbola, like the trigometric functions are related to the circle, they are called *hyperbolic functions*. Three of the hyperbolic functions that appear in heat-transfer problems are defined as

$$\sinh x = \frac{e^x - e^{-x}}{2}$$

$$\cosh x = \frac{e^x + e^{-x}}{2}$$

$$\tanh x = \frac{\sinh x}{\cosh x} = \frac{e^x - e^{-x}}{e^x + e^{-x}}$$

The differentiation formulas for the functions sinh and cosh are

$$\frac{d}{dx}(\sinh x) = \cosh x$$

$$\frac{d}{dx}(\cosh x) = \sinh x$$

A short table of the values for sinh, cosh, and tanh are given in Table C-1.

Table C-1 Values for Hyperbolic Functions

x	$\sinh x$	$\cosh x$	$\tanh x$
0	0.00000	1.0000	0.00000
0.1	0.10017	1.00500	0.09967
0.2	0.20134	1.02007	0.19738
0.3	0.30452	1.04534	0.29131
0.4	0.41075	1.08107	0.37995
0.5	0.52110	1.12763	0.46212
0.6	0.63665	1.18547	0.53705
0.7	0.75858	1.25517	0.60437
0.8	0.88811	1.33743	0.66404
0.9	1.02652	1.43309	0.71630
1.0	1.17520	1.54308	0.76159
1.1	1.33565	1.66852	0.80050
1.2	1.50946	1.81066	0.83365
1.3	1.69838	1.97091	0.86172
1.4	1.90430	2.15090	0.88535
1.5	2.12928	2.35241	0.90515
1.6	2.37557	2.57746	0.92167
1.7	2.64563	2.82832	0.93541
1.8	2.94217	3.10747	0.94681
1.9	3.26816	3.41773	0.95624
2.0	3.62686	3.76220	0.96403
2.5	6.05020	6.13229	0.98661
3.0	10.01787	10.06766	0.99505
3.5	16.5426	16.5728	0.99818
4.0	27.2899	27.3082	0.99933
4.5	45.0030	45.0141	0.99975
5.0	74.2032	74.2099	0.99991

Appendix D

GAUSS ERROR FUNCTION

The *Gauss error function* is defined by the equation

$$\operatorname{erf} x = \frac{2}{\sqrt{\pi}} \int_0^x e^{-\eta^2} d\eta$$

The derivative of the error function is

$$\frac{d}{dx}(\operatorname{erf} x) = \frac{2}{\sqrt{\pi}} e^{-x^2}$$

A useful series expansion for the error function is

$$\operatorname{erf} x = \frac{2}{\sqrt{\pi}} \left(x - \frac{1}{3} x^3 + \frac{1}{5 \cdot 2!} x^5 - \frac{1}{7 \cdot 3!} x^7 \pm \cdots \right)$$

Values for the error function are given in Table D-1.

Table D-1 Error Function

x	$\operatorname{erf} x$	x	$\operatorname{erf} x$	x	$\operatorname{erf} x$
0.00	0.00000	0.76	0.71754	1.52	0.96841
0.02	0.02256	0.78	0.73001	1.54	0.97059
0.04	0.04511	0.80	0.74210	1.56	0.97263
0.06	0.06762	0.82	0.75381	1.58	0.97455
0.08	0.09008	0.84	0.76514	1.60	0.97635
0.10	0.11246	0.86	0.77610	1.62	0.97804
0.12	0.13476	0.88	0.78669	1.64	0.97962
0.14	0.15695	0.90	0.79691	1.66	0.98110
0.16	0.17901	0.92	0.80677	1.68	0.98249
0.18	0.20094	0.94	0.81627	1.70	0.98379

Table D-1. Error Function (Continued)

x	erf x	x	erf x	x	erf x
0.20	0.22270	0.96	0.82542	1.72	0.98500
0.22	0.24430	0.98	0.83423	1.74	0.98613
0.24	0.26570	1.00	0.84270	1.76	0.98719
0.26	0.28690	1.02	0.85084	1.78	0.98817
0.28	0.30788	1.04	0.85865	1.80	0.98909
0.30	0.32863	1.06	0.86614	1.82	0.98994
0.32	0.34913	1.08	0.87333	1.84	0.99074
0.34	0.36936	1.10	0.88020	1.86	0.99147
0.36	0.38933	1.12	0.88079	1.88	0.99216
0.38	0.40901	1.14	0.89308	1.90	0.99279
0.40	0.42839	1.16	0.89910	1.92	0.99338
0.42	0.44749	1.18	0.90484	1.94	0.99392
0.44	0.46622	1.20	0.91031	1.96	0.99443
0.46	0.48466	1.22	0.91553	1.98	0.99489
0.48	0.50275	1.24	0.92050	2.00	0.995322
0.50	0.52050	1.26	0.92524	2.10	0.997020
0.52	0.53790	1.28	0.92973	2.20	0.998137
0.54	0.55494	1.30	0.93401	2.30	0.998857
0.56	0.57162	1.32	0.93806	2.40	0.999311
0.58	0.58792	1.34	0.94191	2.50	0.999593
0.60	0.60386	1.36	0.94556	2.60	0.999764
0.62	0.61941	1.38	0.94902	2.70	0.999866
0.64	0.63459	1.40	0.95228	2.80	0.999925
0.66	0.64938	1.42	0.95538	2.90	0.999959
0.68	0.66278	1.44	0.95830	3.00	0.999978
0.70	0.67780	1.46	0.96105	3.20	0.999994
0.72	0.69143	1.48	0.96365	3.40	0.999998
0.74	0.70468	1.50	0.96610	3.60	1.000000

Appendix E

THERMODYNAMIC PROPERTIES OF SOLIDS

Use of Tabular Properties

The following tables provide the thermophysical properties for materials used in heat-transfer problems. The properties are given in a consistent set of SI units. The conversion factors in Table A-6 may be used if the property values are desired in another unit system.

Whenever a property value is much less than or greater than 1, it is convenient to express it as a numerical value times a power of 10. In this way a great deal of space can be saved without a loss in numerical accuracy. However, this practice can lead to a misunderstanding in the actual value listed in the tables. In the tables that follow, a column heading such as

$$\alpha \times 10^6 \ m^2/s$$

implies that the numerical value for thermal diffusivity occurring in the table has been *multiplied by* 10^6. Therefore, if the value in the table is listed as 5.06, the actual value for thermal diffusivity is

$$\alpha \times 10^6 = 5.06$$

or

$$\alpha = 5.06 \times 10^{-6} \ m^2/s$$

If a column heading is

$$\rho \times 10^{-3} \ kg/m^3$$

the numerical value for density in the table has been *multiplied by* 10^{-3}. For example, if the value in the table is 1.84, the actual value for ρ is

$$\rho \times 10^{-3} = 1.84$$

or

$$\rho = 1.84 \times 10^3 \ kg/m^3$$

The reader should always check the column headings when using the tabular values and be certain that the multiplying factors are applied correctly.

Table E-1 Metallic Elements[a]

Element	Thermal Conductivity k (W/m·K)[b]							Properties at 293 K or 20°C				Melting Temperature
	200 K / −73°C	273 K / 0°C	400 K / 127°C	600 K / 327°C	800 K / 527°C	1000 K / 727°C	1200 K / 927°C	ρ (kg/m³)	c_p (J/kg·K)	k (W/m·K)	$\alpha \times 10^6$ (m²/s)	(K)
Aluminum	237	236	240	232	220			2,702	896	236	97.5	933
Antimony	30.2	25.5	21.2	18.2	16.8			6,684	208	24.6	17.7	904
Beryllium	301	218	161	126	107	89	73	1,850	1750	205	63.3	1550
Bismuth[c]	9.7	8.2						9,780	124	7.9	6.51	545
Boron[c]	52.5	31.7	18.7	11.3	8.1	6.3	5.2	2,500	1047	28.6	10.9	2573
Cadmium[c]	99.3	97.5	94.7					8,650	231	97	48.5	594
Cesium	36.8	36.1						1,873	230	36	83.6	302
Chromium	111	94.8	87.3	80.5	71.3	65.3	62.4	7,160	440	91.4	29.0	2118
Cobalt[c]	122	104	84.8					8,862	389	100	29.0	1765
Copper	413	401	392	383	371	357	342	8,933	383	399	116.6	1356
Germanium	96.8	66.7	43.2	27.3	19.8	17.4	17.4	5,360		61.6		1211
Gold	327	318	312	304	292	278	262	19,300	129	316	126.9	1336
Hafnium	24.4	23.3	22.3	21.3	20.8	20.7	20.9	13,280		23.1		2495
Indium	89.7	83.7	74.5					7,300		82.2		430
Iridium	153	148	144	138	132	126	120	22,500	134	147	48.8	2716
Iron	94	83.5	69.4	54.7	43.3	32.6	28.2	7,870	452	81.1	22.8	1810
Lead	36.6	35.5	33.8	31.2				11,340	129	35.3	24.1	601
Lithium	88.1	79.2	72.1					534	3391	77.4	42.7	454
Magnesium	159	157	153	149	146			1,740	1017	156	88.2	923
Manganese	7.17	7.68						7,290	486	7.78		1517
Mercury[c]	28.9							13,546			2.2	234
Molybdenum	143	139	134	126	118	112	105	10,240	251	138	53.7	2883
Nickel	106	94	80.1	65.5	67.4	71.8	76.1	8,900	446	91	22.9	1726
Niobium	52.6	53.3	55.2	58.2	61.3	64.4	67.5	8,570	270	53.6	23.2	2741

Table E-1 (Continued)

| ELEMENT | THERMAL CONDUCTIVITY k (W/m·K)[b] | | | | | | | PROPERTIES AT 293 K or 20°C | | | | MELTING TEMPERATURE |
	200 K −73°C	273 K 0°C	400 K 127°C	600 K 327°C	800 K 527°C	1000 K 727°C	1200 K 927°C	ρ (kg/m³)	c_p (J/kg·K)	k (W/m K)	$\alpha \times 10^6$ (m²/s)	(K)
Palladium	75.5	75.5	75.5	75.5	75.5	75.5		12,020	247	75.5	25.4	1825
Platinum	72.4	71.5	71.6	73.0	75.5	78.6	82.6	21,450	133	71.4	25.0	2042
Potassium	104	104	52					860	741	103	161.6	337
Rhenium	51	48.6	46.1	44.2	44.1	44.6	45.7	21,100	137	48.1	16.6	3453
Rhodium	154	151	146	136	127	121	115	12,450	248	150	48.6	2233
Rubidium	58.9	58.3						1,530	348	58.2	109.3	312
Silicon	264	168	98.9	61.9	42.2	31.2	25.7	2,330	703	153	93.4	1685
Silver	403	428	420	405	389	374	358	10,500	234	427	173.8	1234
Sodium	138	135						971	1206	133	113.6	371
Tantalum	57.5	57.4	57.8	58.6	59.4	60.2	61	16,600	138	57.5	25.1	3269
Tin[c]	73.3	68.2	62.2					5,750	227	67.0	51.3	505
Titanium[c]	24.5	22.4	20.4	19.4	19.7	20.7	22	4,500	611	22.0	8.0	1953
Tungsten[c]	197	182	162	139	128	121	115	19,300	134	179	69.2	3653
Uranium[c]	25.1	27	29.6	34	38.8	43.9	49	19,070	113	27.4	12.7	1407
Vanadium	31.5	31.3	32.1	34.2	36.3	38.6	41.2	6,100	502	31.4	10.3	2192
Zinc	123	122	116	105				7,140	385	121	44.0	693
Zirconium[c]	25.2	23.2	21.6	20.7	21.6	23.7	25.7	6,570	272	22.8	12.8	2125

[a]Purity for all elements exceeds 99%.

[b]The expected percent errors in the thermal conductivity values are approximately within ±5% of the true values near room temperature and within about ±10% at other temperatures.

[c]For crystalline materials, the values are given for the polycrystalline materials.

Source: E. R. G. Eckert and R. M. Drake, Analysis of Heat and Mass Transfer, McGraw-Hill Book Company, New York, 1972; Raznjevič, Handbook of Thermodynamic Tables and Charts, 3rd ed., McGraw-Hill Book Company, New York, 1976; Y. S. Touloukian, ed., Thermophysical Properties of Matter, IFI/Plenum Publishing Corporation, New York, 1970.

Table E-2   Alloys

METAL	COMPOSITION (%)	PROPERTIES AT 293 K OR 20°C			
		ρ (kg/m³)	c_p (J/kg·K)	k (W/m·K)	$\alpha \times 10^5$ (m²/s)
Aluminum					
Duralumin	94–96 Al, 3–5 Cu, trace Mg	2787	833	164	6.676
Silumin	87 Al, 13Si	2659	871	164	7.099
Copper					
Aluminum Bronze	95 Cu, 5 Al	8666	410	83	2.330
Bronze	75 Cu, 25 Sn	8666	343	26	0.859
Red brass	85 Cu, 9 Sn, 6 Zn	8714	385	61	1.804
Brass	70 Cu, 30 Zn	8522	385	111	3.412
German silver	62 Cu, 15 Ni, 22 Zn	8618	394	24.9	0.733
Constantan	60 Cu, 40 Ni	8922	410	22.7	0.612
Iron					
Cast iron	≈4 C	7272	420	52	1.702
Wrought iron	0.5 CH	7849	460	59	1.626
Steel					
Carbon steel	1 C	7801	473	43	1.172
	1.5 C	7753	486	36	0.970
Chrome steel	1 Cr	7865	460	61	1.665
	5 Cr	7833	460	40	1.110
	10 Cr	7785	460	31	0.867
Chrome-nickel steel	15 Cr, 10 Ni	7865	460	19	0.526
	20 Cr, 15 Ni	7833	460	15.1	0.415
Nickel steel	10 Ni	7945	460	26	0.720
	20 Ni	7993	460	19	0.526
	40 Ni	8169	460	10	0.279
	60 Ni	8378	460	19	0.493
Nickel-chrome steel	80 Ni, 15 C	8522	460	17	0.444
	40 Ni, 15 C	8073	460	11.6	0.305
Manganese steel	1 Mn	7865	460	50	1.388
	5 Mn	7849	460	22	0.637
Silicon steel	1 Si	7769	460	42	1.164
	5 Si	7417	460	19	0.555
Stainless steel	Type 304	7817	461	14.4	0.387
	Type 347	7817	461	14.3	0.387
Tungsten steel	1 W	7913	448	66	1.858
	5 W	8073	435	54	1.525

Source: E. R. G. Eckert and R. M. Drake, *Analysis of Heat and Mass Transfer*, McGraw-Hill Book Company, New York, 1972; F. Kreith, *Principles of Heat Transfer*, 3rd ed., Crowell, New York, 1973.

Table E-3 Insulations and Building Materials

Material	Properties at 293 K or 20°C			
	ρ (kg/m^3)	c_p $(J/kg \cdot K)$	k $(W/m \cdot K)$	$\alpha \times 10^5$ (m^2/s)
Asbestos	383	816	0.113	0.036
Asphalt	2120		0.698	
Bakelite	1270		0.233	
Brick				
Common	1800	840	0.38–0.52	0.028–0.034
Carborundum	2200		5.82	
(50% SiC)				
Magnesite	2000		2.68	
(50% MgO)				
Masonry	1700	837	0.658	0.046
Silica	1900		1.07	
(95% SiO_2)				
Zircon	3600		2.44	
(62% ZrO_2)				
Cardboard			0.14–0.35	
Cement, hard			1.047	
Clay	1545	880	1.26	0.101
(48.7% moisture)				
Coal, anthracite	1370	1260	0.238	0.013–0.015
Concrete, dry	500	837	0.128	0.049
Cork, boards	150	1880	0.042	0.015–0.044
Cork, expanded	120		0.036	
Diatomaceous earth	466	879	0.126	0.031
Earth, clayey	1500		1.51	
(28% moisture)				
Earth, sandy	1500		1.05	
(8% moisture)				
Glass fiber (insulation)	220		0.035	
Glass, window	2800	800	0.81	0.034
Glass, wool	50		0.037	
	100		0.036	
	200	670	0.040	0.028
Granite	2750		3.0	
Ice (0°C)	913	1830	2.22	0.124
Kapok	25		0.035	
Linoleum	535		0.081	
Mica	2900		0.523	
Pine bark	342		0.080	
Plaster	1800		0.814	
Plexiglas	1180		0.195	
Plywood	590		0.109	
Polystyrene	1050		0.157	
Rubber, Buna	1250		0.465	
Hard (ebonite)	1150	2009	0.163	0.0062
Spongy	224		0.055	
Sand, dry			0.582	

Table E-3. Insulations and Building Materials (Continued)

MATERIAL	PROPERTIES AT 293 K OR 20°C			
	ρ (kg/m^3)	c_p (J/kg·K)	k (W/m·K)	α (m^2/s × 10^5)
Sand, moist	1640		1.13	
Sawdust	215		0.071	
Wood				
Oak	609–801	2390	0.17–0.21	0.0111–0.0121
Pine, fir, spruce	416–421	2720	0.15	0.0124
Wood fiber sheets (celotex)	200		0.047	
	400		0.055	
Wool	200		0.038	

Source: E. R. G. Eckert and R. M. Drake, *Analysis of Heat and Mass Transfer*, McGraw-Hill Book Company, New York, 1972; K. Raznjevič, *Handbook of Thermodynamic Tables and Charts*, McGraw-Hill Book Company, New York, 1976; F. Kreith, *Principles of Heat Transfer*, 3rd ed., Crowell, New York, 1973.

Appendix F

THERMODYNAMIC PROPERTIES OF LIQUIDS

Table F-1 Water at Saturation Pressure

Temperature, T		Density, ρ	Coefficient of Thermal Expansion, $\beta \times 10^4$	Specific Heat, c_p	Thermal Conductivity, k	Thermal Diffusivity, $\alpha \times 10^6$	Absolute Viscosity, $\mu \times 10^6$	Kinematic Viscosity, $\nu \times 10^6$	Prandtl Number, Pr	$\dfrac{g\beta}{\nu^2} \times 10^{-9}$
K	°C	(kg/m^3)	$(1/K)$	$(J/kg \cdot K)$	$(W/m \cdot K)$	(m^2/s)	$(N \cdot s/m^2)$	(m^2/s)		$(1/K \cdot m^3)$
273	0	999.3	−0.7	4226	0.558	0.131	1794	1.789	13.7	—
293	20	998.2	2.1	4182	0.597	0.143	993	1.006	7.0	2.035
313	40	992.2	3.9	4175	0.633	0.151	658	0.658	4.3	8.833
333	60	983.2	5.3	4181	0.658	0.159	472	0.478	3.00	22.75
353	80	971.8	6.3	4194	0.673	0.165	352	0.364	2.25	46.68
373	100	958.4	7.5	4211	0.682	0.169	278	0.294	1.75	85.09
473	200	862.8	13.5	4501	0.665	0.170	139	0.160	0.95	517.2
573	300	712.5	29.5	5694	0.564	0.132	92.2	0.128	0.98	1766.

Source: K. Raznjević, *Handbook of Thermodynamic Tables and Charts,* McGraw-Hill Book Company, New York, 1976.

Table F-2   Refrigerants Freon 12 (CCl$_2$F$_2$), Saturated Liquid

TEMPERATURE, T		DENSITY, ρ	COEFFICIENT OF THERMAL EXPANSION, $\beta \times 10^3$	SPECIFIC HEAT, c_p	THERMAL CONDUCTIVITY, k	THERMAL DIFFUSIVITY, $\alpha \times 10^8$	ABSOLUTE VISCOSITY, $\mu \times 10^4$	KINEMATIC VISCOSITY, $\nu \times 10^6$	PRANDTL NUMBER, Pr	$\dfrac{g\beta}{\nu^2} \times 10^{-10}$
K	°C	(kg/m³)	(1/K)	(J/kg·K)	(W/m·K)	(m²/s)	(N·s/m²)	(m²/s)	Pr	(1/K·m³)
223	−50	1547	2.63	875.0	0.067	5.01	4.796	0.310	6.2	26.84
233	−40	1519		884.7	0.069	5.14	4.238	0.279	5.4	
243	−30	1490		895.6	0.069	5.26	3.770	0.253	4.8	
253	−20	1461		907.3	0.071	5.39	3.433	0.235	4.4	
263	−10	1429		920.3	0.073	5.50	3.158	0.221	4.0	
273	0	1397		934.5	0.073	5.57	2.990	0.214	3.8	
283	10	1364		949.6	0.073	5.60	2.769	0.203	3.6	
293	20	1330		965.9	0.073	5.60	2.633	0.198	3.5	
303	30	1295		983.5	0.071	5.60	2.512	0.194	3.5	
313	40	1257		1001.9	0.069	5.55	2.401	0.191	3.5	
323	50	1216		1021.6	0.067	5.45	2.310	0.190	3.5	

Source: E. R. G. Eckert and R. M. Drake, *Analysis of Heat and Mass Transfer*, McGraw-Hill Book Company, New York, 1972.

Table F-2 Refrigerants (Continued) Ammonia (NH_3), Saturated Liquid

TEMPERATURE, T		DENSITY, ρ	COEFFICIENT OF THERMAL EXPANSION, $\beta \times 10^3$	SPECIFIC HEAT, c_p	THERMAL CONDUCTIVITY, k	THERMAL DIFFUSIVITY, $\alpha \times 10^8$	ABSOLUTE VISCOSITY, $\mu \times 10^4$	KINEMATIC VISCOSITY, $\nu \times 10^6$	PRANDTL NUMBER, Pr	$\dfrac{g\beta}{\nu^2} \times 10^{-10}$
K	°C	(kg/m³)	(1/K)	(J/kg·K)	(W/m·K)	(m²/s)	(N·s/m²)	(m²/s)		(1/K·m³)
223	−50	703.7		4463	0.547	17.42	3.061	0.435	2.60	
233	−40	691.7		4467	0.547	17.75	2.808	0.406	2.28	
243	−30	679.3		4476	0.549	18.01	2.629	0.387	2.15	
253	−20	666.7		4509	0.547	18.19	2.540	0.381	2.09	
263	−10	653.6		4564	0.543	18.25	2.471	0.378	2.07	
273	0	640.1		4635	0.540	18.19	2.388	0.373	2.05	
283	10	626.2		4714	0.531	18.01	2.304	0.368	2.04	
293	20	611.8	2.45	4798	0.521	17.75	2.196	0.359	2.02	18.64
303	30	596.4		4890	0.507	17.42	2.081	0.349	2.01	
313	40	581.0		4999	0.493	17.01	1.975	0.340	2.00	
323	50	564.3		5116	0.476	16.54	1.862	0.330	1.99	

Source: E. R. G. Eckert and R. M. Drake, *Analysis of Heat and Mass Transfer,* McGraw-Hill Book Company, New York, 1972.

Table F-3 Organic Compounds at 20°C

LIQUID	CHEMICAL FORMULA	DENSITY, ρ (kg/m³)	COEFFICIENT OF THERMAL EXPANSION $\beta \times 10^4$ (1/K)	SPECIFIC HEAT, c_p (J/kg·K)	THERMAL CONDUCTIVITY, k (W/m·K)	THERMAL DIFFUSIVITY, $\alpha \times 10^9$ (m²/s)	ABSOLUTE VISCOSITY, $\mu \times 10^4$ (N·s/m²)	KINEMATIC VISCOSITY, $\nu \times 10^6$ (m²/s)	PRANDTL NUMBER, Pr	$\dfrac{g\beta}{\nu^2} \times 10^{-8}$ (1/K·m³)
Acetic acid	$C_2H_4O_2$	1049	10.7	2031	0.193	90.6				
Acetone	C_3H_6O	791	14.3	2160	0.180	105.4	3.31	0.418	3.97	802.6
Aniline	C_6H_7N	1022	8.5	2064	0.172	81.5	44.3	4.34	53.16	4.43
Benzene	C_6H_6	879	10.6	1738	0.154	100.8	6.5	0.739	7.34	190.3
n-Butyl alcohol	$C_4H_{10}O$	810	8.1	2366	0.167	87.1	29.5	3.64	41.79	5.99
Chloroform	$CHCl_3$	1489	12.8	967	0.129	89.6	5.8	0.390	4.35	825.3
Ethyl acetate	$C_4H_8O_2$	901	13.8	2010	0.137	75.6	4.49	0.498	6.59	545.7
Ethyl alcohol	C_2H_6O	789	11.0	2470	0.182	93.4	12.0	1.52	16.29	46.7
Ethylene glycol	$C_2H_6O_2$	1113		2382	0.258	97.3	199	17.9	183.7	
Glycerine	$C_3H_8O_3$	1260	5.0	2428	0.285	93.2	14,800	1175	12,609	0.0000355
n-Heptane	C_7H_{14}	684	12.4	1884	0.140	108.6	4.09	0.598	5.50	340.1
Isobutyl alcohol	$C_4H_{10}O$	804	9.4	2303	0.134	72.4	39.5	4.91	67.89	3.82
Methyl alcohol	CH_4O	792	11.9	2470	0.212	108.4	5.84	0.737	6.80	214.9
n-Octane	C_8H_{18}	702	11.4	2177	0.147	96.2	5.4	0.769	8.00	189.1
n-Pentane	C_5H_{12}	626	16.0	2177	0.136	99.8	2.29	0.366	3.67	1171
Toluene	C_7H_8	866	10.8	1675	0.151	104.1	5.86	0.677	6.50	231.1
Turpentine	$C_{10}H_{16}$	855	9.7	1800	0.128	83.2	14.87	1.74	20.91	31.4

K. Raznjević, *Handbook of Thermodynamic Tables and Charts*, McGraw-Hill Book Company, New York, 1976; F. Kreith, *Principles of Heat Transfer*, 3rd ed., Crowell, New York, 1973.

Table F-4 Oils Unused Engine Oil, Saturated Liquid

TEMPERATURE, T		DENSITY, ρ	COEFFICIENT OF THERMAL EXPANSION $\beta \times 10^3$	SPECIFIC HEAT, c_p	THERMAL CONDUCTIVITY, k	THERMAL DIFFUSIVITY, $\alpha \times 10^{10}$	ABSOLUTE VISCOSITY, $\mu \times 10^3$	KINEMATIC VISCOSITY, $\nu \times 10^6$	PRANDTL NUMBER, $Pr \times 10^{-2}$	$\dfrac{g\beta}{\nu^2}$
K	°C	(kg/m^3)	$(1/K)$	$(J/kg \cdot K)$	$(W/m \cdot K)$	(m^2/s)	$(N \cdot s/m^2)$	(m^2/s)		$(1/K \cdot m^3)$
273	0	899.1	0.70	1796	0.147	911	3848.	4280.	471.	8475
293	20	888.2		1880	0.145	872	799.	900.	104.	
313	40	876.1		1964	0.144	834	210.	240.	28.7	
333	60	864.0		2047	0.140	800	72.5	83.9	10.5	
353	80	852.0		2131	0.138	769	32.0	37.5	4.90	
373	100	840.0		2219	0.137	738	17.1	20.3	2.76	
393	120	829.0		2307	0.135	710	10.3	12.4	1.75	
413	140	816.9		2395	0.133	686	6.54	8.0	1.16	
433	160	805.9		2483	0.132	663	4.51	5.6	0.84	

Source: E. R. G. Eckert and R. M. Drake, *Analysis of Heat and Mass Transfer*, McGraw-Hill Book Company, New York, 1972.

Table F-4 Oils (Continued) Transformer Oil (Standard 982-68)

Temperature, T °C	K	Density, ρ (kg/m³)	Coefficient of Thermal Expansion $\beta \times 10^3$ (1/K)	Specific Heat, c_p (J/kg·K)	Thermal Conductivity, k (W/m·K)	Thermal Diffusivity, $\alpha \times 10^{10}$ (m²/s)	Absolute Viscosity, $\mu \times 10^3$ (N·s/m²)	Kinematic Viscosity, $\nu \times 10^6$ (m²/s)	Prandtl Number, $Pr \times 10^{-2}$	$\dfrac{g\beta}{\nu^2}$ (1/K·m³)
−50	223	922		1700	0.116	742	29,320	31,800	4,286	
−40	233	916		1680	0.116	750	3,866	4,220	563	
−30	243	910		1650	0.115	764	1,183	1,300	170	
−20	253	904		1620	0.114	778	365.6	404	52	
−10	263	898		1600	0.113	788	108.1	120	15.3	
0	273	891		1620	0.112	778	55.24	67.5	8.67	
10	283	885		1650	0.111	763	33.45	37.8	4.95	
20	293	879		1710	0.111	736	21.10	24.0	3.26	
30	303	873		1780	0.110	707	13.44	15.4	2.18	
40	313	867		1830	0.109	688	9.364	10.8	1.57	

Source: N. B. Vargaftik, Tables on the Thermophysical Properties of Liquids and Gases, 2nd ed., Hemisphere Publishing Corporation, Washington, D.C., 1975.

Appendix G

THERMODYNAMIC PROPERTIES OF GASES

Table G-1 Dry Air at Atmospheric Pressure

Temperature, T		Density, ρ	Coefficient of Thermal Expansion, $\beta \times 10^3$	Specific Heat, c_p	Thermal Conductivity, k	Thermal Diffusivity, $\alpha \times 10^6$	Absolute Viscosity, $\mu \times 10^6$	Kinematic Viscosity, $\nu \times 10^6$	Prandtl Number, Pr	$\dfrac{g\beta}{\nu^2} \times 10^{-8}$
K	°C	(kg/m³)	(1/K)	(J/kg·K)	(W/m·K)	(m²/s)	(N·s/m²)	(m²/s)		(1/K·m³)
273	0	1.252	3.66	1011	0.0237	19.2	17.456	13.9	0.71	1.85
293	20	1.164	3.41	1012	0.0251	22.0	18.240	15.7	0.71	1.36
313	40	1.092	3.19	1014	0.0265	24.8	19.123	17.6	0.71	1.01
333	60	1.025	3.00	1017	0.0279	27.6	19.907	19.4	0.71	0.782
353	80	0.968	2.83	1019	0.0293	30.6	20.790	21.5	0.71	0.600
373	100	0.916	2.68	1022	0.0307	33.6	21.673	23.6	0.71	0.472
473	200	0.723	2.11	1035	0.0370	49.7	25.693	35.5	0.71	0.164
573	300	0.596	1.75	1047	0.0429	68.9	39.322	49.2	0.71	0.0709
673	400	0.508	1.49	1059	0.0485	89.4	32.754	64.6	0.72	0.0350
773	500	0.442	1.29	1076	0.0540	113.2	35.794	81.0	0.72	0.0193
1273	1000	0.268	0.79	1139	0.0762	240	48.445	181	0.74	0.00236

Source: K. Raznjević, *Handbook of Thermodynamic Tables and Charts*, McGraw-Hill Book Company, New York, 1976.

Table G-2 Carbon Dioxide at Atmospheric Pressure

Temperature, T		Density, ρ	Coefficient of Thermal Expansion, $\beta \times 10^3$	Specific Heat, c_p	Thermal Conductivity, k	Thermal Diffusivity, $\alpha \times 10^4$	Absolute Viscosity, $\mu \times 10^6$	Kinematic Viscosity, $\nu \times 10^6$	Prandtl Number, Pr	$\dfrac{g\beta}{\nu^2} \times 10^{-6}$
K	°C	(kg/m^3)	(1/K)	(J/kg·K)	(W/m·K)	(m^2/s)	(N·s/m^2)	(m^2/s)		(1/K·m^3)
220	−53	2.4733		783	0.010805	0.05920	11.105	4.490	0.818	
250	−23	2.1657		804	0.012884	0.07401	12.590	5.813	0.793	
300	27	1.7973	3.33	871	0.016572	0.10588	14.958	8.321	0.770	472
350	77	1.5362	2.86	900	0.02047	0.14808	17.205	11.19	0.755	224
400	127	1.3424	2.50	942	0.02461	0.19463	19.32	14.39	0.738	118
450	177	1.1918	2.22	980	0.02897	0.24813	21.34	17.90	0.721	67.9
500	227	1.0732	2.00	1013	0.03352	0.3084	23.26	21.67	0.702	41.8
550	277	0.9739	1.82	1047	0.03821	0.3750	25.08	25.74	0.685	26.9
600	327	0.8938	1.67	1076	0.04311	0.4483	26.83	30.02	0.668	18.2

Source: E. R. G. Eckert and R. M. Drake, *Analysis of Heat and Mass Transfer*, McGraw-Hill Book Company, New York, 1972; F. Kreith, *Principles of Heat Transfer*, 3rd ed., Crowell, New York, 1973.

Table G-3   Carbon Monoxide at Atmospheric Pressure

TEMPERATURE, T		DENSITY, ρ	COEFFICIENT OF THERMAL EXPANSION, $\beta \times 10^3$	SPECIFIC HEAT, c_p	THERMAL CONDUCTIVITY, k	THERMAL DIFFUSIVITY, $\alpha \times 10^4$	ABSOLUTE VISCOSITY, $\mu \times 10^6$	KINEMATIC VISCOSITY, $\nu \times 10^6$	PRANDTL NUMBER, Pr	$\dfrac{g\beta}{\nu^2} \times 10^{-6}$
K	°C	(kg/m³)	(1/K)	(J/kg·K)	(W/m·K)	(m²/s)	(N·s/m²)	(m²/s)		(1/K·m³)
220	−53	1.554		1043	0.01906	0.1176	13.88	8.90	0.758	
250	−23	0.841		1043	0.02144	0.1506	15.40	11.28	0.750	
300	27	1.139	3.33	1042	0.02525	0.2128	17.84	15.67	0.737	133
350	77	0.974	2.86	1043	0.02883	0.2836	20.09	20.62	0.728	65.9
400	127	0.854	2.50	1048	0.03226	0.3605	22.19	25.99	0.722	36.3
450	177	0.758	2.22	1055	0.04360	0.4439	24.18	31.88	0.718	21.4
500	227	0.682	2.00	1064	0.03863	0.5324	26.06	38.19	0.718	13.4
550	277	0.620	1.82	1076	0.04162	0.6240	27.89	44.97	0.721	8.83
600	327	0.569	1.67	1088	0.04446	0.7190	29.60	52.06	0.724	6.04

Source: E. R. G. Eckert and R. M. Drake, *Analysis of Heat and Mass Transfer*, McGraw-Hill Book Company, New York, 1972; F. Kreith, *Principles of Heat Transfer*, 3rd ed., Crowell, New York, 1973.

Table G-4 Helium at Atmospheric Pressure

Temperature, T		Density, ρ	Coefficient of Thermal Expansion, $\beta \times 10^3$	Specific Heat, c_p	Thermal Conductivity, k	Thermal Diffusivity, $\alpha \times 10^4$	Absolute Viscosity, $\mu \times 10^6$	Kinematic Viscosity, $\nu \times 10^6$	Prandtl Number, Pr	$\dfrac{g\beta}{\nu^2} \times 10^{-6}$
K	°C	(kg/m³)	(1/K)	(J/kg·K)	(W/m·K)	(m²/s)	(N·s/m²)	(m²/s)		(1/K·m³)
3	−270			5200	0.0106		0.842		0.74	
33	−240	1.466		5200	0.0353	0.04625	5.02	3.42	0.70	
144	−129	0.3380	6.94	5200	0.0928	0.5275	12.55	37.11	0.694	49.4
200	−73	0.2435	5.00	5200	0.1177	0.9288	15.66	64.38	0.70	11.8
255	−18	0.1906	3.92	5200	0.1357	1.3675	18.17	95.50	0.71	4.22
366	93	0.1328	2.73	5200	0.1691	2.449	23.05	173.6	0.72	0.888
477	204	0.1020	2.10	5200	0.197	3.716	27.50	269.3	0.72	0.284
589	316	0.08282	1.70	5200	0.225	5.215	31.13	375.8	0.72	0.118
700	427	0.07032	1.43	5200	0.251	6.661	34.75	494.2	0.72	0.0574
800	527	0.06023	1.25	5200	0.275	8.774	38.17	634.1	0.72	0.0305
900	627	0.05286	1.11	5200	0.298	10.834	41.36	781.3	0.72	0.0178

Source: E. R. G. Eckert and R. M. Drake, *Analysis of Heat and Mass Transfer*, McGraw-Hill Book Company, New York, 1972; F. Kreith, *Principles of Heat Transfer*, 3rd ed., Crowell, New York, 1973.

Table G-5   Hydrogen at Atmospheric Pressure

Temperature, T (K)	T (°C)	Density, ρ (kg/m³)	Coefficient of Thermal Expansion, β×10³ (1/K)	Specific Heat, c_p (J/kg·K)	Thermal Conductivity, k (W/m·K)	Thermal Diffusivity, α×10⁴ (m²/s)	Absolute Viscosity, μ×10⁶ (N·s/m²)	Kinematic Viscosity, ν×10⁶ (m²/s)	Prandtl Number, Pr	$\frac{g\beta}{\nu^2}$ ×10⁻⁶ (1/K·m³)
50	−223	0.50955		10,501	0.0362	0.0676	2.516	4.880	0.721	
100	−173	0.24572	10.0	11,229	0.0665	0.2408	4.212	17.14	0.712	333.8
150	−123	0.16371	6.67	12,602	0.0981	0.475	5.595	34.18	0.718	55.99
200	−73	0.12270	5.00	13,540	0.1282	0.772	6.813	55.53	0.719	15.90
250	−23	0.09819	4.00	14,059	0.1561	1.130	7.919	80.64	0.713	6.03
300	27	0.08185	3.33	14,314	0.182	1.554	8.963	109.5	0.706	2.72
350	77	0.07016	2.86	14,436	0.206	2.031	9.954	141.9	0.697	1.39
400	127	0.06135	2.50	14,491	0.228	2.568	10.864	177.1	0.690	0.782
450	177	0.05462	2.22	14,499	0.251	3.164	11.779	215.6	0.682	0.468
500	227	0.04918	2.00	14,507	0.272	3.817	12.636	257.0	0.675	0.297
600	327	0.04085	1.67	14,537	0.315	5.306	14.285	349.7	0.664	0.134
700	427	0.03492	1.43	14,574	0.351	6.903	15.89	455.1	0.659	0.0677
800	527	0.03060	1.25	14,675	0.384	8.563	17.40	569	0.664	0.0379
1000	727	0.02451	1.00	14,968	0.440	11.997	20.16	822	0.686	0.0145
1200	927	0.02050	0.833	15,366	0.488	15.484	22.75	1107	0.715	0.00667

Source: E. R. G. Eckert and R. M. Drake, *Analysis of Heat and Mass Transfer,* McGraw-Hill Book Company, New York, 1972; F. Kreith, *Principles of Heat Transfer,* 3rd ed., Crowell, New York, 1973.

Table G-6 Nitrogen at Atmospheric Pressure

| Temperature, T | | Density, ρ | Coefficient of Thermal Expansion, $\beta \times 10^3$ | Specific Heat, c_p | Thermal Conductivity, k | Thermal Diffusivity, $\alpha \times 10^4$ | Absolute Viscosity, $\mu \times 10^6$ | Kinematic Viscosity, $\nu \times 10^6$ | Prandtl Number, Pr | $\dfrac{g\beta}{\nu^2} \times 10^{-6}$ |
K	°C	(kg/m³)	(1/K)	(J/kg·K)	(W/m·K)	(m²/s)	(N·s/m²)	(m²/s)		(1/K·m³)
100	−173	3.4808		1072	0.00945	0.0253	6.86	1.97	0.786	
200	−73	1.7108	5.00	1043	0.01824	0.1022	12.95	7.57	0.747	855.6
300	27	1.1421	3.33	1041	0.02620	0.2204	17.84	15.63	0.713	133.7
400	127	0.8538	2.50	1046	0.03335	0.3734	21.98	25.74	0.691	37.00
500	227	0.6824	2.00	1056	0.03984	0.5530	25.70	37.66	0.684	13.83
600	327	0.5687	1.67	1076	0.04580	0.7486	29.11	51.19	0.686	6.25
700	427	0.4934	1.43	1097	0.05123	0.9466	32.13	65.13	0.691	3.31
800	527	0.4277	1.25	1123	0.05609	1.1685	34.84	81.46	0.700	1.85
900	627	0.3796	1.11	1146	0.06070	1.3946	37.49	91.06	0.711	1.31
1000	727	0.3412	1.00	1168	0.06475	1.6250	40.00	117.2	0.724	0.714
1100	827	0.3108	0.909	1186	0.06850	1.8591	42.28	136.0	0.736	0.482
1200	927	0.2851	0.833	1204	0.07184	2.0932	44.50	156.1	0.748	0.335

Source: E. R. G. Eckert and R. M. Drake, *Analysis of Heat and Mass Transfer*, McGraw-Hill Book Company, New York, 1972; F. Kreith, *Principles of Heat Transfer*, 3rd ed., Crowell, New York, 1973.

Table G-7 Oxygen at Atmospheric Pressure

TEMPERATURE, T		DENSITY, ρ	COEFFICIENT OF THERMAL EXPANSION, $\beta \times 10^3$	SPECIFIC HEAT, c_p	THERMAL CONDUCTIVITY, k	THERMAL DIFFUSIVITY, $\alpha \times 10^4$	ABSOLUTE VISCOSITY, $\mu \times 10^6$	KINEMATIC VISCOSITY, $\nu \times 10^6$	PRANDTL NUMBER, Pr	$\dfrac{g\beta}{\nu^2} \times 10^{-6}$
K	°C	(kg/m³)	(1/K)	(J/kg·K)	(W/m·K)	(m²/s)	(N·s/m²)	(m²/s)		(1/K·m³)
100	−173	3.992		948	0.00903	0.0239	7.768	1.946	0.815	
150	−123	2.619	6.67	918	0.01367	0.0569	11.49	4.387	0.773	3398
200	−73	1.956	5.00	913	0.01824	0.1021	14.85	7.593	0.745	850.5
250	−23	1.562	4.00	916	0.02259	0.1579	17.87	11.45	0.725	299.2
300	27	1.301	3.33	920	0.02676	0.2235	20.63	15.86	0.709	129.8
350	77	1.113	2.86	929	0.03070	0.2968	23.16	20.80	0.702	64.8
400	127	0.9755	2.50	942	0.03461	0.3768	25.54	26.18	0.695	35.8
450	177	0.8682	2.22	957	0.03828	0.4609	27.77	31.99	0.694	21.3
500	227	0.7801	2.00	972	0.04173	0.5502	29.91	38.34	0.697	13.3
550	277	0.7096	1.82	988	0.04517	0.6441	31.97	45.05	0.700	8.79
600	327	0.6504	1.67	1004	0.04832	0.7399	33.92	52.15	0.704	6.02

Source: E. R. G. Eckert and R. M. Drake, Analysis of Heat and Mass Transfer, McGraw-Hill Book Company, New York, 1972; F. Kreith, Principles of Heat Transfer, 3rd ed., Crowell, New York, 1973.

Table G-8 Steam (H₂O Vapor) at Atmospheric Pressure

TEMPERATURE, T		DENSITY, ρ	COEFFICIENT OF THERMAL EXPANSION, $\beta \times 10^3$	SPECIFIC HEAT, c_p	THERMAL CONDUCTIVITY, k	THERMAL DIFFUSIVITY, $\alpha \times 10^4$	ABSOLUTE VISCOSITY, $\mu \times 10^6$	KINEMATIC VISCOSITY, $\nu \times 10^6$	PRANDTL NUMBER, Pr	$\dfrac{g\beta}{\nu^2} \times 10^{-6}$
K	°C	(kg/m³)	(1/K)	(J/kg·K)	(W/m·K)	(m²/s)	(N·s/m²)	(m²/s)		(1/K·m³)
380	107	0.5863		2060	0.0246	0.204	12.71	21.6	1.060	
400	127	0.5542	2.50	2014	0.0261	0.234	13.44	24.2	1.040	41.86
450	177	0.4902	2.22	1980	0.0299	0.307	15.25	31.1	1.010	22.51
500	227	0.4405	2.00	1985	0.0339	0.387	17.04	38.6	0.996	13.16
550	277	0.4005	1.82	1997	0.0379	0.475	18.84	47.0	0.991	8.08
600	327	0.3652	1.67	2026	0.0422	0.573	20.67	56.6	0.986	5.11
650	377	0.3380	1.54	2056	0.0464	0.666	22.47	66.4	0.995	3.43
700	427	0.3140	1.43	2085	0.0505	0.772	24.26	77.2	1.000	2.35
750	477	0.2931	1.33	2119	0.0549	0.883	26.04	88.8	1.005	1.65
800	527	0.2739	1.25	2152	0.0592	1.001	27.86	102.0	1.010	1.18
850	577	0.2579	1.18	2186	0.0637	1.130	29.69	115.2	1.019	0.872

Source: E. R. G. Eckert and R. M. Drake, *Analysis of Heat and Mass Transfer*, McGraw-Hill Book Company, New York, 1972; F. Kreith, *Principles of Heat Transfer*, 3rd ed., Crowell, New York, 1973.

Appendix H

THERMODYNAMIC PROPERTIES OF LIQUID METALS

Table H-1   Bismuth

Temperature, T (K)	T (°C)	Density, ρ (kg/m³)	Coefficient of Thermal Expansion, $\beta \times 10^3$ (1/K)	Specific Heat, c_p (J/kg·K)	Thermal Conductivity, k (W/m·K)	Thermal Diffusivity, $\alpha \times 10^5$ (m²/s)	Absolute Viscosity, $\mu \times 10^4$ (N·s/m²)	Kinematic Viscosity, $\nu \times 10^7$ (m²/s)	Prandtl Number, Pr	$\dfrac{g\beta}{\nu^2} \times 10^{-9}$ (1/K·m³)
589	316	10,011	0.117	144.5	16.44	1.14	16.22	1.57	0.014	46.5
700	427	9,867	0.122	149.5	15.58	1.06	13.39	1.35	0.013	65.6
811	538	9,739	0.126	154.5	15.58	1.03	11.01	1.08	0.011	106
922	649	9,611		159.5	15.58	1.01	9.23	0.903	0.009	
1033	760	9,467		164.5	15.58	1.01	7.89	0.813	0.008	

Source: F. Kreith, *Principles of Heat Transfer*, 3rd ed., Crowell, New York, 1973.

Table H-2 Mercury (Saturated Liquid)

Temperature, T (K)	°C	Density, ρ (kg/m³)	Coefficient of Thermal Expansion, $\beta \times 10^4$ (1/K)	Specific Heat, c_p (J/kg·K)	Thermal Conductivity, k (W/m·K)	Thermal Diffusivity, $\alpha \times 10^7$ (m²/s)	Absolute Viscosity, $\mu \times 10^4$ (N·s/m²)	Kinematic Viscosity, $\nu \times 10^6$ (m²/s)	Prandtl Number, Pr	$\dfrac{g\beta}{\nu^2} \times 10^{-10}$ (1/K·m³)
273	0	13,628		140.3	8.20	42.99	16.90	0.124	0.0288	13.73
293	20	13,579	1.82	139.4	8.69	46.06	15.48	0.114	0.0249	
323	50	13,506		138.6	9.40	50.22	14.05	0.104	0.0207	
373	100	13,385		137.3	10.51	57.16	12.42	0.0928	0.0162	
423	150	13,264		136.5	11.49	63.54	11.31	0.0853	0.0134	
473	200	13,145		157.0	12.34	69.08	10.54	0.0802	0.0116	
523	250	13,026		135.7	13.07	74.06	9.96	0.0765	0.0103	
588.7	315.5	12,847		134.0	14.02	81.50	8.65	0.0673	0.0083	

Source: E. R. G. Eckert and R. M. Drake, Analysis of Heat and Mass Transfer, McGraw-Hill Book Company, New York, 1972.

Table H-3 Sodium

TEMPERATURE, T		DENSITY, ρ	COEFFICIENT OF THERMAL EXPANSION $\beta \times 10^3$	SPECIFIC HEAT, c_p	THERMAL CONDUCTIVITY, k	THERMAL DIFFUSIVITY, $\alpha \times 10^5$	ABSOLUTE VISCOSITY, $\mu \times 10^4$	KINEMATIC VISCOSITY, $\nu \times 10^7$	PRANDTL NUMBER, Pr	$\dfrac{g\beta}{\nu^2} \times 10^{-9}$
K	°C	(kg/m³)	(1/K)	(J/kg·K)	(W/m·K)	(m²/s)	(N·s/m²)	(m²/s)		(1/K·m³)
367	94	929	0.27	1382	86.2	6.71	6.99	7.31	0.0110	4.96
478	205	902	0.36	1340	80.3	6.71	4.32	4.60	0.0072	16.7
644	371	860		1298	72.4	6.45	2.83	3.16	0.0051	
811	538	820		1256	65.4	6.19	2.08	2.44	0.0040	
978	705	778		1256	59.7	6.19	1.79	2.26	0.0038	

Source: F. Kreith, *Principles of Heat Transfer*, 3rd ed., Crowell, New York, 1973.

Appendix I

NORMAL EMISSIVITIES

Table I-1 Normal Emissivity of Metals

SUBSTANCE	STATE OF SURFACE	TEMPERATURE (K)	NORMAL EMISSIVITY, ϵ_n^a
Aluminum	Polished plate	296	0.040
		498	0.039
	Rolled, polished	443	0.039
	Rough plate	298	0.070
Brass	Oxidized	611	0.22
	Polished	292	0.05
		573	0.032
	Tarnished	329	0.202
Chromium	Polished	423	0.058
Copper	Black oxidized	293	0.780
	Lightly tarnished	293	0.037
	Polished	293	0.030
Gold	Not polished	293	0.47
	Polished	293	0.025
Iron	Oxidized smooth	398	0.78
	Ground bright	293	0.24
	Polished	698	0.144
Lead	Gray oxidized	293	0.28
	Polished	403	0.056
Molybdenum	Filament	998	0.096
Nickel	Oxidized	373	0.41
	Polished	373	0.045
Platinum	Polished	498	0.054
		898	0.104
Silver	Polished	293	0.025
Steel	Oxidized rough	313	0.94
	Ground sheet	1213	0.520
Tin	Bright	293	0.070
Tungsten	Filament	3300	0.39
Zinc	Tarnished	293	0.25
	Polished	503	0.045

^{a}Hemispherical emissivity values, ϵ, may be approximated by: $\epsilon = 1.2\epsilon_n$ for bright metal surfaces; $\epsilon = 0.95\epsilon_n$ for other smooth surfaces; $\epsilon = 0.98\epsilon_n$ for other rough surfaces.
Source: K. Raznjevič, *Handbook of Thermodynamic Tables and Charts*, McGraw-Hill Book Company, New York, 1976.

Table I-2. Normal Emissivity of Nonmetals

SUBSTANCE	STATE OF SURFACE	TEMPERATURE (K)	NORMAL EMISSIVITY, E_N
Asbestos board		297	0.96
Brick	Red, rough	293	0.93
Carbon filament		1313	0.53
Glass	Smooth	293	0.93
Ice	Smooth	273	0.966
	Rough	273	0.985
Masonry	Plastered	273	0.93
Paper		293	0.80
Plaster, lime	White, rough	293	0.93
Porcelain	Glazed	293	0.93
Quartz	Fuzed, rough	293	0.93
Rubber			
Soft	Gray	297	0.86
Hard	Black, rough	297	0.95
Wood			
Beech	Planed	343	0.935
Oak	Planed	294	0.885

Source: K. Raznjevič, *Handbook of Thermodynamic Tables and Charts,* McGraw-Hill Book Company, New York, 1976.

Table I-3 Normal Emissivity of Paints and Surface Coatings

SUBSTANCE	STATE OF SURFACE	TEMPERATURE (K)	NORMAL EMISSIVITY, ϵ_n
Aluminum bronze		373	0.20–0.40
Aluminum enamel	Rough	293	0.39
Aluminum paint	heated to 325°C	423–588	0.35
Bakelite enamel		353	0.935
Enamel			
White	Rough	293	0.90
Black	Bright	298	0.876
Oil paint		273–473	0.885
Red lead primer		293–373	0.93
Shellac,	Bright	294	0.82
black	Dull	348–418	0.91

Source: K. Raznjevič, *Handbook of Thermodynamic Tables and Charts,* McGraw-Hill Book Company, New York, 1976.

Appendix J

MASS DIFFUSIVITIES

Table J-1. Binary Gas-Phase Mass Diffusivities at Atmospheric Pressure

BINARY GAS MIXTURE	TEMPERATURE		$D_{AB} \times 10^5$ (m²/s)
	K	°C	
Air–ammonia	273	0	1.98
Air–aniline	298	25	0.726
Air–benzene	298	25	0.962
Air–carbon dioxide	273	0	1.36
Air–carbon disulfide	273	0	0.883
Air–chlorine	273	0	1.24
Air–ethyl alcohol	298	25	1.32
Air–hydrogen	273	0	5.472
Air–iodine	298	25	0.834
Air–mercury	614	314	4.73
Air–napthalene	298	25	0.611
Air–oxygen	273	0	1.75
Air–sulfur dioxide	273	0	1.22
Air–toluene	298	25	0.844
Air–water	298	25	2.60
CO_2–benzene	318	45	0.715
CO_2–carbon disulfide	318	45	0.715
CO_2–ethyl alcohol	273	0	0.693
CO_2–hydrogen	273	0	5.50
CO_2–nitrogen	298	25	1.58
CO_2–water	298	25	1.64
Hydrogen–nitrogen	286	13	7.376
Oxygen–ammonia	293	20	2.53
Oxygen–benzene	296	23	0.39
Oxygen–hydrogen	285	14	7.748
Oxygen–nitrogen	285	12	2.025

Source: R. D. Reid and T. K. Sherwood, *The Properties of Gases and Liquids*, McGraw-Hill Book Company, New York, 1966; *Handbook of Chemistry and Physics*, 39th ed., CRC Press, Cleveland, Ohio, 1957–58.

Table J-2 Binary Liquid-Phase Mass Diffusivities (Dilute Solutions; A is Solute, B is Solvent)

BINARY LIQUID MIXTURE	TEMPERATURE K	°C	$D_{AB} \times 10^9$ (m²/s)
Bromine in water	285	12	0.90
CO_2 in water	291	18	1.71
Chlorine in water	285	12	1.40
Glucose in water	288	15	0.52
Hydrogen in water	298	25	3.36
Iodine in water	298	25	1.25
Methonal in water	288	15	1.28
Nitrogen in water	295	22	2.02
Oxygen in water	298	25	2.60
Aniline in methanol	288	15	1.49
CCl_4 in methanol	288	15	1.70
Chloroform in methanol	288	15	2.07
Iodoform in methanol	288	15	1.33
Lactic acid in methanol	288	15	1.36
Acetic acid in benzene	288	15	1.92
Bromine in benzene	285	12	2.00
CCl_4 in benzene	298	25	2.00
Chloroform in benzene	288	15	2.11
Iodine in benzene	293	20	1.95

Source: R. D. Reid and T. K. Sherwood, *The Properties of Gases and Liquids*, McGraw-Hill Book Company, New York, 1966.

Table J-3 Binary Solid-Phase Mass Diffusivities

BINARY SOLID MIXTURE	TEMPERATURE K	°C	$D_{AB} \times 10^{10}$ (m²/s)
Aluminum in copper	293	20	1.30×10^{-24}
Antimony in silver	293	20	3.51×10^{-15}
Bismuth in lead	293	20	1.10×10^{-10}
Cadmium in copper	293	20	2.71×10^{-9}
Helium in silicon dioxide	293	20	$2.40 - 5.50 \times 10^{-4}$
Helium in Pyrex	773	500	2.00×10^{-2}
Helium in Pyrex	293	20	4.49×10^{-5}
Hydrogen in silicon dioxide	773	500	$0.573 - 2.10 \times 10^{-2}$
Hydrogen in nickel	358	85	1.16×10^{-2}
Hydrogen in nickel	438	165	0.105
Mercury in lead	293	20	2.50×10^{-9}

Source: R. M. Barrer, *Diffusion in and Through Solids*, Macmillan Publishing Co., Inc., New York, 1941.

Appendix K

TEMPERATURE CONVERSIONS

Table K-1   Temperature Conversion Table

	In Terms of K	In Terms of °C	In Terms of °F	In Terms of °R
Kelvin, K =	—	°C + 273.15	$\frac{5}{9}$(°F − 32) + 273.15	$\frac{5}{9}$°R
Celsius, °C =	K − 273.15	—	$\frac{5}{9}$(°F − 32)	$\frac{5}{9}$(°R − 491.67)
Fahrenheit, °F =	$\frac{9}{5}$(K − 273.15) + 32	$\frac{9}{5}$°C + 32	—	°R − 459.67
Rankine, °R =	$\frac{9}{5}$K	$\frac{9}{5}$°C + 491.67	°F + 459.67	—

Table K-2   Equivalence of K, °C, and °F to Nearest Degree

K	°C	°F	K	°C	°F	K	°C	°F
220	−53	−63	335	62	144	450	177	351
225	−48	−54	340	67	153	455	182	360
230	−43	−45	345	72	162	460	187	369
235	−38	−36	350	77	171	465	192	378
240	−33	−27	355	82	180	470	197	387
245	−28	−18	360	87	189	475	202	396
250	−23	−9	365	92	198	480	207	405
255	−18	0	370	97	207	485	212	414
260	−13	9	375	102	216	490	217	423
265	−8	18	380	107	225	495	222	432
270	−3	27	385	112	234	500	227	441
275	2	36	390	117	243	505	232	450
280	7	45	395	122	252	510	237	459
285	12	54	400	127	261	515	242	468
290	17	63	405	132	270	520	247	477
295	22	72	410	137	279	525	252	486
300	27	81	415	142	288	530	257	495
305	32	90	420	147	297	535	262	504
310	37	99	425	152	306	540	267	513
315	42	108	430	157	315	545	272	522
320	47	117	435	162	324	550	277	531
325	52	126	440	167	333	555	282	540
330	57	135	445	172	342	560	287	549

Appendix L

DIMENSIONLESS GROUPS

Table L-1 Dimensionless Groups Used in Heat Transfer

GROUP	SYMBOL	FORMULA^a	PHYSICAL SIGNIFICANCE
Biot	Bi	$Bi = \dfrac{h_c L}{k_s}$	Ratio of solid thermal resistance to fluid thermal resistance
Fourier	Fo	$Fo = \dfrac{\alpha t}{L^2}$	Ratio of conduction of heat to storage of heat; used in transient heat-transfer problems
Graetz	Gz	$Gz = RePr\left(\dfrac{d}{L}\right)$ $= \dfrac{\rho V d^2 c_p}{kL}$	Used in forced-convection problems
Grashof	Gr	$Gr = \dfrac{g\beta \Delta T L^3}{\nu^2}$	Ratio of buoyancy to viscous forces
Lewis	Le	$Le = \dfrac{\alpha}{D_{AB}}$	ratio of thermal diffusivity to molecular diffusivity; used in mass-transfer problems
Nusselt	Nu	$Nu = \dfrac{h_c L}{k}$	Basic dimensionless convective heat-transfer coefficient
Peclet	Pe	$Pe = RePr$ $= \dfrac{\rho V L c_p}{k}$	Ratio of heat transfer by convection to conduction; used in forced-convection problems

GROUP	SYMBOL	FORMULA[4]	PHYSICAL SIGNIFICANCE	
Prandtl	Pr	$Pr = \dfrac{\mu c_p}{k} = \dfrac{\nu}{\alpha}$	Ratio of momentum to thermal diffusion	
Rayleigh	Ra	$Ra = GrPr$ $= \dfrac{g\beta \Delta T c_p \rho^2 L^3}{k\mu}$	Used in free-con- vection problems	
Reynolds	Re	$Re = \dfrac{\rho VL}{\mu} = \dfrac{VL}{\nu}$	Ratio of inertia to viscous forces	
Schmidt	Sc	$Sc = \dfrac{\mu}{\rho D_{AB}} = \dfrac{\nu}{D_{AB}}$	Ratio of momentum to mass diffusion	
Sherwood	Sh	$Sh = \dfrac{h_m L}{D_{AB}}$	Ratio of mass diffusivity to molecular diffusivity	
Stanton	St	$St = \dfrac{Nu}{RePr}$	Ratio of heat	
		$= \dfrac{h_c}{\rho V c_p}$	transfer at	*transfer at surface to that transported by fluid*

c_p = specific heat at constant pressure

D_{AB} = mass diffusivity

d = diameter

g = acceleration of gravity

h = convective-heat-transfer coefficient

h_m = convective-mass-transfer coefficient

k = thermal conductivity of a fluid

k_s = thermal conductivity of a solid

L = characteristic length

t = time

V = velocity

α = thermal diffusivity

β = coefficient of thermal expansion

ΔT = temperature difference

μ = dynamic or absolute viscosity

ν = kinematic viscosity

ρ = density

Appendix M

SUBPROGRAM FOR MATRIX INVERSION

The listing below is a subroutine called MATINV which will invert an $N \times N$ matrix called AA. The input to MATINV is the $N \times N$ elements of AA and the number N, of rows (or columns) in the square matrix. The output of MATINV is the N^2 values for the inverse elements called AINV.

The subroutine in its present form is limited to a 50×50 matrix, although the user can increase the size limit of the program so that it can invert larger matrices by simply increasing the storage locations in the DIMENSION statements. MATINV calls a separate subroutine named EXCH whose purpose is to exchange the rows and columns in the matrix AA and make the largest element appear in the location AA(K,K). A listing of MATINV and EXCH taken from Pennington* are as follows:

```
        SUBROUTINE MATINV (AA,N,AINV)
        DIMENSION AA(50,50),AINV(50,50),A(50,100),ID(50)
        NN=N+1
        N2=2*N
        DO 100 I=1,N
        ID(I)=I
        DO 100 J=1,N
100     A(I,J)=AA(I,J)
        DO 200 I=1,N
        DO 200 J=NN,N2
200     A(I,J)=0.
        DO 300 I=1,N
300     A(I,N+I)=1.
        K=1
1       CALL EXCH (A,N,N,N2,K,ID)
2       IF (A(K,K)) 3,999,3
3       KK=K+1
```

```
        DO 4 J=KK,N2
        A(K,J)=A(K,J)/A(K,K)
        DO 4 I=1,N
        IF (K-I) 41,4,41
 41     W=A(L,K)*A(K,J)
        A(I,J)=A(I,J)-W
        IF (ABS(A(I,J))-0.0001*ABS(W)) 42,4,4
 42     A(I,J)=0.0
  4     CONTINUE
        K=KK
        IF (K-N) 1,2,5
  5     DO 10 I=1,N
        DO 10 J=1,N
        IF (ID(J)-I) 10,8,10
  8     DO 101 K=1,N
101     AINV(I,K)=A(J,N+K)
 10     CONTINUE
        RETURN
999     PRINT 1000
        RETURN
1000    FORMAT (19H MATRIX IS SINGULAR)
        END


        SUBROUTINE EXCH (A,N,NA,NB,K,ID)
        DIMENSION A(50,100),ID(50)
        NROW=K
        NCOL=K
        B=ABS(A(K,K))
        DO 2 1=K,N
        DO 2 J=K,NA
        IF (ABS(A(I,J))-B) 2,2,21
 21     NROW=I
        NCOL=J
        B=ABS(A(I,J))
  2     CONTINUE
        IF (NROW-K) 3,3,31
 31     DO 32 J=K,NB
        C=A(NROW,J)
        A(NROW,J)=A(K,J)
 32     A(K,J)=C
  3     CONTINUE
        IF (NCOL-K) 4,4,41
 41     DO 42 I=1,N
        C=A(I,NCOL)
        A(I,NCOL)=A(I,K)
 42     A(I,K)=C
        I=ID(NCOL)
        ID(NCOL)=ID(K)
        ID(K)=I
  4     CONTINUE
        RETURN
        END
```

*Ralph H. Pennington, 'Introductory Computer Methods and Numerical Analysis', Second Edition, The Macmillan Co., London, 1970.

Appendix N

HEAT-TRANSFER PROGRAMS

Over 1500 of the computer programs developed in the space effort are available for use by the public. Some of these programs can be used to solve heat-transfer problems. Programs from the Department of Defense, NASA, and other governmental agencies are cataloged and are for sale at reasonable prices.

The Computer Software Management and Information Center (COSMIC) is operated by NASA by the University of Georgia for the purpose of making the programs available to the public. The table below is a partial listing of programs relating to heat transfer abstracted from a COSMIC catalog of available programs. A more detailed list of programs, including language, machine requirements, program size, and price can be obtained by referring to the reference number and writing to

COSMIC
112 Barrow Hall
The University of Georgia
Athens, Georgia 30602

Another series of computer programs in the general area of heating, refrigeration, air conditioning, and ventilating has been cataloged by ASHRAE under a grant from the National Science Foundation. These programs apply to heat-transfer and fluid-flow problems and they are particularly relevant because of the recent emphasis on energy conservation. Details on these programs are given in NSF publication RA-760002(5559), ASHRAE Research Project GRP-153 (1975).

PROGRAM DESCRIPTION	REFERENCE NUMBER
CINDA-3G Chrysler Improved Numerical Differencing Analyzer (CDC 6000 Series Version)	MSC-11653
Transient Thermal Analysis of Fluid Systems	MSC-19502
General Transient Heat-Transfer Computer Program for Thermally Thick Walls	LAR-10794
Variable Boundary II: Heat Conduction	LEW-10679
SHORE-CATCH: A Heat-Transfer Digital Computer Program	MFS-00261
Heat-Exchanger Program	MSF-02129
Heat-Exchanger Performance Routine	MSC-16492
Two-Dimensional Transient Heat-Conduction Program	MSC-17024
One-Dimensional, Steady-State Heat-Transfer Program	MSC-19181
TRACK: Computer Program for Transient and Steady-State Coupled Fluid Flow and Heat Conduction	NUC-10189
Solving Transient Heat-Conduction Problems by Computer	NUC-10325
Heat Transfer	WLP-10036
Nodal Network Thermal Balance Program	GSC-11158
NASTRAN: Thermal Analyzer Theory and Application	GSC-12162
BETA II: Boeing Engineering Thermal Analyzer	MFS-15055
TAP: Thermal Analyzer Program	MFS-18410
TRASYS/SCOPE: Thermal Radiation Analysis System	MFS-23857
Transient Thermal Analysis of Fluid Systems	MSC-19502
FORTRAN: Thermal Analyzer Incorporating Stable Numerical Integration	MSC-19566
TRAC: Transient Radiation Analysis by Computer Program	GSC-11949
TOSS: Modified Transient and/or Steady-State Digital Heat-Transfer Code	NUC-10162
TAP-A: A Program for Computing Transient or Steady-State Temperature Distributions	NUC-10282

Answers
to Odd-Numbered
Problems

CHAPTER 1

1-1   105 W/m²

1-3   91.07°C

1-5   22.31°C, 26.15°C

1-7   Fiberglass—2.86; plaster—0.123; plywood—0.917;
      common brick—0.358. All units m²·K/W

1-9   (a) 31.3 W; (b) 18.7 W; (c) 1.13 W; (d) 8.88×10^{-3} W; (e) 1.96×10^{-2} W
      (f) 2.75×10^{-3} W

1-11 1066 W

1-13 31,089 W/m²

1-15 29,740 W; 397.4°C

1-17 7.109 W/m·K

1-19 48,000 W; 680,000 W

1-21 51,000 W

1-23 500 cm²

1-25 187.5 W/m²·K; 5625 W/m²

1-27 (b) 23,662 W/m²; (c) 236.6°C; (d) 57,931 W/m²

1-29 60.83°C; 15°C

1-31 (a) 45.92 W; (b) 2322 W; (c) 16,191 W

1-33 17.01 W

1-35 37.18 W/m²·K

1-37 0.3397 W/m²·K

1-41 $\Pi_1 = hL/k$, $\Pi_2 = \rho VL/\mu$, $\Pi_3 = \mu c_P/k$

CHAPTER 2

2-1 -43.33 W/m^2; 0.231 m$^2 \cdot$K/W; 30°C

2-3 -197.2°C

2-5   4050 W/m^2; 8100 W

2-7   50.27 W; -100°C

2-9   115.6 W/m; 0.735 K/W

2-11   6283 W; 7.96×10^{-2} K/W

2-13   2.71 cm; 522.8°C; 1058°C

2-15   1319 W; 201.5 W; 84.7%

2-17   0.804 W/m$\cdot$K; 420°C; 121.3°C

2-19   254 kW, $731

2-21 5.648×10^4 W

2-25 $(T - T_1) + \dfrac{\beta}{2}(T^2 - T_1{}^2) = \dfrac{k_m}{k_o}\dfrac{\ln(r/r_i)}{\ln(r_o/r_i)}(T_2 - T_1)$

2-27 k increases with increasing temperature.

2-31   (a) $T = T_1 + \dfrac{x}{L}(T_2 - T_1) + \dfrac{BL^3 x}{12k}\left[1 - \left(\dfrac{x}{L}\right)^4\right]$

   (b) $x = \left[\dfrac{3k}{BL}(T_2 - T_1) + \dfrac{L^3}{4}\right]^{1/3}$

   (c) $q'' = \dfrac{1}{L}(T_2 - T_1) - \dfrac{BL^3}{4k}$

2-33 $T = \dfrac{A}{9k} r_o{}^3\left[1 - \left(\dfrac{r}{r_o}\right)^3\right] + T_1$

2-35   50°C

2-37   693 W/m

2-39   0.9526 W; 173.8°C; 1.006 W; 173.6°C

2-41 $\theta = \dfrac{\cosh\left[(\text{Bi})^{1/2}(1 - \xi)\right]}{\cosh(\text{Bi})^{1/2}}$; $\xi = \dfrac{x}{L}$; $\text{Bi} = \dfrac{hPL^2}{kA}$

   $P = \pi D$; $A = \pi D^2/4$

   $q = 2(\text{Bi})^{1/2}\dfrac{kA}{L}(T_b - T_\infty)\tanh(\text{Bi})^{1/2}$

2-43   130.9 W; 27 fins;

2-47 $-2.5 \times 10^{-2}\,\hat{i}$W/m^2; $-(10\hat{i} + 7.5\hat{j}) \times 10^2$ W/m^2; $-3 \times 10^3\,\hat{i}$W/m^2

2-49   118.1 W/m; 2.27; 6.77 m

2-51   574 W/m

2-53   403 W/m

2-55   17.0 W/m; 17.6 km

2-57   364 W

2-59 $T_1 = 250$°C; $T_2 = 300$°C; $T_3 = 350$°C; $T_4 = 400$°C

2-61 $\frac{1}{2}(T_4 + T_2) + T_7 + T_8 + (\text{Bi})T_\infty - (3 + \text{Bi})T_1 = R_1$

   $\frac{1}{2}(T_1 + T_9) + (\text{Bi})T_\infty - (1 + \text{Bi})T_2 = R_2$

   $T_7 + T_6 + T_5 + T_4 - 4T_3 = R_3$

2-63 $x = 3$; $y = 8$; $z = 7$; $u = 0.6$; $w = 3$

2-65   (a) $T_1 = 287.5$°C, $T_2 = 312.5$°C, $T_3 = 237.5$°C, $T_4 = 262.5$°C

   (b) $T_1 = 290.0$°C, $T_2 = 315.0$°C, $T_3 = 240.0$°C, $T_4 = 265.0$°C

2-67 $T_1 = 160.0$°C, $T_2 = 140.7$°C, $T_3 = 140.5$°C, $T_4 = 162.4$°C

   $T_5 = 142.0$°C, $T_6 = 137.0$°C, $T_7 = 147.6$°C, $T_8 = 128.2$°C, $T_9 = 123.4$°C

2-69 $T_1 = 170°C$, $T_2 = 178°C$, $T_3 = 180°C$, $T_4 = 178°C$, $T_5 = 172°C$
$T_6 = 158°C$, $T_7 = 162°C$, $T_8 = 164°C$, $T_9 = 160°C$, $T_{10} = 150°C$
$T_{11} = 140°C$, $T_{12} = 150°C$, $T_{13} = 152°C$, $T_{14} = 152°C$, $T_{15} = 150°C$

2-71 $T_1 = 171.53°C$, $T_2 = 178.28°C$, $T_3 = 180.24°C$, $T_4 = 178.75°C$,
$T_5 = 171.80°C$, $T_6 = 158.43°C$, $T_7 = 162.99°C$, $T_8 = 163.94°C$,
$T_9 = 161.39°C$, $T_{10} = 154.53°C$, $T_{11} = 143.05°C$, $T_{12} = 148.80°C$,
$T_{13} = 151.16°C$, $T_{14} = 150.84°C$, $T_{15} = 148.96°C$

2-73 $T_1 = 208.48°C$, $T_2 = 226.54°C$, $T_3 = 227.95°C$, $T_4 = 237.24°C$
$T_5 = 241.63°C$, $T_6 = 243.21°C$, $T_7 = 154.83°C$, $T_8 = 169.76°C$,
$T_9 = 163.85°C$, $T_{10} = 149.63°C$, $T_{11} = 171.96°C$, $T_{12} = 186.08°C$,
$T_{13} = 189.59°C$, $T_{14} = 121.68°C$, $T_{15} = 133.82°C$, $T_{16} = 131.73°C$
$T_{17} = 129.24°C$, $T_{18} = 135.59°C$, $T_{19} = 141.16°C$, $T_{20} = 142.98°C$

CHAPTER 3

3-1   20.43 min; 4.08 min; 2.04 min
3-3   9.12 s; 73.1 W
3-7 7.62×10^{-3} s
3-9   1 hr–35 min
3-11 79°C, 65°C
3-13 40.8 min; 26°C
3-15 41.7 min; 166.6 min
3-17 1.06 hr
3-19 5.88 hr
3-21 33.3 min
3-23 84°C
3-25 5 min, 13.3 min
3-35 15.15 min
3-37 at $t = 64$ s; $T_1 = 130.0°C$, $T_2 = 80.0°C$, $T_3 = 55.0°C$
at $t = 116$ s; $T_1 = 140.6°C$, $T_2 = 101.3°C$, $T_3 = 75.0°C$
at $t = 181$ s; $T_1 = 146.5°C$, $T_2 = 117.2°C$, $T_3 = 87.8°C$

CHAPTER 4

4-1   (a) 1.705×10^4; (b) 3.31×10^4; (c) 4.56×10^5; (d) 1.25×10^3
4-3   15 m/s
4-5   (a) 0.02 m, 0.05 m; (b) 0.0023 m, 0.0057 m
4-7   488.6
4-9 2.01×10^6
4-11 1.19×10^{-4} N/m²; 2.15×10^{-4} N
4-13 $\bar{h} = 0.036 \left(\dfrac{\mu c_p}{k} \right)^{1/3} \left(\dfrac{V_\infty L}{\nu} \right)^{4/5} \dfrac{k}{L}$
4-15 $\dfrac{T - T_s}{T_\infty - T_s} = \dfrac{3}{2} \dfrac{y}{\delta_t} - \dfrac{1}{2} \left(\dfrac{y}{\delta_t} \right)^3$
4-17 (a) $\dfrac{T - T_s}{T_\infty - T_s} = \dfrac{y}{\delta_t}$; (b) $\dfrac{\delta}{\delta_t} = Pr^{1/3}$; (d) $\overline{Nu} = 0.29\ Re^{1/2} Pr^{1/3}$
4-19 (a) 3.706×10^{-3}; (b) 2.55×10^3 N; (c) 5.63 W/m²·K; (d) 1082 W
4-21 (a) 9.16×10^{-4}; (b) 22.11 W/m²·K
4-23 37.3°C

4-27 $\dfrac{h}{\rho V_\infty c_p}$

4-29 722 W

4-31 15.5°C

4-33 100°C

4-35 101.7 W/m²·K; -95.8 W/m

4-37 (a) 448.6 W/m²·K; (b) 538.8 W/m²·K; (c) 296.1 W/m²·K

4-39 (a) $-28{,}763$ W; (b) 101,183 N/m²

4-41 34.23°C; 366 W

4-43 2567 W/m

CHAPTER 5

5-1 $\pi = \left(\dfrac{d}{L}\right)\left(\dfrac{V d \rho c_p}{k}\right)$

5-3 $\pi_1 = \dfrac{\tau_s}{\rho V_\infty^{\,2}}$; $\pi_2 = \dfrac{\rho V_\infty D}{\mu}$

5-5 6.62 m

5-7 3912 W/m²·K; 32.9°C

5-9 2.27 m³/s

5-11 2.67 m

5-13 20.98 kW/m²·K

5-15 550.3 K; 349 kW

5-17 $\overline{Nu}_L = 1.5\, Re_L^{1/3} Pr^{1/3}$

5-19 (a) 1.417×10^5 W; (b) 3.611×10^5 W; (c) 6.39×10^5 W

5-21 (a) 0.0683 W/m²·K; (b) 17.5 W/m²·K; (c) 36.1 W/m²·K

5-23 186 W/m²·K

5-25 142 kW; 5.5°C

5-27 0.64

5-29 6.36×10^5

5-31 (a) 79.7 W/m²·K; (b) 2.25 kW per tube; (c) 101,239 N/m²

5-33 8306 W/m²·K

5-35 (a) 406.8 W/m²·K; (b) 883 kW; (c) 8.4°C; (d) 8284 N/m²

5-37 9.79×10^3 kW; 787 N/m²

5-39 (a) 7.33×10^7; (b) 2.13×10^{10}; (c) 9.14×10^5; (d) 1.61×10^{11}

5-41 125.4 W; 140.9 min

5-43 52.7°C

5-45 3.35 kW

5-47 17.28 W/m

5-49 45.7 kW

5-51 1305 W

5-53 233 W

5-55 620 W

5-57 17.1 kW

5-59 792 W/m²·K; 648 W/m²·K

5-61 (a) 682 W/m²·K; (b) 1001 W/m²·K

5-63 3173°C

5-65 (a) 111°C; (b) 1.7 cm; (c) 155.4 W/m²·K; (d) 3.52×10^{-3}; (e) 8.09 kW/m²

CHAPTER 6

6-1 (a) 1.287×10^5 W/m^3; (b) 4.022×10^8 W/m^3; (c) 1.287×10^{10} W/m^3;
     (d) 4.022×10^{13} W/m^3

6-5 6245 K

6-7 (a) Ultraviolet—0.58%, visible—19.0%, infrared—80.4%;
     (b) Ultraviolet—0.024%, visible—5.28%, infrared—94.7%

6-9 $\sigma = \left(\dfrac{\pi}{C_2}\right)^4 \dfrac{C_1}{15}$

6-11 Sun—83.7%; plant—0.0015%

6-13 (a) 0.0118; (b) 0.00292

6-15 (a) 0.2365; (b) 0.2123

6-17 (1) 0.2; (2) 0.6

6-19 $F_{1\to2} = \sqrt{1 + \left(\dfrac{b}{a}\right)^2} - \dfrac{b}{a}$

6-21 0.8183

6-23 $F_{1\to2} = \dfrac{R^2}{L^2 + R^2}$

6-25 0.0253

6-27 $F_{1\to4} = 0.204$, $F_{1\to3} = 0.414$, $F_{1\to5} = 0.089$, $F_{1\to6} = 0.293$

6-29 (a) -1.27×10^7 W; (b) -1.269×10^7 W; (c) 4.535×10^7 W
     (d) 204.1 W; (e) 3.265×10^7 W

6-31 (a) 114°C; (b) 40,000 W

6-33 (b) 440.1 W; (c) 1589 W; -390 W; -1199 W; (d) 1589 W; 50 W

6-35 104°C

6-37 50°C

6-39 398.4 W/m^2

6-41 $J_1 = 4.048 \times 10^5$ W/m^2; $J_2 = 6.373 \times 10^5$ W/m^2; $J_3 = 8.685 \times 10^5$ W/m^2
     $q_1'' = -3.480 \times 10^5$ W/m^2; $q_3'' = 3.469 \times 10^5$ W/m^2; $T_2 = 1831$ K

6-43 244.1 W/m

6-45 (a) 1095 K; (b) 995 K; (c) 10^5 W; (d) 59136 W

6-47 $\dfrac{dT_2}{dt} = 0.1651 - 5.647 \times 10^{-12}\, T_2^4$, T in K, t in s

6-49 (a) 20°C; (b) 11,640 W

6-51 105°C

6-53 1.0 m

6-55 (a) 962 K; (b) 2.88 kg/hr

6-59 (a) 22310 W; (b) -3130 W; (c) $-19,180$ W; (d) $q_{1\rightleftharpoons2} = 8.703 \times 10^3$ W;
     $q_{1\rightleftharpoons3} = 1.361 \times 10^4$ W; $q_{2\rightleftharpoons3} = 5.572 \times 10^3$ W

6-61 -30.1 W/m^2, 409 W/m^2, -14.5 W/m^2, 613 W/m^2, -482 W/m^2, -496
     W/m^2

6-63 (a) 757 K, 759 K; (b) 729 K; (d) 3300 W/m^2

6-65 0.38

6-67 0.66

6-69 (a) $-16,612$ W/m^2, 919 K; (b) $-19,326$ W/m^2

6-71 (a) $-25,538$ W/m^2; (b) $-26,573$ W/m^2

6-73 $-39,734$ W/m^2

CHAPTER 7

7-1   (a) 47.3 $W/m^2 \cdot K$; (b) 40.5 $W/m^2 \cdot K$

7-3   (a) 46.9 $W/m^2 \cdot K$; (b) 40.2 $W/m^2 \cdot K$

7-5   4.95%

7-7   47.1°C

7-9   (a) 233 kW; (b) 218 kW; 6.52%

7-11   (a) 53.7 m^2; (b) 51.4 m^2

7-13   (a) 4.31 m; (b) 108.5°C; (c) 109.3°C

7-15   (a) 81%; (b) 82%

7-17   (a) $T_w = 54.5$°C; $T_A = 120$°C
       (b) $T_w = 54.7$°C; $T_A = 115.8$°C

7-19   (a) 67%; (b) 48.8 kW; (c) 44.0°C

7-21   (a) 357 kW; (b) 313 kW

7-23   (a) 90.5%; (b) 80.6 kW
       (c) $T_w = 68.1$°C; $T_A = 64.8$°C

INDEX

Absorbed irradiation, 292, 293–294, 295, 297–301, 350, 351, 357
Aerodynamic heating, 268
Altitude of the sun, 361, 362, 365
Analog methods, 93–94
Analogy between the flow of heat and electricity, 7–10
Analytical methods, 85–87
Annular-flow regime, 476
Approximate integral method, 208–212
Average friction coefficient, 202
Axially constant wall heat flux, 264
Azimuth of the sun, 362, 369

Backward-difference technique, 177
Binary diffusion, 481
Binder-Schmidt method, 174–176
Biot number, 63–64, 73, 75, 76, 81, 83, 137, 138–139, 148, 536
Blackbody, 19, 284–285
Black surfaces, radiative exchange between, 315–325
Blocks, convection heat transfer from, 259–261

Boiling heat transfer, 464–480
  boiling regimes, 464–467
  boiling and vaporization in forced convection, 474–478
  correlation of data, 467
  film boiling, 465–467, 479–480
  heat exchangers, 447–448
  nucleate boiling, 465, 467
    with forced convection, 471–474
    maximum heat flux in, 474
  pool boiling, 464, 465, 468–471
  transition boiling, 479
Boundary layer, 199
Boundary-layer thickness, 199
Bubble Reynolds number, 467
Bubbly-flow regime, 475–476
Buckingham (pi) Theorem, 27–29
Bulk boiling, 464
Burnout, 467

Cavity effect, 294
Chart solutions to transient conduction problems, 147–164
  one-dimensional, 147–160

two-and three-dimensional, 160–164
Collector efficiency factor, 432–433
Collector-heat-loss conductance, 424–432
Collector-heat-removal factor, 433–436
Composite wick, 447
Computer programs, 538–541
  steady-state problems, 84, 94–95, 108–112
  radiation problems, 338–346
Condensation heat transfer, 454–464
  dropwise condensation, 463–464
  filmwise condensation, 455–463
    condensation of superheated vapor, 463
    high vapor velocity, 462
    turbulence, 460–462
  mixtures of vapors and noncondensable gases, 464
Conduction equation, 43–51
  cylindrical coordinates, 49–50, 54–58, 70–73
  dimensionless form, 46–48
  rectangular coordinates, 44–46, 68–70
  spherical coordinates, 50–51, 58–59
Conduction heat transfer, 2–14
  combined with other heat transfer mechanisms, 21–25
  contact resistance, 10–11
  defined, 2
  electric analog for, 7–10
  Fourier's law of, 3–7, 43, 44–45, 48, 52, 65, 87, 89, 93, 96
  sign convention for, 2–3
  steady state. See Steady-state conduction
  thermal conductivity, 11–14
  transient. See Transient conduction
Conduction shape factor, 90–93
Configuration factor, 307
Constant wall temperature, 264
Contact resistance, 10–11
Convection heat transfer, 14–18
  analysis of, 197–229
    analogy between heat and momentum transfer in turbulent flow over a flat surface, 212–218
  conservation of mass, momentum, and energy equations for laminar flow over a flat plate, 202–206
  convection process, 197–202
  evaluation of heat-transfer and friction coefficients in laminar flow, 208–212
  integral momentum and energy equations for laminar boundary layer, 206–208
  laminar forced convection in a tube, 221–226
  Reynolds analogy for turbulent flow over a flat plate, 218–221
  Reynolds analogy for turbulent flow in a tube, 226–229
  boiling and, 471–478
  combined with other heat transfer mechanisms, 21–25
  defined, 14
  engineering relations for, 237–269
    blocks, 259–261
    combined free and forced convection, 262–265
    cylinders and spheres, 249–253, 259–261
    dimensionless parameters for correlating data, 237–240
    flat plates, 248–249
    forced convection, 248–255, 262–265
    high-speed flow, 265–269
    natural convection, 255–265
    tube banks, 253–255
    tubes and ducts, 240–248
    vertical planes and cylinders, 259
  forced. See Forced convection
  free. See Free convection
  Newton's law of cooling, 15–16
  thermal resistance, 17–18
Convection-heat-transfer coefficient, 237–239
Convective-mass-transfer, 481, 486–491
Critical excess temperature, 466
Critical radius, 63
Crossed-string method, 312–315
Cylinders
  conduction heat transfer in, 48–50,

54–58, 62–64, 66–67, 70–73, 160–164
convection heat transfer in, 249–253, 259–261
Cylindrical coordinate system, 49–50
one-dimensional conduction with generation, 70–73
one-dimensional conduction without generation, 54–58

Declination of the sum, 362–363, 366, 368
Differential equations for laminar flow over a flat plate, 202–206
Diffusely reflecting surface, 306
Diffuse surface, 304–305
Diffusion mass transfer, 481
Digital computers, 42–43, 84, 94–95
Dimensional analysis, 26–32
Dimensional formulas, 27
Dimensionless groups, 536
Biot, 63–64, 73, 75, 76, 81, 83, 137, 138–139, 148, 536
Fourier, 47–48, 138–139, 148, 169, 536
Graetz, 536
Grashof, 256–257, 259, 262, 489, 536
Lewis, 536
Nusselt, 223, 225, 239, 246, 248–251, 259, 261, 262, 263, 467, 488, 536
Peclet, 246, 537
Prandtl, 202, 205–206, 217, 218, 239, 262, 488, 537
Rayleigh, 259, 260, 537
Reynolds, 32, 200, 201, 239, 248, 251, 262, 460–462, 537
Schmidt, 488, 537
Sherwood, 487, 489, 490, 537
Stanton, 227, 537
Dimensions, 25–26
Directional radiation properties, 291, 301–307
Dropwise condensation, 463–464
Ducts, convection heat transfer in, 240–248
laminar flow, 200–201, 243–246
liquid metals, 246–247
turbulent flow, 200–201, 241–243

Dynamic viscosity, 199

Ecliptic plane, 363
Eddy diffusivity of heat, 217
Eddy heat conductivity, 217
Eddy viscosity, 215
Electric analog for conduction, 7–10
Electromagnetic spectrum, 283–284
Emissivity of a surface, 293–297, 299–301, 350–357, 531–533
Emphemeris, solar, 367
Enclosed spaces, free convection in, 261–262
Enclosure relationship, 309, 311
Energy equation
laminar boundary layer, 207–208
laminar flow over a flat plate, 204–205
Equimass counterdiffusion, 482
Equimolar counterdiffusion, 483
Excess temperature, 464
Explicit numerical method, 165–173

Fick's first law of diffusion, 481–482
Film boiling, 465–467, 479–480
Film condensation, 455–463
condensation of superheated vapor, 463
high vapor velocity, 462
turbulence, 460–462
Film temperature, 264
Fin efficiency, 79–84
Finite-difference technique, 95–100
Finite-element method, 180
Fins, heat transfer from, 74–84
Flat-plate collectors. *See* Solar energy collectors
Flat plates, convection heat transfer in, 248–249
laminar flow, 199–206, 248–249
turbulent flow, 199–200, 218–221, 249
Fluctuating component, 213
Fluidized beds, 437
Forced convection, 14–17, 198–202
boiling and vaporization in, 474–478
combined with free convection, 262–265

defined, 15
in external flow, 248–255
laminar, in a tube, 221–226
nucleate boiling with, 471–474
in transition flow, 248
FORTRAN, 84, 109–112, 182–188
Forward-difference method, 166
Fouling factors, 403–404
Fourier number, 47–48, 138–139, 148, 169, 536
Fourier's law of heat conduction, 3–7, 43, 44–45, 48, 52, 65, 87, 89, 93, 96
Free convection, 14–17, 198, 199, 255–265
  combined with forced convection, 262–265
  defined, 14
  in enclosed spaces, 261–262
  in horizontal cylinders, spheres, and blocks, 259–261
  in vertical planes and cylinders, 259
Free-molecule regime, 269
Friction coefficient, 202, 208–212
Friction factor, 31–32, 228, 264

Gases
  radiative properties of, 350–357
  thermodynamic properties of, 520–527
Gauss error function, 143, 506–507
Gauss-Siedel method, 113
Geometric resistance, 317
Gradient, 503
Graetz number, 536
Graphical methods
  steady-state conduction problems, 87–93
  transient conduction problems, 174–176
Grashof number, 256–257, 259, 262, 489, 536
Gray bodies, 19–20, 299–301
Gray surfaces, radiative exchange between, 325–333
Greenhouse effect, 298

Heat exchangers, 23, 395–448
  effectiveness, 413–422

heat pipes, 438–448
  boiling limitations, 447–448
  entrainment limitation, 444
  sonic limitation, 442–444
  wicking limitation, 444–447
heat transfer and flow on packed beds, 436–438
log-mean temperature difference, 405–413
overall heat-transfer coefficient, 401–404
solar energy collectors, 422–436
  collector efficiency factor, 432–433
  collector-heat-loss conductance, 424–432
  collector-heat-removal factor, 433–436
  energy balance, 423–424
types of, 396–401
Heat-flux vector, 85–86
Heat pipes, 438–448
  boiling limitations, 447–448
  entrainment limitation, 444
  sonic limitation, 442–444
  wicking limitation, 444–447
Heat transfer, principles of, 1–32. *See also* Boiling heat transfer; Condensation heat transfer; Convection heat transfer; Mass transfer; Radiation heat transfer
  combined mechanisms, 21–25
  conduction, 2–14
  convection, 14–18
  dimensional analysis, 26–32
  dimensions, 25–26
  radiation, 19–21
  units, 26
Heat-transfer coefficient, 23, 59–62, 202, 208–212, 401–404
High-speed flow, heat transfer in, 265–268
Hopscotch methods, 177
Hyperbolic functions, 504–505

Imperfect thermal contact, 10
Implicit numerical method, 176–189
Incidence angle, 368
Insolation, 358, 365, 368–369, 372

Insulation thickness for a cylinder, 62–64

Integral equations
evaluation of heat-transfer and function coefficient, 208–212
momentum and energy, 206–208

Interface resistance, 10–11

International System of Units (SI), 26, 497–501

Isotherms, 85, 87, 93, 94

Iteration method, 112–118

Kinetic theory, 12

Kirchhoff's law, 293–295, 319, 325, 326, 346

Knudsen number, 269

Laminar boundary layer, integral momentum and energy equations for, 206–208

Laminar flow
in ducts, 200–201, 243–246
evaluation and heat-transfer and friction coefficients in, 208–212
over flat plates, 199–206, 248–249
in tubes, 201, 221–226, 243–246
velocity profiles, 201

Laminar sublayer, 200

Laplace equation, 46, 85, 93

Laplacian operator, 49, 502–503

Latitude, 363

Lewis number, 536

Liquid metals
flow of, 246–247, 252–253
thermodynamic properties of, 528–530

Liquids, thermodynamic properties of, 514–519

Local friction coefficient, 202

Log-mean Nusselt number, 246

Log-mean temperature difference (LMTD), 405–413

Longitude, 363

Mass, conservation of, 203

Mass diffusivities, 533–534

Mass transfer, 480–491
by convection, 481, 486–491
defined, 480
by diffusion, 481–486

MATINV Subprogram, 110–112, 182–183, 338–346, 538–539

Matrix methods
radiation heat transfer problems, 333–346
computer program, 338–346
surfaces with known net heat flux, 337–338
surfaces with known temperatures, 334–337
steady-state conduction problems, 106–112
transient conduction problems, 182–183

Mean beam length, 351, 352

Mean value, 213

Metals, liquid. See Liquid metals

Mist-flow regime, 476

Mixing cup temperature, 225

Molecular mass transfer, 481–486

Momentum diffusivity, 205

Momentum equation
laminar boundary layer, 206–207
laminar flow over a flat plate, 204, 205

Momentum transfer in turbulent flow, 212–218

Monochromatic radiation properties, 291, 295–298

Newton's law of cooling, 15–16

Nodes, 94, 100–106

Nucleate boiling, 465, 467
with forced convection, 471–474
maximum heat flux in, 474

Number of heat-transfer units, 415–416

Numerical methods
steady-state conduction problems, 94–100
transient conduction problems, 165–189
explicit method, 165–173
graphical interpretation, 174–176
implicit method, 176–189

Nusselt number, 223, 225, 239, 246, 248–251, 259, 261, 262, 263, 467, 488, 536

Ohm's law, 7, 53, 54, 93, 317, 319, 326
One-dimensional and transient problems, 44
One-dimensional conduction, 51–84
   with generation, 68–73
   without generation, 51–64
   heat transfer from fins, 74–84
   variable thermal conductivity, effect of, 64–68
One-dimensional problems, 43
One-dimensional solutions to transient conduction problems, 147–160
Overall heat-transfer coefficient, 23, 59–62, 401–404
Overall transmittance, 23
Overall unit conductance, 23
Overrelaxation, 102

Packed beds, heat transfer and flow in, 436–438
Parallel circuit, 54, 56
Particle theory, 282–283
Peclet number, 246, 537
Perfect thermal contact, 10
Photons, 282–283
Planck's constant, 282
Planck's law, 284–285
Plane walls, conduction heat transfer through, 3–7, 44–46, 51–54, 64–70
Poiseuille's equation, 445–446
Pool boiling, 464, 465, 468–471
Potential flow regime, 199
Prandtl number, 202, 205–206, 217, 218, 239, 262, 488, 537
Prandtl's mixing length, 215
Primary dimensions, 27
Ptolomaic view, 364

Radiation functions, 287–291
Radiation heat transfer, 19–21, 282–378
   combined with other heat transfer mechanisms, 21–25
   defined, 19
   electromagnetic spectrum, 283–284
   matrix methods, 333–346
      computer program, 338–346
      surfaces with known net heat flux, 337–338

surfaces with known temperatures, 334–337
   particle theory, 282–283
   physics of, 284–291
      blackbodies, 19, 284–285
      gray bodies, 19–20
      Planck's law, 284–285
      radiation functions, 287–291
      Stefan-Boltzmann law, 19, 287
      Wien's displacement law, 285–287
   radiative exchange, 315–333
      between black surfaces, 315–325
      between gray surfaces, 325–333
      through absorbing, transmitting media, 346–350
   radiation properties, 291–307
      directional, 291, 301–307
      of gases, 350–357
      gray bodies, 299–301
      Kirchhoff's law, 293–295, 319, 325, 326
      monochromatic, 291, 295–298
      total, 291, 292–293
   shape factor, 307–315
      algebra, 311–312
      crossed-string method, 312–315
   solar, 358–378
      on earth, 372–378
      tilted surfaces, 368–372
   thermal conductance, 20–21
   thermal resistance, 20
   wave theory, 282
Radiosity, 326
Rayleigh number, 259, 260, 537
Rayleigh scattering, 360
Reciprocity relationship, 308–309, 311
Recovery factor, 267
Rectangular coordinate system, 44–46
   one-dimensional conduction with generation, 68–70
   one-dimensional conduction without generation, 51–54
Reflected irradiation, 292
Refractory surface, 320
Relaxation techniques, 94, 100–106
Reradiating surface, 320
Residuals, 100–106
Reynolds analogy, 217–229, 489

for turbulent flow over a flat plate, 218–221
for turbulent flow in a tube, 226–229
Reynolds number, 32, 200, 201, 239, 248, 251, 262, 460, 462, 537
Reynolds stress, 214
Rohsenow equation, 468–471

Saturated boiling, 464
Saturation temperature, 463
Schmidt number, 488, 537
Semi-infinite solids, transient conduction in, 142–147
Series thermal circuit, 53–54, 56
Shape factor, radiation, 307–315
    algebra, 311–312
    crossed-string method, 312–315
Shell-and-tube heat exchanger, 396–401
Sherwood number, 487, 489, 490, 537
SI System (Système International d'Unitès), 26, 497–501
Slip-flow regime, 269
Smithsonian Meteorological Tables, 369, 370
Solar energy collectors, 422–436
    collector efficiency factor, 432–433
    collector-heat-loss conductance, 424–432
    collector-heat-removal factor, 433–436
    energy balance, 423–424
Solar hour angle, 363
Solar maps, 372–374
Solar radiation, 358–378
    on earth, 372–378
    tilted surfaces, 368–372
Solids, thermodynamic properties of, 508–513
Specular reflector, 306–307
Spheres
    conduction heat transfer in, 50–51, 58–59, 66, 67
    convection heat transfer in, 249–253, 259–261
Spherical coordinate system, 50–51, 58–59
Stability limit, 169
Stanton number, 227, 537

Steady-state conduction, 42–135
    conduction equation, 43–51
        cylindrical coordinates, 49–50, 54–58, 70–73
        dimensionless form, 46–48
        rectangular coordinates, 44–46, 68–70
        spherical coordinates, 50–51, 58–59
    one-dimensional, 51–84
        with generation, 68–73
        without generation, 51–64
        heat transfer from fins, 74–84
        variable thermal conductivity, effect of, 64–68
    two-and three-dimensional, 84–118
        analog methods, 93–94
        analytical methods, 85–87
        graphical methods, 87–93
        iteration methods, 112–118
        matrix methods, 106–112
        numerical methods, 94–100
        relaxation methods, 94, 100–106
Steady-state problems, 43
Stefan-Boltzmann constant, 19, 287
Stefan-Boltzmann law, 287
Subcooled boiling, 464
Sun time, 363
Superheated vapor, condensation of, 463
Superimposed free convection, 264–265
Symbols, 28

Temperature conversions, 535
Thermal circuits, 53–54, 56, 317–333, 348–350
Thermal conductance, 7
    conduction heat transfer, 11–14
    radiation heat transfer, 20–21
    variable effect of, 64–68
Thermal diffusivity, 45
Thermal radiation, 283
Thermal resistance
    combined heat transfer mechanisms, 22, 23, 25
    conduction heat transfer, 10–11
    convection heat transfer, 17–18
    radiation heat transfer, 20

Three-dimensional conduction.
See Two-and three-dimensional conduction
Tilt factor, 376
Total emissive power, 287
Total radiation properties, 291, 292–293
Total solar radiation, 358
Transient conduction, 136–189
  chart solutions, 147–164
    one-dimensional, 147–160
    two-and three-dimensional, 160–164
  defined, 136
  with negligible internal resistance, 137–142
  numerical solutions, 165–189
    explicit method, 165–173
    graphical interpretation, 174–176
    implicit method, 176–189
  in a semi-infinite solid, 142–147
Transition boiling, 479
Transmitted irradiation, 292
Tridiagonal matrix, 183–184
Tube banks, convection heat transfer in flow over, 253–255
Tubes, convection heat transfer in, 200–202, 240–248
  combined free and forced convection, 263–264
  laminar flow, 201, 221–226, 243–246
  liquid metals, 246–247
  superimposed free convection, 264–265
  turbulent flow, 201, 226–229, 241–243
Tube-within-a-tube counterflow heat exchanger, 396
Turbulent exchange coefficient for temperature, 217
Turbulent flow
  condensation heat transfer, 460–462
  in ducts, 200–201, 241–243

over flat plates, 199–200, 218–221, 249
over flat surfaces, analogy between heat and momentum transfer in, 212–218
in tubes, 201, 226–229, 241–243
velocity profiles, 201
Turbulent rate of heat transfer, 216
Two-and three-dimensional conduction, 84–118
  analog methods, 93–94
  analytical methods, 85–87
  graphical methods, 87–93
  iteration methods, 112–118
  matrix methods, 106–112
  numerical methods, 94–100
  relaxation methods, 94, 100–106
  transient conduction problems, 160–164
Two-dimensional and steady problems, 44
Two-dimensional problems, 43

Undisturbed flow regime, 199
Units, 26
Unit surface conductance, 24–25
Unit thermal conductance, 7, 20
Unit thermal resistance, 25

Vaporization in forced conduction, 474–478
Vector operations, 502–503
Vertical planes, convection heat transfer in, 259
View factor, 307
Visible radiation, 283

Wave theory, 282
Weber number, 444
Wicks, heat exchanger, 444–447
Wien's displacement law, 285–287

Zenith angle, 363